THE INSECTS AND ARACHNIDS OF CANADA

PART 6

AF772294

The Mosquitoes of Canada

Diptera: Culicidae

D. M. Wood, P. T. Dang,[1] and R. A. Ellis[2]

Biosystematics Research Institute
Ottawa, Ontario

Research Branch
Agriculture Canada

Publication 1686 1979

[1] Research Associate, Carleton University, Ottawa, Ontario.
[2] National Research Council Postdoctorate Fellow; presently Entomologist, Parks and Recreation Department, Winnipeg, Manitoba.

© Minister of Supply and Services Canada 1979

Available in Canada through

your local bookseller

or by mail from
Canada Communication Group — Publishing

Ottawa, Canada K1A 0S9

Reprinted 1993

Catalogue No. A42-42/1979-6
ISBN 0-660-10402-4

The Insects and Arachnids of Canada

Contents

Introduction

Throughout Canada mosquitoes are familiar to everyone as biting pests. In addition to their importance as a nuisance, a few species occasionally create problems out of all proportion to their numbers by transmitting disease organisms. Such pests have been the reason for costly abatement programs. At times, Canada seems to have been afflicted with a greater than average share of mosquitoes. Though the number of species (so far, 74 have been found) is not particularly large compared with the fauna of tropical countries, mosquitoes are often exceptionally numerous, and almost every part of the country experiences periodic outbreaks of one species or another.

An important step in successful mosquito abatement, as in dealing with any pest problem, is the correct identification of the offending species. Entomological history contains many examples of wasted money and effort because some similar but innocent species was mistaken for the pest species. Some of the most abundant widespread species of mosquitoes exhibit variation in color, and even in structure, across their ranges. This intraspecific variation along with extreme similarity of closely related species, especially among some *Aedes* and *Culex*, makes correct identification difficult, particularly of adult females. In difficult cases, adults reared from previously identified larvae can be named with much greater assurance than either larvae or adults identified separately. For example, adult females of *Aedes euedes*, *fitchii*, and *mercurator* are extremely similar, as are adults of both sexes of *Aedes aboriginis* and *punctor*, although their larvae are readily distinguishable. However, the larvae of *Aedes communis*, *implicatus*, and *stimulans* are not readily separable, although the adults are more easily distinguished.

This publication has been prepared for the identification of adults and fully grown larvae. Although eggs, early instar larvae, and pupae of most North American species are known, they are less easily identified than adults and mature larvae. Also, sufficient material for every species and enough experience to offer identification aids for these early stages are lacking. Rearing eggs and pupae to larvae and adults is probably easier and less time consuming than attempting to identifiy them in their early stages. In this publication, the keys to the larvae differ somewhat from the traditional approach, because they are specifically designed for identifying whole specimens in fluid, either alive or preserved, with the aid of a dissecting microscope. The identification of living larvae can be confirmed by rearing them to the adult stage.

Eggs of many species of *Aedes* have been illustrated by Kalpage and Brust (1968). For the identification of first-instar larvae the reader is referred to Bohart (1954), Price (1960), and Dodge (1966); for second and third instars to Dodge (1963) and Smith (1969*a*); and for pupae to Penn (1949) and Darsie (1949, 1951, 1955, 1957).

The following keys describe the usual, or "normal," condition based on a series of each species in the Canadian National Collection. However, every structural character in both adults and larvae shows some individual or regional variation; in most cases this variation has not been described in the literature. Such variations are usually minor, but the examination of enough specimens of any species usually shows a few individuals whose taxonomic characters do not conform to those used in the keys. These specimens, which can best be identified by direct comparison with more typical specimens, are most likely aberrant individuals of species already recorded from Canada, but they could also belong to species known from the northern United States that have not yet been discovered in Canada. These latter species, which are listed in a later chapter, can be identified by referring to Carpenter and LaCasse (1955). This standard work has been brought up to date by a series of three supplements (Carpenter 1968, 1970*b*, 1974) and a summary of taxonomic changes (Darsie 1973, 1978).

When a specimen has been keyed out, it should be compared with the description of that species. The descriptions contain more information than can be included in the key and are therefore provided as a cross-check. The descriptions of the adults follow a standard format. The larval descriptions are abbreviated to include only those characters that, taken together, characterize a particular species; more complete information on larval characters is summarized in Table 3 (foldout).

Generalizations have been necessary in the keys and in the descriptions of species as well as in the presentation of biological information. A given species may behave predictably for many years, but then it "breaks the rules" as a result of some unusual set of environmental circumstances. Some of the common widespread species may be complexes of more than one species, not yet separable by external structures, but having different habits or behavior. Widespread species also encounter differing environmental conditions across Canada, and as a result they may behave differently in different parts of the country. The amount of literature on mosquitoes is enormous, and some reports are even contradictory, which makes generalizations hazardous.

General biology

Life cycle

The mosquito, as in most other flies, passes through seven stages during its life cycle: the egg, the four successively larger larval instars, the pupa, and the adult. Most Canadian species of mosquitoes complete only one generation each year; others complete two or more.

Egg. Depending on the species to which they belong eggs are laid either on water or on moist soil, such as along the margins of ponds and marshes where water is likely to collect or in roadside ditches where snowmelt or excess rainwater will eventually accumulate to form temporary pools. In order to hatch, eggs must be in contact with standing water, although those of some species can survive in dry soil for several years before hatching. Although some species may deposit several hundred eggs, natural selection works against the progeny of females that broadcast their eggs randomly. Little is known about the oviposition sites chosen by each species. The sites determine where larvae will develop, and the weather largely determines when and whether they will develop successfully.

Embryonic development begins immediately after oviposition in all species, but in the case of floodwater mosquitoes, hatching does not take place until an adequately long period of cold or dryness has passed. Eggs that have ceased development during adverse conditions of drought or cold are said to be in diapause. There are two types of hatching stimuli: oxygen concentration and temperature. The species whose eggs hatch in early spring respond to decreasing oxygen tension as bacteria begin to multiply around them, whereas eggs of species hatching in late spring and summer are stimulated to hatch by higher water temperatures, and hatching may take place soon after eggs have been reflooded (see "Overwintering Stage," p. 12).

Larva. The first-instar larva is usually 1–2 mm long, but it may be 4 mm in the largest species. It is always recognizable by its egg burster, a pointed conical protuberance on the top of its head. The fourth, or last, larval instar is 7–15 mm long. Differences among instars are described in a subsequent section, "Anatomy." Except for the two predatory species *Toxorhynchites rutilus* and *Psorophora ciliata* that seize their prey with their sharply toothed mandibles, all larval stages of all species feed on detritus, algae, and other microorganisms, which they strain out of the water. The larvae of some species feed mainly at or near the surface, whereas others feed mainly at or near the bottom, although most species are adaptable enough to feed wherever food is plentiful. Thus *Anopheles* larvae, which are highly adapted for surface feeding, take food from the bottom when necessary. Low-protein diets produce stunted adults, whereas high-protein diets tend to produce larger and longer lived adults. However, no diets have been discovered that produce adults appreciably larger than those that develop under natural conditions.

The mouthparts of mosquito larvae are among the most elaborately evolved of any insect mouthparts. Food particles are gathered by a pair of labral brushes, each of which is a large bundle of long curved setae arranged in closely packed rows on the underside of the labrum. The labral brushes open by internal blood pressure and are closed by two pairs of large muscles. The brushes open and close rapidly, several times per second, and this motion generates a current of water that helps to bring more particles into range and may also move the larva slowly over the substrate. At each closing of the brushes, food particles that have become entrapped by the setae are combed out by a row of stiff parallel setae along the curved dorsal edge of each mandible. As the mandible opens, particles are combed from it by backwardly projecting hairs on the midventral region of the labrum, assisted by the maxillae. Finally, a bundle of setae at the base of the mandible scrapes the food from the midventral labral region into the pharynx. This complex maneuver, in which the mandible plays a dual role by merely opening and closing, is not restricted to mosquitoes but is characteristic of the larvae of other related families, notably dixid midges (Dixidae) and black flies (Simuliidae). If particles are too large to be handled in the previously described manner, they may be rasped into smaller pieces by the labral brushes, whose apices are often saw-toothed. Particles may also be broken up between the sharp tips of the mandibles and the hypostomial teeth (Snodgrass 1959).

Larvae have no legs, or pseudopods, but they swim by lashing the abdomen from side to side, a movement from which their common name of "wrigglers" was derived. The ventral brush (Fig. 16), a dense row of long stiff setae well braced at their bases to prevent them bending from side to side, serves as a paddle. The long setae of the body, which are so useful taxonomically, may also be sensory in function. Although respiration in minute larvae is mainly through the skin, older larvae of all species respire through a single pair of openings, or spiracles, on the dorsum of abdominal segment VIII. In the subfamily Anophelinae the spiracles are flush with the surface, as in dixid midges and the chaoborid genus *Eucorethra*, but in all other mosquitoes they are elevated at the apex of a conical siphon. Larvae are often seen suspended from the surface film by the apex of their siphon. One of the oldest control methods still widely used is to disrupt the surface tension with an oily substance in an attempt to asphyxiate the larvae.

Osmoregulation takes place in the rectal wall in two pairs of conical anal papillae (one pair in *Wyeomyia*) that project from it. These papillae have often been referred to as "gills," but they apparently have no respiratory function. The papillae are greatly reduced in size in all the species living in water that has a high content of dissolved salts, such as coastal salt marshes or inland alkali sloughs. The two pairs of papillae are about the same length, except in seven species of *Aedes*, namely *campestris, cantator, dorsalis, increpitus, melanimon, schizopinax,* and *triseriatus*, in which the ventral pair is shorter than the dorsal pair. The ultrastructure of the rectum and papillae of some brackish water species has been investigated by Phillips and Meredith (1973).

Pupa. The pupa is somewhat suggestive of a tadpole with its abdomen dorsoventrally flattened and ventrally curled. It swims rapidly downward and backward, as does a lobster, by scooping water with a pair of paddle-like flanges at the tip of its abdomen. Unless it can anchor itself on the bottom beneath or among vegetation, it gradually bobs to the surface again. It breathes atmospheric oxygen through a pair of funnel-shaped "respiratory trumpets," each of which arises on the anterolateral side of the thorax behind the head and connects internally to the anterior thoracic spiracles of the developing adult. When at rest, pupae remain motionless, floating at the surface. By this behavior, the pupae of mosquitoes can be distinguished from the pupae of Chironomidae (especially the subfamily Tanypodinae) and Chaoboridae (except *Eucorethra*, whose pupa is very similar to that of the mosquito), which resemble those of mosquitoes but remain suspended below the surface until the adult is about to emerge. Pupae of Dixidae leave the water to rest on emergent vegetation. As in most other aquatic Diptera, the pupal case of the mosquito ruptures middorsally along the thorax and the adult emerges into the air, the surface tension of the water preventing it from getting wet.

Adult. The adult mosquito, especially the female, needs no introduction. No other fly has a long slender densely scaled proboscis as long as the thorax, long slender 15-segmented antennae, and densely scaled wing veins. The male usually has long tufted palpi (Fig. 1). With only one exception (*Wyeomyia*) in Canada, the antennae of the male have many more and longer setae than those of the female. These setae are believed to serve as sound receptors, which respond to the frequency of the female's wingbeat before mating. Males usually emerge a day or two before females, and they soon assemble, sometimes in hundreds, at sunset and sunrise in a characteristic hovering flight called the "mating swarm." In spite of some early controversy, it is now a well-established fact, not only for mosquitoes but for many other Diptera, that the "swarm" is a device for assembling the sexes and that mating takes place after a female enters the swarm (Downes 1958, McAlpine and Munroe 1968). Females need to mate only once, early in their lives, whereas males presumably continue swarming for the rest of their lives. Therefore, mating is observed usually only during the first few days after emergence (Frohne and Frohne 1952).

Before swarming, the long setae of the males' antennae, which are arranged in whorls on the antennae, are erected by internal blood pressure.

In this state, the antennae are sensitive to the frequency of the female's wingbeat and presumably they begin to vibrate when she approaches. The frequencies may differ among species, which may assist in specific recognition. The location of the swarm in relation to surrounding landmarks is important in maintaining conspecific mating (Downes 1958). The female must be attracted to the swarming site by the same visual cues that brings the males to the site, because unlike the male she appears to be insensitive to sounds. Undoubtedly, pheromones play an essential role as soon as the sexes are within close range of one another.

After mating, which usually takes place within a day or so of emergence, the female begins to seek a blood meal. In the search the female may travel several kilometres. Some species have been shown to disperse for great distances, even 300–500 km. Like the male, the female must also seek a carbohydrate meal at regular intervals, but the blood meal is required to initiate and develop each batch of eggs. Carbon dioxide plays an important role in long-range attraction of a blood-seeking mosquito, and the female flies toward increasing concentrations of it. At close range, the female mosquito begins to detect heat if the victim is warm-blooded, and finally she sees the host. The host odor also may be involved in the attraction, but its role is controversial. After the female lands, she may probe the skin a few times, probably searching for a capillary. Only the central stylets, the labrum, the two long hair-like mandibles, the two maxillary blades, and the single median hypopharynx are inserted into the skin; the trough-like supporting labium remains outside. The tips of the mandibles are needle sharp and are driven into the tissue, followed by the maxillae, whose saw-toothed apices serve to anchor the proboscis for another deeper thrust. Engorgement usually takes only a few minutes, during which she may imbibe her own weight of blood (Woodward and Chapman 1965). A few species may even triple their weight. While the blood is digesting and the first batch of eggs is developing, the female remains inactive, but soon she must seek a suitable place to deposit her eggs. In the selection of an oviposition site she may be guided by an attractant left behind by the developing immature stages of her own species (Hudson and McLintock 1967, Kalpage and Brust 1973). After oviposition, the female immediately begins to search for a second blood meal; she may complete several such cycles in her lifetime. Five ovarian cycles in one year are probably the most a mosquito can achieve in Canada, although there are records of as many as 10 cycles per female in the southern United States.

Overwintering stage

Throughout Canada, with the possible exception of southwestern British Columbia, a critical event in the life cycle of every species of mosquito is the initiation and termination of diapause, which makes it possible for the insect to survive the winter. Each species in Canada seems to have evolved

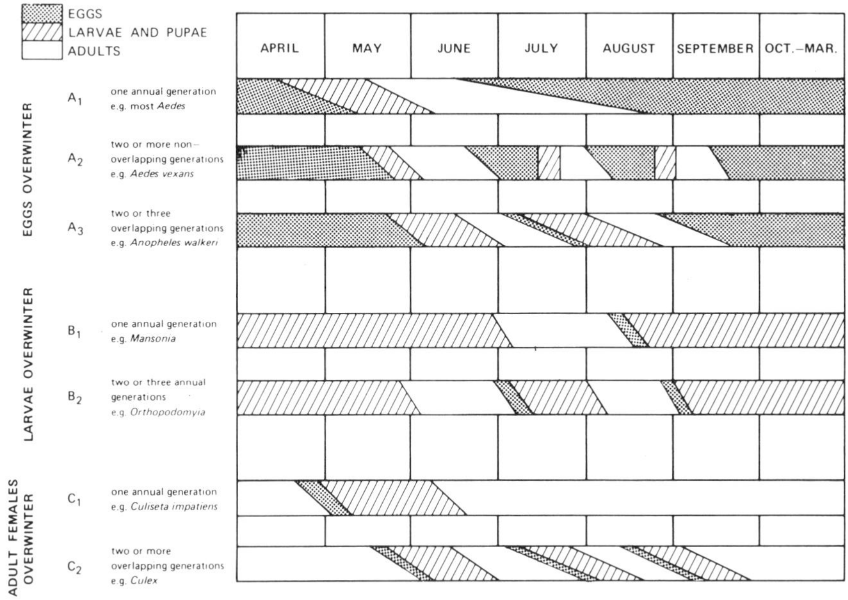

Table 1. Approximate correlation between seasons and developmental stages of Canadian mosquitoes

the ability to overwinter in only one particular stage. Depending on the species, the stage may be an egg, a larva, or an inseminated nulliparous nonblood-fed female. Table 1 shows three basic types of life cycle: A, B, and C, which are distinguished by the stage—egg, larva, or adult female—in which overwintering is accomplished. Each of these three life cycles is further subdivided into subgroups, such as A_1, A_2, and A_3, depending on whether there is a single generation each year or more than one. Several factors, particularly the latitude and the amount of summer rainfall, determine whether a multivoltine species might have two or three or even more generations each year. Cold or drought may reduce a bivoltine species to a univoltine one in a particular season or latitude. For example, *Anopheles earlei* is at least bivoltine in southern Ontario, but is probably univoltine in northern Canada (Happold 1965*a*). Therefore, Table 1 is only a rough guide to the stage that may be expected for a given species at a given time. In the warmest parts of Canada where winters are less severe, especially in southwestern British Columbia, a few species can even overwinter in more than one stage, that is, egg and larva, or larva and adult.

Group A. The species of this group overwinter in the egg stage.

Subgroup A_1: Hatching time is closely related to snowmelt. Beckel (1953) discovered that when eggs had been subjected to a suitably long period of cold, the decomposition of the carcasses of the mosquitoes that had laid those eggs the previous year stimulated hatching. This effect has since been shown to be a result of low oxygen tension, which occurs as bacteria become active in the debris surrounding the eggs. Nutrient broth is used to stimulate hatching in the laboratory (Brust and Kalpage 1967, Kalpage and Brust 1974). Because of the wide variation in spring temperatures from year to year, larval development may be prolonged for several weeks or completed in a week or two. Eggs laid by species in this group enter an obligatory diapause and will hatch only after being subjected to several weeks of cold temperatures. *Culiseta morsitans* and all species of *Aedes* except the 16 species mentioned under subgroups A_2 and B_2 belong to this group.

Subgroup A_2: Species of this group overwinter in the egg stage as do those of subgroup A_1, but they are also capable of additional generations during the same summer if their eggs have dried and been subsequently reflooded. All species of *Psorophora* and 15 species of *Aedes*, namely *atropalpus, campestris, canadensis, cantator, cinereus, dorsalis, hendersoni, nigromaculis, sollicitans, spencerii, sticticus, togoi, triseriatus, trivittatus,* and *vexans,* are capable of producing more than one generation during one summer. The time of first hatching in the spring is also much more varied. *Aedes spencerii* is the earliest species to hatch, whereas some other species in this group, for example, *trivittatus,* hatch relatively late in the season.

Eggs of *Psorophora* and of many of the *Aedes* in this group are laid on the mud at the edges of partly dried-up temporary pools, and they may hatch abruptly anytime during the summer provided they are flooded. However, if conditions remain dry, they may not hatch that year, whereas in wet summers there may be several generations, and adults may be present more or less continuously. Several centimetres of rain, leaving ditches and low-lying areas full of water or causing river levels to rise, are sufficient to initiate hatching of these species. Larval development can be completed in as short a time as 4 or 5 days, and pupal development is even faster; therefore, the cycle from egg to adult requires only a week in hot weather. The best-known member of this group is *Aedes vexans*, although the rock-pool, tree-hole, and salt- and alkaline-marsh breeders listed previously also belong here. *Aedes cinereus* was previously considered to be univoltine (Carpenter and LaCasse 1955), but Brust (1968) showed that about half the eggs laid by this species were nondiapausing eggs that hatched without cold conditioning. Similarly, Brust showed that *Aedes campestris, canadensis*, and *spencerii* all laid both diapausing and nondiapausing eggs.

Subgroup A_3: As in subgroups A_1 and A_2, species of this group overwinter in the egg stage and hatch in late spring. However, eggs of the first generation are laid on permanent water, so that development proceeds immediately. Thus, larvae appear while their mothers are still seeking blood for a second ovarian cycle. As the summer progresses, the second and third generations overlap and all stages may be encountered. *Anopheles walkeri* is the only member of this group.

Group B. In this group, the larval stage enters diapause in preparation for overwintering.

Subgroup B_1: The group has been further subdivided into a univoltine subgroup, B_1, to which *Mansonia perturbans* and *Wyeomyia smithii* belong.

Subgroup B_2: A multivoltine subgroup, B_2, includes *Anopheles barberi, Toxorhynchites rutilus, Aedes sierrensis, Culiseta melanura, Orthopodomyia alba, O. signifera*, and southern populations of *Wyeomyia smithii*. Also *Aedes sierrensis* may overwinter in the egg stage in the northern part of its range, and Loor and DeFoliart (1969) have found a few overwintering eggs of *Orthopodomyia*. The particular larval stage at which overwintering can take place is probably specific for most species, but has not been studied in every case. Smith and Brust (1971) demonstrated that development of *Wyeomyia smithii* was arrested in the third instar by shorter day length, and it did not begin again until temperature and day length were raised above critical levels. In early spring, second instars of *Orthopodomyia alba* that had survived the winter were found in the Ottawa area (and after they were collected they continued to develop to the adult stage), but later instars in the same tree hole had died, presumably as a result of cold. Similarly, larvae of *Anopheles barberi* collected in October and April were all in the second instar. However, *Toxorhynchites rutilus* is considered to be cold hardy only as a mature larva (Horsfall 1955). Larvae of *Mansonia perturbans* may not actually enter a diapause, but may merely experience a greatly reduced growth rate and therefore overwinter in whatever stage they have reached as winter approaches. Synchronization of individuals appears to occur in the last instar before pupation (Hagmann 1952). In most of its Canadian range, *Wyeomyia smithii* is univoltine, although in southern Ontario it may have a second generation. In Canada, the summer behavior of *Culiseta melanura* remains unknown, although larvae are present in May and September.

Group C. Species in this group overwinter as inseminated nonblood-fed females.

Subgroup C_1: Frohne (1953*b*, 1954*a*, 1954*b*) was the first to show that the two most northerly species of *Culiseta, alaskaensis* and *impatiens*, have an extremely long life-span, of a year or more. Newly emerged females do not seek blood in their first summer. Most of their lives are spent in a torpid condition, aestivating or hibernating. They are among the earliest to seek blood the following spring, often before the snow has disappeared. Later, seeking blood for their second or third ovarian cycles, their cycles may overlap with their own progeny. At the latitude of Ottawa, adult *impatiens* have never been collected after April or early May. This species is probably univoltine at that latitude, although there is ample time for a second generation. The two species of *Culiseta* appear to be the only mosquitoes with this type of life cycle in Canada.

Subgroup C_2: Members of this group also overwinter as nonblood-fed adult females, but they undergo successive summer generations. The remaining species of *Anopheles*, namely, *earlei*, *freeborni*, *punctipennis*, and *quadrimaculatus*; three species of *Culiseta*, namely, *incidens*, *inornata*, and *minnesotae*; all species of *Culex*; and *Uranotaenia sapphirina* belong to this group. The species of *Anopheles* emerge in early spring, on the first warm days in April. *Culiseta inornata* appears somewhat later (Rempel 1953), whereas *Culex* spp. and *Uranotaenia sapphirina* appear still later. All of these species inhabit semipermanent marshes, and they breed continuously during the summer, which results in overlapping generations. During warm wet weather substantial populations can build up as the summer progresses. The last generation of adult females in the autumn does not feed on blood before going into hibernation; therefore these females can only be collected in light traps or in damp dark places, such as caves, hollow trees, and unheated buildings, while they are hibernating.

Pest species

More than 60 of the 74 species of mosquitoes occurring in Canada are known to bite man or domestic animals and birds. However, relatively few species are usually numerous enough to be considered serious pests solely on the basis of the discomfort created by their biting activities. *Aedes vexans*, which occurs throughout southern Canada, develops rapidly when summer rainfall is excessive, sometimes achieving enormous populations. This species is probably Canada's most severe mosquito pest. In the Prairie Provinces *Aedes spencerii* and *dorsalis* are regular and abundant nuisances. The salt-marsh species, *Aedes cantator* and *sollicitans*, are pests from spring to late fall along the coastal regions of the Maritime Provinces. In southern Ontario, *Mansonia perturbans* is habitually a problem during July in the vicinity of large inland marshes, whereas *Aedes stimulans* is probably the most numerous and aggressive species in spring (Twinn 1931, Benedict 1962, Belton and Galloway 1966), especially near forested areas on Paleozoic sediments. *Aedes punctor* and *communis* are the dominant species of the Canadian shield and throughout the Boreal forest. The former species prefers acidic water lying over peat, whereas the latter is usually associated with pools on alluvial clay along stream valleys (Maire and Aubin 1976). The subarctic region, particularly near the tree line, is dominated by *Aedes hexodontus*, whereas in the arctic, *impiger* and *nigripes* may be troublesome on warm days.

Adult feeding

In addition to frequently imbibing water, both sexes of all species probably feed regularly on nectar. Males feed only on nectar. A carbohydrate food source is important in prolonging life, although it is apparently not necessary for development of the first batch of eggs of some species (Smith and Brust 1970, Hudson 1970). The species that do not require a blood meal in order to mature eggs are called autogenous. Many workers have observed mosquitoes feeding at flowers of various species (Sandholm and Price 1962, West and Jenkins 1951). The pollinia of a small northern orchid, *Habenaria obtusata*, have often been found attached to the mouth-parts of species of northern *Aedes* (Hocking et al. 1950, Thien 1969*a*, 1969*b*). Haeger (1955) observed *Aedes sollicitans* feeding on aphid honeydew.

Females of most Canadian species require a blood meal in addition to a carbohydrate meal in order to develop a sizeable batch of eggs; these females are called anautogenous. Unlike the carbohydrate meal, which passes into the crop, the blood meal is pumped directly into the midgut where it is enclosed in a transparent sheath of thin cuticle, the peritrophic membrane. Distension of the gut wall helps initiate the production of hormones that stimulate egg development, which proceeds as the blood is digested and assimilated. Digestion may be completed in 24–36 hr at 25°C, 4–5 days at 15°C, and about 15 days at 10°C (Horsfall 1955). Egg development takes somewhat longer. After the first batch of eggs has been laid, the female may again seek blood to repeat the process.

Females of *Orthopodomyia* spp., *Uranotaenia sapphirina*, and *Wyeomyia smithii* have not been observed feeding; they are presumed to reproduce autogenously. Also, females of *Aedes churchillensis* and *rempeli* cannot be induced to take blood, and these species may partly develop a second batch of eggs without it (Ellis and Brust 1973, Smith and Brust 1970). A few females of *Aedes atropalpus* may be induced to take blood before laying eggs, but most of them do not, because blood is not needed for development of the first batch of eggs (Hudson 1970). Only parous females, that is, those that have already laid their first clutch of eggs, require a blood meal in order to develop a second batch of eggs. The number of eggs laid by an autogenous female is determined by the quality and the quantity of the food ingested by her while she was in the larval stage.

Many more species may possess facultative autogeny, which is the ability to mature a small number of eggs by utilizing the food reserves that are present. Corbet (1964) demonstrated that in both *Aedes impiger* and *nigripes*, a few eggs, sometimes only one, were matured after the food reserves from the degeneration of most of the oocytes had been translocated to only a few. Unlike *atropalpus*, both *impiger* and *nigripes* require a source of carbohydrate before ovarian development can proceed. This ability to develop some eggs facultatively, in the absence of blood and at the expense of the rest of the follicles, may be more common than is presently realized. Among large numbers of reared but nonblood-fed adults of 17 species from Nevada, Chapman (1962) found individuals of 10 species that were able to develop some eggs autogenously. Among the 10 were *Aedes campestris*, *communis*, possibly *churchillensis*, *dorsalis*, *melanimon*, *nigromaculis*, *schizopinax*, and *Culex tarsalis*. The first- and last-listed of these species laid the greatest number of eggs.

Host preferences

Unlike black flies (Simuliidae), which are usually rather specific as to the host animal they feed on and the environment that the animal is in, mosquitoes, especially *Aedes*, seem to be much less particular. Hayes (1961), with the use of serological methods, found that females of *Aedes canadensis* were attracted to and fed on birds, mammals, amphibians, and reptiles, although there was a slight preference for mammals. *Mansonia perturbans* also was found to be quite polyphagous (Downe 1962). The high arctic species also appeared to obtain blood from whatever warm-blooded animal was available (Corbet and Downe 1966). In general, however, *Aedes* obtained blood meals more often from mammals than from birds, perhaps because there were more mammals than birds available to them (Downe 1960, Shemanchuk et al. 1963).

Mosquito-borne disease

In addition to their importance as pests, several species of mosquitoes have been implicated in the transmission of the causative organisms of several diseases in Canada. Malaria, transmitted only by *Anopheles*, was a serious problem in southern Ontario during the last century, when hundreds of people died as a result of the infection. At that time, the role of mosquitoes was not known and human carriers of malaria were undoubtedly readily accessible to hungry *Anopheles*. The disease is no longer present in Canada, but five of the six species of *Anopheles* found in Canada have been experimentally infected with malaria in the laboratory, and the sixth, *earlei*, may also be capable of infection, but this has not been documented (Brust, personal communication). Presumably these species could transmit malaria again if they were given the opportunity. However, the possibility of transmission of malaria from an infected person returning from a malarious region is unlikely.

Although members of the genus *Anopheles* are the only vectors of human malaria, several species of malaria in birds are transmitted by culicine mosquitoes. These diseases may play a decisive role in regulating populations of wild birds, although to date they have not threatened domestic birds.

Much more serious are the arboviruses, or arthropod-borne viruses, of various encephalitides. The history of the diseases caused by these viruses in Canada and of the research and related activities has been thoroughly documented by McLean (1975), McLintock and Iversen (1975), and McLintock (1976). The most important of these diseases are caused by the viruses of western equine encephalitis (WEE), eastern equine encephalitis (EEE), and St. Louis encephalitis (SLE). Encephalitides are mainly diseases of birds, therefore, the bird-feeding species of *Culex* and *Culiseta* are important as vectors. Because these viruses appear in the mosquito population each summer in various parts of the continent, researchers consider that epizootics in the wild bird population are probably common, if unpredictable, events. Birds do not seem to be adversely affected by the viruses; only in such unnatural hosts as horses and man do these viruses produce unpleasant symptoms.

In Canada, WEE was first diagnosed in 1935 in horses, and later in 1941 in a major epidemic in the Prairie Provinces, attacking humans (McLintock and Iversen 1975). In subsequent years, WEE virus was isolated from several naturally infected species of mosquitoes in the USA, and more mosquitoes were infected under experimental conditions. Gradually, *Culex tarsalis*, a species of rather omnivorous biting habits, emerged as the principal vector of WEE, although several other species, including spring populations of *Aedes*, remain implicated in transferring the virus. In spite of a prodigious research effort, the mode of overwintering of the virus remains unknown. For a long time it has been assumed that birds returning on their spring migrations carry the virus, but this theory has not been demonstrated. Females of *Culex tarsalis* and of another potential vector, *Culiseta inornata*, do not take a blood meal before they hibernate; therefore they cannot become infected until spring. However, overwintering mosquito eggs have been shown to harbor another arbovirus (Watts et al. 1973), and WEE may also overwinter in this fashion.

In 1975 in Manitoba, the outbreak of WEE that threatened Winnipeg was counteracted by a massive mosquito control program in Winnipeg and several surrounding municipalities. This outbreak was extensively documented in Special Supplement I to the May–June issue of the *Canadian Journal of Public Health* (1976).

Only one outbreak of EEE has been recorded in Canada; in horses, near Knowlton, Que., in 1972 (Harrison and Cousineau 1973). The principal vector of the virus in the USA, *Culiseta melanura*, is a rare and local species in Eastern Canada, probably not capable of supporting an epizootic. Therefore, it is possible that another species such as *Mansonia perturbans*, from which EEE virus has been recovered, may also transmit the disease.

SLE virus was first isolated in Canada in 1971 from Saskatchewan, but the first epidemic occurred in southwestern Ontario in 1975, concurrent with a widespread epidemic in the USA. The principal vector of this virus is believed to be *Culex pipiens*.

The above-mentioned viruses have probably always been present in Canada, and their apparently recent prevalence may be no more than a result of more extensive surveillance. The stationing of sentinel chicken flocks in strategic locations in the Prairie Provinces and the periodic trapping of mosquitoes attracted to them have been standard monitoring procedures for detecting viral presence for many years. In the case of WEE, the infection of horses has always preceded human infections and has warned of a potential epidemic. This warning could also be taken for EEE, but not for SLE, since the latter does not affect horses.

A fourth group of related viruses, collectively labeled California encephalitis (CE) viruses, cause diseases of small mammals and are transmitted predominantly by species of *Aedes*. Three of the CE strains have produced clinical symptoms in humans. One strain, the LaCasse virus, has been found in larvae and in males of its principal vector, *Aedes triseriatus*, suggesting that it can be transmitted transovarially by an infected female mosquito to her progeny (Watts et al. 1973). The most common virus of the California group, the Snowshoe Hare strain, has been isolated from larvae of *Aedes implicatus* by McLintock et al. (1976).

Geographic distribution

The treatment of each species in this publication is accompanied by a map showing the site of collection of all the specimens we have examined (shown by solid circles), locality records from the literature (solid triangles), and those literature records that we believe may be based on misidentified specimens that require verification (open triangles). Most of the specimens indicated by solid circles are in the Canadian National Collection. Each map is intended as only a rough guide to the distribution of a species; the total ranges of Canadian mosquitoes are unknown, and many regions of the country have been inadequately collected.

Table 2 lists the known occurrence of each species in Canada; the distribution in Eurasia and Alaska is included for comparison. The open circles represent literature records that we consider incorrect. These provincial lists are only a rough guide and should not be considered complete; for example, *Culiseta alaskaensis* must occur in Ontario, but we have not seen any records, undoubtedly because the region where it occurs is sparsely collected.

The distribution of each Canadian species outside Canada is outlined briefly in the text dealing with that species. Further details may be obtained from Carpenter and LaCasse (1955), Carpenter (1968, 1970*b*, 1974), Stone et al. (1959), and Stone (1965). Maps showing the distribution of each species in Alaska are given by Gjullin et al. (1961) and in Washington, Oregon, and Idaho by Gjullin and Eddy (1972).

Munroe (1956) has classified insect distribution patterns in Canada into nine categories, and, as might be expected, some mosquito distributions may be readily assigned to several of these categories. However, most of the species in Canada are wide ranging, transcontinental or nearly so, and therefore even overextend some of Munroe's "combined ranges." Almost all of these species also occur across Eurasia (see Table 2). Perhaps because of compression of isotherms south of Hudson Bay, the ranges of these wide-ranging species occupy more degrees of latitude in Western Canada than they do in Eastern Canada, for example, the ranges of *Aedes diantaeus, punctor,* and *pionips* and *Culiseta impatiens* and *morsitans*. Common and often abundant throughout the boreal forest, or taiga, these mainly northern species become less common and more local in southeastern Canada. An extreme example is *Culiseta alaskaensis,* which is found from Alaska to southern British Columbia and southern Manitoba, where it occupies all life

	Eurasia	Alaska	Y.T.	N.W.T.	B.C.	Alta.	Sask.	Man.	Ont.	Que.	N.B.	P.E.I.	N.S.	Nfld.
Anopheles														
barberi									●	●				
earlei		●	●	●	●	●	●	●	●	●	●	●	●	●
freeborni					●									
punctipennis					●			●	●	●	●		●	
quadrimaculatus									●	●				
walkeri					○		●	●	●	●	●		●	
Toxorhynchites														
rutilus									●					
Aedes														
aboriginis		●			●		○							
abserratus								●	●	●		●	●	●
aloponotum					●									
atropalpus									●	●				●
aurifer								○	●	●	●			
campestris		●	●		●	●	●	●	●	●				
canadensis			●	●	●	●	●	●	●	●	●	●	●	●
cantator										●	●	●	●	●
cataphylla	●	●	●		●	●	●	○						
churchillensis						●		●						
cinereus	●	●	●	●	●	●	●	●	●	●	●	●	●	●
communis	●	●	●	●	●	●	●	●	●	●	●	●	●	●
decticus		●	●			●		●	●	●				●
diantaeus	●	●	●	●	●	●	●	●	●	●			●	●
dorsalis	●				●	●	●	●	●	●	●			
euedes	●	●		●	●	●	●	●	●	●				
excrucians	●	●	●	●	●	●	●	●	●	●	●	●	●	●
fitchii	●	●	●	●	●	●	●	●	●	●	●	●		●
flavescens	●	●	●	●	●	●	●	●	●	●				●
grossbecki									●					
hendersoni					●		●	●	●	●				
hexodontus	●	●	●	●	●	●		●	●	●				●
impiger	●	●	●	●	●	●	●	●	●	●				●
implicatus		●	●	●	●	●	●	●	●	●				●
increpitus	●				●	●	●	●						
intrudens	●	●			●	●	●	●	●	●	●	●	●	●
melanimon					●	●	●							
mercurator	●	●	●	●	●	●	●	●	●					
nigripes	●	●	●	●	●			●		●				●
nigromaculis						●	●	●						
pionips	●	●	●	●	●	●	●	●	●	●				●
provocans				●	●	●	●	●	●	●	●	●	●	
pullatus	●	●	●	●	●	●	○			●				●

Table 2. Distribution of Canadian mosquitoes (open circles represent literature records that we believe are incorrect)

	Eurasia	Alaska	Y.T.	N.W.T.	B.C.	Alta.	Sask.	Man.	Ont.	Que.	N.B.	P.E.I.	N.S.	NFLD.
Aedes (cont'd)														
punctor	●	●	●	●	●	●	●	●	●	●	●	●	●	●
rempeli	●			●					●	●				
riparius	●	●	●	●	●	●	●	●	●	●			●	
schizopinax						●								
sierrensis					●									
sollicitans									●		●	●	●	
spencerii					●	●	●	●	●					
sticticus	●				●	●	●	●	●	●	●			●
stimulans		○	○		○	○	○	●	●	●	○	●	●	○
thibaulti	○								●					
togoi	●				●									
triseriatus					○		○	○	●	●	●			
trivittatus								●	●	●			○	
vexans	●		●		●	●	●	●	●	●	●	●	●	
Culex														
pipiens	●				●	○		○	●	●	●		●	
restuans					○	●	●	●	●	●	●		●	
tarsalis				●	●	●	●	●	●					
territans	●	●	●	●	●	●	●	●	●	●	●		●	
Culiseta														
alaskaensis	●	●	●	●	●	●	●	●		●				●
impatiens		●	●	●	●	●	●	●	●	●	●			●
incidens		●	●	●	●	●	●	●						○
inornata			●	●	●	●	●	●	●					
melanura									●	●				
minnesotae					●	●	●	●	●	●				
morsitans	●	●	●	●	●	●	●	●	●	●	●	●	●	●
Mansonia														
perturbans					●	●	●	●	●	●	●	●	●	
Orthopodomyia														
alba									●	●				
signifera									●					
Psorophora														
ciliata									●	●				
columbiae									●					
ferox									●					
signipennis							●							
Uranotaenia														
sapphirina									●	●				
Wyeomyia														
smithii							●	●	●	●	●	●	●	●

zones between the subarctic and the prairie environments, and yet in Eastern Canada it seems to be limited to the taiga.

Another common type of distribution is the "eastern range," which extends at its maximum from southern Manitoba to Nova Scotia, and at its minimum to include only southwestern Ontario and southernmost Quebec. Species with the widest distributions in this category include *Aedes atropalpus, stimulans,* and *trivittatus.* Species with the most restricted distributions include species with southern affinities that are widespread in the USA, but are at their northern limits in southern Canada, such as *Anopheles barberi* and *quadrimaculatus; Toxorhynchites rutilus;* the three eastern species of *Psorophora; Uranotaenia sapphirina; Orthopodomyia* spp.; *Aedes aurifer, thibaulti,* and *triseriatus;* and *Culiseta melanura.* Most of these species tend to be from uncommon to rare in occurrence, possibly because of their more exacting habitat requirements, and the true northern limits of their distribution have probably not been determined. About half of these species were discovered in Canada only within the last 10 years. Another type of "eastern range" with a more northerly extension is shown by *Aedes abserratus, Culex restuans,* and *Wyeomyia smithii.*

A third group of species occupies a "central range," which in Canada is the prairies and the surrounding aspen parkland. *Aedes nigromaculis* and *schizopinax* are found only in the prairies, whereas others such as *Aedes melanimon* and *increpitus* have also invaded prairie-like situations in central British Columbia. *Aedes spencerii* and *dorsalis* have invaded the prairie-like farming areas of Eastern Canada; the latter has also colonized the marshy brackish coast of James Bay. *Aedes campestris* has included in its range the low arctic tundra of Hudson Bay as well as the prairies, but has not yet become established in eastern farmland.

Rather few species occupy a purely "western range." *Aedes aloponotum* and *aboriginis* both seem to be confined to coastal British Columbia. *Anopheles freeborni* seems to be found only in the dry interior of British Columbia, whereas *Aedes sierrensis* ranges widely in both the coastal and interior regions of that province.

Several southern species, with transcontinental ranges in the USA, reach their northern limit in three isolated areas of southern Canada: southern British Columbia, southern Manitoba, and southeastern Canada. This combined eastern and western range is shown by *Anopheles punctipennis, Aedes hendersoni, Mansonia perturbans,* and *Culex pipiens.*

Only three species may be considered as having an "arctic range." These species are *Aedes impiger, nigripes,* and *hexodontus*, and of these only *nigripes* is confined to the tundra. The other two are also subarctic and cordilleran, ranging well into the taiga. *Aedes rempeli* is also low arctic and boreal in range, although too few stations have been found to permit generalization. *Aedes impiger* may have the widest range in Canada of any species of mosquito, and it is probably the most highly adapted arctic species because it is indigenous in several high arctic localities, such as Melville Island and Peary Land, northern Greenland, where even *nigripes* has not been collected.

Finally, the distributions of *Culiseta inornata* and *Aedes pullatus* show an unusual composite range that suggests the possibility of two sibling species in the former and dispersal from two separate glacial refugia in the latter. *Culiseta inornata* is found abundantly throughout the west, extending even to the arctic coast of the Yukon, but in Eastern Canada it is confined to southern Ontario, where it is uncommon. Perhaps it has only recently spread into southwestern Ontario from the west. *Aedes pullatus* has two separate ranges: one in and west of the Rocky Mountains and the other in northern Quebec, Newfoundland, and southern Baffin Island. The Grand Banks, Sable Island, and other offshore areas that were extensive land areas during the fullest extent of the last glacial period could have functioned as refugia for many species (Howden et al. 1970), including *Aedes pullatus.* The western populations of *pullatus* could have spread from Beringia, northwest of the ice sheet.

In summary, few Canadian mosquitoes occupy ranges that conform, even approximately, with the botanists' classification of the major Canadian vegetative regions (Rowe 1972). *Aedes aboriginis* in the wet Pacific Coast forest region, *nigromaculis* on the prairies, *pionips* in the boreal forest region (with a few stations in the aspen parkland and the Great Lakes – St. Lawrence forest region), and *nigripes* confined to the arctic tundra and coastal grasslands of Newfoundland (with a few stations in Alpine forest–tundra as well) are perhaps the only species whose distribution more or less coincides with a major life zone and that could be considered as characteristic of the zone.

Phenology

The time of first appearance of mosquitoes in the spring varies greatly from year to year and from one part of Canada to another. Depending on the onset of spring, the year-to-year variation may be by as much as 3 weeks. This annual variation makes assessment of the latitudinal variation rather difficult. Nevertheless, Map 1 shows the approximate time that mosquitoes are expected to emerge in numbers in any given locale. The isotherms are based on the mean daily temperatures for April, which seemed to correspond more closely than annual mean temperatures, degree-days, or any other measurement to the earliest records of mosquitoes in the literature (Hearle 1926, Happold 1965*b*, Graham 1969*b*, Rempel 1953, Hocking et al.

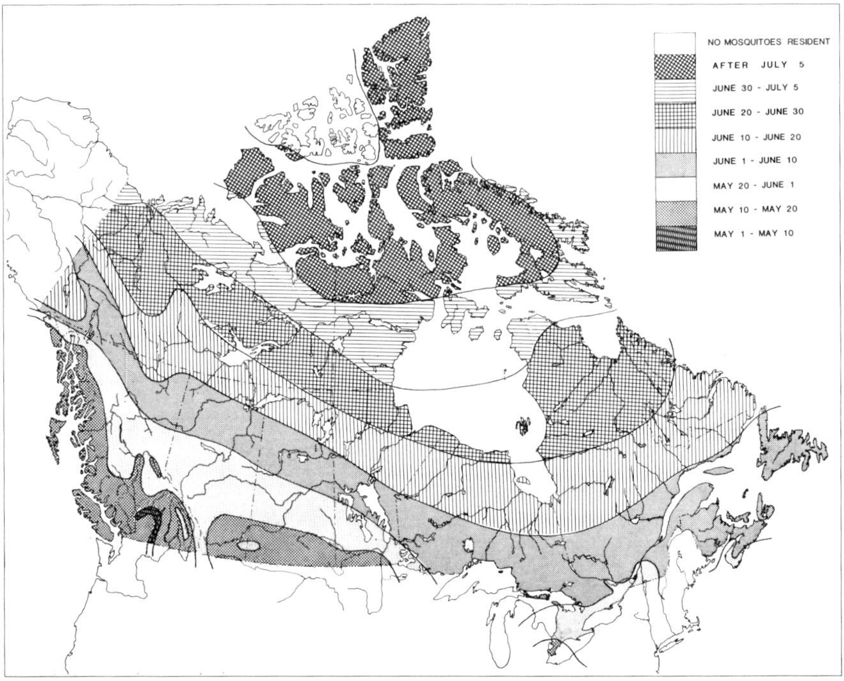

Map 1. Approximate time of first appearance of numbers of *Aedes* mosquitoes. Zone boundaries are based on spring temperature data from several sources, with local modifications to account for different species.

1950, Jenkins and Knight 1950, 1952, Maire and Aubin 1976, and others), as well as to the earliest dates recorded on material in the Canadian National Collection and to our own experience.

Different species also greatly affect the earliest dates of mosquito abundance. The presence of *Aedes cataphylla* in the vicinity of Kamloops, B.C., and of *spencerii* on the prairies advances the mosquito season to the same time as, or even ahead of, that of southern Ontario, where the common species are rather late in emerging, although the mean temperatures in April at Kamloops, B.C., and at Windsor, Ont., are warmer than those in the Prairie Provinces.

Anatomy

Adult

All Canadian mosquitoes, both males and females, can be distinguished from all other midge-like insects by their long, slender scaly proboscis, which in all cases is longer than the length of the head and thorax combined. A few crane flies (Tipulidae) have an elongated rostrum, the region of the head between the antennae and mouthparts, but they can be distinguished from mosquitoes by their maxillary palpi, which arise near the apex of this rostrum instead of at the base of the proboscis. A few other Diptera also have a similarly elongated proboscis, but it is not covered with scales. The proboscis of the mosquito is straight or nearly so, narrowest near the base, widening only slightly near the apex, except in *Toxorhynchites* in which it is thickest basally and tapers to a pointed strongly decurved apex.

Males of all Canadian mosquitoes except *Wyeomyia smithii* can be distinguished from females by their plumose antennae (Fig. 1). Each of the 13 segments, or flagellomeres, of the flagellum bears a whorl of setae in both sexes, but in the female these setae are each scarcely longer than twice as long as their corresponding flagellomere (Fig. 2). However, in all males except of *Wyeomyia*, the setae on all but the last flagellomere are several times as long as their corresponding segment and are much more numerous than in the female. Males of Chaoboridae, Ceratopogonidae, and Chironomidae, which also have somewhat similar plumose antennae, all lack an elongate slender scaly proboscis.

Few other Diptera are as extensively clothed with scales as are mosquitoes. The scales are usually either pale or dark colored. The dark-colored scales are seldom truly black and are called dark scales because the actual color may vary from reddish brown to dark brown, depending on the species and to a lesser extent on the age of the specimen. In some lights, dark scales may appear faintly bluish metallic and thus seem paler than they do in other lights. In a few species, the dark scales are brilliantly and unmistakably metallic blue, green, or purple. The pale-colored scales are white, silvery, yellow, or gold and are called pale by comparison with the darker scales, which may be brown or black. Thus some species, for example, *Aedes aurifer* or *thibaulti*, have areas of gold scales that are distinctly paler than adjacent areas of dark brown scales, whereas members of the *Aedes excrucians* group may have areas of whitish scales contrasting with areas of reddish brown scales. Often pale and dark scales are mixed on the proboscis, palpi, wing veins, legs, and abdomen. The color of the integument also varies interspecifically, and it may show through between the scales, unless they are particularly dense, thereby influencing the overall color.

The antenna is divided into 15 segments (Figs. 1, 2). The basal segment, or scape, is hidden behind the greatly enlarged second segment, or pedicel (sometimes called torus in the literature). The pedicel bears a group of setae on its medial surface, which usually also bears either pale or dark scales that may extend out over its dorsal surface but are usually absent from its ventrolateral surface. The first flagellomere usually also bears a median patch of scales, but the remaining flagellomeres are usually without scales (in female *Orthopodomyia*, a conspicuous median line of white scales is present on the first three to five flagellomeres).

The palpus is typically five-segmented, which is the standard number in the Nematocera. Each segment is called a palpomere. In both sexes of *Anopheles*, and in the males of nearly all other genera, the palpus is

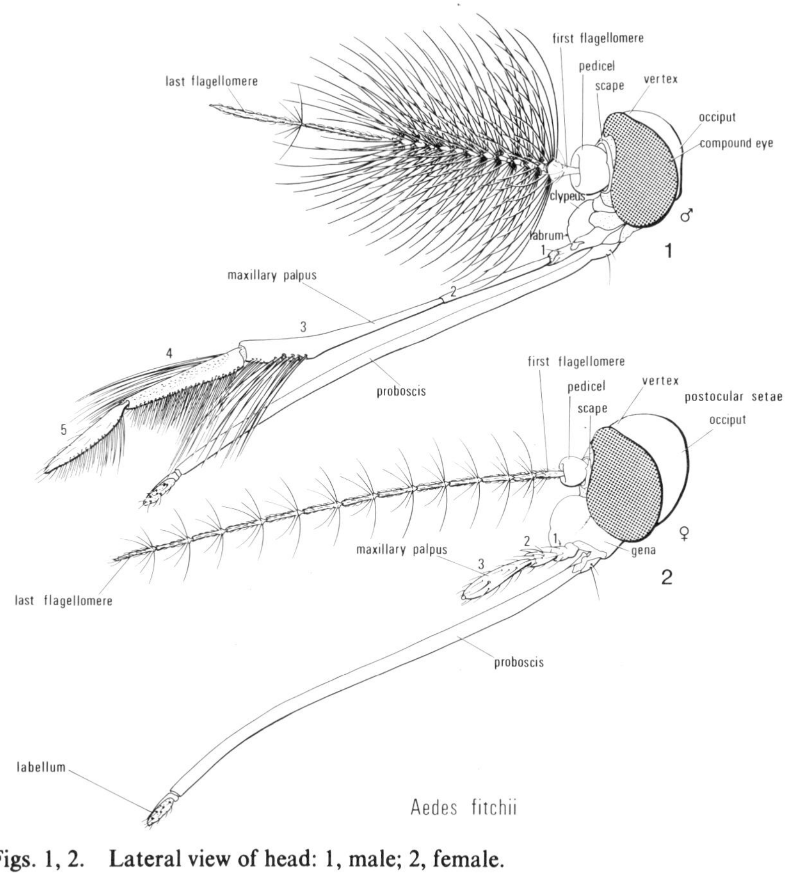

Figs. 1, 2. Lateral view of head: 1, male; 2, female.

elongate and all five palpomeres are distinct. The first (basal) palpomere is partly subdivided, which gives the palpus of males a six-segmented appearance. Because of reduction of the apical segments, the female palpus appears to be two- or three-segmented. The basal half of the first palpomere in both sexes lacks scales and hairs, and in dried material it looks like a basal segment. The distal half of the first palpomere is fully scaled, like the remaining palpomeres, but it can scarcely be distinguished from the second palpomere because the junction is hidden by scales. The third palpomere in both sexes is the longest and often the thickest; in males of most species of Culicinae, it bears a fringe or tuft of long setae near its apex. The fourth and fifth palpomeres of the male are subequal in length and are usually also fringed ventromedially and laterally. To locate a particular palpomere in the female, start counting from the base, but remember that the basal half of the first palpomere is bare of scales and the second half is not. The longest segment, the apparent apical segment, is the third palpomere. In the male, however, and in the female *Anopheles*, it is easier to start counting from the apical, or fifth, segment. In male *Anopheles*, the last two palpomeres are broadened and swollen, and the intersegmental boundary is visible only from below. The apex of the palpus is usually conical and more or less obscured by dense scales and setae; a broken palpus shows a small, round central opening.

The compound eyes are nearly contiguous both dorsally, above the antennae, and ventrally, behind the base of the proboscis. The vertex, or dorsal part of the head behind the eyes, blends indistinguishably into the back of the head, the postcranium or occiput, and is more or less covered with distinctively colored erect and decumbent scales. The erect scales on the vertex are usually truncate or forked apically and often differ in color from the narrow, curved decumbent scales at their bases. A row of forwardly directed (proclinate) setae called the postocular setae arises on the vertex, parallel to the hind margin of the eye; these setae may be reduced to a single median pair in some genera (sometimes called the interocular setae).

Each of the three segments of the thorax (Figs. 3-6) is divided horizontally into a dorsal notum and a ventral sternum. The lateral pleural plate makes up the side of each thoracic segment. In winged insects, a vertical ridge, or thickening, divides this pleural plate into an anterior episternum and a posterior epimeron; the episternum may be further subdivided longitudinally into a dorsal anepisternum and a ventral katepisternum.

Most of the thorax belongs to the second, or middle, segment, the mesothorax. The prothorax and metathorax are smaller, especially middorsally, where they are almost obliterated. What remains of the dorsal tergum of the prothorax, the pronotum, is subdivided into laterally displaced halves. It is divided transversely into a lobe-like antepronotum and a flattened postpronotum, which is flush with the mesothorax and appears to be part of that segment. The postpronotum typically has a vertical row of setae near its

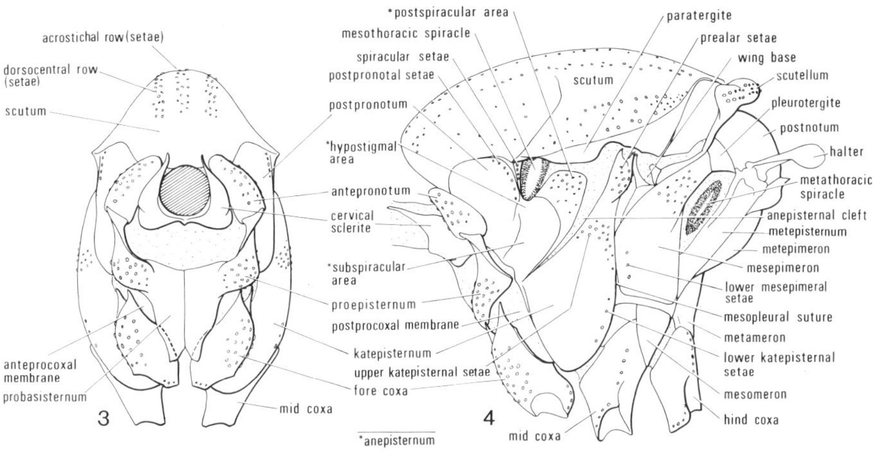

Figs. 3, 4. Thorax of female: 3, anterior view; 4, left lateral view.

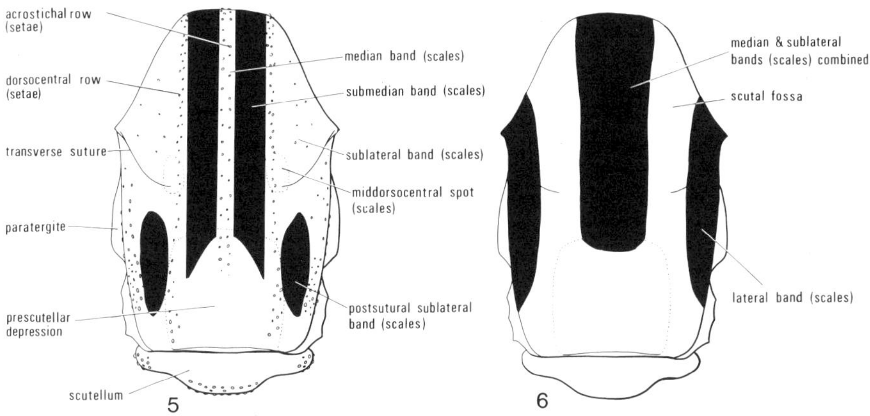

Figs. 5, 6. Scutal pattern of female: 5, *Aedes communis*; 6, *Aedes trivitattus*.

posterior edge that may overlap the spiracle behind it; in arctic species, additional setae may be present along the dorsal part of the postpronotum. Anteriorly, between the fore coxae and below the neck, is a V- or W-shaped sclerite, the probasisternum, occasionally with hairs and scales. The probasisternum is connected above the fore coxa to the proepisternum, which lies between the antepronotum and the fore coxa. The membrane lying between the probasisternum and the fore coxa, the anteprocoxal membrane, and the membrane between the fore coxa and the mesothorax, the postprocoxal membrane, may each bear scales in some species of *Aedes*.

The notum of the mesothorax, or mesonotum, is divided transversely into three sections: the mesoscutum, the mesoscutellum, and the mesopostnotum. Because these subdivisions do not occur in the prothorax and metathorax, it is customary to omit the prefix "meso"; therefore the terms used in this publication are scutum, scutellum, and postnotum. All three make up the mesonotum, a term often incorrectly applied to the scutum alone.

The scutum occupies most of the dorsum of the thorax and is rather imprecisely divided by an incomplete transverse suture into a presutural and a postsutural portion. Setae may arise scattered over much of the scutum, but they are typically more concentrated into a median, or acrostichal, row; a marginal, or supraalar, row; and a row between these two, the dorsocentral row. Scales occur over most or all of the scutum, and though they are easily rubbed off, their arrangement, pattern, and color are usually important for species determination. Although a variety of patterns occurs, two patterns are the most common. In one pattern (Fig. 5), the median area, between the two rows of dorsocentral setae, is divided longitudinally into three stripes, a median stripe of paler scales flanked on each side by a submedian stripe of darker scales. This darker submedian stripe is flanked laterally by paler scales that are concolorous with the median stripe. This pattern is found in *Aedes communis*, *diantaeus*, and *pionips*. Behind the transverse suture, there is usually a second, shorter, dark-scaled postsutural sublateral stripe concolorous with the presutural submedian stripe, either separated from it by pale scales as in *Aedes pionips*, or contiguous with it, as in *Aedes thibaulti*. In another common pattern, the median and submedian stripes are concolorous and contiguous, forming a single broad middorsal stripe (Fig. 6). This pattern is found in *Aedes fitchii*, *punctor*, *sticticus*, *triseriatus*, and others. In *Aedes campestris*, *dorsalis*, and *trivittatus*, the scales of the upper half of the postpronotum and the adjacent edge of the scutum match those of the single middorsal stripe, giving a tripartite appearance to the thorax, as shown in Fig. 6. Finally, the presutural sublateral area, or scutal fossa, which is somewhat triangular and slightly depressed, is usually pale scaled, but it may enclose a patch of scales concolorous with those of the midlongitudinal stripe, thus confining the pale scales to the region of the dorsocentral setae. In some *Culex* and *Culiseta*, the pale scales are most evident as a pair of middorsocentral spots. The central region of the postsutural part of the scutum between the wings is the prescutellar depression. It is usually surrounded by paler scales. The posterior margin of the scutellum may be evenly convex, with a uniform row of posteriorly directed setae, or it may be tripartite, with setae grouped on the lobes. The postnotum has a small group of scales only in *Wyeomyia smithii* (and its tropical relatives), whereas in other species it is bare.

A small oblong strip of integument, the paratergite, lies at the edge of the scutum between the anterior thoracic spiracle and the wing base. The mesothoracic pleural suture, which runs from the mid coxa up to the base of the wing, divides the mesopleuron unequally into a large anterior mesepisternum and a smaller posterior mesepimeron. The mesepisternum is subdivided by a longitudinal anapleural suture, which extends posteriorly or posterodorsally from the proepisternum into a dorsal mesanepisternum and a ventral mesokatepisternum (called mesepisternum by Knight and Laffoon 1970*a*). The prefix "meso" is not used for these two sclerites because this subdivision is used only on the mesepisternum. The anepisternum is further subdivided by a diagonal anepisternal cleft (Fig. 4) into an anterior and a posterior portion. The anapleural suture is distinct in *Uranotaenia*, but indistinct in most other mosquitoes, so that the posterior anepisternum, which bears the prealar setae, appears as a dorsal extension of the katepisternum. The somewhat triangular anterior portion of the anepisternum, with its right angle next to the postpronotum and scutum, bears the anterior (mesothoracic) spiracle. A small taxonomically important group of setae, the spiracular setae, arise just anterior to the sclerotized anterior margin of the spiracle, but posterior to the postpronotal setae, which can be mistaken for the spiracular setae. The spiracular setae are absent in *Aedes, Mansonia,* and *Culex*, but are present in most other genera. The remainder of the anterior anepisternum is arbitrarily divided into three areas according to the three scale patches that may be present. A small area, about half the size of the spiracle and immediately ventral to it, is the hypostigmal area. The subspiracular area is ventral to this area, whereas the postspiracular area, which may bear the taxonomically important postspiracular setae, is posterior to the spiracle. The long, tapering ventral angle of the postspiracular area extending between the subspiracular area and the anepisternal cleft may bear scales in addition to those in the vicinity of the postspiracular setae. The katepisternum is particularly well developed in mosquitoes and related families and it often unites with the opposite katepisternum between the fore and mid coxae, separating them from one another and obscuring the predominantly internal mesosternum. The katepisternum has often been called the sternopleuron, because of the mistaken assumption that the mesosternum has become fused with it. Setae are present in two groups: the upper katepisternal setae in a group along the posterodorsal margin, and the lower katepisternal setae along the posterior margin above the mid coxa. Scales are associated with all three groups of setae, and in most *Aedes* those scales in the midsection associated with the upper katepisternal setae extend varying distances across the katepisternum toward the anterodorsal corner, which is just behind the proepisternum. Behind the mesopleural suture, a rectangular sclerite, the mesepimeron is also divided transversely into a large dorsal mesanepimeron and an insignificant mesokatepimeron. The latter is a narrow ribbon-like, light-colored bare strip along the ventral margin of the mesanepimeron. Because only the mesanepimeron is used taxonomically, it is referred to simply as the mesepimeron. This sclerite always has a group of setae, the upper mesepimeral setae, at its upper end under the base of the wing. In many species, a few lower mesepimeral setae are also present, located along the anterior edge of the mesepimeron at the level of the upper katepisternal setae. Scales that are always present on the upper half of the mesepimeron may extend ventrally for varying distances in various species. Below the mesepimeron and between the mid and hind coxae is a rounded sclerite, the mesomeron, which is considered to be a portion of the mid coxa that has become detached (Crampton 1942).

The metathorax, like the prothorax, has become greatly reduced. However, the metepisternum, the metepimeron, and the metameron are distinguishable; they are in the same relative positions as their homologues of the mesothorax. The metanotum has been obliterated and is nothing more than a narrow ridge separating the mesopostnotum from the first abdominal segment, just above the halter. The halter, the metathoracic spiracle just in front of the halter, and the hind leg are the only prominent metathoracic structures.

The wing (Fig. 7), which belongs to the mesothorax, is long and slender in all mosquitoes. The wing veins are typically covered, dorsally and ventrally, with more or less decumbent scales. Even the two crossveins have scales in *Culiseta alaskaensis*. The posterior margin of the wing is fringed with a row of long scales that is a continuation of the scales at the apex of the costa. The scales on the wing veins are usually dark, but many species have some pale scales intermixed, either scattered or as alternating patches. The scales on the wing veins of the male, though similar in color to those of the female, are sparser and seem easily detached; therefore, their color cannot be used as a distinguishing character in males. The calypter, a rounded scale at the base of the margin of the hind wing, is sometimes fringed along its edge.

Except for CuP, a minor vein of no taxonomic importance lying just behind CuA, all wing veins are covered with scales on both their dorsal (Fig. 7) and ventral surfaces. The relative positions of dark and pale scales are of considerable taxonomic value. Some species have dark or pale scales aggregated into spots, and the positions of these spots are described with reference to the branching points and the apices of the veins. The costa (C) and subcosta (Sc) are unbranched, but the radius (R) branches near its base into a dorsal branch (R_1) and the radial sector (R_s), which branches again near the middle of the wing into R_{2+3} and R_{4+5}. R_{4+5} remains unbranched to the wing margin, but R_{2+3} branches again into R_2 and R_3. A short crossvein (r-m), which connects R_{4+5} to the media (M), is without scales except in *Culiseta*. The next vein behind M is the anterior branch of the cubitus (CuA), ending in two branches CuA_1 and CuA_2. CuA_1 is considered to

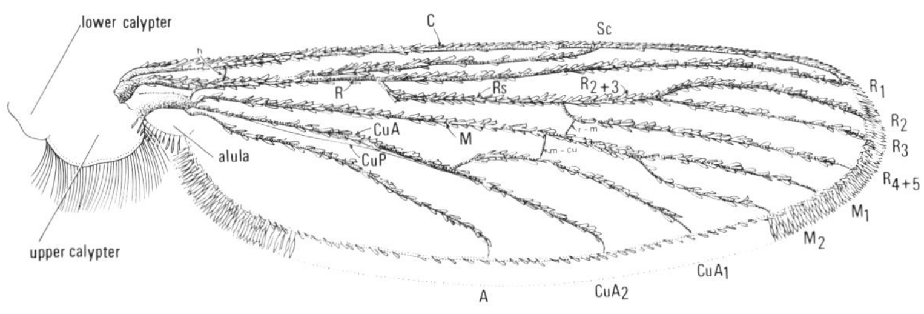

Fig. 7. Dorsal view of wing.

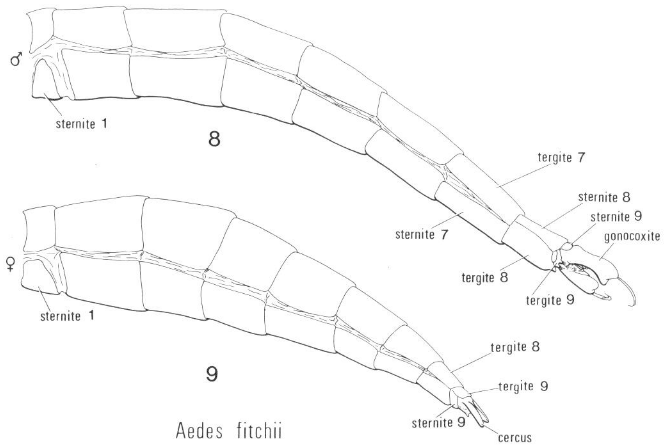

Figs. 8, 9. Lateral view of abdomen: 8, male; 9, female.

contain elements of the media (see Hennig 1973) and therefore it has been referred to sometimes as M_4 or as $M_4 + CuA_1$. However, in mosquitoes CuA_1 seems to be more obviously a part of the cubitus than of the media. The media and cubitus are joined by a short crossvein, shown in Fig. 7 as m-cu. The hindmost vein is the anal vein (A).

Each of the three thoracic segments bears one pair of legs, which are referred to here as forelegs, midlegs, and hind legs. The legs are almost entirely covered with scales. Each leg is divided into nine segments. The coxa and trochanter, the two short basal segments, are usually pale scaled and have not been useful taxonomically. The femur and tibia are the two longest segments. The tarsus is five segmented; the terminal segment, or apical tarsomere, almost always bears a pair of curved tarsal claws. The tarsomeres are numbered from the first (basal) to the fifth (distal). Thus the basal tarsomere of the hind leg is the first hind tarsomere. The surfaces of each leg are described as if they were held straight out to one side; thus there is an anterior surface, an anterodorsal surface, a dorsal surface, and so on. The femur is usually paler scaled basally than apically, but otherwise has not been examined for characters. The tibia and each of the five tarsomeres may be almost entirely dark scaled or they may be characteristically banded with rings of white scales. In the females of most genera, all tarsal claws are simple, that is, without subbasal teeth, and they are more or less alike on all three pairs of legs. In the male the outer fore and mid claws are highly modified for grasping the legs of the female. Only on the hind leg of the male are the two claws similar, and both resemble those of the female. The claws of most female *Aedes* and *Psorophora* have a subbasal tooth, whose length and angle are often characteristic of the species.

The abdomen (Figs. 8 and 9) in both sexes consists of ten segments of which only eight are readily visible. Segments IX and X are reduced, usually telescoped into segment VIII, and are more or less modified as part of the terminalia, or external genitalia. Unless distended with blood or eggs, the mosquito abdomen is strongly flattened dorsoventrally. The sclerite on the dorsum of each segment, or tergite, and that on the venter, or sternite, are each (except for the first sternite) entirely covered with hairs (as in *Anopheles*) or decumbent scales (in all other genera). An hour or so after the adult emerges, the apex of the male abdomen beyond segment VII rotates 180°, either clockwise of counterclockwise, bringing the external genitalia into an inverted position, with the eighth and ninth tergites in a ventral position. Nevertheless, these structures are referred to as though they were in their correct anatomical positions, not their rotated positions. All drawings of male terminalia (Figs. 10 and 11) are of a dorsal view. The male segment VIII is usually undistinguished, but in *Culiseta* subgenus *Culiseta* its tergite is adorned apically with modified setae. The ninth tergite is usually bilobed, with each lobe armed apically with short stout setae. Segment X, usually called proctiger, is a simple membranous cone in *Anopheles*, but is complexly sclerotized in the other genera. In these other genera, the 10th tergite and sternite are each longitudinally subdivided; each half of the 10th tergite is insignificant, whereas each half of the 10th sternite is modified into a more or less S-shaped sclerite, the paraproct, whose strongly sclerotized apex is usually armed with small teeth. In *Culex* this sclerite also has a basal lobe or arm. The aedeagus (also called phallosome in mosquito literature, but a term not widely used for other Diptera) is typically a bulb- or flask-like structure whose apex is partly enclosed by the paraprocts. In *Culex*, the aedeagus develops complex lateral processes. The aedeagus is supported basally by a pair of short simple rods, the parameres. Ventral to the proctiger and aedeagus (but actually above them in the rotated terminalia) are the gonopods (claspers), by far the most conspicuous elements in the male terminalia. Each gonopod is subdivided into a massive basal gonocoxite (basistyle, basimere, or sidepiece) and a more slender distal gonostylus (dististyle, distimere, or clasper) that usually terminates in a short stout peg-like apical seta. In *Aedes*, the gonocoxite, though cylindrical, is sclerotized only ventrally, laterally, and dorsally, and thus appears "open" medially along its full length. The two sclerotized "edges" of this open cylinder are the medioventral and mediodorsal edges. The medioventral edge usually bears a fringe of long setae, whereas the mediodorsal edge is often convoluted, usually with two taxonomically important lobes, a basal lobe and an apical lobe, variously shaped and variously armed with setae of different lengths and thicknesses. In *Culiseta*, the basal and apical lobes are not configurations of this sclerotized "edge," but they arise as tubercles from the "closed" cylinder. In *Culex*, the basal lobe is displaced so far apically, because of extreme median enlargement of the basal foramen of the gonocoxite, that it is called the subapical lobe. This subapical lobe bears a set of complexly modified setae, which may be rod-like or flattened and paddle-like and often have recurved apices. The median dorsal corner of the gonocoxite is elongated anteriorly into a process on which the paramere articulates. The aedeagus is thus supported, by the parameres, from the median dorsal corners of the two gonocoxites. The median ventral corner of the gonocoxite also supports a process, the

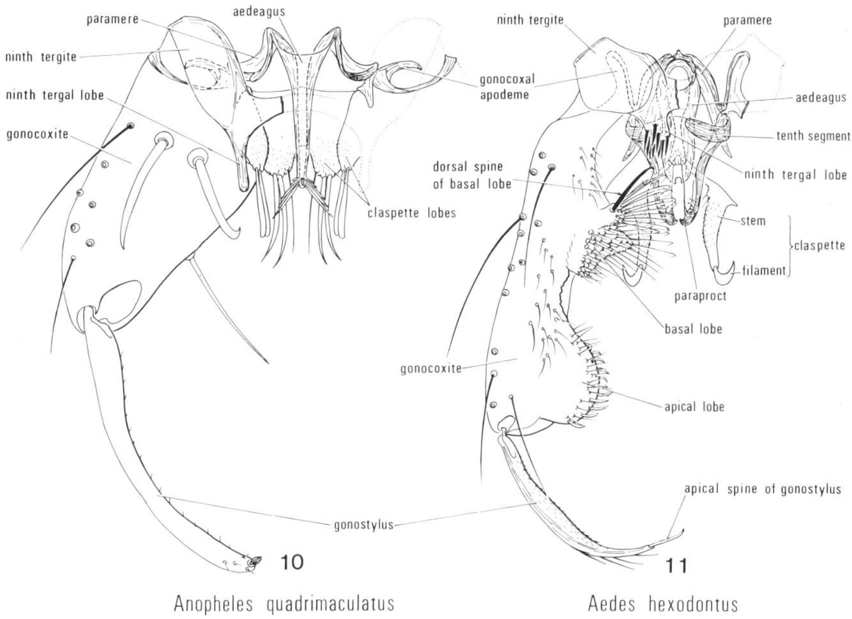

Figs. 10, 11. Dorsal view of male terminalia.

claspette. In *Anopheles*, the claspette is bilobed and armed with setae, whereas in *Psorophora* and most species of *Aedes* it is a complex, dorsally curved finger-like structure of great taxonomic significance. In the other genera it is lacking. The pair of claspettes is ventral to the aedeagus, whereas the proctiger is dorsal. However, the actual position, because of rotation, is the reverse, with the claspettes lying above the aedeagus and the proctiger below it.

The female terminalia are of little taxonomic importance. A pair of apical processes, the cerci are long and slender in *Psorophora* and most species of *Aedes*, and are usually protruding sufficiently to be diagnostic (as in Fig. 9). Otherwise they are short, rounded, and inconspicuous.

Larva

The presence of labral brushes, the fusion and enlargement of the three thoracic segments, and the tubular, cylindrical, or conical respiratory siphon arising from the dorsum of abdominal segment VIII (actually VIII and IX fused) in all genera but *Anopheles* distinguish mosquito larvae from all other larvae. Chaoborid larvae also have the thoracic segments fused and enlarged, but their labral brushes are reduced and are not used for filter feeding. Dixid larvae, which have labral brushes but lack a siphon, could be mistaken for *Anopheles* larvae except that their thoracic segments are discrete and are not broadened.

As in almost all Diptera, the larvae of mosquitoes pass through four instars. The first instar is decidedly different from the other instars, and its species cannot be as easily identified as the others. It is readily recognizable by the presence of an egg burster, a sharply pointed tubercle arising in mid center of the dorsal surface of the head that is assumed to rupture the egg shell. In Culicinae, only the apex of the siphon is sclerotized, with the pecten teeth arising from the transparent membranous basal portion. Several papers have been published that deal extensively with the first instar, and they provide keys for the separation of the species (Bohart 1954, Price 1960, Dodge 1966). Penn (1938) compared instars of *Aedes triseriatus* with respect to an increase in number of pecten teeth and comb scales. Also, Abdel-Malek (1949) compared the four instars of *Aedes trivittatus*. We have not prepared a key to first instars because we lack material and we believe it is easier to rear them to later instars than it is to attempt to identify them.

Identification of mosquito larvae to species is largely dependent on the arrangement and numbers of branches of the setae (Figs. 12–16). All setae are paired and are symmetrically arranged. A seta is referred to in the singular, thus seta 1-C is one of a pair. A seta may have many branches arising from a common base; therefore it is necessary to identify the base of each seta. Even when a seta is broken off, the socket in which it was attached can still be located.

The second, third, and fourth instars resemble one another more closely than they do the first instar. In all three, the egg burster is lacking and the siphon is more completely sclerotized. The branching of the setae increases with subsequent instars, as does the number of comb scales. In genera other than *Anopheles* and *Aedes*, the last instar is easily recognized by the saddle that completely encircles the anal segment. In the second and third instars of all species and in *Anopheles* and most *Aedes*, the saddle is merely an oval sclerite on the dorsal surface of the anal segment, hence the name "saddle." The last instar of those *Aedes* that do not have a completely encircling saddle may be recognized by the position of the saddle seta 1-X, which arises closer to the posterior margin of the saddle than to its ventral margin (Smith 1969*a*, 1969*b*). In all second and third instars, the saddle seta arises at, or very close to, the ventral margin of the saddle, often in the membrane ventral to the saddle. Only in two species of *Aedes, atropalpus* and *togoi*, the saddle seta arises in the membrane below the saddle in the fourth instar as well as in the third and second; these two species have been dealt with accordingly in the key. Dodge (1963) and Smith (1969*a*) provided keys to the second, third, and fourth instars of most of the eastern species.

The external morphology of the fourth instar has been treated in detail (Belkin 1950, Knight and Laffoon 1971, Laffoon and Knight 1973, Harbach and Knight 1978). Only those structures and setae used in this publication are discussed in the following description. The larvae of *Toxorhynchites* are so easily recognized and the arrangement of their setae is so distinctive from that of other mosquitoes that they are excluded from the following description.

The dorsal surface of the cranium, or head capsule, is dominated by a large rather quadrate apotome that occupies most of the center of the head and is separated from the lateral parts of the cranium by an ecdysal line. Anteriorly the clypeolabral suture separates the crescent-shaped labrum from the apotome. Each labral brush consists of hundreds of long sigmoid setae closely and precisely arranged in one of a pair of large patches on the

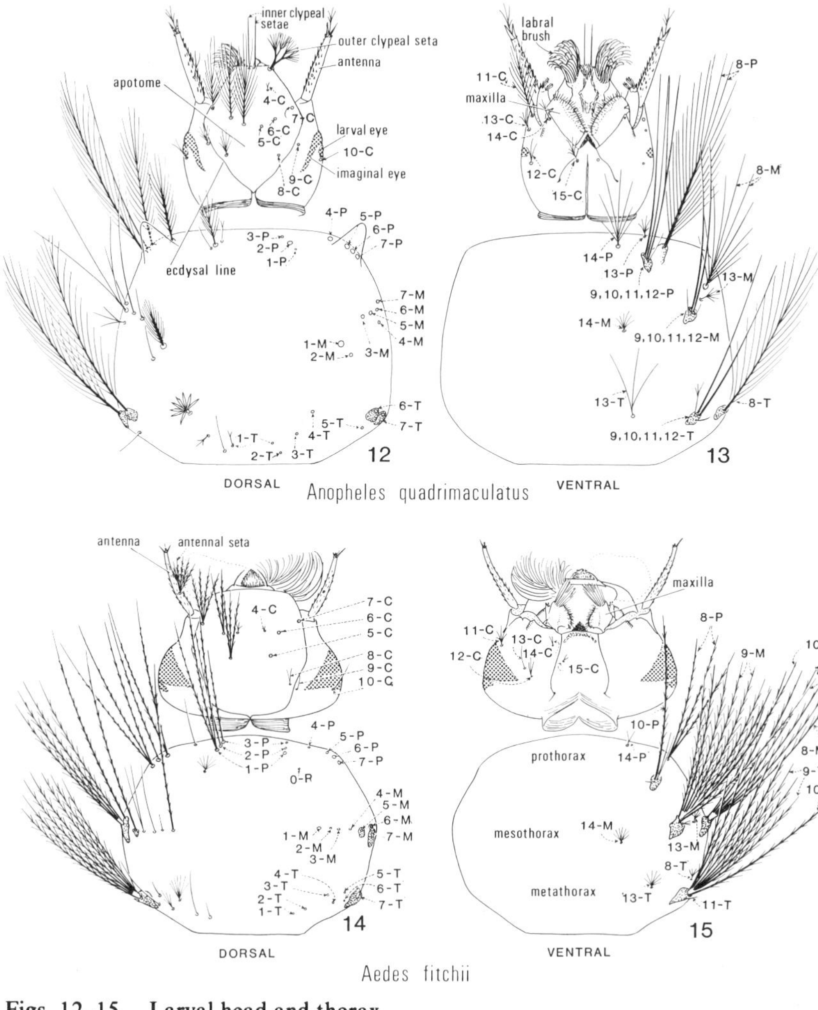

Figs. 12–15. Larval head and thorax.

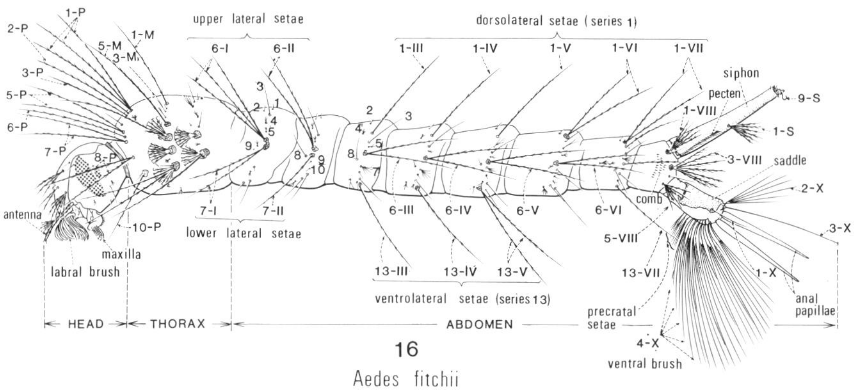

Fig. 16. Left lateral view of larva.

ventral side of the labrum. Each antenna arises outside the apotome at the anterolateral corner of the head. The compound eye of the adult develops precociously in most genera and probably also functions for larval vision along with the smaller simple larval eye, or stemma, behind it. In *Mansonia*, only the larval eye is evident as a pigmented structure, even in the fourth instar.

The cranial setae were previously assigned names but now they are numbered, followed by the letter C. There are 18 pairs of cranial setae, numbered 0-C to 17-C, but only about 6 of these are regularly used. Seta 1-C is a prominent, thick, forwardly directed paired seta arising on the labrum. Setae 2-C to 8-C arise on the apotome. Seta 2-C, the inner clypeal hair, is long and prominent only in *Anopheles*, arising very close to the midline and just beside the other member of its pair. Seta 4-C is usually small and arises far forward on the apotome. In Culicinae, seta 4-C is typically multiple and arises closer to the midline than do setae 5-C to 8-C. Setae 5-C, 6-C, and 7-C are almost always the longest, and they arise more or less side by side in *Anopheles* and in a few *Aedes*, forming a conspicuous transverse row. In most of the culicines, however, seta 6-C is displaced far forward, toward and even in front of 4-C. In this position, seta 6-C often is located in front of seta 5-C, hence the names lower head seta for 6-C and upper head seta for 5-C. The relative position and degree of branching of setae 5-C and 6-C are of utmost importance in identifying species in most culicine genera. Seta 7-C arises near the edge of the apotome, medial to the antennal base. Setae 9-C to 17-C are generally small and have not been fully examined for specific characters.

The antenna is also equipped with setae, which are numbered 1-A to 6-A. All of these, except seta 1-A, which arises from the dorsomedial aspect of the antennal shank, are minute and located at the antennal apex. Seta 1-A may be single or have many branches, and its position is diagnostic. The antennal shank is usually covered with microscopic anteriorly projecting points, or spicules.

The mouthparts vary substantially from species to species, but they have seldom been adequately studied for taxonomic characters. However, the shape of the maxilla varies noticeably among several species of *Aedes* that otherwise look very much alike. The maxilla itself is divided into three parts: the maxillary body, the maxillary palpus, and a small triangular basal sclerite, the palpifer. The maxillary body has an anterior lobe that bears a dense apical tuft of setae and a posterior lobe fringed with shorter setae.

Each of the three thoracic segments and the first seven abdominal segments are partly encircled laterally by a transverse row of paired, symmetrically arranged setae. Each pair of setae is typically numbered from 0 to 14, beginning with the one closest to the middorsal line and ending with the one nearest the midventral line. Some setae have migrated from their usual positions and therefore are difficult to number. The prothorax and abdominal segments II–VII may each have up to 15 pairs of setae; the mesothorax has up to 14; the metathorax, 13; and abdominal segment I, only 12 present at a time, although all 15 are known to occur but not in any one species (Belkin 1950).

The setae of the prothorax are numbered 0-P to 14-P. Seta 0-P is a minute tuft; in culicines it is displaced laterally, lying behind 4-P and 5-P. Setae 1-P to 3-P arise very close to each other and appear at a glance as branches of a single seta. All three setae are close to the midline, and 1-P is almost invariably the longest. Seta 4-P is usually single in *Aedes* subgenus *Ochlerotatus*, but is usually branched in others. Setae 5-P, 6-P, and 7-P are the longest prothoracic setae, each arising separately but side by side on the "shoulder," close to the lateral margin in dorsal view. Seta 8-P is below and usually in front of 7-P and is not shown in any dorsal view in this publication. Setae 9-P to 12-P arise from a small tubercle; according to Belkin (1950), 10-P is the longest of the four.

The setae of the mesothorax are arranged according to the same plan as in the prothorax except that seta 0 is missing. In the Culicinae the first four mesothoracic setae, 1-M to 4-M, are fine, usually minute, and rather regularly spaced in a transverse row. Seta 5-M is typically long and unbranched. Setae 6-M and 7-M share a common base in *Aedes* and in some *Culex*, but they are separate in the remaining genera. Seta 7-M is strongly branched in *Aedes*, but is usually single in other genera. Seta 8-M is long and strongly branched, but like 8-P is not shown in any dorsal view in the plates in this publication.

The metathoracic setae 1-T to 13-T have not been fully examined for characters and are used only to distinguish the species of *Orthopodomyia*.

Belkin (1950) used the arrangement of setae on abdominal segment II as a standard for determining homologies of the setae on segments I and III–VII. Setae on segment VIII cannot be homologized, and they have been numbered 1-VIII to 5-VIII. When numbering setae, the series number is followed by the segment number, for example, 6-III refers to seta 6 on segment III. Series 1, the dorsolateral abdominal setae, and series 13, the ventrolateral abdominal setae, are generally the longest or most conspicuous setae in their respective positions, although they are not always closest to the midline either dorsally or ventrally. These are never elongate on segments I or II. In *Anopheles*, series 1 consists of the palmate "float" setae. The most laterally placed setae, the upper lateral setae (series 6), are the longest setae on segments I–VI and are thus easily recognized. The upper lateral seta (6-VII) on segment VII is short and multiple, unlike those on the preceding segments. On segments I and II, the lower lateral setae (7-I and 7-II) immediately ventral to the upper laterals (6-I and 6-II) are also usually well developed. Because of great variation in the positions of abdominal setae and consequent doubt about their homologies, the terms dorsolateral, upper lateral, lower lateral, and ventrolateral are used throughout, in addition to the supposed series numbers, to refer to the longest seta in the particular part of the segment specified. Lengths of the dorsolateral and ventrolateral setae are given in terms of the adjacent upper lateral setae, except on segment VII, where comparison is made with the upper lateral setae of segment VI.

Segment VIII is entirely different from the preceding segments. The five setae have been arbitrarily numbered. Setae 1-VIII, 3-VIII, and 5-VIII are typically long and strongly branched, whereas setae 2-VIII and 4-VIII, interspersed between them, are finer, shorter, and unbranched, except in *Mansonia*. This pattern does not apply to *Anopheles* or to *Wyeomyia*. Laterally, segment VIII also bears, in most Culicinae, a patch of comb scales. Each scale is thorn-like and does not arise from a socket, as does a seta, but is rigidly attached to the integument. The number of scales varies greatly, from 4 or 5 to more than 100, depending on the species. Intraspecifically there is also considerable variation in numbers. The apex may be long and drawn-out, spine-like, flanked with short subapical spinules, or it may be evenly fringed with short spinules, none of which is markedly longer than the others. The spinules can only be adequately discerned under compound magnification, hence the shape of the comb scales and the length of their spinules are avoided in the larval key. The characteristics of the spinules are illustrated because they are of considerable taxonomic importance. The paired spiracular openings are surrounded, partly in *Anopheles* or completely in Culicinae, with a chitinous ring, which in Culicinae is elongated into a tubular cylindrical siphon. The siphon is so short in *Anopheles* that the spiracles appear to be flush with the surface of the segment, and the siphon is often said to be absent (Fig. 30). The siphon in *Culex* and *Wyeomyia* has a series of paired tufts or setae; no setae are designated 1-S. The spiracles at the apex of the siphon are surrounded by five flaps bearing setae 2-S to 13-S. Only 9-S, at the apex of the largest flap (the posterolateral), is considered here; it is elongate, thickened, curved, and hook-like in some *Aedes* of the *excrucians* group. Laterally, in *Anopheles* and most Culicinae, a row of pecten teeth is present on each side of the siphon. The teeth may alternate long and short in *Anopheles,* but in Culicinae they usually increase uniformly in size toward the apex. Each pecten tooth may have a single basal cusp, or a series of cusps. The last two or three pecten teeth may, in some *Aedes,* be more widely and irregularly spaced than the rest of the teeth; most authors refer to these teeth as being "detached." A single paired seta, the siphonal seta, 1-S, is usually present, either at the base, as in *Culiseta,* or between the mid length and the apex, as in other Culicinae.

The last, or anal, segment is believed to be segment X, with the obliterated remnant of segment IX at its base. In all instars, it bears an ovoid or rectangular dorsal sclerite, the saddle, which in the last instar may completely encircle the segment like a cylinder. The saddle is probably a tergite, and similar tergite-like sclerites appear also on other abdominal segments in *Anopheles* and some *Orthopodomyia.* The saddle bears a single lateral saddle seta, 1-X, which in the first three instars (and also the last instars of *Aedes atropalpus* and *togoi*) arises either ventral to the saddle on the membrane, at the edge of the saddle, or if within the saddle, closer to its ventral margin than to its posterior margin (Smith 1969*b*). In the last (or fourth) instar, the saddle seta arises well within the saddle (except in *Aedes atropalpus* and *togoi* and *Wyeomyia*) and closer to its posterior margin than to its ventral margin. Dorsally, near the posterior border of the saddle, are two long paired setae, 2-X and 3-X. Seta 2-X is the more dorsal of the two and is always branched (except in *Aedes abserratus* and *Culex restuans*). Seta 3-X is ventral to, and almost always longer than, seta 2-X; seta 3-X is branched in *Anopheles, Culiseta, Mansonia, Uranotaenia,* and *Wyeomyia*, but unbranched in *Aedes, Culex, Psorophora,* and *Orthopodomyia.*

Along the midventral line near the apex of segment X arises a series of long setae numbered 4-X that constitutes the ventral brush. *Wyeomyia smithii* has only one seta 4-X, but all other genera have several. All setae in the ventral brush are branched, and all but a few of the smallest and most proximally placed (the precratal setae) have a Y-shaped base, presumably for greater support basally. The apices of the Y-shaped bases are linked together on each side by a bar-like sclerite; all the bases together resemble a ladder, or grid. Setae are staggered alternately on either side of the midline to allow closer crowding of the bases. Setae thus linked together are called cratal setae. The precratal setae are not staggered; they arise from the midline and they lack Y-shaped bases. They arise behind the saddle when it fully encircles the anal segment, except in *Psorophora* and in most *Culiseta*, in which they arise from the saddle. At the apex of segment X are two pairs of anal papillae, except in *Wyeomyia*, which has only one pair. These anal papillae are more or less transparent conical organs, which are involved in osmoregulation; they are extremely short in species that are associated with brackish or alkaline water. Some evidence suggests that the papillae increase or decrease in length from one instar to another in response to different environmental conditions. The differences in length within any species have not been studied, and the papillae are often found to be broken off. However, the papillae sometimes offer useful taxonomic information.

Pupa

Interspecific and even intergeneric variations seem to be much less obvious among pupae than among larvae or adults. No attempt has been made in this publication to provide a means of identifying pupae that is any easier or less time-consuming than rearing the pupae to the adult stage. Darsie (1949, 1951) described and illustrated pupae of the species occurring in the northeastern USA. Additional descriptions of pupae are given by Darsie (1955, *riparius*, *pionips*; 1957, the *punctor* subgroup) and Darsie et al. (1962, *melanura*).

Egg

Eggs of most species have been described, but as with the pupae, it is easier and more certain to hatch and rear the eggs than to try to identify them. Kalpage and Brust (1968) described and illustrated the eggs of 27 species of *Aedes* in Manitoba; they also provided details on the techniques of obtaining, conditioning, and hatching eggs and on preparing them for study. The scanning electron microscope is a powerful instrument for studying and identifying eggs (Brust 1974).

Methods of collecting, rearing, and preserving mosquitoes

Collecting larvae and pupae

Mosquito larvae can be found in almost every type of nonflowing aquatic habitat from the marshy edges of large lakes, provided that fish are not present; through swamps and marshes of all types and sizes; along sections of rivers, streams, or ditches where the water is not flowing; to small collections of water in rock crevices, tree or stump holes, leaves of pitcher-plants, and artificial containers. To determine the temporal and spatial distribution of larvae in a certain area, sample each type of habitat separately and make regular collections from each, preferably at weekly intervals, beginning as soon as the snow melts and continuing until September. Make a preliminary survey of mosquito breeding in any locality by carefully examining every possible type of breeding place. Sandy or rocky lakeshores, especially those exposed to wave action, do not support larvae.

Larvae are usually collected from large ground pools and marshes by using a fine-mesh net or a white-enameled dipper. Some species of larvae seek the protective cover of floating or emergent vegetation, where they are difficult to collect. Aluminum or plastic window screening, stretched tightly across the rim of an insect net, is effective for sieving a few scarce larvae out of a large volume of water. Larvae of *Anopheles* rest at the surface of the water, but they are hard to see among vegetation because they are cryptically colored and often sparsely distributed. When they have been disturbed, most larvae swim rapidly to the bottom, where they may remain for several minutes. Consequently, it is most productive to make only one rapid skimming pass with the net in any one place, then to collect elsewhere for a few minutes before revisiting the first area for a second time. The net may be inverted over a shallow pan of water and the larvae removed with a pipette (Fig. 17*a*). When larvae swim to the bottom, you may have to wait quietly for a while until they come to the surface again. Stirring up the bottom mud to make the larvae more visible has at times been helpful, especially when larvae are wary (as are species of *Culiseta*) or are sparsely distributed.

In smaller ground pools, rock crevices, rock pools, and large artificial containers, a white-enameled dipper with a wooden dowel as a handle extender is most useful. Larvae that live in rock pools seem to burrow in the detritus on the bottom for long periods, and they may have to be stirred up and sieved out if they do not readily rise to the surface.

Several feet of transparent vinyl tubing are necessary to siphon out the liquid from tree holes. An internal diameter of at least a centimetre ensures a rapid flow with minimum blockage so that the larvae are less likely to escape to crevices in the cavity. Siphon the liquid into a pail or a clean polyethylene bag. Because there is a marked seasonal fluctuation in species, return some liquid to the tree hole to ensure continuity in sampling. Small containers, such as the leaves of pitcherplants, can be sampled directly with a pipette.

Special efforts are required to obtain the larvae and pupae of *Mansonia perturbans*, which are buried in the bottom mud attached to the roots of emergent vegetation. The various methods devised have been based either on floating the larvae from the debris at the bottom with the use of concentrated salt solutions or on forcing larvae to detach themselves and swim about by removing the emergent parts of the plant and thus curtailing their oxygen supply (Bidlingmayer 1954). We have not attempted the latter method, but we have suceeded in obtaining a few larvae by uprooting cattails and *Calla* and soaking their roots in a strong solution of KOH. The larvae are reported to be less inclined to detach in cold water (Hagmann 1952).

Complete observations and field notes on all collections provide valuable information on mosquito biology. It is essential to give each collection a number that corresponds with a record kept on a collection data sheet. However, because such information may become separated from the specimens in the course of time, be sure to record the locality, date, habitat, and name of the collector in pencil or India ink on a paper label and to place it inside every larval container.

Transporting living larvae

Larvae are most easily identified when they are in the fourth instar. However, a few immature larvae should also be collected, and, when it is feasible, they can be reared to maturity in the laboratory for easier identification. To obtain adults, especially males, in good condition, it is also necessary to transport mature larvae to the laboratory. Mosquito larvae, especially young larvae intended for rearing, should be chilled before being subjected to much jostling. Therefore, an ice chest is an essential piece of equipment, especially in hot weather. Larvae can be cooled suddenly by adding colder fresh water to their container (Weathersby 1963), but they should never be warmed suddenly. Insulated food containers with tight-fitting lids make useful transporting and rearing containers. Also, plastic bags are useful for large samples. Leave some air above the water in the bag or container and open the container to admit fresh air once a day, or more often in warm weather. Never leave samples in direct sunlight. Collect an extra container of pond water for topping up your samples later. Food should not be necessary for a day or two, provided leaves from the larval habitat are included with your larvae. For longer periods, powdered or paste fish food or powdered liver maintains an adequate growth rate. Overfeeding, as with fish in aquariums, results in foul water, which is detrimental to some species.

Rearing immature stages

Many different techniques have been developed for rearing various species from egg to adult in the laboratory (McLintock 1952, Tremblay 1955, Beckel and Copps 1953, Beckel 1958, Weathersby 1962, 1963, Brust and Kalpage 1967, Chapman and Barr 1969, Gerberg 1970, Ellis and Brust 1973). Various rearing methods were summarized by Gerberg (1970).

Collected eggs of most non-*Aedes* hatch within a few days after being laid, and they need only to be kept floating on water. The eggs of *Aedes* and of other genera that undergo diapause in the egg stage, namely, *Culiseta morsitans* and *Anopheles walkeri*, must be conditioned by storage at 20±1°C for 16 hr of light and 8 hr of darkness for 2 months, then at 5±1°C in total darkness for 3 months (Horsfall and Fowler 1961, Brust and Horsfall 1965). Soil samples containing eggs conditioned in this way need only to be immersed in dechlorinated tap water at 10±1°C for 16 hr of light and 8 hr of darkness. Eggs laid in the laboratory may be conditioned on moist nylon pads sealed in petri dishes. To initiate hatching, they must be placed in dilute nutrient broth (1:1000 by weight of powdered broth in tap water) at 10±1°C for 16 hr of light and 8 hr of darkness. The beginning of bacterial action in the soil or broth quickly lowers the dissolved oxygen in the water and stimulates hatching.

Larvae may be reared in any clean glass, plastic, or porcelain container, but photographic trays are preferable for larger numbers. The 300 mL insulated food containers are satisfactory for small samples of larvae. A density of about 75–100 larvae per litre of distilled water in a shallow pan should minimize the detrimental effects of overcrowding. The previously fed larval diets of ground-up rabbit and dog chow with added yeast and other supplements have largely been supplanted by powdered liver, powdered fish food, Tetra Min Staple, and Tetra Min Tube Food 66. The last-mentioned diet was preferred by Ellis and Brust (1973), because it tended not to foul the rearing medium if offered in excess of larval requirements. Liver powder is particularly difficult to use in the correct amount, although for pollution-tolerant species it is ideal. Brust (personal communication) found that Tetra Min Staple powdered with peat moss was effective for northern species of *Aedes*. A change or two of water may be necessary, depending on accumulation of excess food or on crowding of larvae. The rearing temperature should be maintained at that which the larvae might encounter under natural conditions. In general, 20°C is suitable for spring species, but is too cold for summer species, especially those of *Anopheles*.

Pupae should be removed with a pipette daily or separated from larvae with the use of cold water (Weathersby 1963), and transferred to a styrofoam food container half-filled with distilled water (50 pupae/300 mL container). This container should then be placed in a rearing cage fitted with a sleeve to facilitate its removal after the adults have emerged. Adults can be killed in a deep freezer at suitable intervals, usually 1 or 2 days after emergence (see the following section).

Killing and preserving larvae

Only living larvae should be preserved, because dead ones seldom make good specimens. When a live larva is placed directly into ethanol it usually curls up and later may turn black. The usual method of avoiding this has been to drop the larvae into hot, but not boiling, water for a few minutes before transferring them to 70–80% ethanol for storage. Not only is hot water seldom readily available in the field, but the larvae so treated tend to be flaccid if the water is too hot, and our results have not been consistent. To avoid the use of hot water, one of us (P.T.D.) developed a preservative, KA_3A_{30}, consisting of 1 part kerosene (reagent grade), 3 parts glacial acetic acid, and 30 parts 95% ethanol that produces firm straight larvae with erect mouthparts and setae. As much water as possible must be removed from the larvae before transferring them to the preservative, either by decanting off the water before adding the preservative, or preferably, by partly drying the larvae on absorbent paper. Leave the larvae in this preservative for about 15 min, then decant off the preservative, leaving the larvae in the vial. The preservative can be reused until it becomes cloudy. Wash the larvae in 95% ethanol, decant them, and reflood them with 95% alcohol for permanent storage. Throughout the process, do not handle the larvae; leave them in the same vial. The KA_3A_{30} preservative must be refrigerated or made up fresh periodically, especially in hot weather, because when the ingredients are left standing, acetic acid and ethanol combine to form ethyl acetate, which can be detected by its characteristic odor. The preservative may be destructive to neoprene stoppers; therefore the larvae should be rinsed thoroughly with 95% ethanol before permanent storage.

Transporting and storing preserved larvae

Rough treatment will quickly denude preserved larvae of their setae, regardless of how carefully they have been preserved. Ideally, place larvae in a vial filled with 95% ethanol and stopper it with cotton so that all air bubbles are excluded. Then place this vial in a larger neoprene-stoppered alcohol-filled vial with additional cotton to support the smaller vial, especially during shipment by mail. For permanent storage, avoid cork or polyethylene stoppers and screw caps. Only neoprene bungs have been shown to consistently prevent evaporation of 95% ethanol. Every container must contain a good-quality paper label, giving the locality, date, habitat, and name of the collector, written in pencil or India ink. The locality must be clearly indicated so that it is universally intelligible, and the month should be written in Roman numerals. Inadequately or cryptically labeled material is of no permanent value and will eventually be discarded.

Shipping

Pack the vials in a small stiff-walled inner box. Separate the vials from one another and from the sides of the box by wrapping each one in facial tissue or paper toweling. Gaps in the box can be filled with more tissue or toweling. Tape or tie this box securely and then pack it in a larger outer box so that the inner box is firmly cushioned on all sides by packing material (for example, more toweling, tightly balled newspaper, excelsior, and so on). Securely seal the outer box with tape and affix the address label (with the sender's and the addressee's name and address in full) and another label stating "preserved specimens for scientific study—of no commercial value," and then tie the box with strong string.

Mounting larvae on permanent slides

Before larvae can be mounted, they must first be dehydrated in 95% ethanol (if they have not already been preserved in ethanol), then in 100% ethanol, then in xylene (cedarwood oil can also be used but it takes longer to penetrate), and finally in a 1:1 mixture of xylene and Canada balsam, with at least 15 min in each reagent. On a clean microscope slide, place one large drop of viscous Canada balsam slightly to the right of center. Transfer one larva from the xylene–balsam mixture and cut it with optical scissors or break it in two with fine forceps at the sixth abdominal segment. Orient the anterior section, dorsal side up, with the head toward the top edge of the slide (Fig. 17*b*) and the posterior section, left side up, with the cut end also toward the top edge of the slide. *Anopheles* larvae do not need to be cut in half. When the larva has been positioned, moisten the surface of the drop of balsam with a minute amount of xylene and place a clean cover slip on top of the balsam. The specimen may be oriented after placing the cover slip in position by inserting a No. 000 pin between the cover slip and the slide. If the balsam is viscous enough and you have used the correct amount, the larva will not be crushed when the balsam dries, and therefore cover slip supports such as pieces of glass rod are not necessary.

Collecting and preserving adults

The identification of adults, especially females, depends to such an extent on the color and arrangement of scales that, regardless of how the

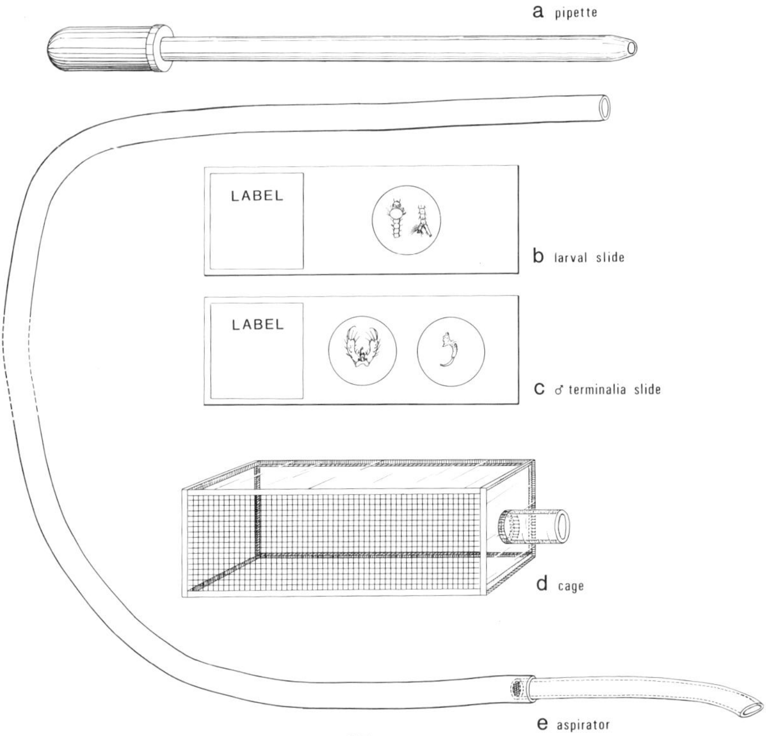

Fig. 17. *a, d, e,* Collecting equipment; *b,* microscope slide of larva; *c,* microscope slide of male terminalia.

specimens are to be collected, the most important consideration is preventing them from becoming denuded. The best way to obtain a series of adults in good condition for study is to rear them from identified larvae. Adults caught in the wild have some advantages: their integument has hardened properly so they do not collapse too badly when they are dried, and they are often taken while swarming or feeding and, therefore, they may contribute useful data. However, if you rear the larvae to adults, you must allow them to live for a day or so, long enough to allow the integument to harden, but not so long that they become denuded. In the laboratory, adults may be easily and conveniently killed by placing the entire cage containing them in a deep freezer for an hour or two. When killed this way, they usually die with their wings folded and their legs outstretched, in a convenient position for pinning. When they are frozen, they are extremely brittle and easily broken. They soon begin to dry out, even in the freezer, and after a couple of weeks it may no longer be possible to manipulate their legs. Therefore, pinning is best done before they have begun to dry. Some workers prefer double mounts, but to do this is an extra time-consuming step, and the specimens require more space in the collection. We prefer gluing the specimen by its right side directly to the side of a No. 1 pin, about 1 cm below the head of the pin, by means of a small droplet of shellac gel.* The dorsum should always be up, and the right side of the specimen glued to the pin so that the left side is exposed for study when the pin is held in the right hand. Return the specimens to the deep freezer as soon as possible after pinning, and leave them there for 2 or 3 months to freeze-dry. By this method shrinkage is minimal. In the field, reared adults may have to be kept alive for 2 or 3 days and fed on fruit or sugar-water so that the integument hardens sufficiently.

Males are usually collected in the field when they are swarming. Arctic species swarm in full sun, either in protected hollows or above a prominent landmark such as a tent. Below the tree line, swarming usually does not start until sundown. Dyar (1922*b*) has described the swarming habits of many species. Swarms may form in clearings over the shoreline or near prominent trees or shrubs. A long-handled net such as one marketed by Bio Quip, Box 61, Santa Monica, California, for tropical collecting is indispensible for netting males from the higher swarms. Such specimens are inevitably denuded of their scales, but this is not a handicap because the palpi, hind tarsi, and terminalia are usually sufficient for determination.

Females of most species are usually collected as they settle on the observer and prepare to bite. A clean, dry, transparent plastic or glass vial, 2–3 cm in diam, can usually be placed over even the most wary individual, once she has begun to feed. Less easily disturbed specimens can be collected directly into a killing tube if they are not to be kept alive. A simple killing tube can be made by placing a few drops of ethylene dichloride (or ethylene tetrachloride, which is less volatile but may be more dangerous to the collector) on a small wad of sponge rubber (not plastic) in the bottom of a wide-mouthed glass (not plastic) vial that is several centimetres long and narrow enough for your thumb to prevent a mosquito from escaping. Separate the specimens from the solvent-impregnated rubber with a soft paper plug and stopper the top of the vial with a cork. A killing tube lasts for 2 or 3 weeks before requiring recharging.

Adults of both sexes can be collected in Malaise traps. This device has the advantage of sampling the various species present without attracting some in preference to others (Breeland and Pickard 1965). However, the effectiveness of a Malaise trap for catching females may be greatly enhanced by the use of CO_2 in the form of dry ice (Graham 1969*a*). A small piece of Vapona strip can be used as a killing agent so that the adults can be kept dry. The trap should be emptied as often as necessary to prevent the

---

* To make shellac gel, boil 125 mL of pure white shellac for about 15 min, stirring continuously, until it turns white and foams; add 10 mL of 70% ethanol and boil for another 4 min until it foams again; pour into small screw-cap vials and keep them tightly closed when not in use.

catch from drying out. Some females of *Culex* and *Culiseta*, which are not attracted to man, can be collected with a Malaise trap. Both sexes of some species are attracted to light; our specimens of *Uranotaenia* adults have only been collected in light traps. Female mosquitoes may also be captured with the use of an insect net, but they lose their scales readily. Dead mosquitoes cannot be carried any distance in your pocket or in a vehicle without becoming unrecognizably denuded, unless you take special precautions to prevent them from shifting around in their container. This problem can be alleviated by bringing them back to the laboratory alive in a cage.

Do not preserve adults in liquids such as ethanol or formalin, because the scales cannot be seen easily in fluid, and they soon become detached when the container is transported. If preservation in fluid is inescapable, Vockeroth (1966) has devised a method of removing small Diptera from alcohol before they are pinned.

Label each pinned adult with its locality, date, and collector, similar to the labeling described for larvae. If the locality is unintelligible, the specimen has no future value to others. The use of numbers alone means nothing to anyone but the collector, and numbers are easily incorrectly transcribed. Many otherwise fine collections have been discarded because the key to the numbering system was lost.

Adult females that are to be kept alive in order to collect eggs from them can be transferred to small cages. Horsfall has developed a simple cage constructed of transparent acrylic sheet plastic and fiber glass window screening (Fig. 17*d*). Two sides and two ends are acrylic; one end is bored to hold a neoprene stopper. The mosquitoes can be transferred to the cage from a clear plastic vial or aspirator (Fig. 17*e*). Wrap the cage in damp toweling, and place a raisin or a wad of cotton soaked in a sucrose solution touching the screening. To provide a blood meal, you can strap the cage to your arm. Eggs will be laid on the surrounding toweling.

Macerating and mounting of terminalia and tarsal claws

Remove the tip of the abdomen by breaking it off with fine forceps or by cutting it with optical scissors just in front of segment VIII. If the specimen is dry, you can soften the abdomen somewhat by placing it in a 100% humid atmosphere for a few hours or by moistening the apical segments with a drop of water. Place the terminalia in a solution of NaOH (1 pellet in 20 mL of water) and heat just to the boiling point for 5–10 min. Remove the terminalia in the solution from the heat and while the solution is still hot draw 1–2 cm^3 of it into a hypodermic syringe, pin down the end of the abdomen with the point of the needle, and with the orifice of the needle directed away from the terminalia eject the contents of the syringe. The terminalia may burst if the stream of solution is directed into the terminalia, but when it is directed away from it, any partly dissolved tissues are pulled out by suction. When they are free of internal tissue, transfer the terminalia to glacial acetic acid in order to neutralize the NaOH and to dissolve any remaining fat droplets. You can examine the terminalia in glycerin, but for permanent preparations they should be dehydrated for 5–10 min each in successively more concentrated changes of ethanol, 75%, 85%, 95%, and 100%, before being transferred to xylene (or cedarwood oil), then to a 1:1 mixture of xylene (or cedarwood oil) and Canada balsam, and finally mounted in balsam on a glass microscope slide beneath a cover slip. A common, though unsatisfactory, procedure used by previous workers was to squash the terminalia under the cover slip, deflecting the claspettes laterally. The claspettes arise ventrally and curve dorsally; therefore they are not usually visible in profile in a dorsal view. Many drawings in the literature, however, show the claspettes splayed to the side, and presumably the shape of the gonocoxite and its lobes is also distorted in such a preparation. All illustrations in this publication were made from unflattened terminalia. To obtain a good lateral view of the claspette, remove it from the terminalia and mount it under a separate smaller cover slip to one side of the remainder of the terminalia. With the use of two pairs of watchmakers' forceps, grasp the base of one gonocoxite with one pair and the base of the opposite claspette stem with the other pair, and pull off the claspette. Mount it immediately in a small drop of balsam, placing it well to the right of center (Fig. 17*c*). Then transfer the remainder of the terminalia to a larger drop of balsam in the center of the same slide, and orient it dorsal side up and anterior end toward you (because it will be inverted when viewed under most compound microscopes). The balsam should not be as viscous as that used for mounting larvae, but it must be thick enough and in sufficient quantity that when the cover slip is finally settled it does not flatten the specimen. If the remaining claspette (or both in an undissected specimen) is deflected laterally, probably you used too little balsam or too much xylene.

Tarsal claws can be removed from the dried specimen and, without further treatment, they can be mounted directly in balsam or they can be cleared first in NaOH, then dehydrated, and mounted, as described previously. Use a very small amount of balsam. After you have placed the cover slip in position, press it down to flatten the claws and to separate them from one another.

Synonymy and authorship

Some species have been named more than once, and these extra names are called synonyms. The oldest, the senior synonym, is the one that must be used, except under exceptional circumstances; the junior synonyms are not used, but are listed in chronological order at the beginning of each species treatment. In this publication, only synonyms based on North American specimens are listed; those based on specimens originating outside of North America have been omitted, but all synonyms can be found in Stone et al. (1959), Stone (1961, 1963, 1967), and Knight and Stone (1977). For example, *Aedes vexans* has two synonyms based on North American specimens, *sylvestris* Theobald and *euochrus* Howard, Dyar & Knab; both of these synonyms are given for *vexans*. The name *sylvestris* was preoccupied, that is, already used, and therefore was not available for use again in *Aedes*, so Blanchard (1905) proposed *montcalmi* Blanchard to replace *sylvestris* Theobald. However, *vexans* also occurs widely in the Old World, where it has been named at least five other times, but these five names have not been listed in this publication. The same situation is true of the other holarctic species, as well as those such as *Psorophora* with an extensive range in Central and South America.

As mentioned previously, it is mandatory that the senior, that is, the oldest, synonym be used in preference to younger names, but this is occasionally not possible. For example, *Aedes melanimon* Dyar, 1924, is really a junior synonym of *mediolineatus* Ludlow, 1907, but *mediolineatus* Theobald, 1901, a still earlier name, is the valid name for an oriental species of *Aedes*; therefore, the name *mediolineatus* is preoccupied in *Aedes* and cannot be used a second time. The species must therefore be called by its second oldest name, *melanimon*.

There are various reasons why synonyms have been, and still are, created. Usually the author of a name was either unaware of an earlier name or he thought that the earlier name referred to something else. In earlier times, when mosquito descriptions bore few important details, it is not surprising that common species were described several times. Another rich source of synonyms was authors such as Dyar, who translated every individual peculiarity into species differences and gave each a name, but which have been found by subsequent authors to belong to one species only. This consolidation can go too far, however; the name *melanimon* was believed for a long time to be a synonym of *dorsalis* until the two species were more carefully studied. As a result of improved taxonomic techniques, some of the common species of today may yet be shown to be composed of more than one similar species. If this happens, all the synonyms will have to be reexamined. Because of their potential importance, synonyms cannot be discarded and forgotten.

It is easy to understand from the foregoing explanation how it is possible for the same name to be coined twice independently, or in different genera that become subsequently united, for example, *mediolineatus* Theobald, 1901, and *mediolineatus* Ludlow, 1907. They cannot both be used in the same genus, because there can only be one *Aedes mediolineatus*. Because *mediolineatus* Ludlow was described subsequent to *mediolineatus* Theobald, the former is therefore a homonym of the latter and must be renamed or replaced by the next oldest synonym (in this case *melanimon*). Because of the possibility of duplication among the thousands of mosquito names, it is customary to write the name of the author after a specific name at least once in a text.

Names were seldom originally described in the genus to which they now belong. For example, the two species with the name *mediolineatus* mentioned previously were proposed as *Culex mediolineatus* Theobald and *Grabhamia mediolineata* Ludlow respectively. Because both species are now considered to belong to *Aedes*, parentheses have been introduced around the authors' names to show that they are no longer in the genus in which they were originally described. They are now written *Aedes mediolineatus* (Theobald) and *Aedes mediolineatus* (Ludlow), the ending of the latter changing to the masculine form to agree with *Aedes* (*Grabhamia* is feminine, *Culex* is masculine).

Acknowledgments

We are deeply grateful to C. F. Wood, Toronto, Ont., and to the National Research Council of Canada for financial assistance to P. T. Dang and R. A. Ellis, respectively, without which our collaboration would not have been possible. We wish to thank the Department of Biology, Carleton University, Ottawa, Ont., for administration of this assistance.

For the donation of specimens, the loan of type material, and the opportunity to examine material under their care, we gratefully acknowledge the help of Dr. R. Brust, University of Manitoba, Winnipeg, Man.; Dr. J. Burger, University of New Hampshire, Durham, N.H.; Dr. H. C. Chapman, Gulf Coast Mosquito Research Laboratory, Lake Charles, La.; Dr. R. A. Costello, British Columbia Department of Agriculture, Cloverdale, B.C.; Mr. M. Enfield, formerly at the University of Calgary, Calgary, Alta.; Dr. B. Helson, Dr. G. Surgenor and Mr. L. Shipp, University of Guelph, Guelph, Ont.; Dr. J. McLintock, Research Station, Agriculture Canada, Saskatoon, Sask.; Dr. S. M. Smith and Mr. R. M. Trimble, University of Waterloo, Waterloo, Ont.; Dr. R. A. Ward and his colleagues at the Medical Entomology Project, Smithsonian Institution, Washington, D.C.; and Dr. G. B. White, Diptera Section, British Museum (Natural History), London, England. Dr. A. Maire and his colleagues, Université de Québec à Trois-Rivières, shared with us their unpublished collection records from Quebec and Dr. J. Shemanchuk, Research Station, Agriculture Canada, Lethbridge, Alta., kindly made available his maps of published distributions. Mr. L. Forster, Biosystematics Research Institute, Agriculture Canada, Ottawa, identified large numbers of unsorted mosquitoes in the Canadian National Collection, plotted their distributions, and prepared photographs of tarsal claws. Mr. T. Stovell, Graphics Section, Research Program Service, Agriculture Canada, Ottawa, provided the photographs of the wings. We also thank Dr. J. R. Vockeroth, Biosystematics Research Institute, for his interest and helpful advice. We are especially indebted to Dr. R. Brust, University of Manitoba, Winnipeg, Man., for critically reading the manuscript and for encouraging us to use many of his unpublished observations.

Key to the genera of mosquitoes of Canada—adults

Males can be distinguished from females by their long plumose antennae (Figs. 1, 2).

1. Posterior margin of scutellum evenly rounded and uniformly setose (Fig. 18); palpus of female longer than antenna (Figs. 20, 21) 2
   Posterior margin of scutellum trilobed, with setae confined to the lobes (Fig. 19); palpus of female much shorter than antenna (Culicinae) 3
2. Proboscis relatively straight (sometimes curved dorsally when dry); palpus of both sexes about as long as proboscis (Fig. 20); apex of palpus (last 2 segments) of male inflated, club-like; abdomen hairy but almost devoid of scales; tarsi dark-scaled (Anophelinae) .... ***Anopheles***
   Proboscis broadest at base and strongly bent ventrally beyond middle; palpus of female straight, as long as unbent basal portion of proboscis (Fig. 21); palpus of male strongly bent dorsally, tapering apically, without long setae, and longer than proboscis; abdomen and tarsi covered with blue green or purple metallic scales (Toxorhynchitinae) .... ***Toxorhynchites***

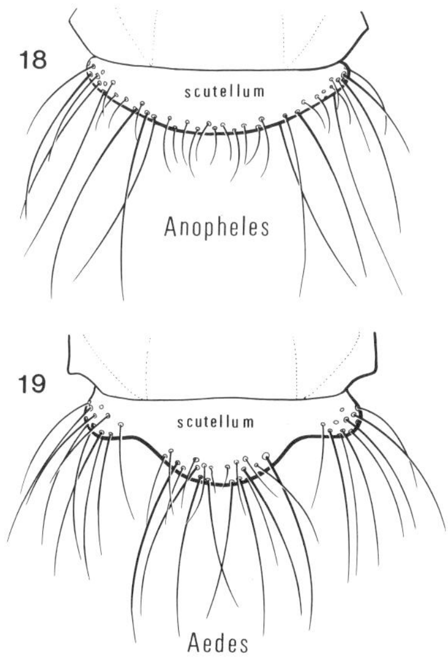

Figs. 18, 19. Dorsal view of scutellum.

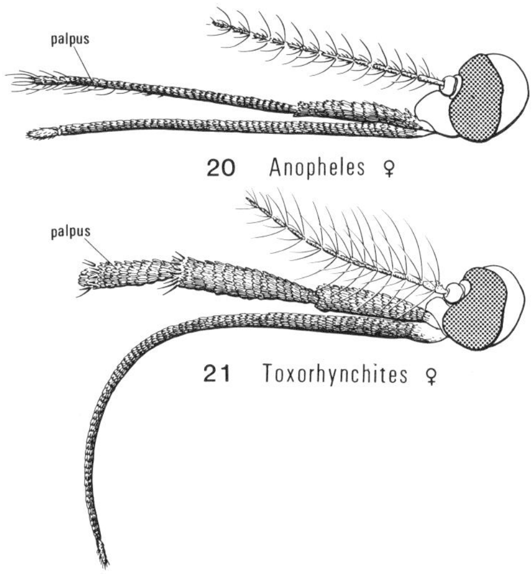

Figs. 20, 21. Left lateral view of head of female.

3. Postspiracular setae present (Figs. 22, 23: *ps*); abdomen of female tapering evenly to apex; cercus long, slender (except in *Aedes togoi* and *atropalpus*), usually conspicuously exserted (Fig. 24) 4
   Postspiracular setae absent (Figs. 26, 27); abdomen of female parallel-sided, rounded apically (Fig. 25); cercus short, rounded, inconspicuous 5
4. Spiracular area with setae (sometimes only one or two present in *Psorophora signipennis*) (Fig. 22, *s*); paratergite (Fig. 22, *pt*) and subspiracular area of anepisternum (Fig. 22, *sa*) bare or with a few scattered scales .... ***Psorophora***
   Spiracular area (Fig. 23, *s*) without setae, although setae on hind margin of postpronotum may be mistaken for them; paratergite (Fig. 23, *pt*) and subspiracular area (Fig. 23, *sa*) usually with scales (except in *Aedes cinereus, atropalpus*, and *trivittatus*) .... ***Aedes***

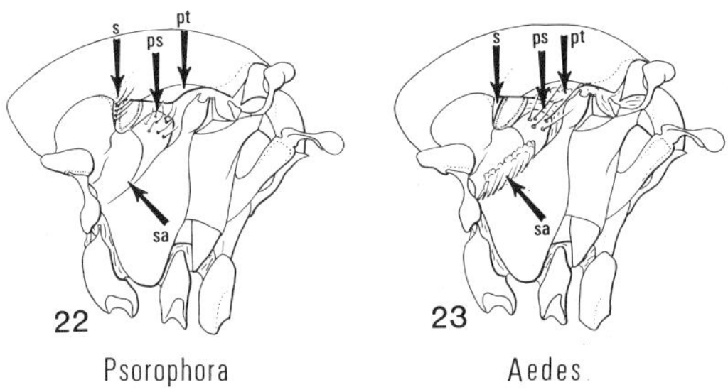

Figs. 22, 23. Left lateral view of thorax.

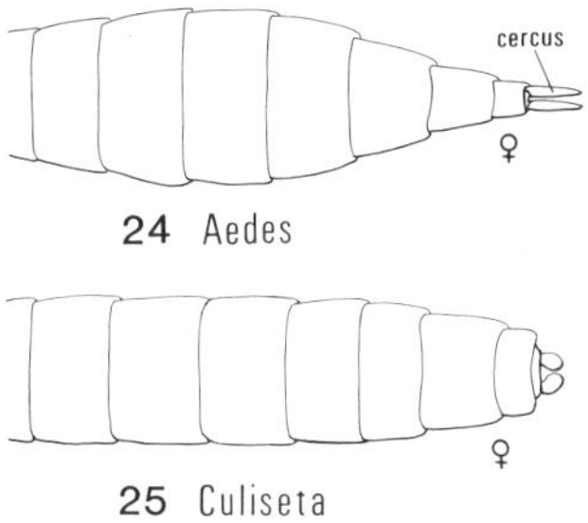

Figs. 24, 25. Dorsal view of abdomen of female.

5. Spiracular area (Figs. 26, 27: *s*) with setae 6
   Spiracular area without setae 8
6. Scales on ventral surface of subcosta at base anterior to humeral crossvein, erect, hair-like (Fig. 26, *sc*); upper calypter with fringe of long fine setae (Fig. 26, *c*); paratergite with scales or hairs on ventral margin (Fig. 26, *pt*); abdominal scales without metallic reflections .... ***Culiseta***
   Scales on ventral surface of subcosta at base broad, flat, and recumbent (Fig. 27, *sc*); upper calypter lacking fringe (Fig. 27, *c*); paratergite barc (Fig. 27, *pt*); scales on dorsum of abdomen dark brown with greenish blue metallic reflections 7
7. Entire scutum covered with broad flattened brown scales; entire postpronotum, most of pleuron, and venter of abdomen covered with silvery white scales; postnotum with a small median group of anteriorly directed setae (Fig. 27, *pn*); found in bogs, closely associated with the pitcherplant (*Sarracenia purpurea*) .... ***Wyeomyia***
   Scutum sparsely clothed with narrow brown scales except for a narrow middorsal stripe of rounded flat metallic blue scales and a patch of similar scales above paratergite; metallic blue scales also present on vertex, antepronotum, and katepisternum; usually found near semipermanent marshes, not bogs .... ***Uranotaenia***

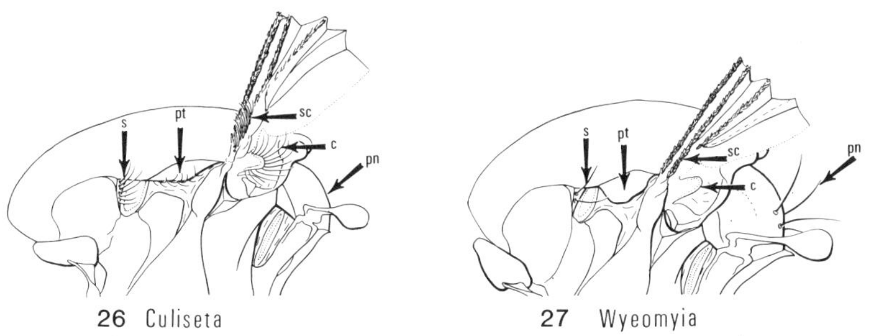

Figs. 26, 27. Left lateral view of thorax.

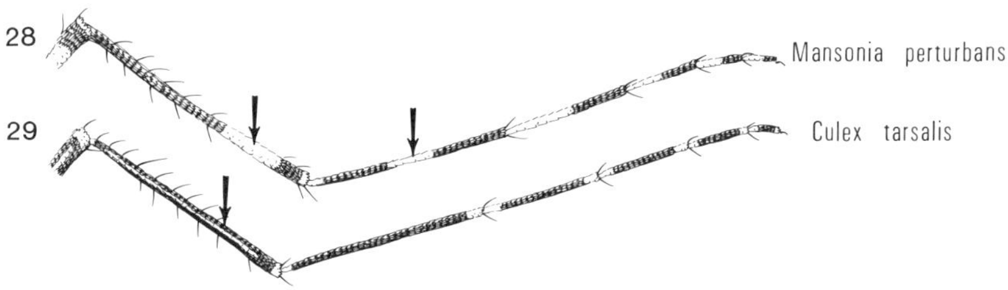

Figs. 28, 29. Hind tibia and tarsus: arrows show different patterns of banding.

8. Hind tibia with a wide ring of yellow scales in distal third and another similar ring at apex; first tarsal segments each with basal and middle rings of yellowish scales; remaining tarsal segments white-scaled on basal half, dark-scaled on apical half (Fig. 28) .... ***Mansonia***
   Hind tibia and basal tarsal segments with scattered pale scales, but lacking well-defined rings at their mid lengths; midtarsal and hind tarsal segments with apices and bases white-scaled, or all tarsi dark-scaled (Fig. 29) 9
9. White tarsal rings conspicuous on hind legs, but scarcely present (reduced at most to a small dorsoapical spot) on fore- and mid-legs; scutum with narrow submedian and lateral lines, and on posterior half also sublateral lines of white scales contrasting with dark brown integument, which is otherwise largely devoid of scales except for acrostichal and dorsocentral stripes of dark brown scales .... ***Orthopodomyia***
   All tarsi, or none, with white rings; scutum uniformly clothed with pale brown scales, sometimes with a pair of small paler median spots .... ***Culex***

Key to the genera of mosquitoes of Canada—larvae

1. Siphon absent, with spiracles opening between large valves nearly flush with dorsal surface (Fig. 30); flattened fan-like palmate hairs (series 1) present on abdominal segments III–VII .... ***Anopheles***
   Siphon present as a sclerotized cylinder elevating spiracles above dorsal surface (Figs. 31, 32); flattened fan-like palmate hairs absent (one pair of palmate hairs belonging to pupal abdominal tergite II may be seen through the larval cuticle of larvae ready to molt) 2
2. Siphon conical, nearly as wide at base as long, its apical third with strongly sclerotized pointed apex for piercing underwater parts of plants (Fig. 31); antenna long, whip-like, with apical portion distal to the antennal tuft and twice as long as basal portion .... ***Mansonia***
   Siphon more than twice as long as wide, blunt-tipped (Fig. 32); basal portion of antenna equal to or longer than apical portion 3
3. Predaceous forms very large (more than 12 mm) when fully grown; apoteme very broad and somewhat parallel-sided, widest posteriorly, occupying nearly all of dorsum of head, with concave anterior margin (Figs. 33, 34) 4
   Nonpredaceous forms medium to small (less than 10 mm) when mature; apotome rounded or oval, broadest in middle, with convex or truncate anterior margin (Figs. 35, 36) 5
4. Siphon lacking pecten teeth but with a prominent multiple siphonal seta 1-S; setae of first 7 abdominal segments grouped on 3 pairs of convex sclerotized plates, with those of segment VIII on one minute and one large plate (Fig. 32) .... ***Toxorhynchites***
   Siphon with numerous pecten teeth, each with long hair-like apex; siphonal seta 1-S a single long hair (Fig. 37); abdominal setae each arising separately, not grouped on sclerotized plates .... ***Psorophora* subgenus *Psorophora***

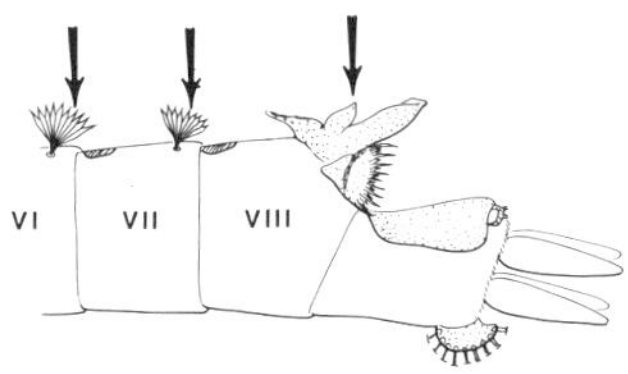

30

Anopheles

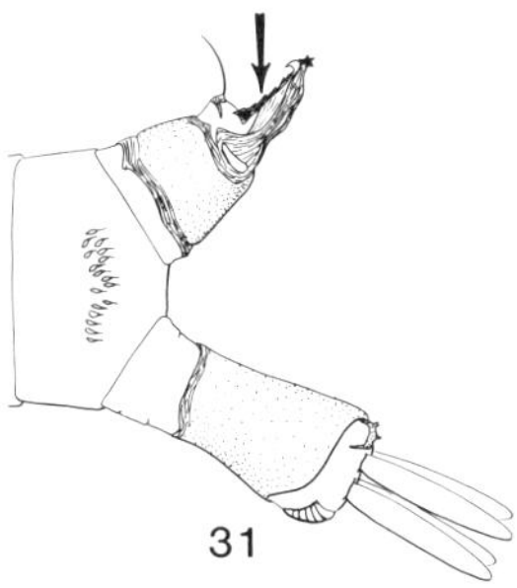

31

Mansonia

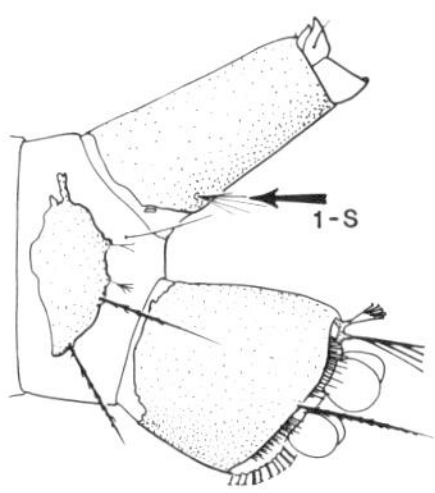

32

Toxorhynchites

Figs. 30–32. Left lateral view of terminal segments of larvae.

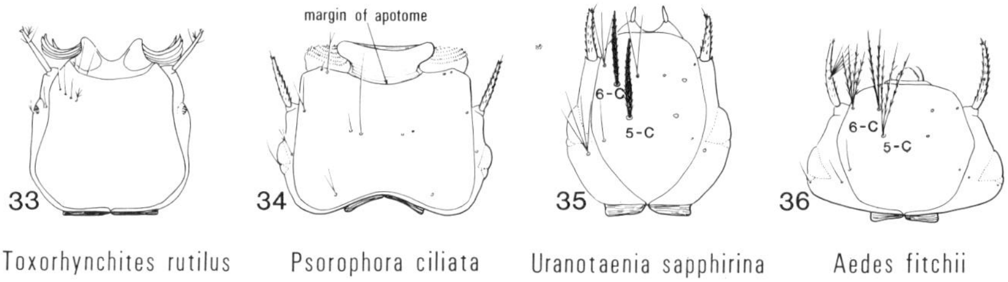

Figs. 33–36. Dorsal view of head of larva.

5. Head oval, as long as wide; head setae 5-C and 6-C each single, inflated, strongly sclerotized (Fig. 35), and much more prominent than other head setae; upper lateral setae (series 6) of abdominal segments III–VII each short, with numerous stellately arranged branches; pecten teeth pale, obscure (easily overlooked), fringed on both sides, with numerous subequal spinules .... ***Uranotaenia***

   Head semicircular, broader than long (Fig. 36); head setae 5-C and 6-C often branched, but when single not conspicuously different in thickness and sclerotization from other head setae; upper lateral setae (series 6) of abdominal segments III–VII each long and at most triple, more or less the same length as those of segments I and II; pecten teeth strongly sclerotized, toothed only on one side, usually only at base or lacking 6
6. Siphon with a basal row of pecten teeth (Figs. 38–44) 7

   Siphon without pecten teeth (Figs. 45, 46) 13
7. Siphon with a single seta 1-S on each side arising distal to most of the pecten teeth (Figs. 38, 39) 8

   Siphon with either a single seta 1-S on each side arising at the base of siphon proximal to the pecten teeth or with 4 or more pairs of setae or tufts distal to the pecten teeth (Figs. 40–44) 9
8. Siphon moderately to greatly inflated, widest near middle; pecten teeth 6 or fewer; siphonal seta 1-S minute, arising remote from last pecten tooth; saddle completely encircling anal segment, enveloping 8 or more precratal setae (Fig. 38) .... ***Psorophora* subgenus *Janthinosoma* and *P.* subgenus *Grabhamia***

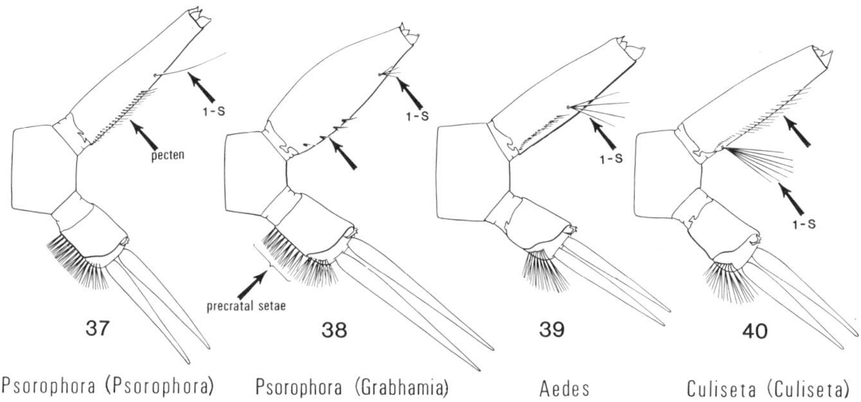

Figs. 37–40. Left lateral view of terminal segments of larva.

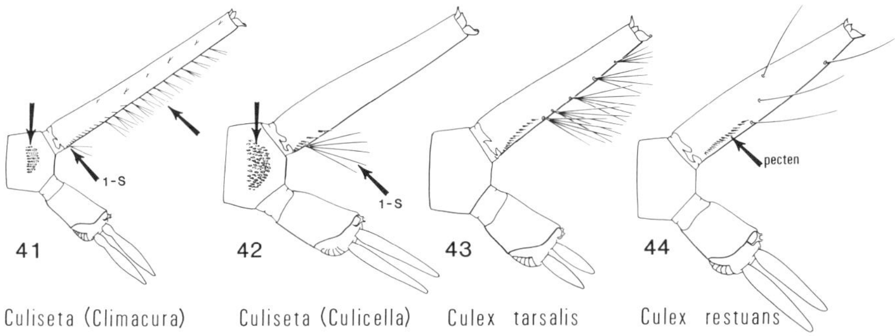

Figs. 41–44. Left lateral view of terminal segments of larva.

Siphon not inflated, nearly parallel-sided at middle, widest at or near base; pecten teeth more than 10; siphonal seta 1-S no farther than its own length from one of the last pecten teeth; when completely encircling anal segment, saddle not enclosing any precratal setae (Fig. 39) .... ***Aedes***

9. Pecten immediately followed by a row of closely spaced setae that are virtually a continuation of the pecten; siphonal seta 1-S as conspicuous as the largest seta (3-VIII) on segment VIII; siphon rather short and stout, about 3 times as long as maximum width (Fig. 40) .... ***Culiseta* subgenus *Culiseta***

   Pecten not immediately followed by a row of closely spaced setae, although four or five widely spaced setae or tufts may be present distal to the pecten teeth; siphonal seta 1-S, when present, smaller than seta 3-VIII; siphon slender, 4–8 times as long as maximum width 10

10. Siphon with midventral row of 10–16 unpaired tufts, a small obscure basal seta 1-S, and a few minute scattered two-branched setae; comb of segment VIII with about 25 long narrow scales closely and precisely arranged side by side (Fig. 41) .... ***Culiseta* subgenus *Climacura***

   Siphon with all setae more or less paired, and in a lateral, ventrolateral, or dorsal position rather than a midventral position; comb scales irregularly arranged in a patch (Figs. 42–44) 11

11. Siphon with a single seta 1-S on each side at extreme base, ventral to the pecten (Fig. 42); remainder of siphon devoid of setae or tufts; pecten teeth small, fewer than 10 .... ***Culiseta* subgenus *Culicella***

   Siphon with several pairs of setae, including seta 1-S, none of which arises from base of siphon (Figs. 43, 44); pecten teeth more than 10 12

12. Distal 3–5 pecten teeth unevenly and more widely spaced than basal teeth; siphon with a dorsolateral row of 4 or 5 tufts, a lateral row of 3 or 4 smaller tufts, and a single large tuft ventral to the pecten at about the middle of the pecten series (see Fig. 161 of *Aedes* larval key) .... ***Aedes provocans***

   Pecten teeth evenly spaced; siphon with all tufts or single hairs arising distal to the pecten teeth (Figs. 43, 44) .... ***Culex***

13. Siphon with scattered unbranched setae over its entire length; anal segment with all 4 pairs of setae subequal in length, each seta with 2–4 branches; anal papillae 1 pair (Fig. 45); found in leaves of pitcherplants (*Sarracenia purpurea*) .... ***Wyeomyia***

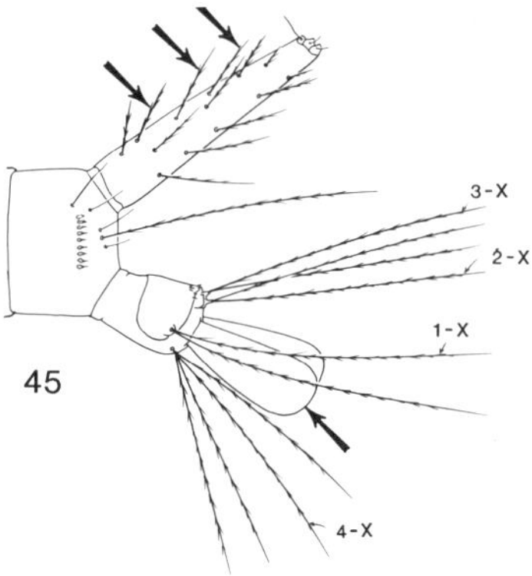

Wyeomyia smithii

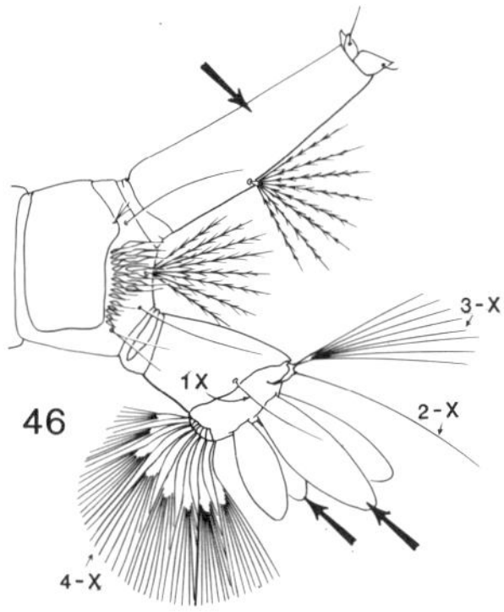

Orthopodomyia signifera

Figs. 45, 46. Left lateral view of terminal segments of larva.

Siphon with a branched seta 1-S on each side; 4 pairs of setae of anal segment varying widely in length and branching; anal papillae 2 pairs (Fig. 46); found in tree holes .... ***Orthopodomyia***

Subfamily Anophelinae

This subfamily, though worldwide in distribution, contains only three genera (Edwards 1932, Stone et al. 1959). *Bironella* Theobald and *Chagasia* Cruz are small genera, restricted to the Australasian and Neotropical regions respectively, whereas most of the species of the subfamily, including all Canadian species, belong to the genus *Anopheles* Meigen. For a diagnosis of the subfamily, see Edwards (1932), who treated it as a tribe, Anophelini, of the subfamily Culicinae (*sensu* Culicidae of this publication). The majority of recent authors, after recognizing dixids and chaoborids as separate families, have upgraded Anophelini, Toxorhynchitini, and Culicini *sensu* Edwards (1932) to subfamilies, but their constituent genera and species remain unchanged.

Adult anophelines, in life, hold their proboscis straight out in front of them in line with the body axis, not at an angle as do the culicines. They also generally have longer legs and stand with the body axis more strongly inclined to the surface and the proboscis down. Females, at least of the Canadian species, have elongated palpi, usually as long as the proboscis; the palpus of the male of nearly all mosquitoes is elongate, but that of an anopheline has the two apical segments swollen and inclined laterally. Perhaps the most important distinguishing trait is the manner of larval feeding: anophelines, like dixids, usually rest at the surface and take in particles from the surface film, but unlike dixids, which tilt their head back to feed, the anophelines rotate their head 180° to bring their mouthparts into contact with the surface. Culicine larvae cannot rotate their head, instead they bend their whole body in order to feed at the surface.

In the early history of Upper Canada, our native *Anopheles* were undoubtedly involved in the transmission of human malaria (*Plasmodium* spp.). Endemic malaria reached its peak in Canada in the middle of the last century, but it declined and disappeared altogether by the beginning of this century. Accounts of the early history of malaria in Canada are given by O'Rourke (1959) and McLintock and Iversen (1975). At least five of the six species of *Anopheles* in Canada have been shown to be potentially capable of transmitting malaria, although *quadrimaculatus* was probably the major vector (O'Rourke 1959).

Genus *Anopheles* Meigen

Adult. Integument medium to dark brown; Canadian species with recumbent scales only on vertex, palpus, proboscis, legs beyond coxae, and wing veins, but absent from most of head and from thorax and abdomen; palpus about as long as proboscis in both sexes, slender throughout, with a few short scattered setae in female, enlarged apically in male, with the last two palpomeres somewhat inflated and fringed (except in *barberi*) and having long setae both mesally and laterally; first palpomere minute, and second and third palpomeres longest, with the second palpomere having somewhat raised scales giving it a shaggy appearance (except in *barberi*) and the last palpomere shorter than the fourth in female, subequal in male; clypeus longer than broad, triangular, narrowing to rounded apex; labium slender throughout, relatively straight and parallel-sided or nearly so; mandible and maxillary blade of female well-developed, toothed apically; frons with tuft of pale setae directed forward between eyes; erect forked scales of head numerous, but restricted to region of occiput behind occipital setae; scutum narrowed laterally and not strongly arched, sparsely pollinose; setae of acrostichal and dorsocentral rows pale, erect, relatively numerous, and conspicuous; spiracular setae present (few, rarely none, in *barberi*); anepisternum otherwise bare; tarsal claws without subbasal tooth; wing scales brown, more or less aggregated to form darker spots (except in *barberi*) at base of R_2, at bifurcations of R_{2+3} and R_{4+5}, of R_2 and R_3, of M_1 and M_2, and on M at junction of crossvein r-m; spots less pronounced in male than in female; calypter with marginal fringe; abdomen parallel-sided, pilose, but without scales in the Canadian species; cercus of female short, rounded, inconspicuous; gonocoxite of male in the Canadian species without apical or basal lobes, but with a pair of prominent spines (parabasal spines) in place of basal lobe; gonostylus long, slender, and curved medially, with short apical spine; claspette divided longitudinally into a dorsolateral, or outer, lobe and a ventromedial inner lobe, each armed apically with long stout setae; aedeagus tubular with (except in *barberi*) one or more pairs of laterally projecting apical spines, or "leaflets"; anal segment conical, membranous, grooved ventrally to partly encircle aedeagus.

Larva. Head as long as or longer than broad, usually somewhat oval in outline; major setae of at least the first three abdominal segments of the thorax (except in *barberi*) and of the head conspicuously pinnately branched or plumose; seta 1 of some or all of the first seven abdominal segments palmately branched, forming a "float hair," each of these hairs consisting of a stout short base and 12–20 laterally directed flattened branches arranged fan-like (Figs. 65–68); abdominal segment VIII bearing a lateral sclerite with posterior row of conspicuous pecten teeth; each plate connected to the corresponding plate on the other side by a narrow sclerotized band encircling respiratory apparatus posteriorly but not anteriorly, thus a true siphon as found in Toxorhynchitinae and Culicinae absent; spiracular apparatus flush with dorsal surface, as in Dixidae, consisting of a small semicircular anteromedian plate, a small lateral triangular plate on each side of paired spiracular openings, and a large complex posteromedian plate; saddle not encircling anal segment (Fig. 30).

Biology. Eggs are laid on water, separately rather than in rafts, usually on permanent or semipermanent ponds, marshes, sloughs, or ditches, wherever emergent aquatic vegetation is growing. The egg is oblong, boat shaped, pointed at each end, and somewhat pinched in at the middle, with a flattened upper surface bordered end to end with a flange-like girdle of air cells or "float chambers," which may be wider in the pinched-in middle region. Eggs of *Anopheles walkeri* overwinter (Hurlbut 1938), but those of the other species are believed not to diapause but to hatch within a few days of oviposition.

Larvae of *Anopheles* live at the surface of semipermanent stagnant bodies of water, where they feed on floating particles. Unlike larvae of the Culicinae, anopheline larvae rotate their head through 180° while feeding so that their mouthparts are uppermost and break the surface film. *Anopheles* larvae are also capable of "grazing," in the same way as culicines, on food particles too large to swallow, either at the surface or on the bottom. A detailed account of their feeding anatomy and behavior is given by Schremmer (1949). Larvae grow rather slowly, requiring a couple of weeks to reach maturity, even in midsummer.

The pupae of nearctic *Anopheles* have been described and illustrated by Penn (1949), and those of the northeastern USA by Darsie (1949).

Adult females usually become active and seek blood in the early evening, and then continue to be active throughout the night. They live a long time and are surprisingly fecund; Aitken (1945) recorded a female of *punctipennis* that laid five batches of eggs, for a total of 500 eggs. O'Rourke (1959) implied that *quadrimaculatus* could lay as many as 10 batches of eggs, with up to 200 in a batch. Mitchell (1907) obtained 1569 eggs in six ovarian cycles from a single female of *punctipennis* in Louisiana. All the species in Canada have more than one generation during the growing season, but in the north *earlei* is univoltine because of the shortness of the season.

Key to the species of *Anopheles* of Canada—adults

1. Wing scales evenly distributed, not aggregated into spots; scutal setae brown, rather long, erect, with those of acrostichal and dorsocentral rows each confined to an irregular single row; aedeagus lacking terminal spines (Fig. 51); outer claspette lobe with three overlapping flattened setae (Fig. 58) .... ***barberi***

   Wing scales denser in some areas than others, forming a pattern of spots; scutal setae yellow, short, fine, curved, with those of acrostichal and dorsocentral rows arising haphazardly over a relatively wide area, especially on posterior half of scutum; apex of aedeagus with three or four pairs of divergent spines (Figs. 52, 53); setae of outer claspette lobe pointed, or if blunt, no more than two present (Fig. 59) 2

2. Wing mostly dark scaled with some creamy yellow scales either as part of fringe at wing apex or forming a pattern of light spots (Figs. 47, 48); aggregations of dark wing scales dense, conspicuous; outer claspette lobe with one or two sharply pointed setae, usually single (Fig. 59) 3

   Wing entirely dark scaled (Figs. 49, 50), although fringe scales of apex may appear paler because of bluish iridescence in some lights; dark wing spots small and vaguely defined; outer claspette lobe with two setae, rounded apically (Fig. 60) (except in *freeborni* of British Columbia, which cannot be readily distinguished from *earlei* or *punctipennis* by terminalia alone) 4

3. Wing with a crescent-shaped spot of cream-colored fringe scales on wing tip between apices of R_1 and R_{4+5} (Fig. 48), and with pronounced dark spots of aggregated dark scales, especially on radial sector; ninth tergal lobe of male usually dilated apically, with a somewhat triangular apex (Fig. 54) .... ***earlei***

   Wing with several cream-colored spots: the largest spot U-shaped, situated at the anterior wing border between the end of the subcosta and the bifurcation of R_{2+3}; the second spot narrower, V-shaped, also situated at the anterior border near the wing apex; each spot including a portion of the costa, R_1, and R_{2+3} (or in the more distal spot, R_2 and R_3) (Fig. 47); remaining wing veins, especially

47

48

49

50

Figs. 47–50. Dorsal view of wing of *Anopheles* species: 47, *punctipennis*; 48, *earlei*; 49, *walkeri*; 50, *freeborni*.

on anterior half of wing, thickly and rather uniformly dark scaled; ninth tergal lobe of male attenuated (Fig. 55) .... ***punctipennis***

4. Occurring only in central British Columbia; scutum of female with fairly well defined acrostichal and dorsocentral gray pollinose stripes, coalescing on posterior half (as in *earlei*); setae of outer claspette lobe sharply pointed; ninth tergal lobes of male rather long, parallel-sided, widely separated from one another (Fig. 56) .... ***freeborni***

   Occurring east of Alberta (except for a specimen of *walkeri* supposedly from Vancouver Island); scutum of female dull brown with scarcely any pollinose pattern when viewed from above; setae of outer claspette lobe flattened with rounded apices (Fig. 60); ninth tergal lobes of male smaller, narrower, enlarged apically, closer to each other (Fig. 57) 5

5. Intersegmental joints of palpus accentuated with whitish scales; scutum dull, reddish to dark brown; aedeagal spines without basal serrations; claspette lobes strongly oriented dorsoventrally, with spines appearing superimposed over one another; outer (dorsal) lobe shorter than inner (ventral) lobe .... ***walkeri***

   Palpus uniformly brown scaled; scutum paler, medium sandy brown; some aedeagal spines with basal serrations; claspette lobes oriented more side by side, with spines readily visible; outer and inner lobes about equal in length (Fig. 60) .... ***quadrimaculatus***

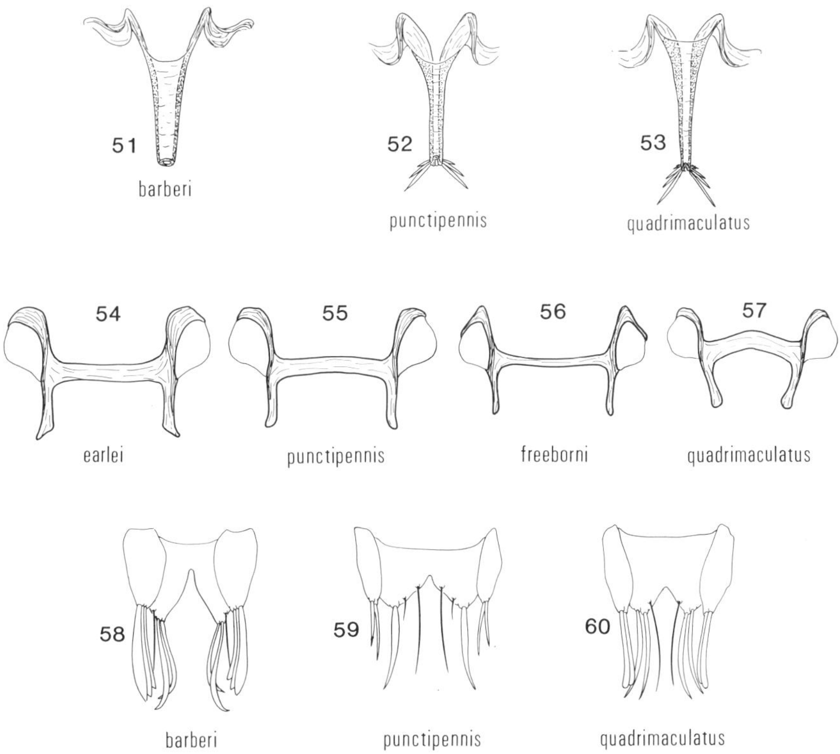

Figs. 51–60. Dorsal view of *Anopheles* species: 51–53, aedeagus; 54–57, ninth tergum of male; 58–60, claspette lobes.

Key to the species of *Anopheles* of Canada—fourth-instar larvae

1. Head setae short and simple; upper lateral setae 6-IV to 6-VI on abdominal segments IV–VI plumose .... ***barberi***

   Head setae 5-C to 7-C long, plumose; upper lateral setae 6-IV to 6-VI on segments IV–VI branched from base but not plumose 2

2. Basal tubercles of inner clypeal setae 2-C separated from each other by a distance at least as great as their own diameter (Fig. 61, *a*, *b*); setae 2-C divergent or parallel .... ***quadrimaculatus***

   Basal tubercles of inner clypeal setae 2-C closely approximated, almost touching each other (Figs. 62–64); setae 2-C parallel or convergent 3

3. Abdominal setae 2-IV and 2-V of segments IV and V each single; abdominal setae 0-II to 0-VII branched (visible only under compound magnification) (Fig. 65); apex of inner clypeal seta 2-C minutely plumose (visible only under compound magnification) (Fig. 63) .... ***walkeri***

   Abdominal setae 2-IV and 2-V multiple; abdominal setae 0-II to 0-VII each single (Fig. 66); inner clypeal seta 2-C either simple or with Y-shaped apical fork (one branch may be shorter than the other, somewhat resembling that of *walkeri*) (Figs. 62, 64) 4

4. Branches of abdominal seta 1-II ("float hair") of segment II hair-like, not flattened into "leaflets" (best seen under compound magnification) (Fig. 66); apex of inner clypeal seta 2-C usually forked (Fig. 64, *a*), sometimes with unequal branches (Fig. 64, *b*), occasionally simple (a rare specimen may be simple on both sides, thus resembling *punctipennis*) .... ***earlei***

   Branches of abdominal seta 1-II of segment II each flattened, blade-like (Figs. 67, 68); inner clypeal seta 2-C simple (Fig. 62). 5

5. Abdominal segments IV–VII with middorsal circular sclerite only (Figs. 65–67); antennal seta 1-A usually arising at or before basal one-quarter of shaft (Freeborn and Bohart 1951); southwestern British Columbia (Fraser Valley) and central and Eastern Canada .... ***punctipennis***

   A pair of small circular sclerites on posterior half of dorsum of abdominal segments IV–VII (Fig. 68); antennal seta 1-A usually arising beyond basal one-quarter of shaft; central British Columbia (Okanagan Valley, B.C., only) .... ***freeborni***

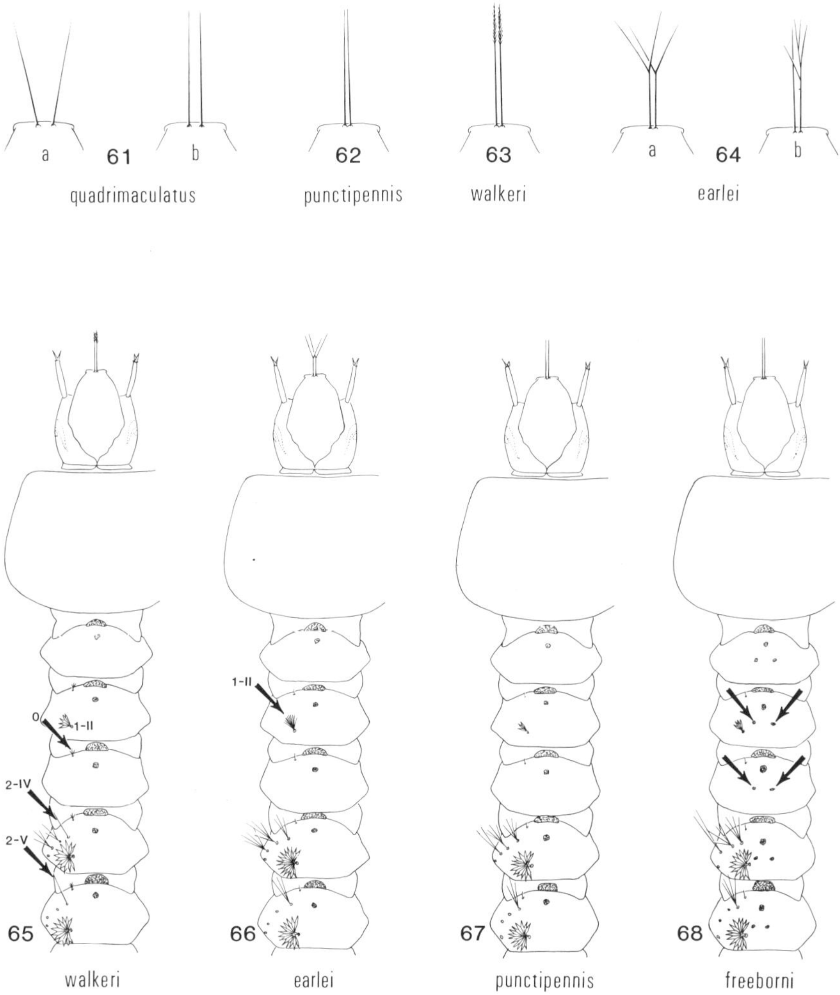

Figs. 61–68. Dorsal view of *Anopheles* larvae: 61–64, clypeal setae (*a*, *b*, two variants); 65–68, abdominal segments I–V.

Anopheles barberi Coquillett

Plate 1; Map 2

Anopheles barberi Coquillett, 1903*b*:310.

Adult. Female: A small unmarked species; palpus brown-scaled, slender throughout, with the second palpomere lacking raised scales (that are present in other species) and thus not apparently any broader than third

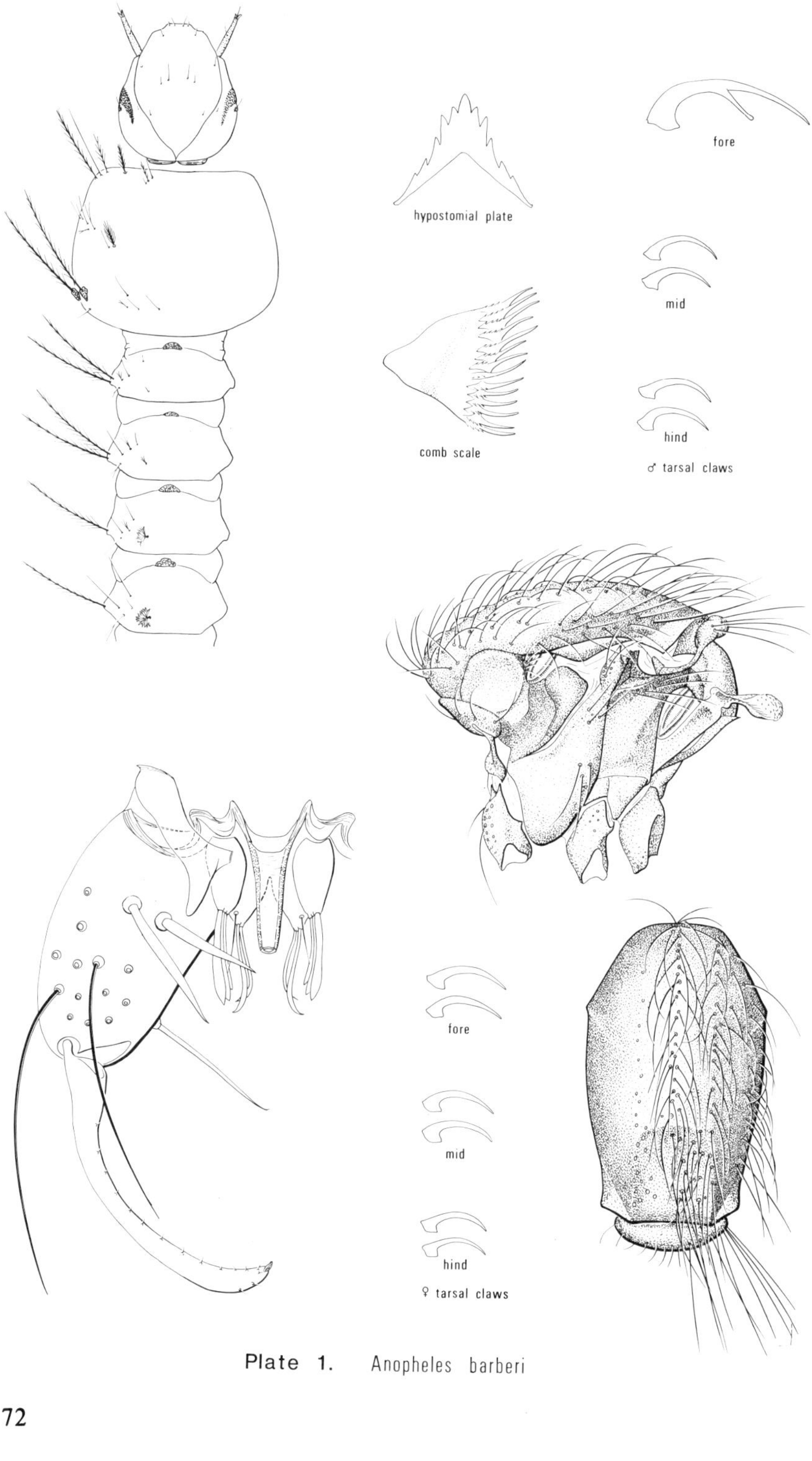

Plate 1. *Anopheles barberi*

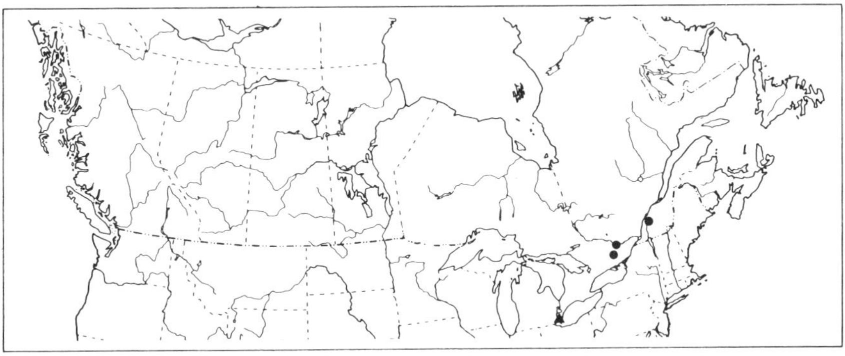

Map 2. Collection localities for *Anopheles barberi* in Canada: • specimens we examined, ▲ literature records.

or fourth palpomeres; proclinate setae and scales of vertex few, sometimes only one or two; erect forked scales of occiput yellow; scutum medium brown, subshining; acrostichal and dorsocentral setae brown, rather long, erect, few in number, with narrower rows and correspondingly broader bare areas than the rows and spaces in other species; spiracular bristles few, sometimes absent; scales of wing veins brown, uniformly distributed along veins, not aggregated into spots.

Male: Last two palpomeres inflated but with few scales along lateral and medial edges; erect forked occipital scales yellowish medially, brown laterally; ninth tergal lobe scarcely evident; aedeagus without apical spines (leaflets); outer lobe of claspette with three broad flattened setae; inner lobe of claspette with two narrower flattened S-shaped setae and one straight cylindrical seta.

Larva. Easily recognized, not only because of its short unbranched head setae and the plumose upper lateral setae of segments IV–VI, but by its presence in tree holes. *A. barberi* is the only *Anopheles* larva in Canada that occurs in such a habitat.

Remarks. A species distinctly different in several respects, especially in larval structure, from the other species of *Anopheles* in Canada. *A. barberi* was at one time placed in another genus, *Coelodiazesis* Dyar & Knab. Recent authors do not recognize this taxon even as a subgenus, but from a phylogenetic point of view it is worth mentioning that *barberi* occupies an isolated position in the subgenus *Anopheles*.

Biology. Larvae of *A. barberi* have been found almost exclusively in rot cavities of trees (Smith and Trimble 1973) and rarely in artificial containers (Jenkins and Carpenter 1946). The species overwinters in the larval stage, probably in the second instar (Matheson 1944). In the Ottawa area, first instars collected in August slowly progress to the second instar, then cease development. In mid-April, collections are still only second instars. In Mississippi and Minnesota, only early instars are collected during the winter (Peterson and Smith 1945, Price and Abrahamsen 1958). However, in Tennessee, all instars overwinter (Breeland et al. 1961). Snow (1949) found that second instars can survive long periods surrounded by ice. In Ontario, larvae of *barberi* are occasionally unassociated with other species; usually, however, larvae of *Aedes hendersoni, A. triseriatus,* and *Orthopodomyia* spp. are also present (Smith and Trimble 1973). Besides filter-feeding in the usual way, larvae of *barberi* are said to be predaceous on other mosquito larvae (Dyar and Knab 1906*b*, Petersen et al. 1969) and ceratopogonid larvae (Baker 1935).

Larvae can be present throughout the summer (Smith and Trimble 1973), which indicates that there are at least two generations. Larval development is rather slow, requiring a month to reach maturity in both spring and summer (Horsfall 1955). At Perth, Ont., tree holes in which *barberi* larvae were not found in early spring nevertheless supported larvae during the summer. Perhaps only certain cavities are suitable as overwintering sites.

Females are timid biters, taking a small amount at a time, even if undisturbed (Thibault 1910). Although they have been experimentally infected with malaria *Plasmodium vivax* and shown capable of transmitting it (Horsfall 1955), they are too uncommon to be of any importance.

Distribution. Eastern North America, from Minnesota and New Mexico east to Quebec and Florida.

Anopheles earlei Vargas

Plate 2; Figs. 48, 54, 64, 66; Map 3

Anopheles earlei Vargas, 1943*a*:9.

Anopheles maculipennis: Hearle (1927*a*) and other authors, not Meigen.

Anopheles maculipennis occidentalis (in part): Aitken (1945).

Adult. Female: Palpus entirely brown-scaled; vertex white-scaled, contrasting with dark-scaled occiput; scutum brown, sparsely gray pollinose (usually greasy, especially specimens collected in the fall), appearance changing with direction of observation, for example, when viewed from above, entire median region between outer edges of dorsocentral band appearing grayish except for a narrow darker stripe between dorsocentral and acrostichal setae, but when viewed laterally, the sublateral and lateral areas appearing gray; a few setae on middle of anterior edge of scutum usually somewhat whiter than the surrounding setae; scales of fringe at apex of wing between apices of R_1 and R_{4+5} creamy white, contrasting with

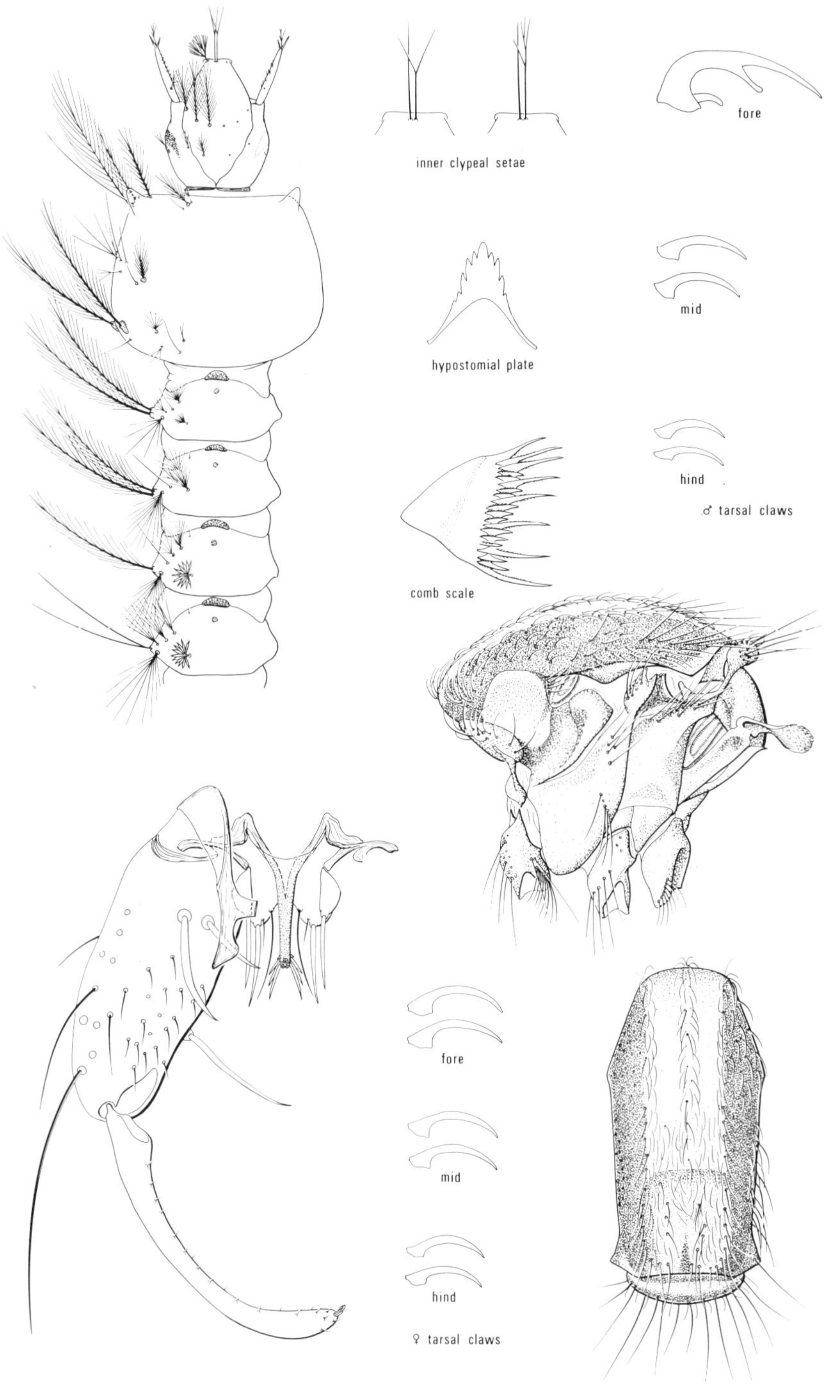

Plate 2. *Anopheles earlei*

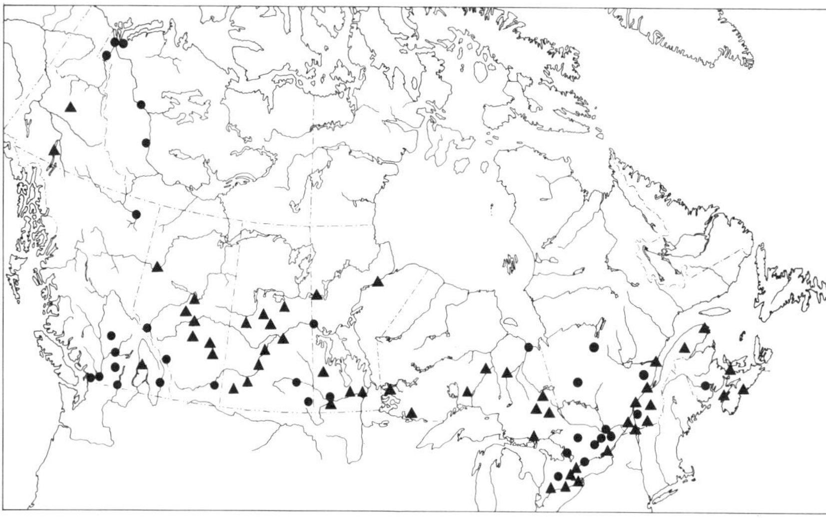

Map 3. Collection localities for *Anopheles earlei* in Canada: • specimens we examined, ▲ literature records.

remaining dark scales of wings, which are strongly aggregated to form conspicuous spots, as outlined in generic description; a conspicuous spot on R_{2+3} at point of bifurcation of R_{4+5}.

Male: Coloration as in female; dark spots on wings much less prominent than in female because of reduced scaling and therefore cannot be as readily used as in female for distinguishing this species from *quadrimaculatus* and *freeborni;* apex of wing with spot of cream-colored scales; ninth tergal lobe with expanded foot-shaped apex.

Larva. Inner clypeal setae 2-C almost contiguous at base, convergent apically, usually with a Y-shaped forked apex, often one forked, the other not, occasionally neither forked, although a minute side branch sometimes present; head setae 5-C to 7-C long, strongly plumose; abdominal setae 2-IV and 2-V multiple; abdominal setae 0-II to 0-VIII single; dorsolateral seta 1-II multiple, its branches each hair-like, not flattened dorsoventrally.

Remarks. Previous authors have implied that the common species of *Anopheles* across most of Canada is *earlei* and that *occidentalis* occurs in British Columbia (Pratt 1952, Carpenter and LaCasse 1955, Curtis 1967). These authors distinguished adults of the two species on the basis of whether the scales on vein R_{2+3} were raised or appressed. However, Vargas (1943*b*, 1944) and Vargas and Matheson (1948) clearly stated that the adult of *occidentalis* could be distinguished from *earlei* by the absence of scales on R_{2+3} between the basal dark spot of dense black scales and the bifurcation of

R_2 and R_3. These authors apparently did not point out that the scales are lacking only on the upper side of the vein. In a series of 10 males and 25 females of *occidentalis,* all in good condition, in the U.S. National Museum, from Stanford University, Calif., the type locality of *occidentalis,* the central portion of R_{2+3} was invariably devoid of scales on the upper side, but the lower side was fully scaled. These latter scales, however, were small, broad, and closely appressed to the vein, identical with the scales on the underside of the same vein in *earlei,* but unlike the long slender divergent scales on the upper side of the vein in *earlei.* Because the wing is transparent, scales on the underside can be seen easily from above; therefore, when examining the wing, avoid direct light and use reflected light. In their key to *Anopheles* females, Carpenter and LaCasse (1955, p. 29, couplet 8) probably mistook which side of the wing had the closely appressed scales, and did not mention that separation should be made on the basis of the presence (*earlei*) or absence (*occidentalis*) of scales on the upper side of the middle portion of R_{2+3}.

Myklebust (1966) recorded *occidentalis* from six counties in the state of Washington, but records from the interior of the state were later changed to *earlei* (Gjullin and Eddy 1972). Only one other record from Washington remained, Dyar's (1917*b*) original record from Whatcom Co., just south of the B.C. border. This record was based on some larvae that were neither reared nor carefully studied and were only suspected of being *occidentalis.* Dyar and Knab's (1906*b*) cotypes from Revelstoke, B.C., which we could not locate, are surely *earlei.* We have not seen any Canadian specimens that could be *occidentalis,* and we assume that Pratt's (1952) map showing *occidentalis* distributed over most of British Columbia and southern Alaska was purely supposition based on Dyar and Knab's mixed type series. *Anopheles occidentalis* appears to be confined to Oregon and California.

Biology. Females of *earlei* overwinter in buildings (McLintock 1944), caves (Price et al. 1960), and mammal burrows (Shemanchuk 1965). Beaver lodges, where we have found *earlei* resting during the summer, are also potential hibernating sites, as well as hollow logs and trees. The female hibernates as a nonblood-fed nullipar (that is, never having laid eggs) and comes out to bite on the first warm days of mid-April at Ottawa, but not until late May in Alaska (Frohne 1951). In the north, the summer season is too short for a second generation, and therefore Frohne (1956) believes that *earlei* is univoltine in Alaska. Graham (1969*b*) found newly emerged females in late June and again in late July in central Alberta, which indicates two generations, but he thought that *earlei* could be univoltine in some years. At Ottawa, larvae can be found from June to September, and three or more generations may be customary. During the summer, however, fewer adults are present. Rempel (1950) reported that the species is rare on open plains, but more common in northern wooded areas.

Happold (1965*a*) observed mating swarms in mid-September in north central Alberta; after sunset one swarm formed 1 m above ground over a Stevenson screen. Rozeboom (1952) succeeded in getting males of *earlei* from Montana to mate in cages of 11 dm^3. However, because males from Alaska would not mate in any of the cages tested, Frohne (1956) speculated that the Alaskan *Anopheles* might be the Eurasian *maculipennis* Meigen rather than *earlei*.

Distribution. North America, from Alaska and Washington east to Newfoundland and Massachusetts.

Anopheles freeborni Aitken

Plate 3; Figs. 50, 56, 68; Map 4

Anopheles maculipennis spp. *freeborni* Aitken, 1939:192.

Anopheles quadrimaculatus: Hearle (1927*a*) and many authors before 1939 (see Aitken 1945:296), not Say.

Adult. Female: Scales of the fringe at apex of wing dark, although paler and iridescent in some light, never creamy yellow as in *earlei*; scales in midsection of R_{2+3}, adjacent to base of R_{4+5}, not as densely aggregated, especially not along anterior edge of vein, as either of the other three spots on the radial sector (in *earlei*, the midsection has a prominent spot of aggregated scales); otherwise similar to *earlei* and distinguished from *quadrimaculatus* by its more pollinose scutum (the midsection of R_{2+3} in *quadrimaculatus* usually lacks an aggregation of scales) and its occurrence in British Columbia.

Male: Coloration as in female, perhaps somewhat paler; wing spots almost nonexistent, with the one at base of R_s being most noticeable; ninth tergal lobe long, slender, straight, extending (in unmounted terminalia) to apex of claspette; outer lobe of claspette superimposed over inner lobe, each with one or two long slender pointed setae; aedeagus with apical spines.

Larva. No reliable characters have been found to distinguish this species from *punctipennis*. Our few records suggest the two species may be allopatric, overlapping only in the Kamloops–Vernon area. Freeborn and Bohart (1951) found that in most larvae of *freeborni* the antennal seta 1-A arises beyond the basal one-quarter of the shaft, but our material is insufficient to be conclusive. Dodge (1963) mentioned the presence of a pair of small circular sclerites on the posterior half of the dorsum of abdominal segments IV–VII.

Biology. Adult females overwinter in various sites, such as abandoned buildings (Beck 1961), abandoned mines (Chapman 1961), root cellars (Gjullin and Yates 1945), talus slopes and marmot burrows, especially when they are among rocks or at the base of a cliff (Harwood 1962), woodrats' nests (Ryckman and Arakawa 1951), and both natural and artificial rock piles 30–90 cm deep (Rush et al. 1958). These authors also found males in

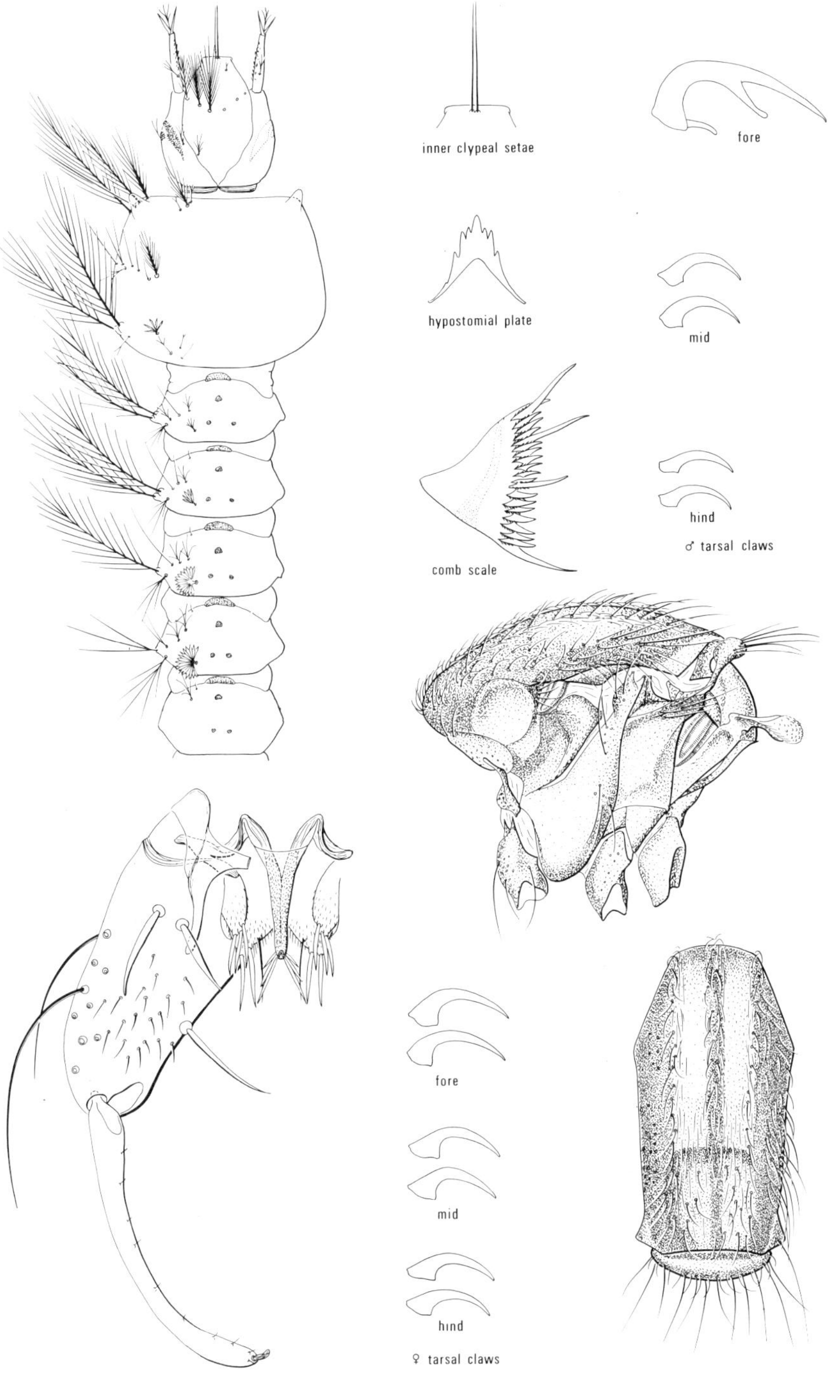

Plate 3. *Anopheles freeborni*

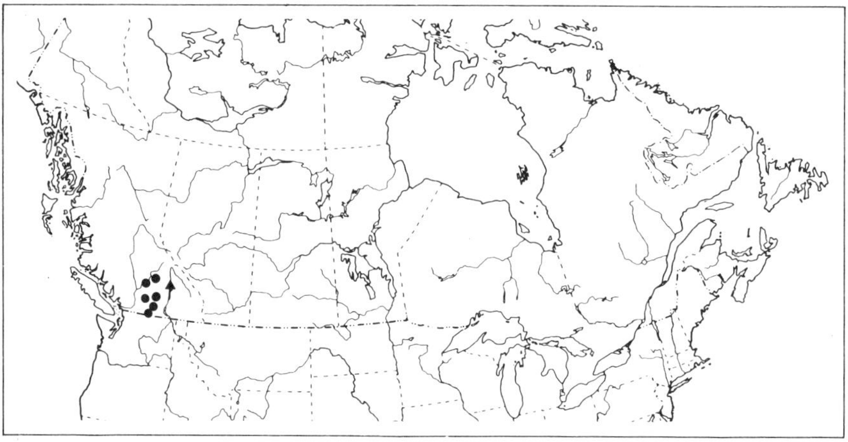

Map 4. Collection localities for *Anopheles freeborni* in Canada: • specimens we examined, ▲ literature records.

overwintering sites in early fall, but as winter progressed males were no longer recovered. In early spring, females actively seek blood and they are most noticeable then; thereafter, females of the summer generation are less often seen (Portman 1954), probably because of the greater number and variety of hosts available.

Larvae (identified as *occidentalis*) in association with *punctipennis* and *earlei* were reported from pools and sloughs formed by creeks, large marshes, and irrigated pastures in Washington (Gjullin and Yates 1945). Rice fields are an important breeding site in California (Portman 1954). Larvae tolerate some salinity, and they are usually found at sites that have emergent vegetation (O'Rourke 1959).

Oviposition takes place while the female is either hovering over the water or resting on it (Wallis 1954*a*). Water containing a concentration of up to 5% NaCl is acceptable to ovipositing females, but is avoided when it is over 6%.

Burgess (1948, 1955) obtained some viable eggs from a cross between *freeborni* and *punctipennis*, but not from a cross between *freeborni* and *quadrimaculatus*.

Distribution. Western North America, from south central British Columbia south to California, Texas, and northern Mexico.

Anopheles punctipennis (Say)

Plate 4; Figs. 47, 52, 55, 59, 62, 67; Map 5

Culex punctipennis Say, 1823:9.
Culex hyemalis Fitch, 1847:281.

Adult. Female: Palpus dark-scaled; second palpomere with somewhat erect and shaggy scales, making the segment appear broader; reclinate setae and scales of vertex white; erect forked scales of median area of occiput pale, contrasting with darker scales laterally; scutum similar to *earlei*, but with slightly denser and more conspicuous gray pollen; wing predominantly dark-scaled, with diagnostic pattern of cream-colored scales forming two large distinctive spots and several smaller more variable spots as follows: a large wedge-shaped spot involving the costa, R_1, and R_{2+3} between the apex of the subcosta and the bifurcation of R_2 and R_3; a narrower wedge-shaped spot near the apex of the wing including the costa, the extreme apex of R_1, the apical one-quarter of R_2, and a small middle portion of R_3; a small spot at the apex of R_{4+5}, which usually also extends to include the fringe of the wing; one or two spots (sometimes combined) on M proximal to the crossvein r-m; another spot just distal to the crossvein r-m; one in the middle third of each of M_1 and M_2; and another in the middle third of A.

Male: Coloration as in female; wing pattern, though not as pronounced as in female, equally distinctive and characteristic; ninth tergal lobe slightly broadened and rounded apically; outer claspette lobe with one straight, flattened, sharply pointed seta; inner lobe with one similar flattened seta and one slender cylindrical seta; apical spines of aedeagus with basal serrations.

Larva. Inner clypeal setae 2-C closely approximated basally and convergent apically; larvae not always distinguishable from those of *freeborni* (characters for separation are given under *freeborni*).

Biology. Adult females overwinter as blood-fed nullipars in the same types of niches as were described for *earlei* and *freeborni*. The females emerge in early spring to take a blood meal and to develop and lay a batch of eggs. The life cycle appears to be the same as that of *earlei* and *freeborni*, with two or three generations annually. Larvae of *punctipennis* are usually associated with those of *quadrimaculatus*, *walkeri*, and *earlei*, wherever their distributions overlap, and in the western USA with *freeborni*. In simulated natural environments, *punctipennis* bred in pools containing floating debris, emergent vegetation, or no vegetation, but they avoided pools that were covered with duckweed (Furlow and Hays 1972). Larvae of *punctipennis* have been found in rock pools associated with *Aedes atropalpus*, *Culex territans*, and *C. restuans* in southern Ontario (James 1964) and have occasionally been collected in tree holes and artificial containers (Dorsey 1944). James (1964) collected larvae as late as 10 Nov. in Ontario.

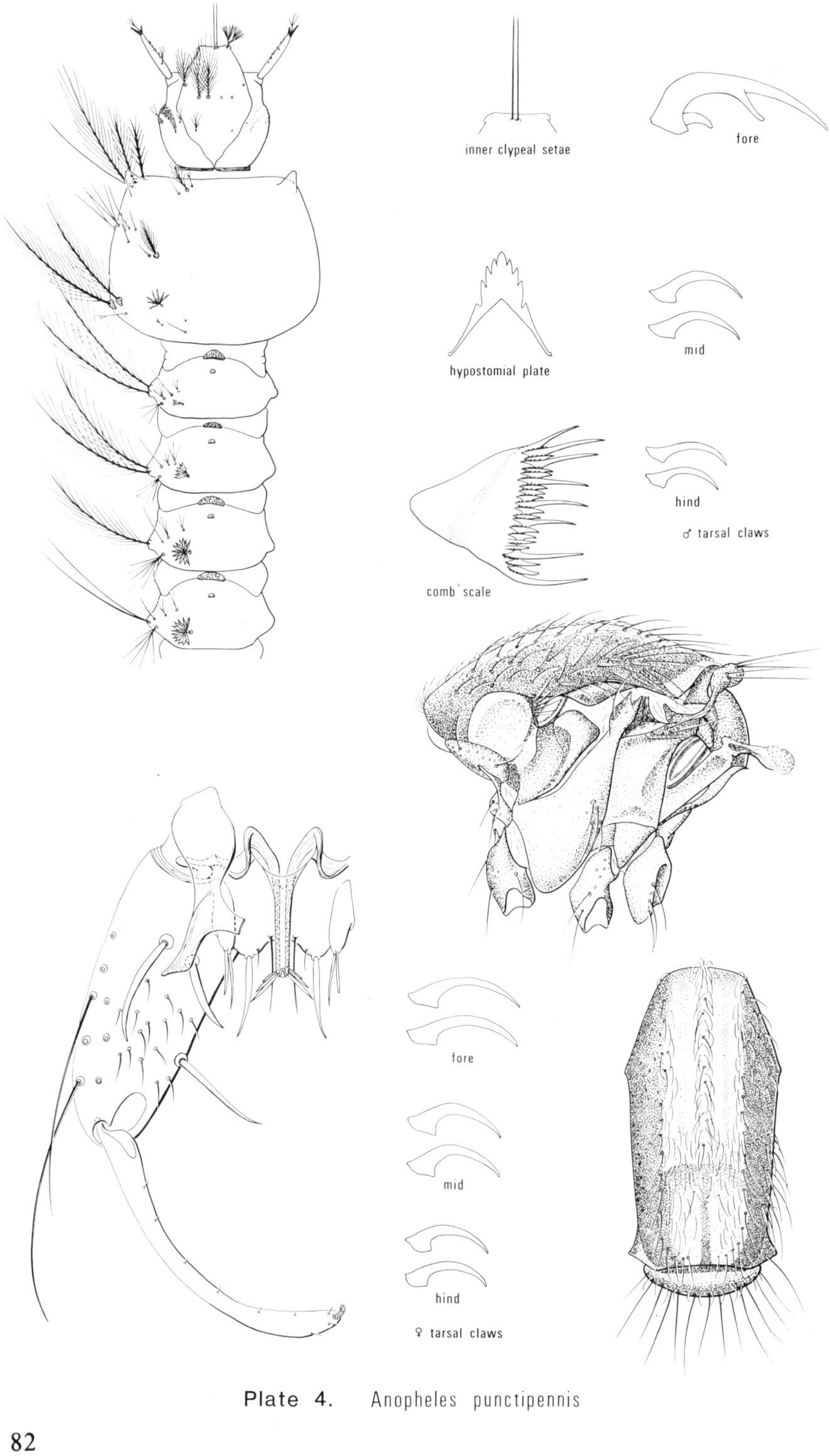

Plate 4. *Anopheles punctipennis*

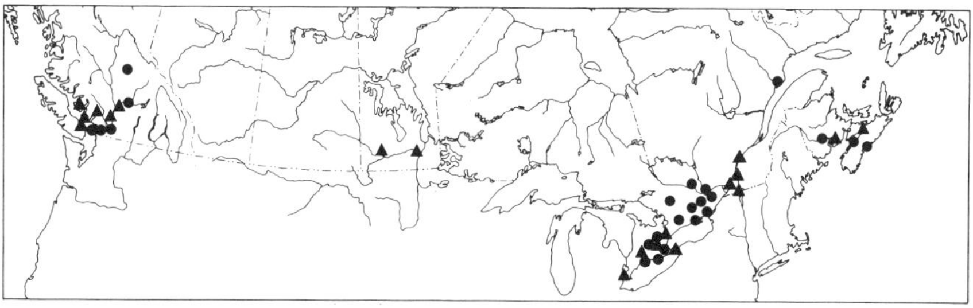

Map 5. Collection localities for *Anopheles punctipennis* in Canada: • specimens we examined, ▲ literature records.

In Nebraska, larvae were collected in brine seepage pits from oil wells that were high in concentration of inorganic salts (Rapp 1960). Breeland et al. (1961) gave a detailed account of the biology of *punctipennis* in Tennessee.

Distribution. North America, from southern British Columbia, southern Manitoba, and southeastern Canada south throughout the USA to northern Mexico.

Anopheles quadrimaculatus Say

Plate 5; Figs. 10, 12, 13, 53, 57, 60, 61; Map 6

Anopheles quadrimaculatus Say, 1824:356.
Anopheles annulimanus Wulp, 1867:129.

Adult. Female: Palpus entirely dark-scaled; upper occiput and scutum sparsely grayish brown pollinose, scarcely visible when viewed from above, at least insufficient to form the pattern characteristic of *earlei* and *freeborni*; scutal vestiture of yellow setae, slightly longer and less dense than in *walkeri*; wing scales entirely dark, with spots (of aggregated scales) weakly defined, those on midsection of R_{2+3} (at origin of R_{4+5}) usually not distinct.

Male: Palpus entirely dark-scaled, but the scales appearing paler in some lights because of bluish iridescence; scutum medium brown, sparsely grayish brown pollinose medially when viewed from above; scutal setae longer and sparser than those of *walkeri*; lobes of claspette aligned more side to side than in *walkeri*, with setae more readily visible; outer lobe with two flattened setae, rounded apically as in *walkeri*; inner lobe with two flattened, sharply pointed setae and one more slender cylindrical third seta arising medially; outer and inner claspette lobes almost equal in width; spines at apex of aedeagus with serrations at their bases.

Larva. Inner clypeal setae 2-C divergent apically and not as closely approximated at base as in other species, with basal tubercles of each seta separated from each other by a distance equal to the width of one tubercle; abdominal setae 0–II to 0–VII on segments II–VII single (as in *earlei*, but not *walkeri*); each branch of dorsolateral seta 1–II (float hair) flattened and

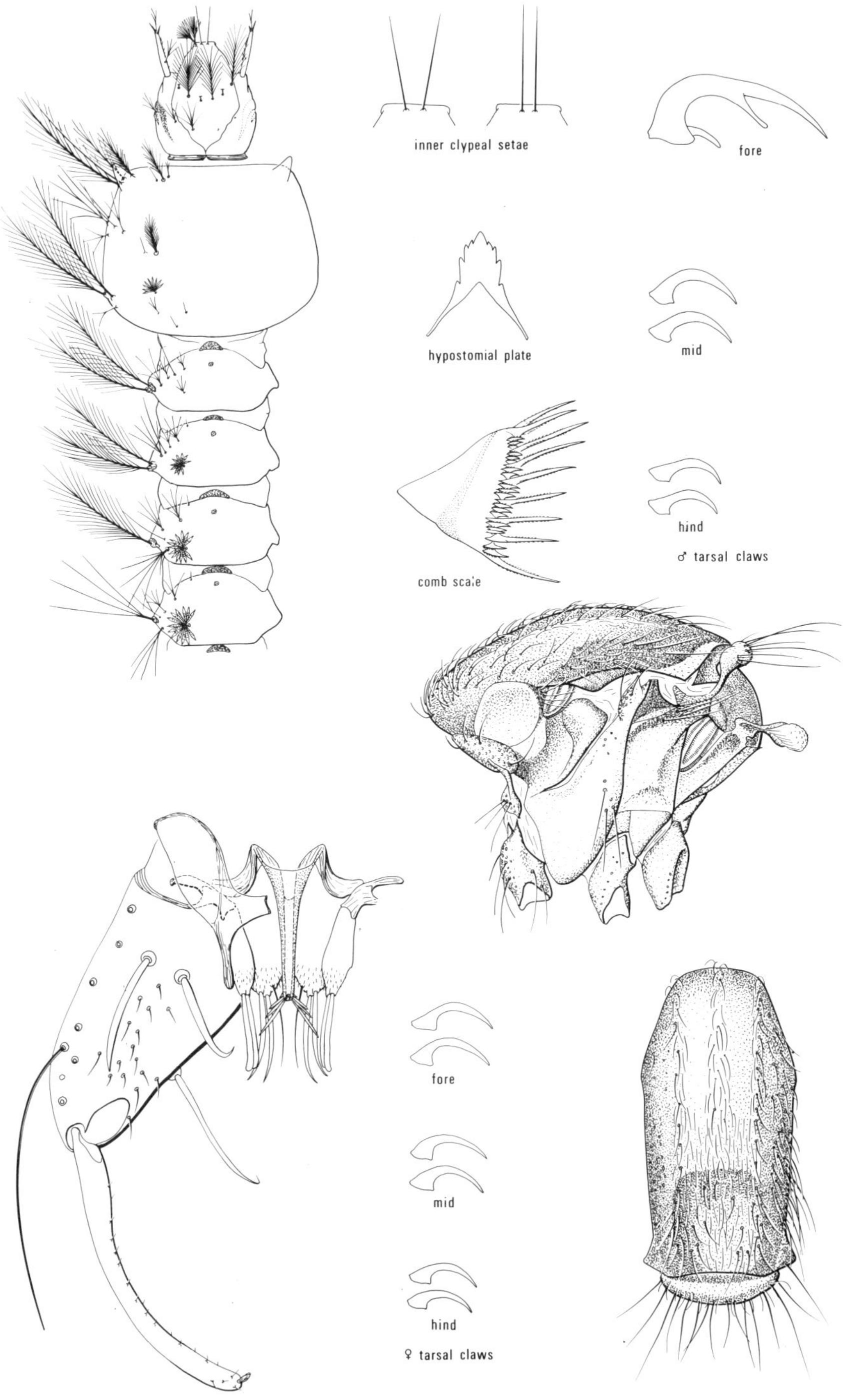

Plate 5. Anopheles quadrimaculatus

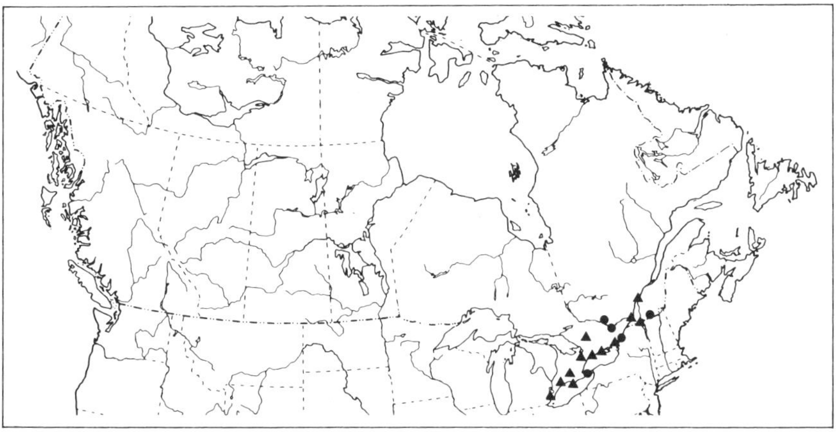

Map 6. Collection localities for *Anopheles quadrimaculatus* in Canada: • specimens we examined, ▲ literature records.

expanded; abdominal seta 2–IV single to triple (usually single in specimens from southern USA); seta 2–V single or double.

Biology. *Anopheles quadrimaculatus*, like *barberi*, is at the edge of its range in Ontario and Quebec. It is the least common of the four largest marsh-breeding species in Canada, and data on its biology at this latitude are lacking. In the eastern USA, however, it is one of the most common species of *Anopheles*, and because of its importance as a pest and its ability to transmit human malaria, it has been extensively studied. Horsfall (1955) gives a thorough review of the extensive literature on this species.

In the USA and presumably also in Canada (O'Rourke 1959), the female overwinters in hollow trees, barns, cellars, caves, and similar protected niches (Breeland et al. 1961). Winter may reduce the population severely; Peterson and Smith (1945) remarked that "a comparatively small number of. . . females" was all that carried the species through the winter, even as far south as Mississippi, and Headlee (1945) found a high mortality among hibernating females in New Jersey.

Data on the role of egg development of *quadrimaculatus* (Huffaker 1944) indicate that development is extremely rapid at high temperatures, about 33 hr at 33°C compared with almost 500 hr at 10°C. Horsfall (1955) found that eggs can withstand a week or more of drying and their development in the southern USA is greatly retarded during the winter. However, retardation would not be sufficient to carry the eggs over a Canadian winter, and an egg diapause such as has been demonstrated for *walkeri* has evidently not been discovered.

Larvae of *quadrimaculatus* have been collected as early as June at London, Ont. (Judd 1954). Larval development takes about 3 weeks at

23°C (Bradley and Fritz 1945). In the laboratory, the fastest development occurred at 31°C, when larval and pupal development together took only 7.6 days, but this would never occur under Canadian conditions. Larvae are tolerant of high salinity and are commonly found in salt marshes in New Jersey (Chapman 1959). Inland they have been found in various habitats, but in Ontario they occur along the marshy edges of rivers, in association with larvae of the other three marsh-breeding *Anopheles* (Wishart and James 1946).

Females hide in dark places during the day, and this predilection has been exploited for sampling the population by setting up empty wooden containers, such as nail kegs, with the opening downward, at strategic positions (Snow 1949). As in *freeborni*, females land on water to drink and also to lay their eggs, either while standing on the surface or when in flight (Wallis 1954*a*). Ovipositing females avoid water containing a concentration of higher than 5% NaCl.

Distribution. Eastern North America, from North Dakota and Texas east to Maine and Florida.

Anopheles walkeri Theobald

Plate 6; Figs. 49, 63, 65; Map 7

Anopheles walkeri Theobald, 1901*a*:199.

Adult. Female: Palpus mostly dark-scaled except for a narrow ring of whitish scales at apices of second, third, fourth, and fifth palpomeres, and a few whitish scales also at bases of third and fourth palpomeres; upper occiput rather conspicuously gray pollinose, with forked scales pale to dark brown; scutum, when viewed from above, appearing dull reddish brown to dark brown with scarcely any pollinosity showing; vestiture of scutum rather dense and woolly, of short yellow recumbent setae; wing scales entirely dark, with spots (of aggregated scales) even more obscure than in *quadrimaculatus*, usually absent on midsection of R_{2+3} (at origin of R_{4+5}).

Male: Second palpomere usually with at least a few white scales at apex; scutum dull reddish brown (similar to *Culex*), with short yellow somewhat recumbent setae; lobes of claspette aligned more dorsoventrally than in other species, so that setae of outer lobe are almost superimposed on those of inner lobe; outer lobe with two flattened, apically rounded setae; inner lobe with three sharply pointed setae, two flattened and the third more slender and cylindrical arising from ventral surface of inner lobe; inner lobe somewhat longer than outer lobe; spines at apex of aedeagus without serrations.

Remarks. Although consistently a darker species than *quadrimaculatus*, with which it usually occurs and is most easily mistaken, this

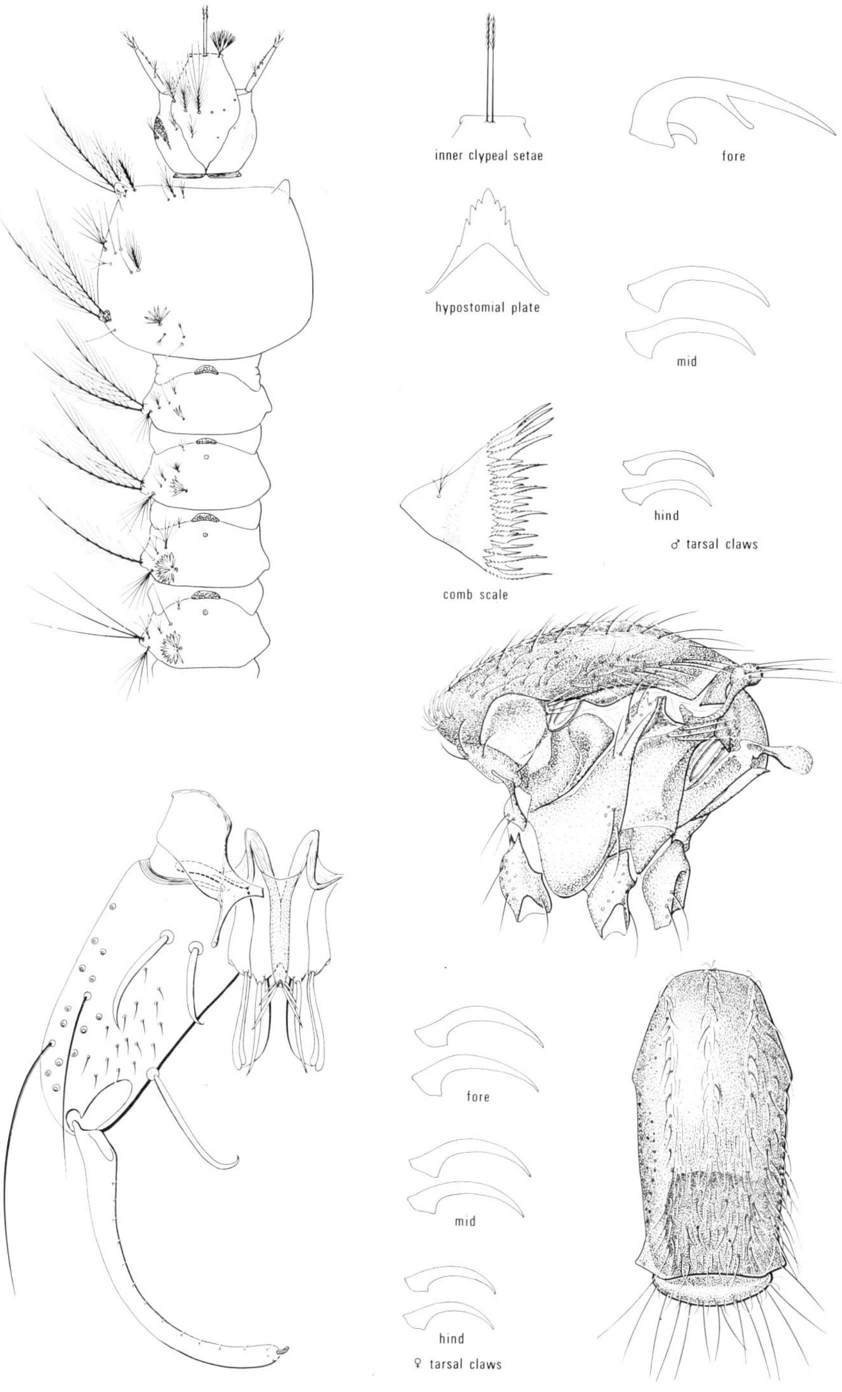

Plate 6. *Anopheles walkeri*

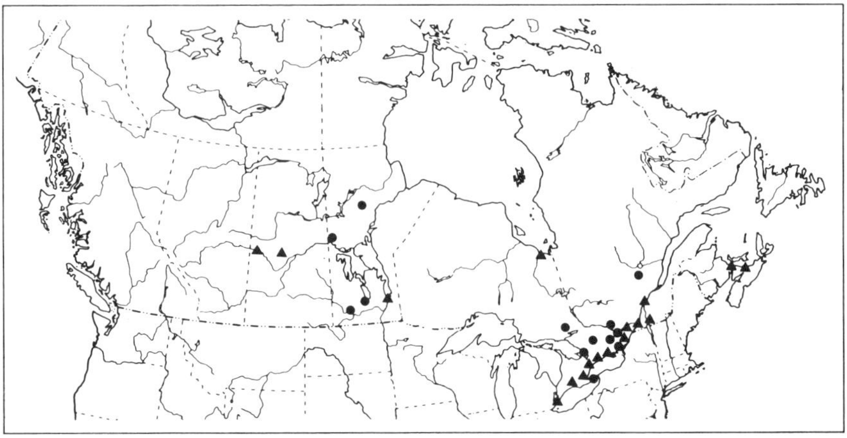

Map 7. Collection localities for *Anopheles walkeri* in Canada: • specimens we examined, ▲ literature records.

color difference can only be appreciated by comparison with known material. Hence the only reasonably reliable character for recognizing *walkeri* is the banded palpus, which is fortunately more obvious in the female than in the male. Some males have scarcely any white scales at the intersegmental junction between the second and third palpomeres, and none on more distal palpomeres, so they can only be separated from male *quadrimaculatus*, without reference to the terminalia, by the more reddish and less pollinose scutum with its shorter woollier vestiture.

Biology. *Anopheles walkeri* is the only species of the genus in North America known to overwinter in the egg stage, at least in the northern part of its range. Eggs laid near the end of August at Ithaca, N.Y., were larger and of a different shape from those laid earlier. These larger eggs did not hatch until subjected to a lengthy period of cold conditioning, and were called "winter eggs" by Matheson and Hurlbut (1937). These eggs are capable of withstanding temperatures as low as −21°C for 72 hr (Peters 1943).

Larvae have been collected in various habitats, usually in association with the other three marsh species, but they are more commonly found in ponds with stable water levels that are overgrown with cattails and other emergent vegetation (Matheson and Hurlbut 1937, Peters 1943).

Adult females show a greater predilection to light than do those of *quadrimaculatus* (Bradley and King 1941 cited by Horsfall 1955). During the day females rest in vegetation near their breeding grounds, rather than seeking secluded quarters, and therefore cannot be collected in the same manner as females of *quadrimaculatus* (Snow and Smith 1956). They occasionally bite during the day if they are disturbed, but they are mainly nocturnal. They are usually the most common *Anopheles* during summer in southern Ontario.

Distribution. Central and eastern North America, from Saskatchewan and Texas east to Nova Scotia and Florida. Ozburn (1945) recorded *walkeri* from Esquimalt, B.C., the only record west of the Rocky Mountains in North America. The specimen, a female in the Canadian National Collection, Ottawa, has white-banded palpi and appears to be correctly identified, but it may not be correctly labeled as to locality. We have decided to consider it mislabeled until additional specimens can be found in British Columbia to substantiate the record.

Subfamily Toxorhynchitinae

The smallest of the three culicid subfamilies, Toxorhynchitinae contains only one genus, *Toxorhynchites*, a mainly tropical group of exceptionally large mosquitoes that reaches its northern limit in the New World at Point Pelee, Ont., in the southernmost part of Canada (Parker 1977). The diagnosis of the genus, which follows, thus applies also to the subfamily. A further division of subfamily and generic characters may be found in Edwards (1932, p. 58) as the genus *Megarhinus* Robineau-Desvoidy. No other mosquito in Canada is so large or so vividly colored with metallic blue green and purple scales.

Genus *Toxorhynchites* Theobald

Adult. Integument dark brown; head and body extensively covered with ovoid, flattened, closely appressed metallic scales; eyes contiguous dorsally; postocular setae confined to vertex; erect forked scales confined to posterior margin of upper occiput; first flagellomere twice as long as second flagellomere in male, one and one-half times as long in female; clypeus trapezoidal, wider than long, widest along rather straight anterior edge; proboscis thickened basally, uniformly tapering to long, slender posteroventrally recurved apex; palpus of male longer and stouter than proboscis, and curving dorsally at about the same level as proboscis curves ventrally, with fifth palpomere long, slender, tapering to pointed apex; palpus of female, in most species of the New World, straight, with fifth palpomere minute, hidden within apical scales of fourth palpomere, therefore appearing only 4-segmented; palpus of both sexes with at most a few short, scattered spine-like setae; anepisternum narrowed, compressed between postpronotum and katepisternum, bare except for spiracular setae; paratergite enlarged, triangular; scutum almost devoid of setae except for a dense crescent-shaped fringe above wing base; posterior edge of scutellum rounded, with a uniformly distributed row of setae; wing long, narrow, thinly scaled; cercus of female small, rounded, not exserted; male terminalia small, with gonocoxite lacking apical lobe, but with a few inner basal spines; gonostylus rather straight, folded mediodorsally, terminating in a short spine-like seta; claspette absent.

The one Canadian species, along with all but three of the New World species placed in the subgenus *Ankylorhynchus* Lutz, belongs to the subgenus *Lynchiella* Lahille. The subgenus *Toxorhychites*, with palpi in the female shorter than in the male and apparently consisting of only two segments, occurs only in the Old World.

Larva. Very large, up to 15 mm long; head subquadrate, about as wide as long; labrum strongly notched in middle of anterior margin; labral brush bases raised and prominently extended anteriorly; labral brush consisting of about 10 stout, curved, flattened setae, each concave anteriorly; head setae inconspicuous; larval eye prominent but adult compound eye not evident; thoracic and abdominal setae short plumose, most major setae arising from sclerotized tubercles or "islets"; thoracic setae rather short, with some major setae spike-like; abdominal segment VIII with roughly circular plate giving rise to two minute branched setae dorsally and two longer spike-like setae ventrally; siphon short, with siphonal seta 1-S near base but without pecten teeth; saddle completely encircling anal segment; saddle seta 1-X spike-like.

Toxorhynchites rutilus (Coquillett)

Plate 7; Figs. 21, 32, 33; Map 8

Megarhinus rutilus Coquillett 1896:44.
Megarhinus septentrionalis Dyar and Knab 1906*d*:294.

Toxorhynchites rutilus and *septentrionalis* are now considered subspecies; the nominal subspecies *T. rutilus rutilus*, with a white band on the fore tarsus of the male, is confined to the southeastern USA. The more northerly populations, including the Canadian specimens, belong to *T. rutilus septentrionalis* and they lack the white band on the male fore tarsus. Females and larvae of the two subspecies have not been distinguished.

Adult. Upper third of occiput metallic blue scaled, lower two-thirds yellow-scaled; pedicel and clypeus pale gray pollinose; second and third palpomeres predominantly yellow-scaled, remaining palpomeres and proboscis mostly metallic purple scaled; postpronotum blue-scaled; median and lateral bands of scutum yellow-scaled, remainder dull metallic blue scaled, which may in some lights appear dark brown; pleura and coxae yellow-scaled; femora mostly yellow-scaled; in male, second and basal half of third tarsomeres of forelegs and midlegs and fourth hind tarsomere and in female, second to fourth tarsomeres including forelegs and midlegs and fourth and fifth tarsomeres of hind leg silvery white scaled, with all other legs purple-scaled; abdomen metallic blue green scaled above, yellow-scaled laterally and ventrally.

Larva. An unmistakable species in all four instars. Easily distinguished from all other species by the peculiar labral brushes, whose few strong stiff setae remain tightly pressed together, curving anteriorly away from the mouth rather than posteriorly as in all other mosquitoes. In addition, the short spike-like setae on thorax, abdominal segment VIII, and saddle are peculiar to *Toxorhynchites*. The only other Canadian culicid for which it might be mistaken, *Psorophora ciliata*, occurs in ground pools and

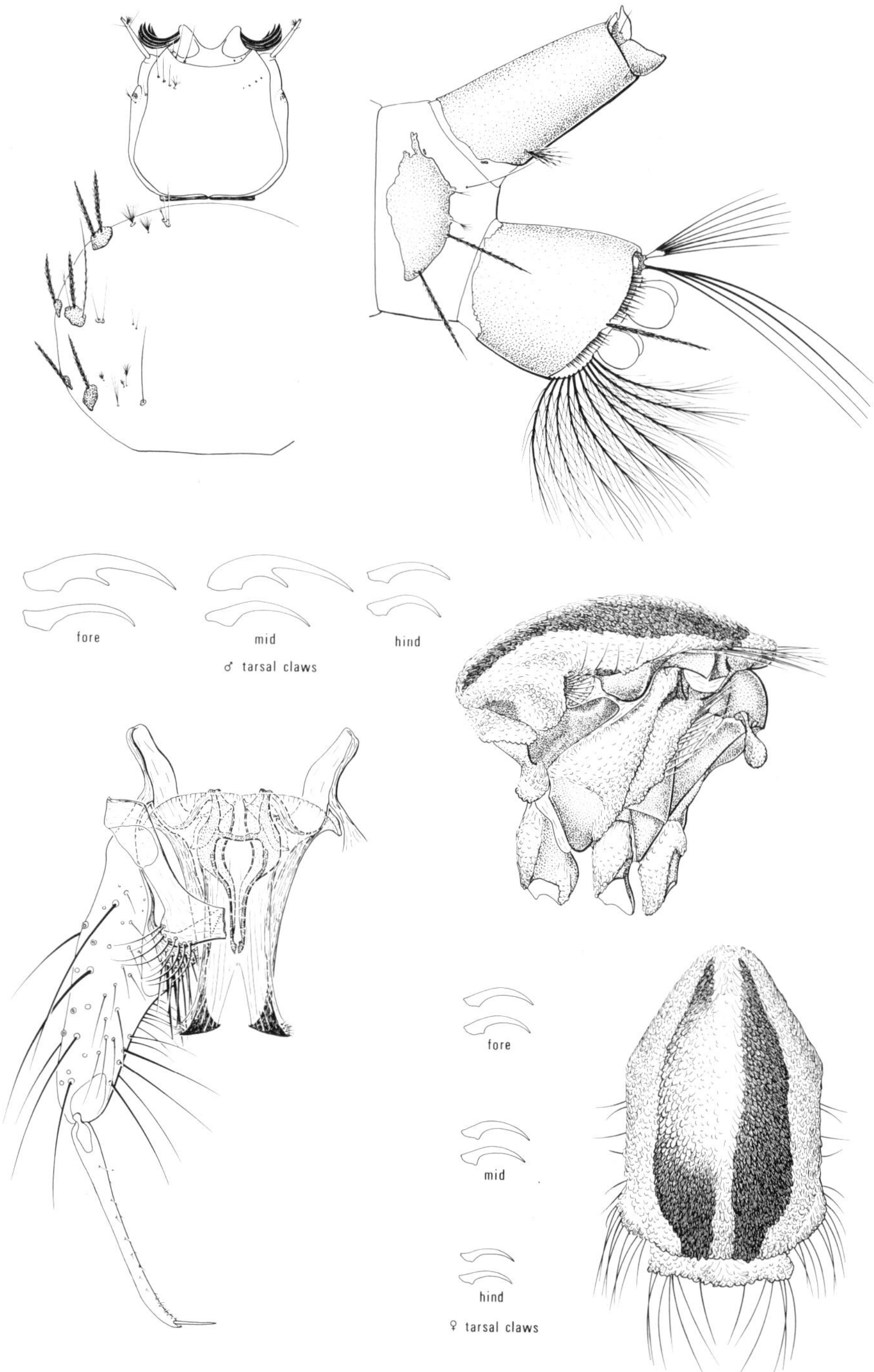

Plate 7. Toxorhynchites rutilus

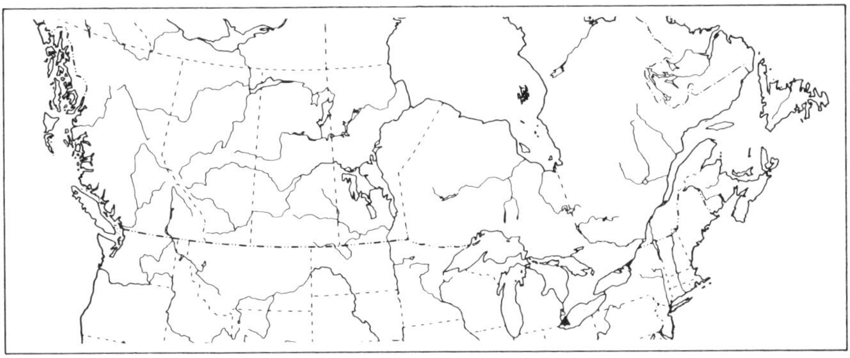

Map 8. Collection localities for *Toxorhynchites rutilus* in Canada: ▲ literature records.

has regular labral brushes and a much longer siphon with a row of pecten teeth. The larva of the large chaoborid *Eucorethra underwoodi* Underwood is also superficially similar, but occurs only in ground pools and lacks a fully developed siphon.

Biology. Only two larvae of *T. rutilus* have been found in Canada, in a large rot cavity in a red oak in Point Pelee National Park, Ont. (Parker 1977). The species usually inhabits rot holes in various hardwoods, but it has also been collected from artificial containers and even from the water contained in leaf bases of bromeliads (Carpenter and LaCasse 1955). The larvae are predaceous, usually feeding on the larvae of other tree-hole mosquitoes, such as *Orthopodomyia* spp., *Aedes triseriatus* and *hendersoni*, and *Anopheles barberi*, as well as on adult flies and other insects captured from the surface. Breland (1949) achieved faster development of the last instar by introducing fruit flies into the diet, whereas larvae fed only on mosquito larvae often remained in the last instar for several months before pupating. They are also cannibalistic and known for their voraciousness; one larva can devour over 100 smaller larvae of *Orthopodomyia* and *Aedes* (Basham et al. 1947) in 16 days. The prey is grasped by the mandibles (Breland 1948), not the labral brushes, whose tines are curved the wrong way for an effective grasping structure. The winter is passed as a last-stage larva. In contrast with the long larval life, the pupal period is of normal duration.

Adult females do not feed on blood, but, as do males, take only nectar. The white eggs are ejected singly onto the surface of the water while the female hovers rhythmically up and down over the surface (Breland 1949).

Distribution. Southeastern North America, from Illinois, southwestern Ontario, and New York south to Texas and Florida.

Subfamily Culicinae

Although some species of *Anopheles* (Anophelinae) can be locally common, most of the mosquito pests in Canada belong to the subfamily Culicinae, the largest and most diverse of the three culicid subfamilies. Some 27 genera and many more subgenera are recognized (Stone et al. 1959), including the large and important genera *Aedes* and *Culex*.

Culicines are most easily recognized by elimination, because the only two nonculicine genera in Canada, *Anopheles* and *Toxorhynchites*, are distinctive and easily recognized with the unaided eye. Female culicines have short palpi, less than half as long as the proboscis. The proboscis, in life, is carried down, at an angle to the body axis, rather than straight out in front as in *Anopheles*. Males of Culicinae usually have long palpi, with brushes of long setae on the apical segments, which are not swollen. The scutellum of both sexes is trilobed, and the scutellar setae are confined to the lobes. The egg is not equipped with wing-like floats as is that of *Anopheles*. Whereas anopheline larvae have small heads compared with the size of the body, the culicine larval head is usually large and broad. All culicine larvae have the respiratory apparatus elevated on a siphon, and although they may feed at or near the surface while suspended from the surface film by the apex of the siphon, most of them feed by grazing on the bottom or on objects in the water.

Genus *Aedes* Meigen

Adult. Small to large in size; female palpus short; male palpus usually about as long as proboscis (short only in *cinereus*); postocular setae numerous, arising close to eye margin; occiput fully covered with both erect and recumbent scales; acrostichal and dorsocentral setae usually well-developed, exceptionally so in some arctic species; spiracular setae absent; postspiracular area with setae and scales, but not as extensively covered as in *Psorophora*; tarsal claws usually each with a subbasal tooth of a different length; calypter with fringe; abdomen of female tapering; cercus of female usually long, slender, exserted; terminalia of male rather larger than those of other genera; gonocoxite usually with elaborately setose basal lobe and rounded apical lobe, often with tufts of setae in addition; gonostylus long, slender, sickle-shaped, with small apical spine; claspette present, typically with curved cylindrical stem bearing a single blade-like filament that is usually keeled on convex side (apex of claspette stem with several setiform filaments in *vexans*, or nothing in *cinereus*).

Larva. Head oval, usually wider than long, with rounded clypeal margin; setae of labral brush each usually with serrate inner margin; antenna rarely longer than length of head; antennal tuft usually arising from near the middle; prothoracic seta 4-P short, fine, single (except in *togoi*); mesothoracic seta 7-M multiple, its base usually combined with that of 6-M; siphon with well-developed row of pecten teeth and siphonal tuft arising near or beyond middle; saddle usually not completely encircling anal segment.

Subgenera of *Aedes*

The genus *Aedes* in Canada has been subdivided into five subgenera. Because the subgenera are based mainly on details of the male terminalia and are not well marked in the larval stage or in the female, they are not used in this text. *Aedes cinereus* is the type species of the genus, and therefore belongs to the subgenus *Aedes*. *Aedes vexans* belongs to the subgenus *Aedimorphus*; *togoi* to *Finlaya* (Stone et al. 1959); *hendersoni* and *triseriatus* to *Protomacleaya* (Zavortink 1972); and all the rest except *atropalpus* and *sierrensis* to *Ochlerotatus*. The subgeneric positions of *atropalpus* and *sierrensis* are still unsettled. Previously both were referred to *Finlaya*, but Zavortink (1972) transferred *atropalpus* to *Ochlerotatus*. Belkin and McDonald (1957) and Zavortink (1972) placed *sierrensis* in *Ochlerotatus*, whereas Cupp and Horsfall (1969), with the use of parasitological and cytochemical as well as morphological and behavioral evidence, argued in favor of retaining *sierrensis* in *Finlaya*. We lack experience with the subgenus *Finlaya* and cannot comment further.

Larvae of *Aedes togoi* and of all our species of *Ochlerotatus* including *atropalpus* have a single saddle seta 1-X, whereas in the other species this seta is branched. Among the last-mentioned group, *Aedes vexans* and *cinereus* have unevenly spaced distal pecten teeth, whereas those of *sierrensis* and of the two species of *Protomacleaya* are evenly spaced. In the males of *Ochlerotatus* the hind claw has a subbasal tooth that is absent in the males of *atropalpus, sierrensis*, and *cantator*. One of each pair of hind claws of *cantator* has a tooth, the other lacks it. Hind claws of males of non-*Ochlerotatus, sierrensis*, and *atropalpus* lack a subbasal tooth. The hind claws of the females of *Protomacleaya*, as well as of *cantator, atropalpus*, and *sierrensis* also lack a subbasal tooth; this tooth is present in females of all other *Ochlerotatus* and in *cinereus* and *vexans*. Thus there is no clear distinction at the subgeneric level that is common to the male, the female, and the larva, and we have therefore not used subgenera here as we have in *Culiseta*.

Key to the species of *Aedes* of Canada—adult females

1. Tarsomeres ringed with bands of white scales (Figs. 69–73) 2

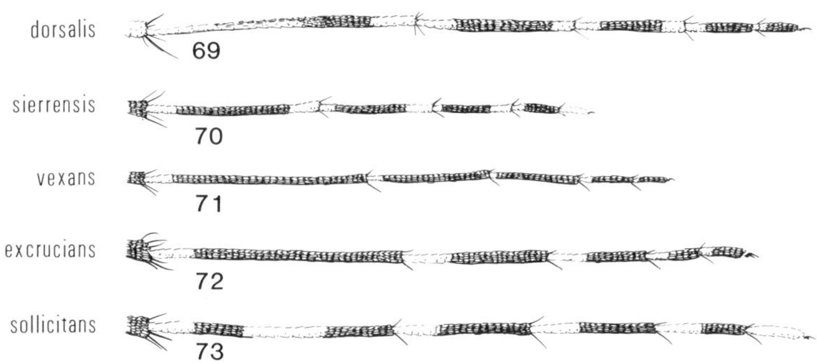

Figs. 69–73. Hind tarsus of female of *Aedes* species.

Tarsomeres brown-scaled, without rings of white scales, although scattered white scales can be present 23

2. Basal four tarsomeres each with base and apex white-scaled (Figs. 69, 70) 3
   Basal four tarsomeres each with white scales at base only (Figs. 71–73) 9
3. Postprocoxal membrane bare (Fig. 74); wing veins (except for base of costa in some species) entirely brown-scaled 4
   Postprocoxal membrane with scales (Fig. 75); wing veins, especially basal portions of vein R, with white and brown scales intermixed (*dorsalis* group) 7
4. Scutum with uniformly colored narrow gold scales on a reddish brown integument; costa entirely dark-scaled .... ***canadensis***
   Scutum with contrasting stripes and patches of gold and brown scales; costa usually with a patch of pale scales at base 5
5. First (basal) tarsomere with equibroad basal and apical white bands; second and third tarsomeres each with apical white band more than three times as broad as basal band (Fig. 70); scales of katepisternum in two separate groups; apices of palpus and of pedicel of antenna extensively white-scaled .... ***sierrensis***
   All tarsomeres (except apical one) with apical and basal white bands subequal, or the basal slightly broader than the apical bands (as in Fig. 69); scales of

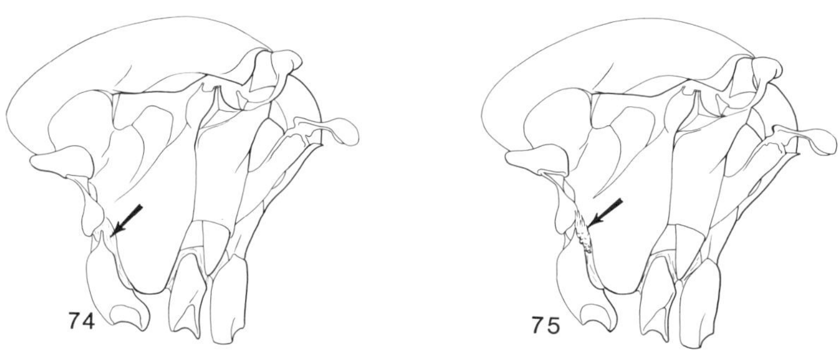

Figs. 74, 75. Left lateral view of thorax showing: 74, bare postprocoxal area; 75, postprocoxal area with scales.

Figs. 76–78. Tarsal claw of female of *Aedes* species: 76, 77, fore claw; 78, hind claw.

katepisternum in a single large group; apices of palpus and of pedicel with few or no white scales 6

6. Hind apical tarsomere entirely white-scaled (as in Fig. 70); palpus entirely brown-scaled; Eastern Canada .... ***atropalpus***

   Hind apical tarsomere brown-scaled; palpus with white scales at apex; Pacific coast .... ***togoi***

7. Tarsal claw moderately and evenly curved (Fig. 76), the subbasal tooth (distance *x*-*y*) short, one-third or less as long as the distance between apices of subbasal tooth and of claw (*y*-*z*) .... ***dorsalis***

   Tarsal claw strongly and abruptly bent just beyond subbasal tooth (Fig. 77), the subbasal tooth (*x*-*y*) about half as long as the distance between apices of tooth and of claw (*y*-*z*) 8

8. All wing veins with pale and dark scales intermixed, the pale scales usually predominant; scales of middorsal stripe of scutum dull yellowish brown, flanked by paler yellowish scales .... ***campestris***

   Wing veins predominantly dark-scaled, veins R_{4+5}, M_1, M_2, M_4, Cu, and A almost entirely so; scales of middorsal stripe of scutum reddish brown, flanked by whitish, more contrasting scales .... ***melanimon***

9. Basal white-scaled bands on hind tarsomeres narrow, each band less than one-quarter as long as respective tarsomere 10

   Basal white-scaled bands on hind tarsomeres broader, each about one-third as long as respective tarsomere 11

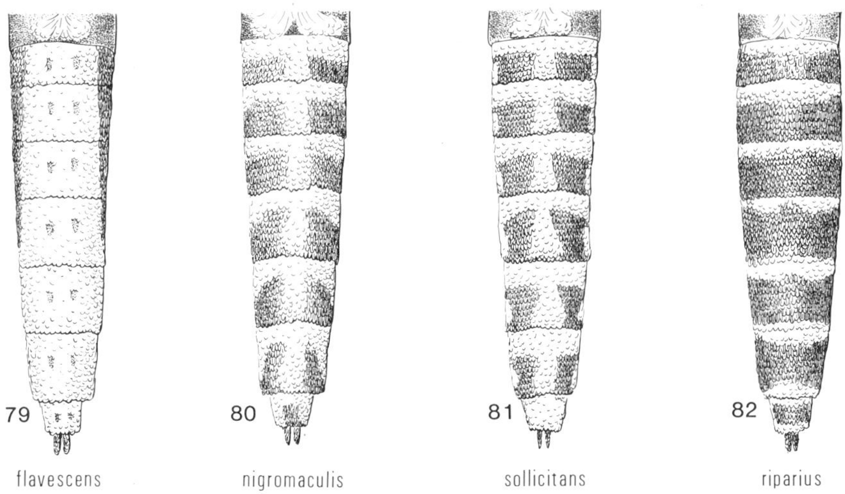

Figs. 79–82. Dorsal view of abdomen of female of *Aedes* species.

10. Hind claw lacking a subbasal tooth (Fig. 78); lower mesepimeral setae usually present; first abdominal sternite with fine hair; katepisternum with scales extending to anterodorsal angle .... ***cantator***

Hind claw with conspicuous basal tooth; lower mesepimeral setae absent; first abdominal sternite bare; katepisternum without scales on anterodorsal angle .... ***vexans***

11. Abdominal tergites each with a median longitudinal stripe of pale yellowish scales in addition to the basal transverse band (Fig. 80), or more or less entirely yellow-scaled (Fig. 79) 12

Abdominal tergites each predominantly dark-scaled with basal transverse band of pale scales only; pale scales, when present behind this band, scattered rather than concentrated into a median longitudinal stripe (Fig. 82) 14

12. Abdominal tergites almost entirely covered with pale yellow scales, except for a pair of small bare spots on each of tergites II–VII (Fig. 79); proboscis dark-scaled, with scattered yellow scales not organized into a discrete band .... ***flavescens***

Abdominal tergites with lateral areas of dark scales (Fig. 80); proboscis dark-scaled, usually with a more or less well-defined ring of white scales near mid length 13

13. Abdominal tergites each with a lateral patch of white scales contrasting in color with the pale yellow scales of the transverse band and middorsal longitudinal stripe (Fig. 81); first (basal) hind tarsomere with discrete ring of pale yellow scales at its midlength, distinctly separated from basal white ring by dark scales (Fig. 73); last hind tarsomere entirely white-scaled; postpronotum mostly bare, with the yellowish brown scales confined to its upper edge; Maritime Provinces and southwestern Ontario .... ***sollicitans***

Abdominal tergites each with all lateral pale scales concolorous with those of the transverse band and middorsal longitudinal stripe; pale yellow scales in middle of basal hind tarsomere scarcely separated from the white scales of basal ring; last hind tarsomere mostly dark-scaled; postpronotum mostly scaled, with the yellowish brown scales occupying its upper third; Prairie Provinces only .... ***nigromaculis***

14. Tarsal claw strongly bent at or before midlength, the apical portion sinuous with a hooked apex and more or less parallel to the subbasal tooth, enclosing an angle of less than 30° (Fig. 85*a*–85*c*); subbasal tooth about half as long as claw beyond fork, its length (x–y) usually greater than distance (y–z) between apices of tooth and claw .... ***excrucians***

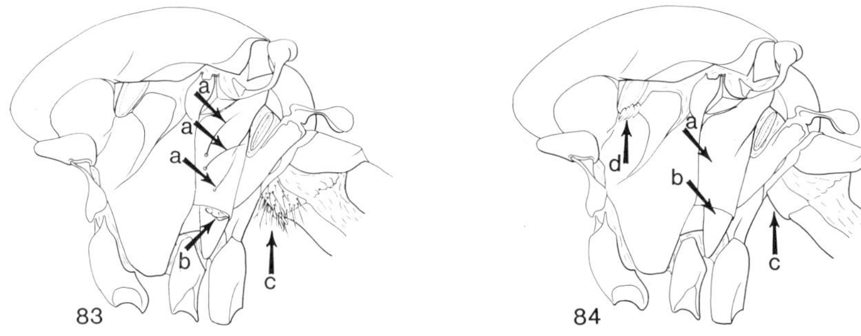

Figs. 83, 84. Left lateral view of thorax and abdominal segments I and II.

Fig. 85. Fore tarsal claw of female of *Aedes excrucians*: *a*, *b*, southern variants; *c*, a northern variant.

Tarsal claw more evenly curved, the apical portion not parallel to subbasal tooth, enclosing an angle of more than 30° (Figs. 88–92); subbasal tooth about one-third as long as claw beyond fork, its length (*x–y*) about half distance (*y–z*) between apices of tooth and claw 15

15. Lower mesepimeral setae present (Fig. 83*a*) mesomeron usually with a few scales on posterodorsal corner, directed posterodorsally (Fig. 83*b*) 16

Lower mesepimeral setae absent (Fig. 84*a*); mesomeron bare (Fig. 84*b*) (loose scales sometimes adhere, but their haphazard position usually reveals their extraneous origin) 20

16. Scales of pedicel all or mostly dark; scutum mostly reddish brown scaled, with pale scales confined to dorsocentral and lateral stripes (Fig. 86) 17

Scales of pedicel mostly white; median and submedian stripes of scutum brown-scaled, forming single middorsal dark stripe contrasting with pale-scaled sublateral and lateral areas (Fig. 87) 18

17. First abdominal sternite usually densely clothed with pale scales and hairs (Fig. 83*c*); intermediate palpomeres with complete basal white-scaled rings; proboscis usually entirely dark-scaled; tarsal claw smaller, more evenly curved beyond subbasal tooth, distal portion not thickened (Fig. 88); Saskatchewan westward .... ***increpitus***

First abdominal sternite bare (Fig. 84*c*); intermediate palpomeres with scattered pale scales; proboscis with scattered white scales; tarsal claw larger, more strongly bent and slightly thickened between bend and apex (Fig. 89); Manitoba eastward, possibly also Saskatchewan .... ***stimulans***

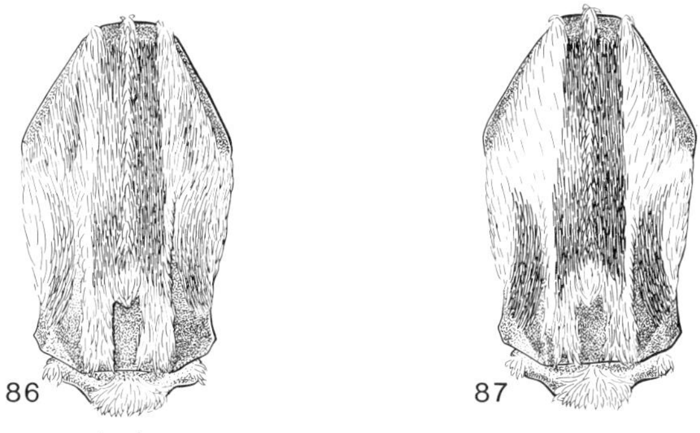

Figs. 86, 87. Scutum of female of *Aedes* species.

88 increpitus 89 stimulans 90 riparius 91 euedes 92 fitchii

Figs. 88–92. Fore tarsal claw of female of *Aedes* species.

18. Wing entirely dark-scaled; dorsal brown-scaled area of postpronotum half as large as ventral pale-scaled area; scutum with dark brown middorsal stripe flanked by pale yellow scales of sublateral and lateral stripes; third fore tarsal segment without complete white basal ring, although a small spot of white scales may be present ventrolaterally .... ***mercurator***
    Wing usually with scattered pale scales intermixed with the predominant dark scales; dorsal brown-scaled area of postpronotum as large as or larger than ventral pale-scaled area; scutum with reddish brown middorsal stripe flanked by whitish-scaled sublateral and lateral stripes; third fore tarsomere with complete basal white ring 19
19. Wing scales large, broad, somewhat triangular in outline, about half as wide as long and usually truncate at apex; pale and dark scales evenly scattered over all wing veins .... ***grossbecki***
    Wing scales much narrower, one-third or less as wide as long, usually rounded apically; pale scales more numerous on veins C, Sc, and R_1 .... ***fitchii***
20. Tarsal claw relatively long, not strongly curved (Fig. 90), with short subbasal tooth not more than one-fifth total length of claw; angle between claw and subbasal tooth 45° or more 21
    Tarsal claw shorter, more strongly curved (Figs. 90, 91), with longer subbasal tooth one-quarter or more total length of claw; angle between claw and subbasal tooth 40° or less; hypostigmal area without scales 22
21. Hypostigmal area usually with a patch of scales; scutum mostly orange-scaled .... ***riparius***
    Hypostigmal area bare; scutum mostly yellowish brown scaled .... ***aloponotum***
22. Proboscis, cercus, and first tarsomeres usually (except for northern specimens) with numerous scattered pale scales, intermixed with predominant black scales; tarsal claw longer and straighter (Fig. 91); large species with mottled reddish brown scutal pattern .... ***euedes***
    Proboscis, cercus, and first tarsal segments beyond basal white ring usually entirely dark-scaled; tarsal claw shorter and more strongly curved (Fig. 92); medium-sized species with distinct middorsal brown scutal stripe and whitish-scaled lateral areas .... ***fitchii***
23. Postprocoxal membrane with scales (Fig. 75) 24
    Postprocoxal membrane bare (Fig. 74) 34
24. Wing veins R_2, R_3, M, and A with pale scales, other veins mostly dark-scaled; abdomen usually (especially in specimens from the Prairie Provinces) with middorsal longitudinal stripe of white scales in addition to transverse white bands .... ***spencerii***
    Wing veins all predominantly dark-scaled, except sometimes at base; abdominal tergites with transverse basal white bands only 25
25. Postpronotum with scattered setae in addition to those along posterior margin usually present in all species (Fig. 93); sublateral area of scutum also with numerous scattered setae 26
    Setae of postpronotum confined to posterior margin, occasionally also with 4 or 5 setae along dorsal margin (Fig. 94); sublateral area of scutum with few scattered setae 27

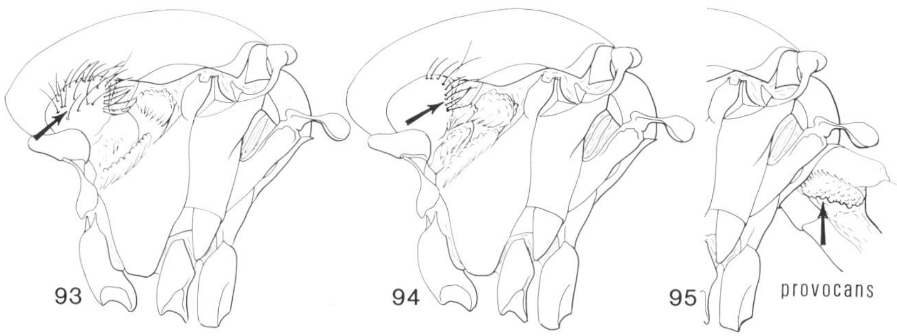

Figs. 93–95. Left lateral view of *Aedes* species: 93, thorax showing extra postpronotal setae; 94, thorax showing usual postpronotal setae; 95, metathorax and base of abdomen.

26. Tarsal claw rather sharply bent near middle, with the portion beyond the bend nearly parallel with long subbasal tooth (Fig. 96); denuded scutum showing paler pollinose band paralleling transverse suture .... ***impiger***
    Tarsal claw evenly and not strongly curved, with its subbasal tooth much shorter than length of claw (Fig. 97); pollen of denuded scutum velvety black, with no pale transverse band .... ***nigripes***
27. Anepisternum (hypostigmal, subspiracular, and postspiracular areas combined) almost completely covered with white scales except for a small central bare patch in middle of subspiracular area (Fig. 94) 28
    Anepisternum not so extensively scaled, with the upper half of subspiracular area including hypostigmal area and anterior half of postspiracular area bare (if a few scales are present on hypostigmal area, wing veins are dark-scaled except at base of costa, and proboscis and palpi are almost entirely dark-scaled) 29
28. Costa, subcosta, and R_1 predominantly dark-scaled, but with scattered white scales; membrane between tergum and sternum of first abdominal segment with at most four or five scales; tarsal claw rather short and strongly curved (Fig. 98) .... ***cataphylla***
    Wing veins dark-scaled, except at base; membrane between tergum and sternum of first abdominal segment with a large patch of white scales (Fig. 95); tarsal claw rather straight, with small subbasal tooth (Fig. 99) .... ***provocans***
29. Anterodorsal corner of katepisternum and ventral fifth of mesepimeron bare of scales (Fig. 100, *a*, *b*); dorsum of head and periphery of scutum white-scaled; palpus with numerous scattered white scales .... ***implicatus***
    Scales along dorsal edge of katepisternum extending to anterodorsal corner (Fig. 101, *a*); scales of mesepimeron extending to its ventral margin (Fig. 101, *b*); dorsum of head and periphery of scutum yellow-scaled (among arctic specimens

Figs. 96–99. Fore tarsal claw of female of *Aedes* species.

these scales sometimes fade to a pale brassy yellow, and then the entire scutum becomes pale, lacking a dark central contrasting area); palpus mostly dark-scaled 30

30. Setae of scutum mostly black; submedian stripes broad, dark brown, separated by a narrow median stripe of yellow scales, and contrasting strongly with yellow-scaled sublateral and lateral areas ..... ***pionips***

Setae of scutum yellow or bronze; scutum with a single, broad, middorsal dark brown stripe (median and submedian stripes combined) or without darker submedian stripes (the scales of this stripe, though concolorous with those of sublateral area, are usually narrower and may appear darker because more dark integument is revealed) 31

31. Probasisternum and adjacent membranes rather extensively scaled, and usually with a few setae; costa of wing with a patch of white scales at base 32

Probasisternum without setae and with at most a few scattered scales at apex; costa of wing usually dark-scaled at base 33

32. Underside of proboscis with some pale scales ..... ***schizopinax***

Underside of proboscis entirely dark-scaled ..... ***hexodontus***

33. Scutum with a single broad middorsal longitudinal stripe of scales that are distinctly darker brown than those of sublateral area (Fig. 87) ..... ***punctor***

Scutum uniformly medium brown (or with darker submedian stripes that may be darker or appear darker because the scales are narrower, revealing more dark ground color) (coastal British Columbia) ..... ***aboriginis***

(Manitoba and east) ..... ***abserratus***

34. Anepisternum bare along ventral margin (Fig. 102, *a*); anterior surface of fore coxa with a patch of brown scales (Fig. 102, *b*) ..... ***cinereus***

Anepisternum densely scaled along ventral margin; fore coxa pale-scaled (except in *aurifer*) 35

35. Paratergite and lateral margin of first abdominal tergite bare; lateral margin of scutum with reddish brown scales, concolorous with those of middorsal (median and submedian) stripe, contrasting with white scales of the sublateral area (Fig. 6) ..... ***trivittatus***

Paratergite and lateral margin of first abdominal tergite with white scales; scales of lateral margin of scutum concolorous with those of adjacent sublateral area 36

36. Abdominal tergites each with broad basal transverse band of white scales 37

Abdominal tergites dark-scaled dorsally, except for a lateral triangular patch of white scales on each side (which may rarely be connected by a narrow basal band of white scales on apical segments) 42

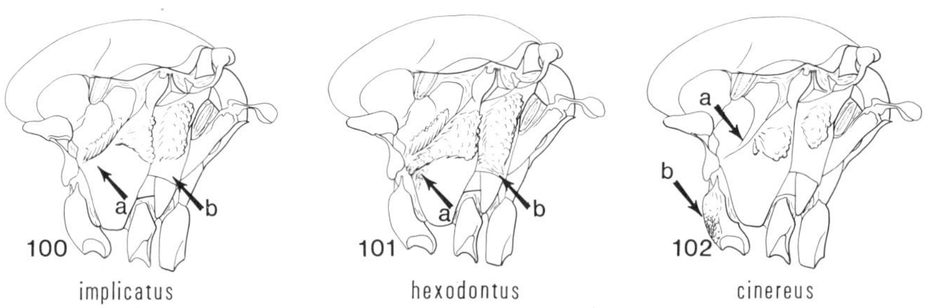

Figs. 100–102. Left lateral view of thorax of female of *Aedes* species showing scale patterns of mesopleuron.

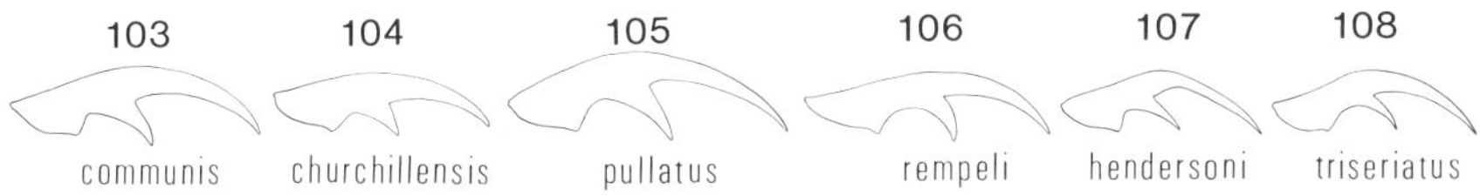

Figs. 103–108. Fore tarsal claw of female of *Aedes* species.

37. Katepisternum with scales extending to anterodorsal corner (Fig. 101, *a*) (scales in this region should all be directed toward wing base, if not, they may be displaced from elsewhere) 38
Anterodorsal corner of katepisternum bare (Fig. 100, *a*) (specimens that have this area denuded will not have any of the combination of characters found in couplets 38–39; however, if the scutum is also denuded, the specimen should be considered unidentifiable unless the basal pit of each scale can be seen after NaOH treatment of the pleuron) 40
38. Scales of median and submedian stripes of scutum concolorous, forming a single broad longitudinal uniformly reddish brown stripe; scales of upper half of postpronotum reddish brown, concolorous with middorsal stripe; lower mesepimeral setae absent........ ***sticticus***
Submedian stripes of scutum usually dark brown scaled, separated by a paler yellowish brown scaled median stripe; scales of the upper half of postpronotum and (except in subarctic specimens) of the lateral and sublateral areas yellowish brown; lower mesepimeral setae present 39
39. Hind tarsal claw with longer narrower subbasal tooth (Fig. 103) .... ***communis***
Hind tarsal claw with shorter subbasal tooth (Fig. 104) .... ***churchillensis***
40. Scales of the scutum uniformly bronzy or yellowish brown, with darker submedian stripe lacking (these scales can be slightly narrower, giving the stripe a darker appearance because of exposure of integument); pedicel yellowish laterally, paler than clypeus; setae of antepronotum and scutum mixed yellow and brown; erect and recumbent scales on vertex yellow, concolorous; integument of head and body medium to light brown........ ***intrudens***
Scales of the median stripe of scutum pale in contrast with narrower, sparser dark brown scales of submedian stripe; pedicel dark brown, concolorous with clypeus; most of the setae of antepronotum and scutum dark brown; erect scales on vertex brown, with whitish recumbent scales among them; integument of head and body dark brown, appearing grayish in some lights because of sparse pollinosity........ 41
41. Mesepimeron with scales extending to anteroventral corner; upper border of postpronotum with scattered setae; hypostigmal scale patch usually present (Fig. 84, *d*); tarsal claw evenly curved with prominent subbasal tooth (Fig. 105)........ ***pullatus***
Mesepimeron bare on lower third; upper border of postpronotum without setae; hypostigmal area with at most two or three scales, usually bare; tarsal claw rather straight basally, curved only near apex, with small subbasal tooth (Fig. 106)........ ***rempeli***
42. Scales of postpronotum and lateral and sublateral areas of scutum silvery white; scales of postpronotum oval or rounded, overlapping one another, obscuring integument; hind tarsal claw without subbasal tooth........ 43
Scales of postpronotum and lateral and sublateral areas of scutum yellow or yellowish brown; scales of postpronotum crescent-shaped, not overlapping; claws of all tarsi each with subbasal tooth 44

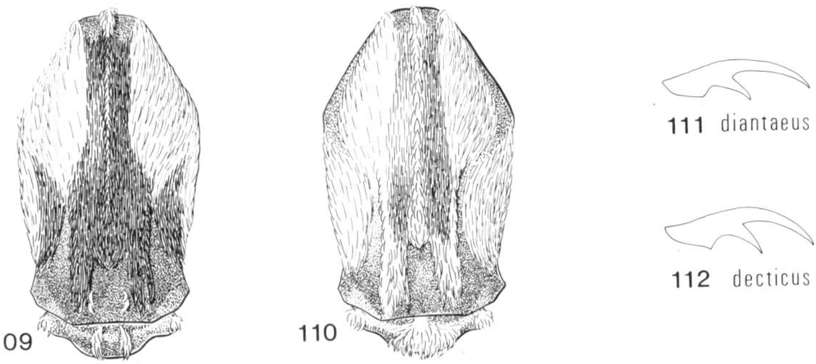

Figs. 109–112. Female of *Aedes* species: 109, 110, dorsal view of scutum; 111, 112, fore tarsal claw.

43. Fore tarsal and midtarsal claws strongly bent just beyond subbasal tooth, with the portion distal to the bend rather straight (Fig. 107); metameron without scales .... ***hendersoni***

Tarsal claw evenly curved (Fig. 108); metameron usually with one or two scales .... ***triseriatus***

44. Two-thirds or more of central area of scutum dark brown scaled; only the lateral and outer margin of sublateral area golden brown scaled; almost all acrostichal and dorsocentral setae on anterior half of scutum lacking; fore coxa mostly brown-scaled .... ***aurifer***

Dark brown scales anterior to transverse suture confined to submedian and median stripes (Figs. 109, 110); lateral and sublateral areas completely pale-scaled (some *decticus* have sublateral dark patches); acrostichal and dorsocentral setae present on anterior half of scutum; fore coxa pale-scaled 45

45. Dark brown submedian stripes of scutum broad, contiguous with each other medially, the median yellow stripe absent or reduced to a few scales anteriorly (Fig. 109); sublateral dark stripes extending from transverse suture to scutellar margin, and contiguous with submedian stripes; prescutellar depression almost completely surrounded anteriorly and laterally with dark scales, metameron bare .... ***thibaulti***

Dark brown submedian stripes of scutum narrower, completely separated medially by a narrow median stripe of yellow scales (Fig. 110); sublateral dark stripes of posterior half of scutum less extensive, separated from adjacent submedian stripe by a line of yellow scales; prescutellar depression surrounded with yellow scales; metameron usually with scales 46

46. Scales of vertex and postpronotum yellow; fore tarsal claw long, slender, and rather straight (Fig. 111) .... ***diantaeus***

Scales of vertex and postpronotum predominantly yellow, usually with patches of brown scales; fore tarsal claw more strongly curved (Fig. 112) .... ***decticus***

Key to the species of *Aedes* of Canada—adult males

Males are most easily recognized by their tarsal, palpal, and genitalic characters. Males and females have similar color and scaling patterns, and similar hind tarsal claws. However, because the scales are sparser or more easily detached from males, many specimens cannot be identified by the key to females. Males have fewer scales in the postprocoxal scale patch, the white basal abdominal bands are not reliable, and the wing scales tend to be pale and easily lost. Consequently, the following key avoids color characters, except those of the leg scales. Tarsal and palpal characters are used first, followed by those of the terminalia. Unfortunately, the external characters of the banded-legged species of the *excrucians* group (couplets 13–21) and most of the black-legged species (couplets 27–41) are too slight to be reliable, and the terminalia must be removed and cleared in NaOH for identification. When terminalia are slide-mounted, future identification can always be assured as long as a hind tarsus and a palpus are also mounted on the slide, even if the rest of the adult is lost or discarded.

1. Tarsomeres ringed with bands of white scales (Figs. 69–73) 2
   Tarsomeres dark-scaled, without rings of white scales, although scattered white scales may be present 22
2. Basal three tarsomeres, especially of hind leg, each with both base and apex white-scaled (white rings need not be of equal width) (Figs. 69, 70) 3
   Basal three tarsomeres each with ring of white scales at base only (Figs. 71–73) 9
3. Palpus longer than proboscis, with dense brushes of long setae on fourth palpomere and apex of third (Fig. 113); apical lobe of gonocoxite present 4
   Palpus equal to or shorter than proboscis, without dense brushes of setae (Fig. 114); apical lobe of gonocoxite absent, with the gonocoxite tapering evenly to apex 7
4. Wing veins entirely dark-scaled; abdomen mostly dark-scaled dorsally, with white scales confined to transverse basal bands; postprocoxal membrane bare; claspette filament narrowly sickle-shaped, without expanded region (Fig. 118 *b*) .... ***canadensis***
   Wing veins, especially basal portion of radius, with white and brown scales intermixed; abdomen mostly white-scaled; postprocoxal membrane with scales; claspette filament expanded (Fig. 120*b*) 5
5. Hind tarsal claw evenly and only slightly curved; subbasal tooth short (distance *x–y*), one-third as long as distance between apices of subbasal tooth and of claw (*y–z*, Fig. 115 *b*); basal lobe of gonocoxite with two enlarged spines: one long, directed medially and the other short, separated by about half the length of the shorter spine (Fig. 115*a*) .... ***dorsalis***
   Hind tarsal claw bent just distal to subbasal tooth; subbasal tooth longer, about half as long as distance between apices of tooth and claw (Figs. 116*b*; 117*b*); basal lobe of gonocoxite either with one large and one small spine arising side by side (Fig. 117*a*) or with one large and two or three small spines (Fig. 116*a*) 6

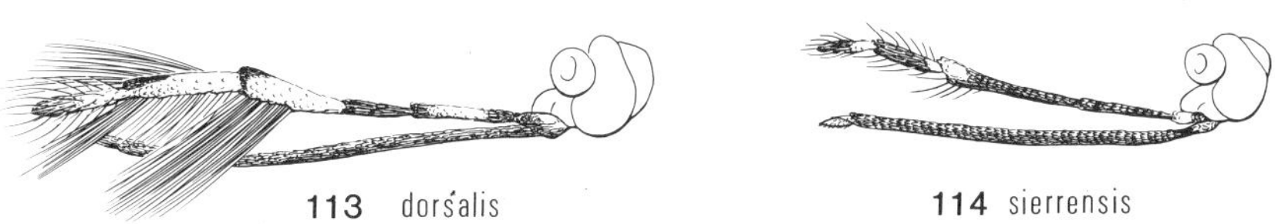

Figs. 113, 114. Left lateral view of head of male of *Aedes* species. Showing differences in the palpus (antenna omitted).

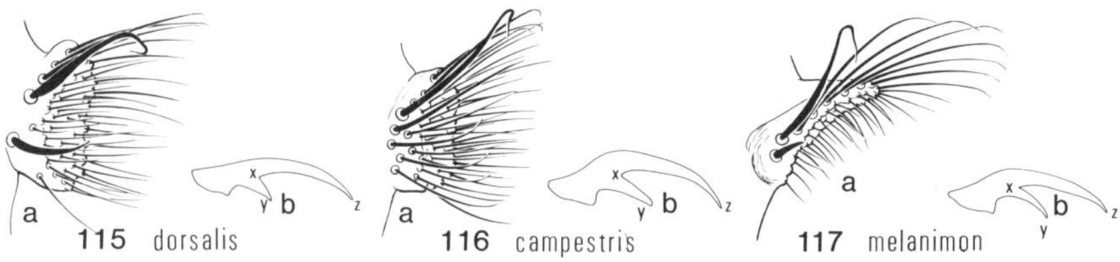

Figs. 115–117. *Aedes* species: *a*, basal lobe of gonocoxite; *b*, hind tarsal claw; *x-y*, length of subbasal tooth; *x-z*, length of main claw.

6. Setose surface of basal lobe of gonocoxite somewhat convex, inclined posterodorsally, with thicker spines at base of dorsal edge of lobe not strongly differentiated in size or sclerotization (Fig. 116*a*) .... ***campestris***
   Setose surface of basal lobe of gonocoxite flattened, inclined posteroventrally, with one short and one long enlarged spine arising side by side at base of dorsal edge (Fig. 117*a*) .... ***melanimon***
7. Last hind tarsomere entirely brown-scaled; hind claw with subbasal tooth; medioventral edge of gonocoxite angulate at level of claspette filament, bearing a dense fringe of long hairs overlapping with those of opposite side .... ***togoi***
   Last hind tarsomere entirely white-scaled; hind claw without subbasal tooth; medioventral edge of gonocoxite lacking basal angle and dense fringe 8
8. Palpus subequal in length to proboscis, with rings of white scales at intersegmental junctions of last three palpomeres (Fig. 114); basal lobe of gonocoxite bearing a dense clump of strong, apically curved setae .... ***sierrensis***
   Palpus entirely dark-scaled, about three-quarters as long as proboscis; basal lobe of gonocoxite with fine setae only .... ***atropalpus***
9. White rings of tarsal segments relatively narrow, those of second or third hind tarsomere occupying less than one-fifth total length of tarsomere (Fig. 71) 10
   White rings of tarsal segments relatively broad, those of second or third hind tarsomere occupying one-third or more of tarsomere (Fig. 72) 11
10. Mesepimeral setae present; hind claw without subbasal tooth; first abdominal sternite with pale hairs; gonostylus with apical spine; claspette stem slender, terminating in sickle-shaped filament .... ***cantator***
    Mesepimeral setae absent; hind claw with subbasal tooth; first abdominal sternite bare; gonostylus with subapical spine; claspette stem broad, terminating in a clump of setae, filament absent .... ***vexans***

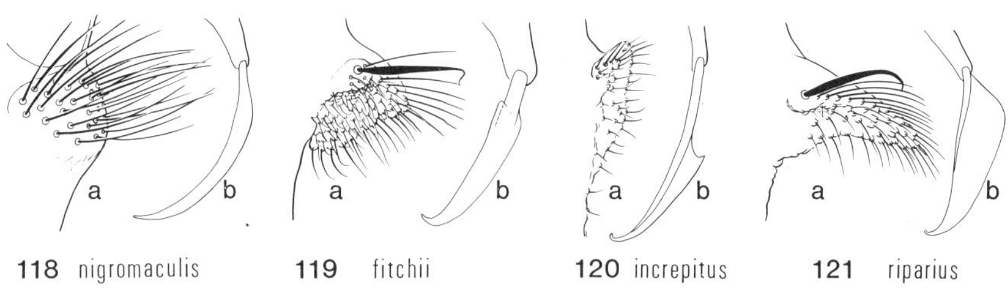

Figs. 118–121. Male of *Aedes* species: *a*, basal lobe of gonocoxite; *b*, lateral view of claspette filament.

11. Basal lobe of gonocoxite a small rounded tubercle (Fig. 118*a*); apical lobe not differentiated; claspette filament a narrow, nearly parallel-sided rod (Fig. 118*b*) 12
Basal lobe of gonocoxite a conical or triangular lobe (Figs. 119*a*–121*a*), often extending part way along mediodorsal margin of gonocoxite (Figs. 124*a*–126*a*); apical lobe prominent; claspette filament expanded at or near middle (Figs. 119*b*; 120*b*) 13
12. Postpronotum mostly scaled, with yellowish brown scales occupying its upper third; western provinces .... ***nigromaculis***
Postpronotum mostly bare, with yellowish brown scales confined to its upper margin; Maritime provinces and southwestern Ontario .... ***sollicitans***
13. Basal lobe of gonocoxite with a strongly differentiated enlarged spine in addition to fine hairs (Fig. 119*a*) 14
Basal lobe of gonocoxite with fine hairs only (Fig. 120*a*) 18
14. Claspette filament with a sharp angle near base of concave side (Fig. 119*b*) .... ***fitchii***
Claspette filament lacking such an angle on concave side (Figs. 120*b*; 121*b*) 15
15. Claspette filament with a long parallel-sided stem-like base, not expanding until near middle (as in Fig. 120*b*); abdominal tergites largely dark-scaled except for basal transverse white bands 16
Claspette filament expanding from base (Fig. 121*b*); abdominal tergites with numerous scattered pale scales 17
16. Integument of scutum reddish brown; middorsal stripe and sublateral area of scutum pale reddish brown, separated by a narrow stripe of yellowish scales along row of dorsocentral setae; apical lobe of gonocoxite extending distally beyond point of attachment of gonostylus; spine of basal lobe scarcely crossing midline (Fig. 122*a*) .... ***stimulans***
Integument of scutum dark brown; middorsal stripe dark brown contrasting with yellow-scaled sublateral area; apical lobe of gonocoxite not extending as far distally as point of attachment of gonostylus; spine of basal lobe long, extending to base of spine on opposite lobe (Fig. 123*a*) .... ***mercurator***

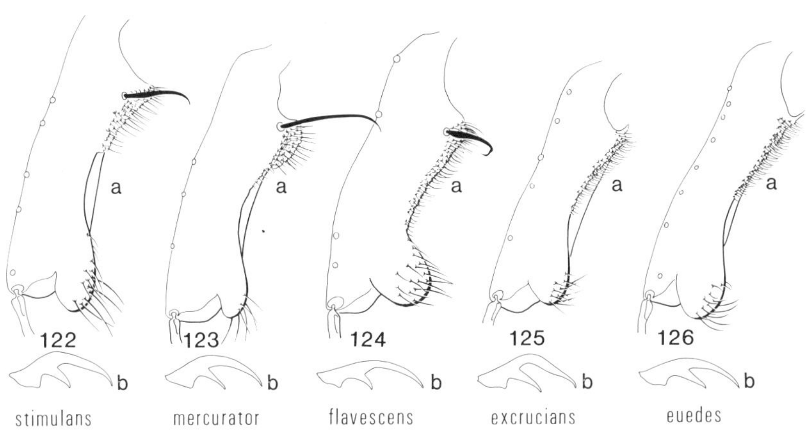

Figs. 122–126. *Aedes* species: *a*, dorsal view of gonocoxite; *b*, hind tarsal claw.

17. Wing scales broad, about half as wide as long and usually truncate at apex; pale and dark scales evenly scattered on all wing veins .... ***grossbecki***

Wing scales much narrower, one-third or less as wide as long, usually rounded apically; pale scales more numerous on veins C, Sc, and R .... ***fitchii***

18. Basal lobe of gonocoxite depressed, triangular, extending half way to apical lobe (Fig. 124*a*); large yellowish species .... ***flavescens***

Basal lobe of gonocoxite conical, strongly pointed medially (Fig. 121*a*); reddish orange species .... ***riparius***

19. Basal lobe of gonocoxite distinct (Fig. 120*a*); cylindrical stem-like base of claspette filament as long as blade (Fig. 120*b*); first abdominal sternite covered with scales and hairs .... ***increpitus***

Basal lobe scarcely differentiated, extending nearly to base of apical lobe (Figs. 125*a*; 126*a*); cylindrical base of claspette filament much less than half as long as blade; first abdominal sternite bare 20

20. Hind claw strongly bent beyond subbasal tooth, which is half as long as claw beyond tooth and almost parallel to it, enclosing an angle of less than 30° (Fig. 125*b*); apical lobe small, ending before base of gonostylus (Fig. 125*a*) .... ***excrucians***

Hind claw less strongly bent; subbasal tooth of hind claw shorter, less than half as long as claw beyond tooth and more strongly divergent from it, enclosing an angle of more than 30° (Fig. 126*b*) 21

21. Apical lobe extending to or beyond base of gonostylus (Fig. 126*a*) .... ***euedes***

Apical lobe smaller, ending before base of gonostylus (as in Fig. 125*a*) .... ***aloponotum***

22. Palpus minute, no larger than that of female (Fig. 127); lower margin of anepisternum without scales; gonostylus twice forked, arising proximal to apex of gonocoxite (Fig. 131) .... ***cinereus***

Palpus more than three-quarters as long as proboscis; anepisternum scaled along ventral margin; gonostylus unforked, arising at apex of gonocoxite 23

23. Apex of third palpal segment scarcely swollen, with, at most, a small group of fewer than 10 setae arising close to apex (no farther from apex than maximum width of segment) (Figs. 128–130) 24

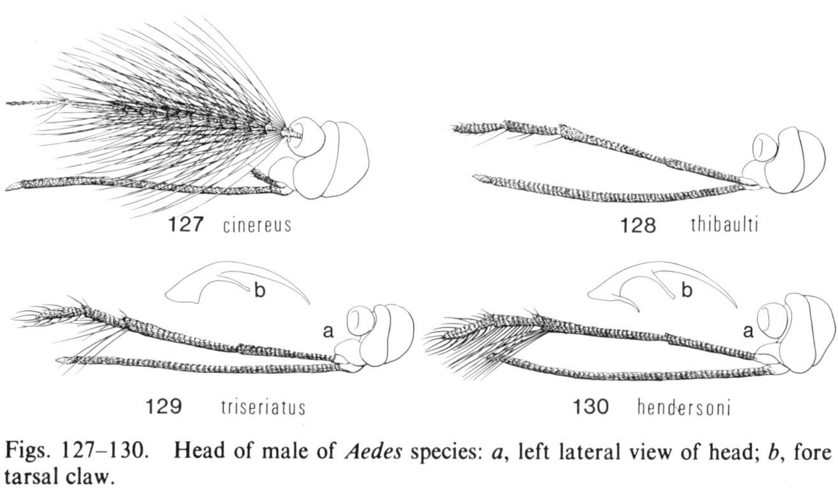

Figs. 127–130. Head of male of *Aedes* species: *a*, left lateral view of head; *b*, fore tarsal claw.

Apical one-quarter or one-third of third palpal segment distinctly enlarged, bearing a conspicuous tuft or fringe of 50 or more long setae arising from ventrolateral surface (as in Fig. 113) 28

24. Scales of postpronotum and sublateral and lateral areas of scutum anterior to transverse suture silvery white; scales of postpronotum rounded or oval, overlapping each other; terminal spine of gonostylus half as long as remainder of gonostylus (Fig. 132); apical lobe of gonocoxite undifferentiated; medioventral edge of gonocoxite with small tuft of setae at mid length; claspette a sickle-shaped unexpanded rod (Fig. 134) 25

Scales of postpronotum and of sublateral and lateral areas of scutum anterior to transverse suture yellow or yellowish brown; scales of postpronotum crescent-shaped, not overlapping; terminal spine of gonostylus one-quarter or less the remaining length of gonostylus (Fig. 133); apical lobe of gonocoxite well-defined; medioventral edge of gonocoxite with either a large dense conspicuous tuft of setae on distal half or no tuft; claspette filament greatly expanded and either fenestrated or elaborately convoluted (Figs. 135–137) 26

25. Apex of third palpal segment with a distinct tuft of about 10 long setae (Fig. 130*a*); ventrolateral edge of fourth palpal segment with a fringe of long setae; fore and mid enlarged tarsal claws each with both basal and subbasal teeth (Fig. 130*b*); mediodorsal margin of gonocoxite concave, and with fewer setae, at mid length .... ***hendersoni***

Apex of third palpal segment with at most two or three long setae (Fig. 129*a*); fourth palpal segment with scattered long setae but lacking a distinct fringe; fore and mid enlarged tarsal claws each with subbasal tooth only (Fig. 129*b*); mediodorsal margin of gonocoxite straight, with setae distributed evenly along its entire length .... ***triseriatus***

26. Fourth and fifth palpal segments without fringes or tufts of long setae (Fig. 128); submedian dark brown stripes of scutum confluent medially, the median yellow stripe absent or narrowly V-shaped and restricted to anterior half of scutum; gonocoxite lacking a dense medioventral tuft of setae; claspette filament arising from a short subapical side branch of claspette stem, and with a flattened median finger-like lobe overlapping its opposite number (Fig. 135) .... ***thibaulti***

Fourth and fifth palpal segments each with medioventral fringe of long setae (as in Fig. 129); narrow yellow median stripe of scutum fully separating dark brown submedian stripes; distal half of medioventral margin of gonocoxite with a conspicuous brush of long setae; claspette filament arising from apex of claspette stem, fenestrated or convoluted, but without median lobe overlapping midline (Figs. 136, 137) 27

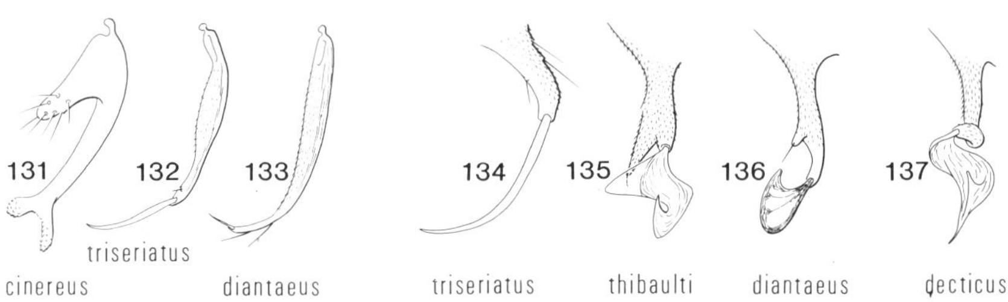

Figs. 131–137. Male of *Aedes* species: 131–133, gonostylus; 134–137, claspette filament, greatly magnified.

27. Claspette filament crescent-shaped, fenestrated with window-like transparent areas (Fig. 136); proctiger enlarged, heavily sclerotized .... ***diantaeus***
Claspette stem twisted spirally; claspette filament further twisted spirally in the same direction, ligulate, with tapered base, subapical dorsally directed pointed conical process, and sharply pointed apex (Fig. 137); proctiger not enlarged .... ***decticus***
28. Basal lobe of gonocoxite bearing distally one or two long, thick, apically directed bristles paralleling median margin of gonocoxite (Figs. 138–141) 29
Basal lobe of gonocoxite without apically directed bristles, although one or more medially directed ones may arise from base of lobe 32
29. Basal lobe of gonocoxite with one apically directly bristle (Fig. 138*a*); a dense pencil-like tuft of long setae (as long as gonostylus) arising from extreme apex of ventral surface of gonocoxite; convex margin of claspette filament with retrorse projection (Fig. 138*b*) .... ***aurifer***
Basal lobe of gonocoxite with two apically directed bristles (Figs. 139*a*–141*a*); setae, if present, at apex of ventral surface of gonocoxite, not confined to extreme apex and not as long as gonostylus 30
30. Gonocoxite lacking a large medially directed spine arising near basal lobe (Fig. 139*a*); claspette filament small, conical, and transversely striated (Fig. 139*b*); membrane between first abdominal tergite and sternite with large patch of white scales, usually visible only from below .... ***provocans***
Gonocoxite with a large medially directed spine arising near basal lobe (Figs. 140*a*; 141*a*); claspette filament blade-like, smooth (Figs. 140*b*; 141*b*); membrane bare between first abdominal tergite and sternite 31
31. Basal lobe of gonocoxite flask-shaped, longer than wide, with the two large apically directed bristles evenly curved, arising on narrow neck-like apex at level of claspette filament (Fig. 140*a*); claspette stem with seta-bearing side branch (Fig. 140*b*) .... ***intrudens***
Basal lobe of gonocoxite thumb-like, with the two large apically directed bristles sinuous, arising near level of base of claspette stem (Fig. 141*a*); claspette stem geniculate or elbowed near middle (Fig. 141*b*) .... ***pullatus***
32. Basal lobe of gonocoxite with uniform fine setae only, lacking a distinctly enlarged spine (Figs. 142, 143) 33
Basal lobe of gonocoxite with the most dorsally placed seta conspicuously longer and thicker than remaining setae (Fig. 145*a*) 34

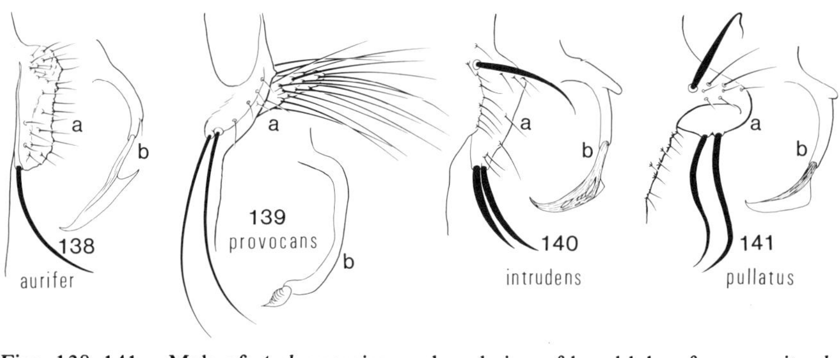

Figs. 138–141. Male of *Aedes* species: *a*, dorsal view of basal lobe of gonocoxite; *b*, lateral view of claspette.

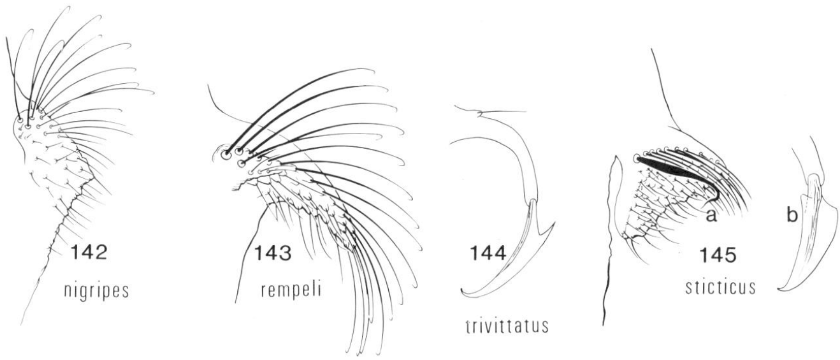

Figs. 142–145. Male of *Aedes* species: 142, 143, dorsal view of basal lobe of gonocoxite; 144, lateral view of claspette; 145*a*, dorsal view of basal lobe of gonocoxite; 145*b*, lateral view of claspette filament.

33. Postprocoxal scale patch present; basal lobe of gonocoxite broadly conical (Fig. 142) .... ***nigripes***
    Postprocoxal scale patch absent; basal lobe of gonocoxite long, narrow, finger-like (Fig. 143) .... ***rempeli***
34. Convex ventral side of claspette filament with sharp retrorse projection (Fig. 144); paratergite bare .... ***trivittatus***
    Convex side of claspette filament evenly rounded; paratergite with patch of scales 35
35. Basal lobe of gonocoxite strongly constricted at base (Fig. 145*a*); concave margin of claspette filament with a small angle at base (Fig. 145*b*) 36
    Basal lobe of gonocoxite conical or triangular, broadest at basal attachment (Figs. 151–153); concave edge of claspette filament tapering evenly to base 37
36. Apical lobe of gonocoxite constricted, with the base as wide as its distance from base of gonostylus (Fig. 146); postprocoxal membrane usually with a few scales .... ***spencerii***
    Apical lobe of gonocoxite broadest at base, with the base twice as wide as its distance from base of gonostylus (Fig. 147); postprocoxal membrane bare .... ***sticticus***

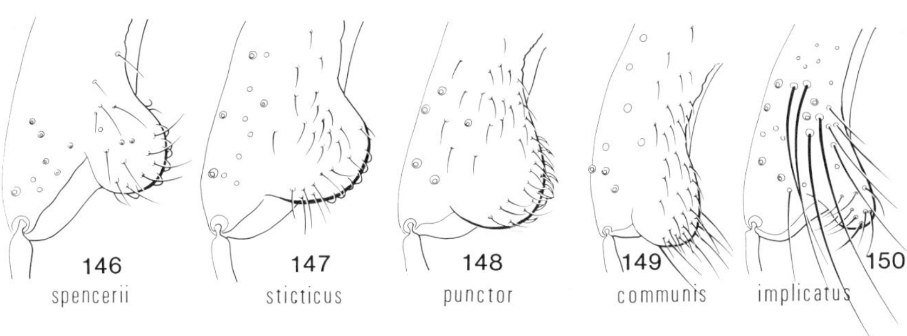

Figs. 146–150. Apical lobe of gonocoxite of male of *Aedes* species.

37. Enlarged seta of basal lobe of gonocoxite arising lateral to a tuft of setae (Figs. 151*a*; 152*a*); medial edge of apical lobe of gonocoxite with short, thick, curved, ventrally directed setae (Fig. 148); claspette filament relatively short, four times longer than wide, tapering rather abruptly distally, inflated, droplet-shaped in section (Fig. 152*b*, *c*), without flange or keel on convex side (*punctor* subgroup) 38

Enlarged seta of basal lobe arising lateral to a row of setae (Figs. 153*a*; 156*a*); median edge of apical lobe with either long straight setae (Fig. 149) or none; claspette filament five or more times as long as wide, not inflated, with thickened concave side, and expanded keel-like flange or flanges on convex side (Figs. 153*c*; 156*c*) 39

38. Basal lobe of gonocoxite thumb-like, with its tuft of setae occupying apex of lobe (Fig. 151); enlarged seta of basal lobe shorter than most of the other finer setae .... ***abserratus***

Basal lobe relatively large and triangular (Fig. 152*a*); enlarged seta of basal lobe longer than all other setae .... ***aboriginis***

.... ***hexodontus***

.... ***punctor***

.... ***schizopinax***

39. Distal surface of basal lobe of gonocoxite concave, with its ventral margin fringed with rather widely spaced, long, curved setae (Fig. 153*a*); apical lobe of gonocoxite with a group of long straight setae projecting ventrally (Fig. 149); claspette filament with keel displaced medially from the normal midventral position, usually also with a second lower keel along ventrolateral edge (Figs. 153*b*, *c*) 40

Distal surface of basal lobe convex, with its ventral margin lacking fringe of long setae (Figs. 156*a*; 157*a*); apical lobe of gonocoxite without long setae on median edge; claspette filament with midventral keel in line with body of filament (Figs. 156*b*, *c*; 157*b*, *c*) 42

40. Postprocoxal membrane with scales; scutal setae mostly dark brown .... ***pionips***

Postprocoxal membrane bare; scutal setae yellow to bronze 41

41. Setae at apex of third and at base of fourth palpomeres extending beyond apex of fourth segment (Fig. 155) .... ***communis***

Setae at apex of third and at base of fourth palpomeres not reaching apex of fourth segment (Fig. 154) .... ***churchillensis***

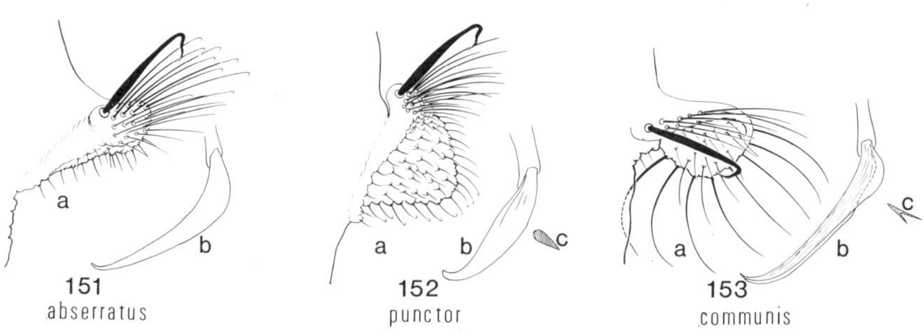

Figs. 151–153. Male of *Aedes* species: *a*, dorsal view of basal lobe of gonocoxite; *b*, lateral view of claspette filament; *c*, cross section of claspette filament.

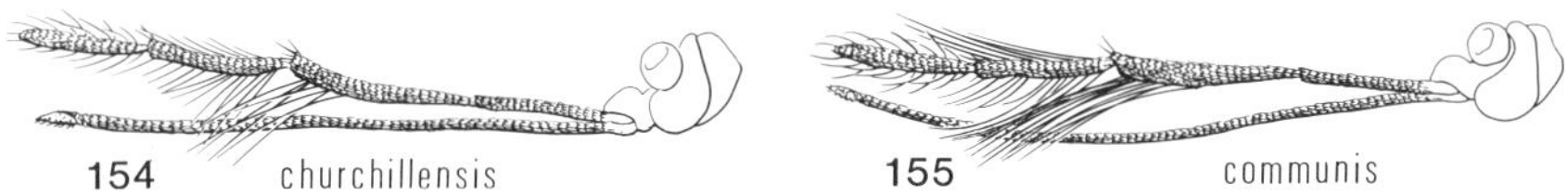

Figs. 154, 155. *Aedes* species: left lateral view of head of male, flagellum omitted.

42. Apical lobe of gonocoxite strongly exserted, medioventrally directed, somewhat parallel-sided; a conspicuous cluster of 6–12 long posteromedially directed setae arising from dorsal surface of gonocoxite lateral to apical lobe (Fig. 150) .... ***implicatus***
    Apical lobe of gonocoxite less prominent, evenly rounded if narrow, parallel-sided, and then distally rather than medially directed; dorsal surface of gonocoxite without cluster of long setae 43
43. Medioventral edge of gonocoxite with numerous long medioventrally directed setae; median edge of basal lobe with a dense row of closely spaced setae ventral to the enlarged seta (Fig. 156*a*); apical lobe at least as wide as adjacent portion of gonocoxite bearing gonostylus; hypostigmal area usually with a patch of scales; basal two-thirds of third palpal segment predominantly white-scaled .... ***cataphylla***
    Medioventral edge of gonocoxite with relatively few long setae; median edge of basal lobe with a sparse fringe of about 10 well-spaced setae (Fig. 157*a*); apical lobe reduced, narrower than adjacent portion of gonocoxite bearing gonostylus; hypostigmal area bare; palpus entirely dark-scaled .... ***impiger***

Key to the species of *Aedes* of Canada—fourth-instar larvae

The third instar resembles the fourth instar in most respects (see chapter on morphology) except that the saddle never encircles the anal segment, the comb scales are fewer, and the setae have fewer branches. The second instar is even less like the fourth, and the first instar is entirely different. Because we lack the third instars of many species, we do not know how well this key would work for them.

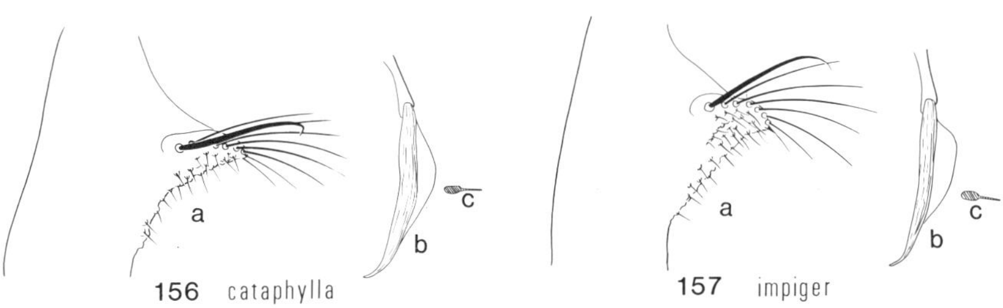

Figs, 156, 157. Male of *Aedes* species: *a*, dorsal view of basal lobe of gonocoxite; *b*, lateral view of claspette filament; *c*, cross section of claspette filament.

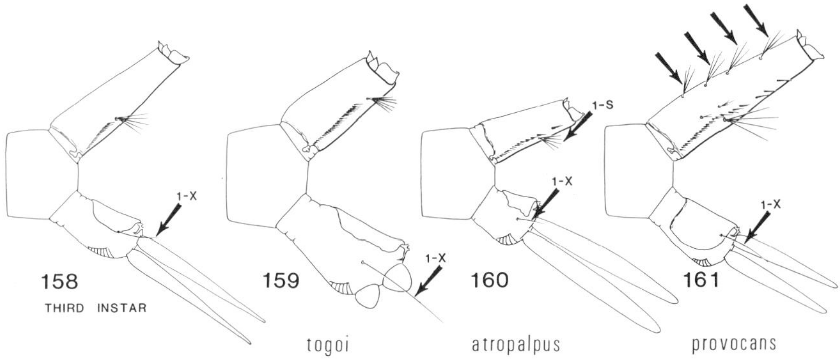

Figs. 158–161. Left lateral view of terminal segments of larva of *Aedes* species.

1. Saddle encircling dorsal one-third or less of anal segment, with seta 1-X arising ventral to the saddle, at its margin, or if from within saddle closer to its ventral margin than to its posterior margin (Figs. 158–160) (second and third instars of all *Aedes* and all four instars of *atropalpus* and *togoi*) 2

   Saddle encircling dorsal half or more of anal segment, with seta 1-X arising from within margin of saddle, closer to its posterior margin than to its ventral margin (Fig. 161) (fourth instars of all but *atropalpus* and *togoi*) 4

2. Siphon short and narrow compared with length of body of larva, twice as long as maximum width or shorter, with the last pecten tooth usually arising close to apex of siphon, well beyond siphonal seta 1-S and second last pecten tooth (Fig. 160); saddle seta 1-X arising from membrane ventral to saddle margin; anal papillae longer than siphon; in rock pools of Eastern Canada .... ***atropalpus***

   Not with the above combination of characters 3

3. Head setae 5-C and 6-C arising side by side far forward on head anterior to 7-C; seta 5-C with 10 or more branches in last instar; prothoracic seta 4-P as long as 1-P or 5-P; siphon short, 2.2 times as long as maximum width or shorter; anal papillae reduced to small tubercles (Fig. 159); in rock pools along southern coast of British Columbia .... ***togoi***

   Head seta 5-C arising posterior to 6-C and 7-C, with fewer than 10 branches; prothoracic seta 4-P usually less than half as long as 1-P or 5-P; siphon usually longer than 2.2 times maximum width; if in Pacific coastal rock pools, not having such reduced anal papillae; for early instars 3–4 mm or longer try couplet 4

4. Siphon with one or more distal pecten teeth more widely spaced than remaining teeth (as in Figs. 160, 161) (if only one tooth appears separated, examine the other side; if both sides have separated teeth, proceed to couplet 5, otherwise try both couplets 5 and 22 and eliminate one of the alternatives with the help of Table 3) 5

   Siphon with all pecten teeth more or less equally and evenly spaced, or if some distal teeth appear slightly more widely separated, the spacing increases regularly distally (as in Figs. 158, 159) 22

5. Siphon with several paired lateral and subdorsal branched setae in addition to siphonal seta 1-S (Fig. 161) .... ***provocans***

   Siphon without branched setae except for siphonal seta 1-S 6

6. Prothoracic seta 6-P with two or more branches; setae 5-P and 7-P multibranched, usually with 4 or 5 and 6 or more branches respectively (Fig. 162); abdominal segments III–VII each with a conspicuous, long, dorsolateral seta (as in Fig. 182); saddle usually completely encircling anal segment in fourth instar (as in Fig. 178*a*); tree line and tundra species .... ***nigripes***
   Prothoracic seta 6-P single; seta 5-P with three or fewer branches; seta 7-P at most quadruple; abdominal segment III lacking long dorsolateral setae (as in Fig. 166); saddle not encircling anal segment (although ventral edges of saddle in fourth-instar *decticus* are closely approximated, and occasional specimens have a narrow connection); boreal and southern species 7
7. Siphonal seta 1-S arising near mid length of siphon, exceeded by several widely and irregularly spaced pecten teeth; last pecten tooth usually close to apex of siphon .... ***cataphylla***
   Pecten teeth confined to basal two-thirds or less of siphon, followed by siphonal seta 1-S (in *nigromaculis,* pecten teeth usually occupy about two-thirds of siphon, but siphonal seta always follows last tooth; in *intrudens,* siphonal seta may arise between last two teeth, but pecten teeth occupy scarcely more than half of siphon) 8
8. Mesothoracic seta 1-M double or triple and as long as head seta 5-C (Fig. 163); saddle encircling only upper half of anal segment; head seta 7-C with 8–12 branches, and seta 6-C usually single (rarely double) .... ***campestris***
   Mesothoracic seta 1-M minute; saddle usually encircling two-thirds or more of anal segment; head seta 7-C usually with fewer than 8 branches, if with more, then seta 6-C at least triple 9
9. Upper caudal seta 2-X unbranched, as long as 3-X just ventral to it (Fig. 183); saddle completely encircling anal segment in fourth instar .... ***abserratus***
   Upper caudal seta 2-X multibranched, much shorter than 3-X (as in Fig. 182); saddle not encircling anal segment except in *nigromaculis* 10

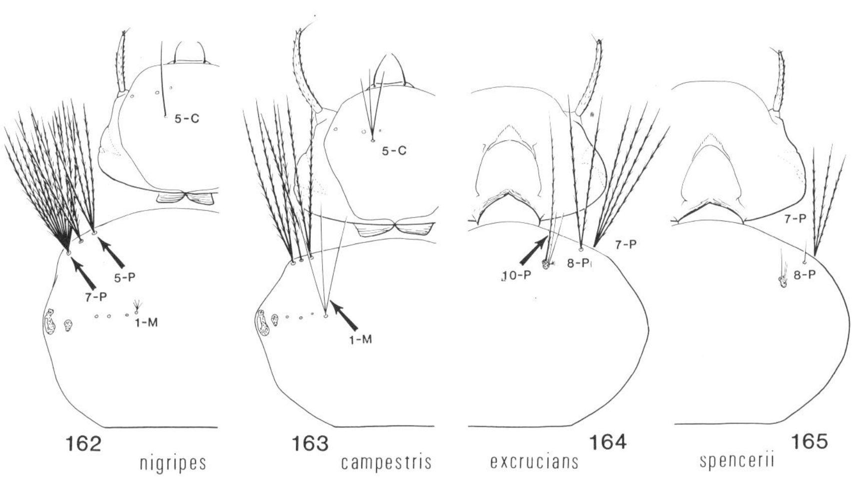

Figs. 162–165. Head and thorax of larva of *Aedes* species: 162, 163, dorsal view; 164, 165, ventral view.

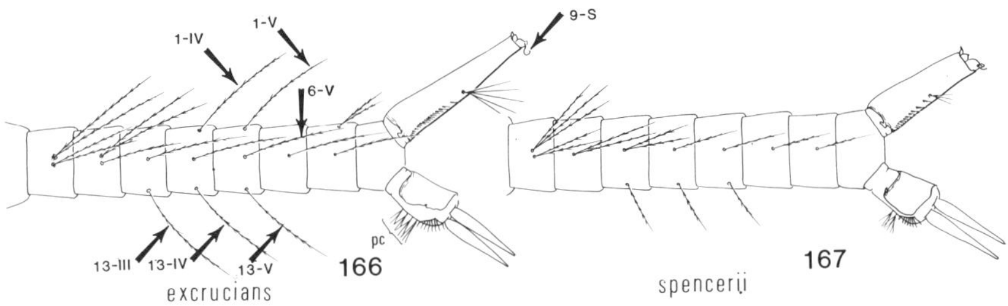

Figs. 166, 167. Left lateral view of abdomen of larva of *Aedes* species.

10. Abdominal segments III–V each with a conspicuous long ventrolateral seta (13-III, 13-IV, and 13-V) as long as height of segment to which it is attached (Fig. 166) and almost as long as lateral seta (6-III to 6-V) of same segment; prothoracic seta 10-P at least two-thirds as long as any of the branches of 7-P (Fig. 164); four or more precratal setae (Fig. 166, *pc*); mature larva about 10 mm long 11

Abdominal segments III–V each with a ventrolateral seta, at most, scarcely half the height of the segment to which it is attached and (except in *spencerii*, Fig. 167) no more than half as long as lateral seta of same segment; prothoracic seta 10-P short, less than one-quarter the length of a branch of 7-P and much finer (Fig. 165); four or fewer precratal setae; mature larva 8 mm or less long 15

11. Abdominal segments IV and V each with a dorsolateral seta (1-IV, 1-V) as long as either lateral seta (6-IV, 6-V) or ventrolateral seta (13-IV, 13-V) of same segment (Fig. 166); siphon long, abruptly narrowed apically, its apical diameter one-seventh as long as siphon; seta 9-S on ventrolateral flap at apex of siphon a strong, curved hook 12

Abdominal segment IV with dorsolateral seta much shorter than corresponding lateral seta, or with none; segment V with dorsolateral seta at most two-thirds as long as lateral seta; siphon tapering more evenly to apex, its apical diameter more than one-sixth as long as siphon (Fig. 168); seta 9-S on ventrolateral flap at apex of siphon relatively weak, scarcely curved 13

12. Integument of thorax and abdomen with a fur-like vestiture of minute hairs; southwestern British Columbia only .... ***aloponotum***

Integument without minute hairs; transcontinental .... ***excrucians***

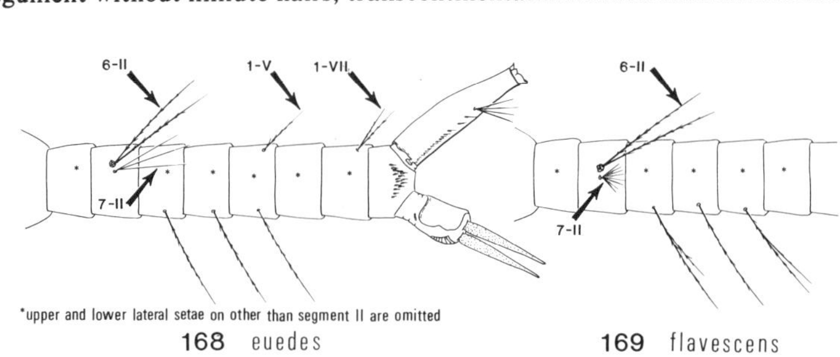

Figs. 168, 169. Left lateral view of abdomen of larva of *Aedes* species. Upper and lower lateral setae are shown only on segment II.

13. Comb scales large, 10 or fewer (usually 6–8); no conspicuous dorsolateral setae (1-IV to 1-VII) on abdominal segments IV, V, or VII .... ***riparius***
Comb scales more than 10; abdominal segments V and VII each with dorsolateral setae (1-V, 1-VII) usually at least half as long as lateral setae (Fig. 168) 14
14. Pecten teeth extending beyond middle of siphon; siphonal seta 1-S arising two-thirds to three-quarters of the distance from base of siphon; anal papillae brown, one and one-half times as long as saddle in life; sublateral seta 7-II on second abdominal segment usually with five or fewer branches (Fig. 168) . ***euedes***
Pecten teeth not extending beyond middle of siphon; siphonal seta 1-S inserted near midpoint; anal papillae transparent, no longer than saddle; sublateral seta 7-II with five or more branches (Fig. 169) .... ***flavescens***
15. Bases of head setae 5-C, 6-C, and 7-C arranged in a straight line, with base of seta 4-C anterior and out of line (Fig. 172); setae 5-C and 6-C each with five or more branches (rarely three or four) .... ***cinereus***
Either bases of head setae 5-C, 6-C, and 7-C not arranged in a straight line (Figs. 173, 176) or 4-C to 7-C all arranged in a straight line and antenna elongate as in *aurifer* and some *diantaeus* (Figs. 174, 175) 16
16. Siphonal seta 1-S short, its length less than apical diameter of siphon and usually no more than twice as long as last pecten tooth (Fig. 178); body of maxilla short, rounded apically, less than twice as long as maxillary palpus (see Fig. 171); prothoracic seta 8-P short, less than half as long as either branch of 7-P (as in Fig. 165) 17
Siphonal tuft longer than apical diameter of siphon, and usually three or more times as long as last pecten tooth (Fig. 179); body of maxilla elongate and pointed apically, more than twice as long as maxillary palpus (Fig. 170*a,b*); prothoracic seta 8-P nearly as long as either branch of 7-P (as in Fig. 164) 19
17. Saddle completely encircling anal segment in fourth instar (Fig. 178, *a*); siphon short and stout, only twice as long as wide; pecten teeth extending well beyond middle of siphon; antenna extremely short, about one-fifth as long as head (Fig. 173) .... ***nigromaculis***
Saddle only partly encircling anal segment; siphon longer, two and one-half times as long as wide or longer; pecten teeth usually confined to basal half of siphon; antenna longer, more than one-third as long as head 18

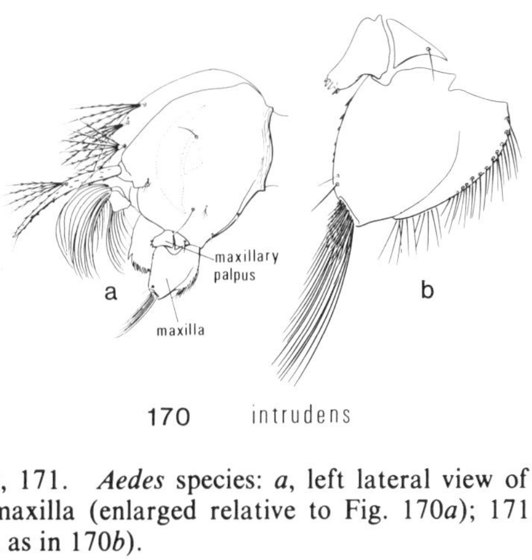

170 intrudens

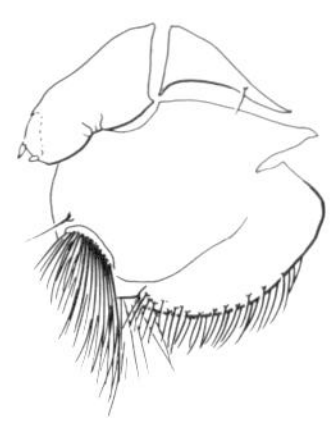

171 spencerii

Figs. 170, 171. *Aedes* species: *a*, left lateral view of head of larva; *b*, left lateral view of maxilla (enlarged relative to Fig. 170*a*); 171, left lateral view of maxilla (enlarged as in 170*b*).

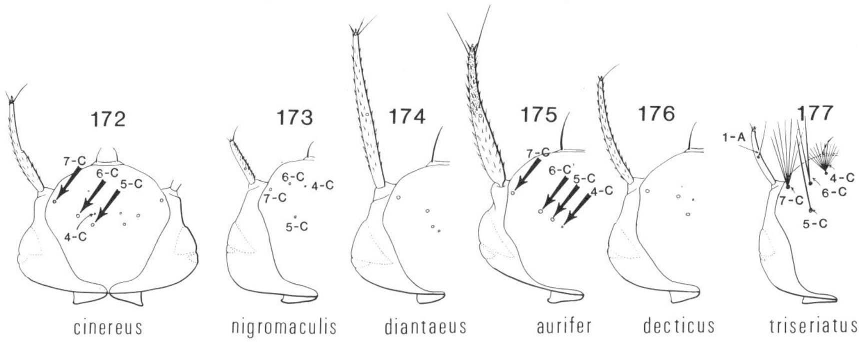

Figs. 172–177. Dorsal view of head of larva of *Aedes* species.

18. Head setae 5-C and 6-C single (5-C occasionally double on one side); integument of thorax and abdomen clothed with minute, regularly spaced hairs, best viewed from the side on a curved part of thorax; prothoracic seta 8-P minute, single, arising ventral to 7-P (Fig. 165); ventrolateral setae (13-III to 13-V) of abdominal segments III–V almost as long as lateral setae (6-III to 6-V) of same segment (Fig. 167) .... ***spencerii***

    Head setae 5-C and 6-C each with two or more branches; integument of thorax and abdomen not clothed with minute hairs; prothoracic seta 8-P double or more, displaced anterior to 7-P; ventrolateral setae of abdominal segments III–V minute .... ***vexans***

19. Antenna white, rather translucent, with black apex, swollen basally, tapering abruptly beyond seta 1-A (Fig. 175); head setae 4-C, 5-C, 6-C, and 7-C arising in a straight line .... ***aurifer***

    Antenna pale to dark brown, evenly tapering from base to apex (Figs. 176, 177); head setae 4-C and 6-C usually arising somewhat anterior to a line between 5-C and 7-C 20

20. Antenna shorter than half maximum head width; saddle incised at its posterolateral corner (Fig. 179); siphonal seta 1-S sometimes arising quite close to base of last pecten tooth, even proximal to it .... ***intrudens***

    Antenna longer than half maximum head width (Figs. 174, 176); saddle not incised at its posterolateral corner; siphonal seta 1-S always arising distal to last pecten tooth, usually separated from it by length of last pecten tooth 21

21. Antenna longer than head (Fig. 174), pale yellowish brown, slightly darker only on apical one-seventh; head and body setae tapering evenly from base; comb scales 6–13 .... ***diantaeus***

    Antenna shorter than head (Fig. 176), medium brown, with apical half darker; head and body setae thick, stiff, parallel-sided or even slightly wider in middle, thickened almost to apex; comb scales 5–7 .... ***decticus***

22. In tree rot cavities, occasionally in artificial containers; antenna smooth, without spicules; antennal seta 1-A usually single (Fig. 177); head seta 6-C usually with more branches than seta 5-C; head seta 4-C exceptionally large, about half as long as seta 6-C; saddle seta 1-X branched 23

    In ground pools; antenna rugose, with spicules; antennal seta 1-A with three or more branches; head seta 6-C with the same number of branches or fewer than seta 5-C; head seta 4-C minute, one-eighth or less the length of seta 6-C; saddle seta 1-X single 25

23. Thoracic setae 1-P, 1-M, and 1-T, and abdominal setae 1-I to 1-VII and 13-I to 13-VII each with three or more stellately arranged branches (Figs. 180, 181); pecten teeth extending beyond basal one-third of siphon 24
Thoracic setae 1-P, 1-M, and 1-T, and abdominal setae 1-I to 1-VII and 13-I to 13-VII each single or double; pecten teeth confined to basal one-third of siphon; British Columbia only .... ***sierrensis***

24. Anal papillae usually less than twice as long as saddle (view against black background), ventral pair shorter than dorsal pair (Fig. 181); saddle rectangular in side view, its anteroventral corner well-developed; acus fused with siphon; thoracic setae 1-P, 1-M, and 1-T, and abdominal setae 1-I to 1-VII and 13-I to 13-VII each with four or more branches; Eastern Canada .... ***triseriatus***
Anal papillae about three times as long as saddle, the ventral pair as long as dorsal pair (Fig. 180); saddle somewhat triangular in side view, its anteroventral corner truncated; acus free from siphon; thoracic setae 1-P, 1-M, and 1-T and abdominal setae of first and thirteeth series each with four or fewer branches (an occasional seta can be five-branched); transcontinental .... ***hendersoni***

25. Mesothoracic seta 1-M as long as head seta 5-C (as in Fig. 163) 26
Mesothoracic seta 1-M minute, less than one-third length of seta 5-C (as in Fig. 162) 35

26. Siphon exceptionally long and slender, four–five times as long as maximum width, and seven–eight times as long as wide at apex; seta 9-S on ventrolateral flap at apex of siphon thick and strongly curved, hook-like; abdominal segments III–VII each with a long, well-developed dorsolateral seta (1-III to 1-VII) that is as long as lateral seta (6-III to 6-VII) of the same segment and almost twice as long as height of that segment (Fig. 182) .... ***fitchii***
Siphon less than four times as long as wide at base and less than six times as long as wide at apex; seta 9-S not thickened or strongly curved; dorsolateral seta (1-III) minute on abdominal segment III (except in *rempeli*, with single head setae); dorsolateral seta 1-VI on segment VI also minute (except in *dorsalis* and *cantator*, in which dorsolateral seta is shorter than lateral) 27

27. Mesothoracic seta 1-M double or triple at least in fourth instar (Fig. 163) (an occasional specimen of *dorsalis* may have 1-M single); anal papillae shorter than saddle 28
Mesothoracic seta 1-M single; anal papillae longer than saddle (except in *cantator*, which is found in eastern coastal salt marshes and has head seta 6-C double or triple) 30

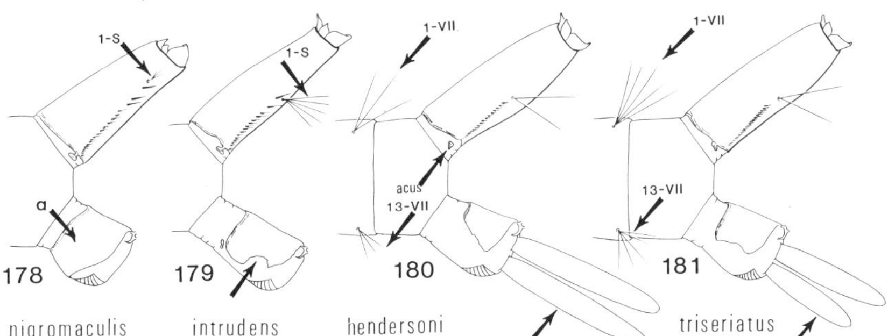

Figs. 178–181. Left lateral view of terminal segments of larva of *Aedes* species.

28. Comb scale with long median spine-like apex and short subapical spinules each less than half its length (as in Fig. 196); prothoracic seta 2-P almost as long as 1-P; dorsolateral abdominal setae 1-IV to 1-VII each as long as upper lateral seta (6-IV to 6-VII) of same segment .... ***schizopinax***

Comb scale without a long median spine-like apex, fringed with subequal spinules (as in Fig. 197); prothoracic seta 2-P less than two-thirds as long as 1-P; dorsolateral abdominal setae 1-IV to 1-VII each no more than two-thirds as long as upper lateral seta of same segment 29

29. Head seta 5-C with two or three branches; pecten teeth extending beyond mid length of siphon; dorsolateral abdominal setae 1-IV to 1-VII and lower lateral seta 7-II minute, less than one-tenth as long as upper lateral seta of same segment .... ***campestris***

Head seta 5-C usually single (sometimes double, rarely triple); pecten teeth confined to basal half of siphon; dorsolateral abdominal setae 1-IV to 1-VII and lower lateral seta 7-II about one-third as long as upper lateral seta of same segment .... ***dorsalis***

30. Head setae 5-C and 6-C each single; abdominal segment III with long dorsolateral seta 1-III (as in Fig. 182); saddle completely encircling anal segment in fourth instar .... ***rempeli***

Head setae 5-C and 6-C each with three or more branches; dorsolateral seta 1-III minute; saddle partly encircling anal segment 31

31. In brackish or salt marshes adjacent to Atlantic coast, Gulf of St. Lawrence and James Bay; anal papillae much shorter than saddle, with the ventral pair shorter than the dorsal pair; dorsolateral setae 1-IV to 1-VII all double or triple and subequal in length .... ***cantator***

In freshwater inland habitats; anal papillae at least as long as saddle, with the ventral pair as long as the dorsal pair; dorsolateral setae 1-IV to 1-VII not as above, either single or double, if double, seta 1-VI markedly shorter than seta 1-V (*aboriginis*, in which these setae appear similar to those in *cantator*, is found on the Pacific coast) 32

32. Prothoracic setae 2-P and 3-P both short, fine, less than half·the length and thickness of 1-P .... ***mercurator***

Prothoracic setae 2-P and 3-P each more than half as long as 1-P (Figs. 184–186) 33

33. Lower lateral seta 7-II of second abdominal segment single or double, half as long as lateral seta 6-II immediately dorsal (as in Fig. 168); seta 1-X usually as long as saddle .... ***aboriginis***

Lower lateral seta 7-II of second abdominal segment with three or more branches, much shorter than half the length of lateral seta 6-II (as in Fig. 169); seta 1-X less than half as long as saddle 34

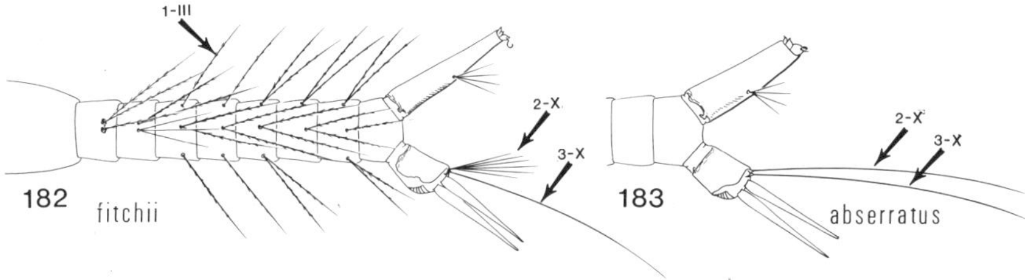

Figs. 182, 183. Left lateral view of larva of *Aedes* species: 182, abdomen; 183, terminal segments.

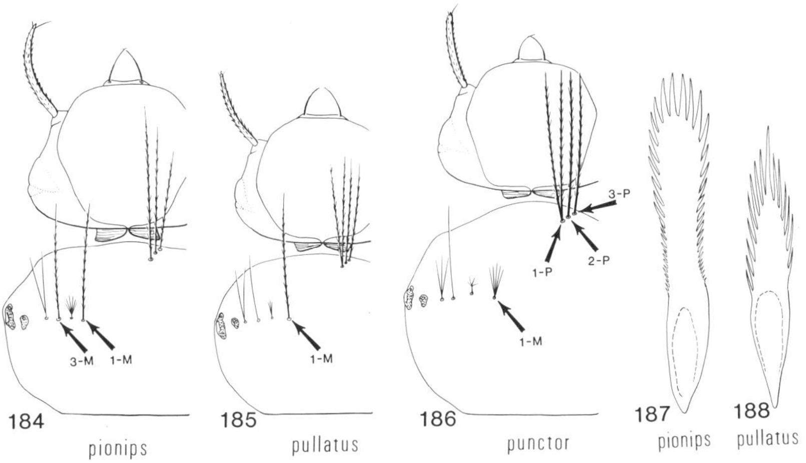

Figs. 184–188. *Aedes* species: 184–186, dorsal view of head and thorax of larva; 187, 188, comb scale (enlarged).

34. Mesothoracic setae 1-M and 3-M subequal in length, both at least half as long and as thick as 1-P (Fig. 184); usually more than 60 comb scales; each scale with small base and long apically broadened apex fringed with subequal spinules (Fig. 187) .... ***pionips***
    Mesothoracic seta 1-M one and one-half times as long as 3-M (Fig. 185); fewer than 60 comb scales; each scale more pointed, with longer median spine; base and apex of scale subequal in length (Fig. 188) .... ***pullatus***
35. Prothoracic setae 2-P and 3-P almost as long and as thick as each branch of 1-P (Fig. 186); seta 1-P usually double; saddle completely encircling anal segment in last instar 36
    Prothoracic setae 2-P and 3-P weak, both usually less than half as long as 1-P; seta 1-P usually single; if saddle complete, siphon less than two and one-half times as long as greatest width 38
36. Upper caudal seta 2-X unbranched, as long as 3-X (Fig. 183); dorsolateral setae 1-IV and 1-V minute, less than half as long as 6-IV and 6-V .... ***abserratus***
    Upper caudal seta 2-X branched, markedly shorter than 3-X; dorsolateral setae 1-IV and 1-V longer than two-thirds length of 6-IV and 6-V 37
37. Prothoracic seta 5-P usually single; comb scales 5–25, usually more than 10, each scale 0.06–0.08 mm long, shorter than last three or four pecten teeth .... ***punctor***
    Prothoracic seta 5-P usually double; comb scales 4–12, usually fewer than 10, each scale longer than 0.1 mm, longer than all pecten teeth except the last .... ***hexodontus***
38. Siphonal seta 1-S shorter than abdominal seta 5-VIII and shorter than apical diameter of siphon (as in Fig. 178) 39
    Siphonal seta 1-S longer than abdominal seta 5-VIII and longer than apical diameter of siphon 42
39. Saddle completely encircling anal segment in fourth instar; siphon two and one-half times as long as maximum width or less 40
    Saddle incomplete; siphon more than two and one-half times as long as maximum width 41
40. Anal papillae bud-like, much shorter than saddle; in saline marshes .... ***sollicitans***
    Anal papillae as long as anal segment; in temporary summer rain pools .... ***trivittatus***

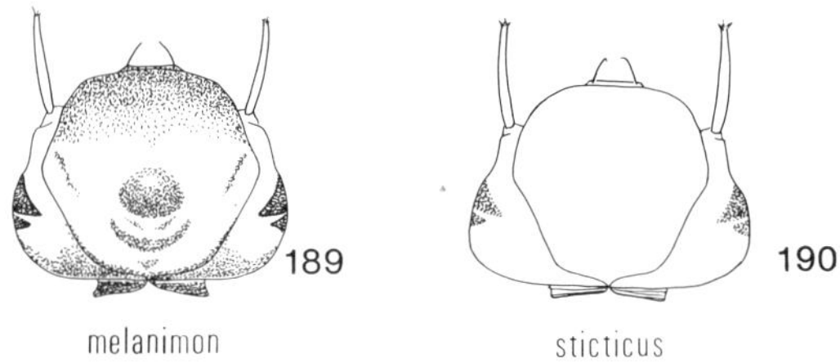

Figs. 189, 190. Dorsal view of head of larva of *Aedes* species.

41. Head capsule with dorsal median circular darkly pigmented spot and a transverse crescent-shaped darkly pigmented bar behind it (Fig. 189); ventral pair of anal papillae shorter than dorsal pair (as in Fig. 199) .... ***melanimon***
 Head capsule without prominent pattern of pigment dorsally (Fig. 190); ventral pair of anal papillae subequal to dorsal pair (as in Fig. 198) .... ***sticticus***
42. Head seta 6-C arising between 5-C and 7-C in a straight line (as in Fig. 172); maxilla long and pointed (as in Fig. 170); clypeal setae small, pale brown, separated from one another by more than their own length; a rare species found in southern Ontario in inundated hollow stumps or logs .... ***thibaulti***
 Head seta 6-C arising anterior to a line between 5-C and 7-C; maxilla short and blunt (Figs. 200–203); clypeal setae larger, darker, separated from one another by less than their own length; in open habitats 43
43. Prothoracic setae 8-P and 10-P both more than two-thirds as long as a branch of 7-P (Fig. 191); dorsolateral seta 1-IV of fourth abdominal segment about half as long as seta 1-V on fifth segment; four or more precratal setae 44
 Either prothoracic seta 8-P or seta 10-P, or both, no more than half as long as 7-P (Figs. 192, 193); dorsolateral setae 1-IV and 1-V on abdominal segments IV and V about the same length; usually fewer than four precratal setae 45
44. Prothoracic seta 2-P less than half as long as seta 1-P; mesothoracic seta 1-M multiple, shorter than seta 3-M; basal cusps of each pecten tooth exceptionally long, the most distal cusp longer than maximum width of that tooth .... ***flavescens***
 Prothoracic seta 2-P more than two-thirds as long as seta 1-P; mesothoracic seta 1-M single, as long as seta 3-M; basal cusps of each pecten tooth all shorter than maximum width of tooth .... ***grossbecki***
45. Head seta 5-C with five or more branches in fourth instar (about three branches in third instar), an occasional mature specimen may have only four on one side but five or more on the other; mesothoracic seta 8-P longer than 10-P (Fig. 192) .... ***canadensis***

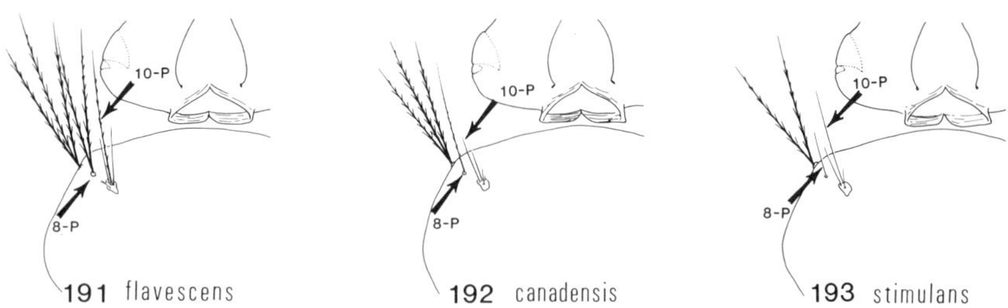

Figs. 191–193. Ventral view of prothorax of larva and adjacent part of head of *Aedes* species.

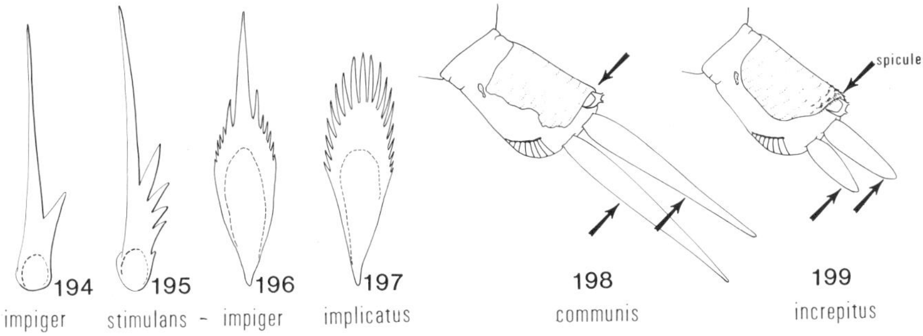

Figs. 194–199. *Aedes* species: 194, 195, pecten tooth (enlarged); 196, 197, comb scale (enlarged); 198, 199, left lateral view of abdominal segment X.

Head seta 5-C with three or fewer branches; mesothoracic seta 8-P shorter than 10-P (Fig. 193) 46

46. Pecten teeth (when viewed under a dissecting microscope) each with a single large basal cusp, occasionally with a minute second basal tooth (Fig. 194); posterior lobe of maxilla narrowly rounded (Figs. 200–201) 47

Pecten teeth each with two or more distinct basal cusps (Fig. 195); posterior lobe of maxilla broadly rounded to truncated (Figs. 202, 203) 48

47. Comb scales 8–16, each with long spine-like apex and short apical spinules (Fig. 196) .... ***impiger***

Comb scales 15–35, each fringed apically with long subequal spinules (Fig. 197) .... ***implicatus***

48. Spicules along posterior edge of saddle apparently absent (when viewed under a dissecting microscope), but actually present though very small (too short to project beyond edge of saddle) (Fig. 198) .... ***communis****

.... ***churchillensis****

Spicules along posterior edge of saddle projecting beyond saddle edge, thus easily visible under magnification of 50X or more (Fig. 199) 49

49. Ventral pair of anal papillae shorter than dorsal pair (Fig. 199); Saskatchewan and west .... ***increpitus***

Ventral pair of anal papillae more or less equal in length to dorsal pair (as in Fig. 198); Manitoba and east .... ***stimulans***

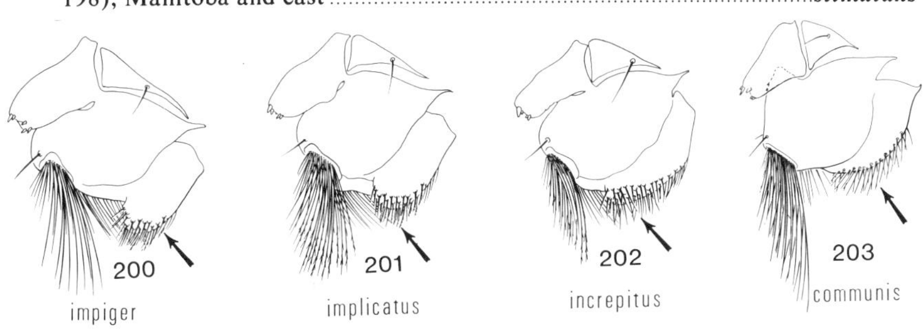

Figs. 200–203. Left lateral view of maxilla of larva of *Aedes* species.

* These two sibling species can be separated only by analysis of populations (Ellis and Brust 1973).

excrucians

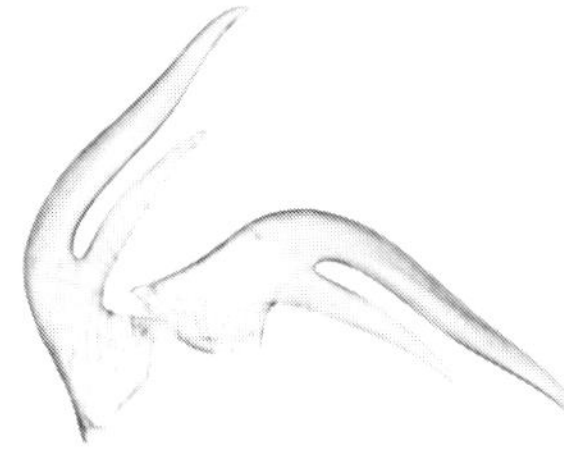

excrucians

excrucians

excrucians

excrucians

euedes

euedes

euedes

Plate 8. Fore tarsal claws of *Aedes* spp.

aloponotum

riparius

fitchii

fitchii

stimulans

stimulans

mercurator

increpitus

Plate 9. Fore tarsal claws of *Aedes* spp.

Aedes aboriginis Dyar

Plate 10; Map 9

Aedes aboriginis Dyar, 1917*b*:99.

Adult. Female: Integument medium brown; proboscis and palpus dark-scaled; pedicel scaled on dorsomedial half, with scales dark medially, pale dorsally; scales of vertex, upper two-thirds of postpronotum, and median stripe, presutural sublateral area, and lateral stripe of scutum yellowish brown to bronzy brown; scales of submedian stripe and postsutural sublateral stripe of scutum coppery brown, narrower and darker than those of remainder of scutum; postprocoxal membrane, anterodorsal corner of katepisternum, and all of mesepimeron with pale yellowish scales; lower mesepimeral setae present; probasisternum without scales; tibiae and first tarsomeres dark-scaled apically, pale-scaled ventrally; remaining tarsomeres entirely dark; tarsal claws moderately and evenly curved, each with a short subbasal tooth; wing veins, including base of costa, entirely dark-scaled; abdominal tergites dark-scaled, each with a transverse basal band of white scales.

Male: Palpus dark-scaled, longer than proboscis by about half the length of last palpomere; apex of third and edges of fourth and fifth palpomeres with dense fringes of long dark setae; basal lobe of gonocoxite large, with its setose surface slightly concave, somewhat triangular or quadrate, oriented mediodorsally, and thus entirely visible in dorsal view; a single enlarged medially directed bristle arising from proximolateral corner of basal lobe; apical lobe well-developed, with short, curved, flattened, ventrally directed setae along medial margin; claspette filament short, inflated, without crest on convex side, droplet-shaped in cross section. Coloration and arrangement of scales as in female.

Larva. Head setae 5-C and 6-C each with two to four branches; prothoracic setae 2-P and 3-P approaching 1-P in length and thickness; metathoracic seta 1-M as long as head seta 5-C; abdominal segment III lacking a well-developed dorsolateral seta (1-III); lower lateral seta (7-II) of second abdominal segment usually double and over half as long as upper lateral seta (6-II); all pecten teeth evenly spaced; other characters as in Table 3. Characters unique to this species seem to be lacking.

Remarks. A member of the *punctor* subgroup (Knight 1951), this species, either male or female, is scarcely separable from *punctor* or *hexodontus*, with which its range overlaps along the coast of British Columbia. However, the larva, having an incomplete saddle and multi-branched head setae, is easily distinguished from *punctor* and *hexodontus*, but resembles *mercurator* or *schizopinax*.

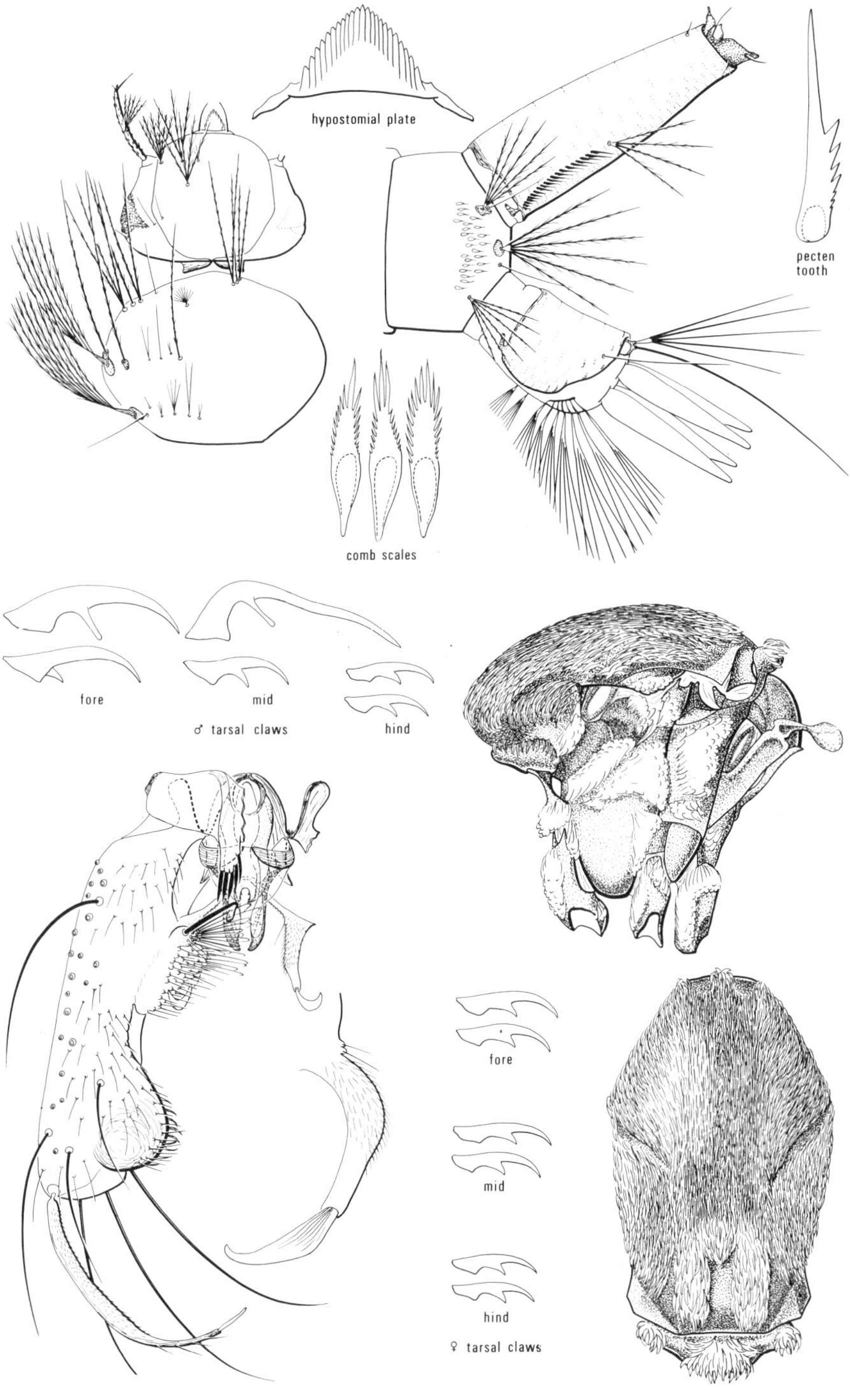

Plate 10. *Aedes aboriginis*

In the female, the submedian stripes of the scutum are bronzy brown, darker than the adjacent sublateral areas, and usually separated medially by a narrow median stripe of lighter brown scales, whereas in *punctor*, the dark brown submedian bands are usually confluent medially (that is, there is scarcely any evidence of a lighter median band, except for the occasional lighter scale). The few specimens we have seen that we believe belonged to this species were larger than *punctor*.

In the male, the seta-bearing portion at the apex of the third palpal segment occupies between one-third and one-half of the segment, and this area may be significantly different from *punctor*.

Biology. An inhabitant of temporary rain and snow pools in the wet Pacific coastal forest, larvae of *aboriginis* are one of the earliest species to appear in spring (Hearle 1926, Boddy 1948). Curtis (1967) did not find the species east of Chilliwack, although it has been recorded from the wet, heavily timbered western slopes of northern Idaho as well as the Cascades of Washington and Oregon (Stage et al. 1952). The species has not been collected in the interior wet belt of British Columbia on the west slopes of the Selkirk Range, although it may occur there. In southern Alaska larvae occurred from May to July in roadside ditches, associated with *Aedes pullatus, A. punctor,* and *Culiseta impatiens* (Frohne and Sleeper 1951). In Washington, larvae were found at the edges of open marshes and grassy-edged temporary pools (Boddy 1948).

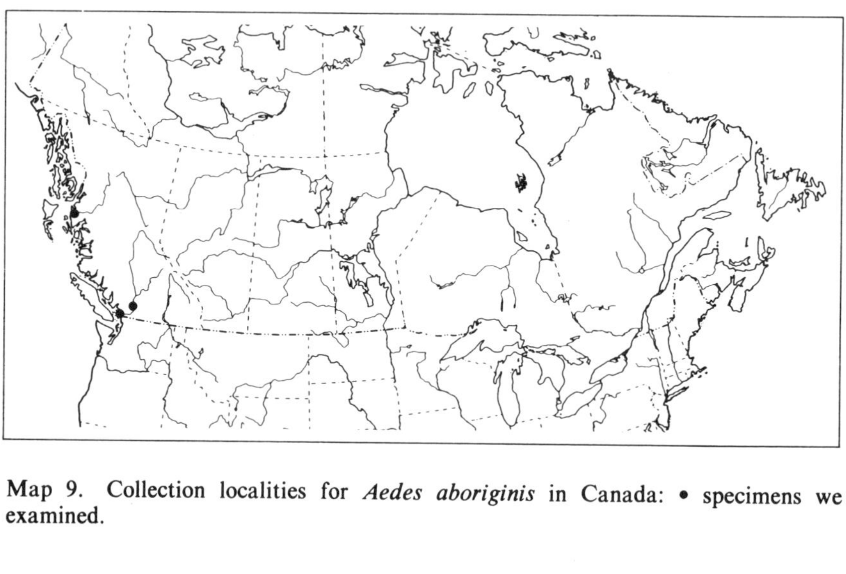

Map 9. Collection localities for *Aedes aboriginis* in Canada: • specimens we examined.

Adults appear in southwestern British Columbia in mid-June (Hearle 1926). The females have seldom constituted a serious problem except in certain situations, such as the Olympic Peninsula, Washington (Stage et al. 1952). Dyar (1917*b*) observed males of this species swarming in the morning beside the trunks of large cedar trees. The swarms were 2–3 m above the ground. Although some rays of the sun penetrated the dense canopy, the swarming areas were heavily shaded.

Distribution. Western North America, from southwestern Alaska south to Oregon and Idaho. Twinn (1949) recorded *aboriginis* from Saskatchewan, the only record east of the Rocky Mountains. We have not located any specimens and have decided to ignore the record as a misidentification.

Aedes abserratus (Felt & Young)

Plate 11; Figs. 151, 183; Map 10

Culex abserratus Felt and Young, 1904:312.
Aedes centrotus Howard, Dyar, and Knab, 1917:747.
Aedes dysanor Dyar, 1921*b*:70.
Aedes implacabilis, authors before 1954, not Walker, 1848.

Adult. Female: Integument medium brown; proboscis and palpus dark-scaled; pedicel scaled on dorsomedial half, with scales dark medially, pale dorsally; scales of vertex, upper two-thirds of postpronotum, and all of scutum bronzy or coppery brown; submedian stripes of scutum, if visible at all, of narrower and sometimes slightly darker scales; probasisternum and membrane between cervical sclerite and dorsolateral corner of probasisternum usually with some scattered pale scales; postprocoxal membrane, anterodorsal corner of katepisternum, and all of mesepimeron pale-scaled; lower mesepimeral setae present; tibiae and first tarsomeres dark-scaled dorsally, pale-scaled ventrally; remaining tarsomeres dark-scaled; wing veins including base of costa dark-scaled; abdominal tergites dark-scaled, each with transverse basal band of white scales.

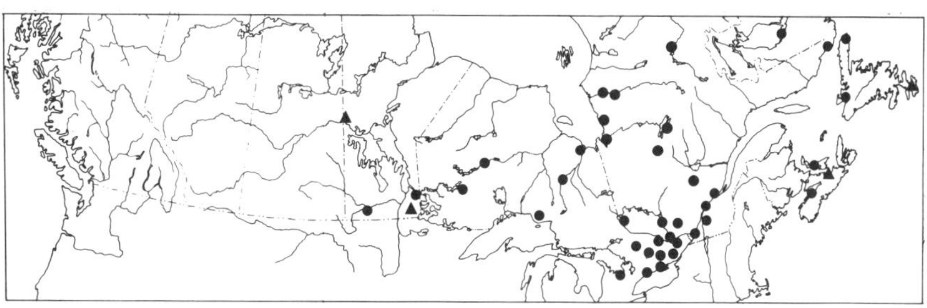

Map 10. Collection localities for *Aedes abserratus* in Canada: • specimens we examined, ▲ literature records.

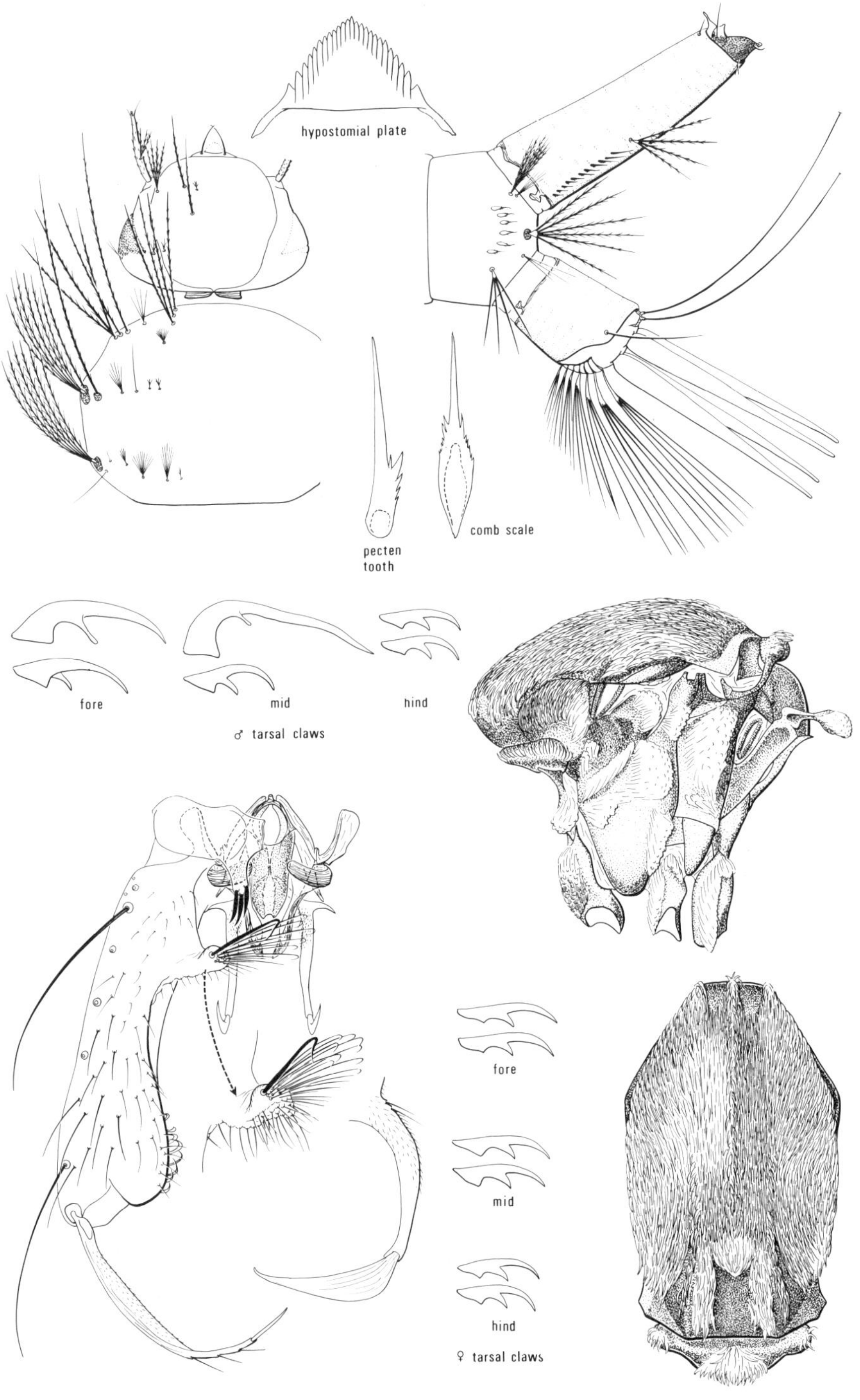

Plate 11. Aedes abserratus

Male: Palpus usually slightly longer than proboscis; apex of third palpal segment with a tuft of long setae that extends beyond apex of fourth segment; fourth and fifth palpal segments with dense lateral and medial fringes of long setae; postprocoxal scale patch sometimes reduced to two or three scales; coloration and scaling as in female; scutum usually uniformly yellowish brown or with poorly defined submedian stripes of slightly narrower and slightly darker scales; dorsomedial surface of gonocoxite strongly excavated distal to basal lobe; basal lobe rather prominent, thumb-like, with its setose surface directed posteriorly, at right angles to dorsomedial margin of gonocoxite, bearing a dense tuft of setae subtended by an enlarged but not more strongly sclerotized seta (because it lacks darker pigment, this enlarged seta is indistinct and could be overlooked; it is best recognized by the presence of a large socket); apical lobe of gonocoxite prominent, evenly rounded, with its median edge fringed with short, flattened, ventrally curved setae; claspette filament short, somewhat inflated, without crest on convex side, droplet-shaped in cross section.

Larva. Prothoracic setae 2-P and 3-P as long and as thick as one of the branches of 1-P; pecten teeth may be evenly spaced or one or two distal teeth may be more widely spaced than the rest; comb scales five to seven, each long and slender with minuscule subapical spinules; saddle completely encircling anal segment; siphon three times as long as wide (about two and one-half times in *hexodontus*); anal papillae no longer than siphon (nearly twice as long in most northern specimens of *hexodontus*); upper caudal seta 2-X single, as long and as thick as seta 3-X immediately ventral to it. Other characters are given in Table 3.

No other species of *Aedes* has the upper caudal seta 2-X single; therefore larvae of *abserratus* can usually be recognized by this character. Because one or both of these upper caudal hairs can be broken off at the base, 3-X can be mistaken for 2-X. If both upper caudal setae are broken, *abserratus* can still be distinguished from *hexodontus*, with which it undoubtedly coexists in the northern fringes of the boreal forest of Ontario and Quebec, by the proportions of the siphon, by the shorter anal papillae, and by the unbranched upper lateral abdominal setae.

Remarks. The male terminalia illustrated by Carpenter and LaCasse (1955: Fig. 115, p. 150) as *abserratus* appear to be those of *punctor*.

Biology. *Aedes abserratus* overwinters in the egg stage and has only one generation in spring. A technique of rearing eggs of this species to adults has been developed by Brust and Kalpage (1967). The larvae of *abserratus* have been taken in a variety of situations, from roadside ditches and the edges of permanent or semipermanent cattail and sedge marshes to the edges of muskeg pools (Price 1963, Pickavance et al. 1970). We also have taken larvae in all these situations in the Ottawa area, but always in or at the edge of wooded or shrubby situations. The larvae were most abundant, however, in boggy situations, usually associated with larvae of *Aedes punctor, canadensis,* and *cinereus*. They were not taken in temporary pools in hardwood forest, where *Aedes communis, diantaeus,* and *intrudens* were abundant.

On the Avalon Peninsula, Nfld., *abserratus* was second only to *punctor* in abundance (Pickavance et al. 1970). However, in Ontario, we have not found it to be a common species. Parkin et al. (1972) have isolated the virus of California encephalitis from females of *abserratus*, and it, along with *intrudens,* has been shown to transmit larvae of the beaver filarioid, *Dipetalonema sprenti* Anderson (Addison 1973).

Distribution. Eastern North America, from Manitoba and Illinois east to Newfoundland, Nova Scotia, and Massachusetts.

Aedes aloponotum Dyar

Plates 9, 12; Map 11

Aedes aloponotum Dyar, 1917*b*:99.

Adult. Female: Integument reddish brown; proboscis mostly dark-scaled; palpus dark-scaled, with same pale scales at base of second palpomere, and fewer pale scales at base of third, not forming basal pale ring as in *excrucians*; pedicel pale-scaled on medial half; vertex, upper two-thirds of postpronotum and scutum predominantly yellowish brown scaled, resembling *riparius* but not so orange and without the whitish scales present in *excrucians* (can only be judged by comparison with known material); postprocoxal membrane with scales; hypostigmal area bare; pleural scales more yellowish than those of *excrucians;* scales of katepisternum extending to its anterodorsal corner; lower third of mesepimeron bare; lower mesepimeral setae absent; tibiae predominantly dark-scaled dorsally, pale-scaled ventrally; first fore tarsomere mostly pale-scaled basally, mostly dark-scaled apically but without definite basal pale ring; first mid and hind tarsomeres similar, but with distinct basal ring; remaining tarsomeres dark-scaled, each with broad basal ring of white scales except last fore tarsomere; tarsal claw straight basally, moderately and evenly curved distally, subbasal tooth relatively small, usually intermediate between that of *riparius* and of *euedes*; wing veins with mixed pale and dark scales throughout; abdominal tergites dark-scaled, each with narrow yellowish-scaled transverse basal band and a few scattered pale scales.

Male. Palpus longer than proboscis by more than half the length of last palpomere, dark-scaled, with transverse basal pale-scaled band on second, third, and fourth tarsomeres; middorsal region of third palpomere with scattered pale scales; apex of third palpomere with dense tuft of long dark setae extending beyond base of last palpomere; fourth palpomere with medial fringe of long setae and lateral and ventral fringes of shorter setae; coloration and scaling of thorax, legs, and wings and shape of hind tarsal claw as in female; basal lobe of gonocoxite not differentiated, with long

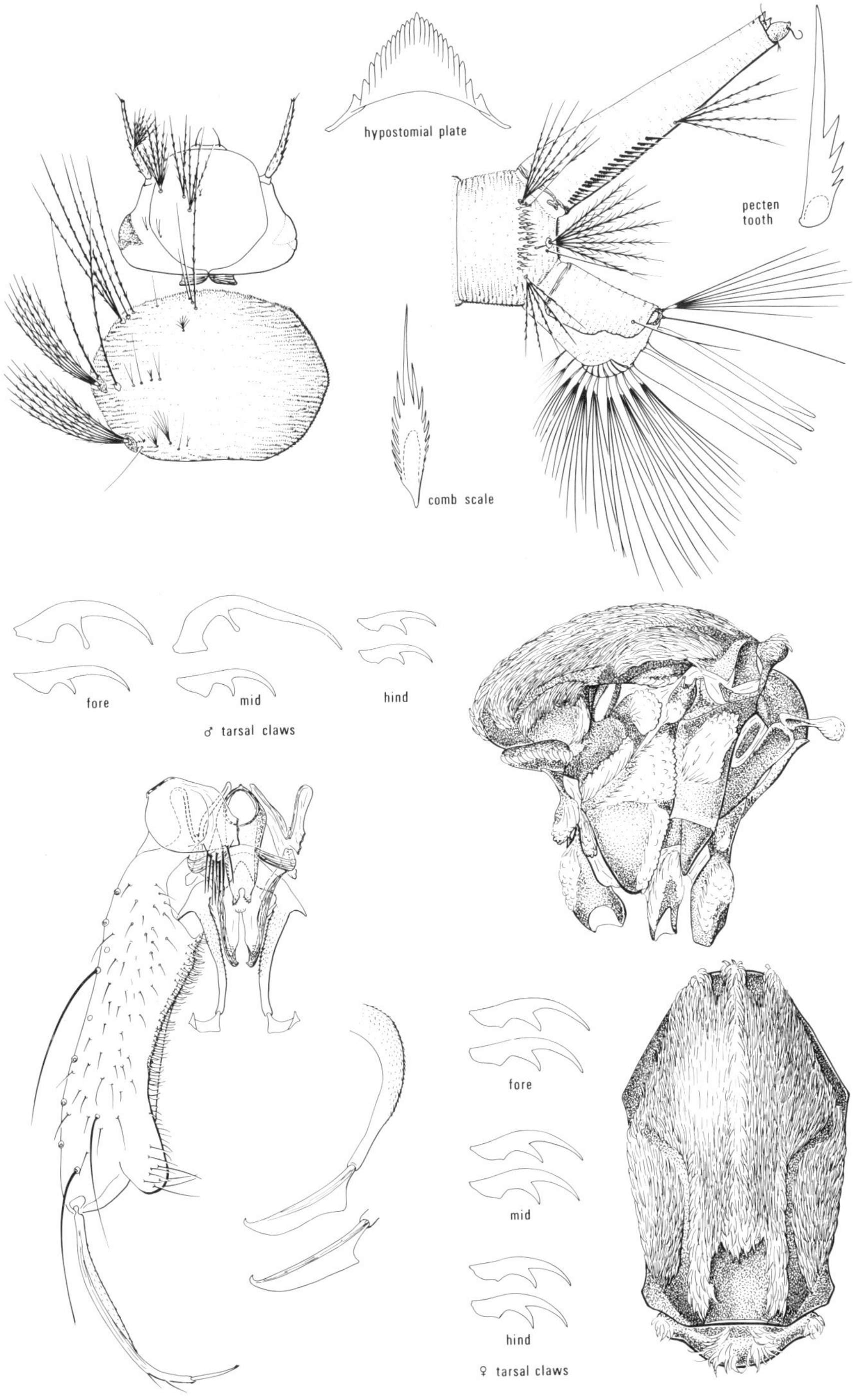

Plate 12. Aedes aloponotum

medially directed seta, setose area extending from base of gonocoxite along mediodorsal edge nearly to apical lobe; apical lobe not extending distal as far as base of gonostylus (that is, not as long as in *euedes*); claspette filament with flange on convex side.

Larva. Head seta 5-C with two to four branches; seta 6-C with two or three branches; integument of thorax and abdomen furry with small microtrichiae that are longer and denser than those of *spencerii*, hence readily visible under a dissecting microscope at 40× or more; otherwise apparently indistinguishable from *excrucians*.

Boddy (1948) noted, in addition to the differences given above, that the upper lateral abdominal setae, series 6, of segments III–VI were double in *aloponotum*, whereas they were single in *excrucians*. Many of our *excrucians* larvae do have the upper lateral setae 6-III to 6-VI single, but others have varying numbers of these setae double; in a few larvae these setae are all double, and the character is not entirely reliable.

Biology. Larvae of *aloponotum* were collected from a pond heavily overgrown with *Veronica scutellata* (Dyar 1924*b*), in a flooded semipermanent pond, in nearby marshy areas (Boddy 1948, Gjullin et al. 1968), and in wooded and semiwooded areas (Gjullin and Eddy 1972). The species has been mistaken for *excrucians* for so long that nothing more definite can be said about its biology. Hearle (1927*a*), in referring to large numbers of biting *aloponotum*, was probably dealing with an assortment of related *Aedes*.

Distribution. Western North America, from southern British Columbia south to Oregon and Idaho.

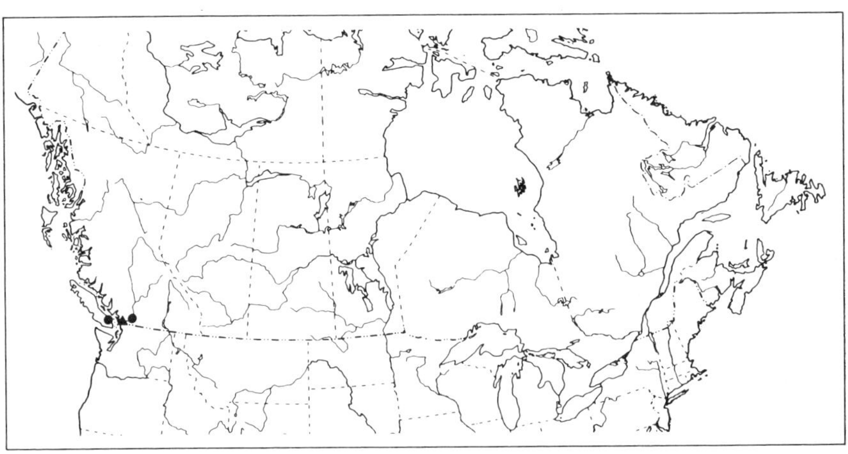

Map 11. Collection localities for *Aedes aloponotum* in Canada: • specimens we examined, ▲ literature records.

Aedes atropalpus (Coquillett)

Plate 13; Fig. 160; Map 12

Culex atropalpus Coquillett, 1902*a*:292.

Adult. Female: Integument dark brown; proboscis and palpus dark-scaled; pedicel with a small group of pale scales on dorsomedial corner; vertex with yellow recumbent scales only, the erect forked scales confined to occiput; postocular setae few, rather widely and regularly spaced; median and submedian stripes of scutum confluent, entirely dark brown scaled; sublateral area anterior to transverse suture entirely yellow-scaled, brown-scaled posterior to suture, separated from submedian stripe by narrow line of yellow scales; lateral stripe yellow-scaled; postprocoxal membrane, hypostigmal area, paratergite, and metameron without scales; scales of katepisternum in two patches: a posterodorsal and a posteroventral one; lower third of mesepimeron bare; tibiae and first and second tarsomeres (and third and fourth hind tarsomeres as well) each dark-scaled except for narrow basal and apical ring of white scales; last hind tarsomere entirely white-scaled; tarsal claws rather straight, curving ventral on apical third, those of hind leg each lacking subbasal tooth; costa basal to humeral crossvein white-scaled; remaining wing veins dark-scaled; abdominal tergites dark-scaled, each with narrow transverse basal band of white scales; cercus short, rounded, inconspicuous.

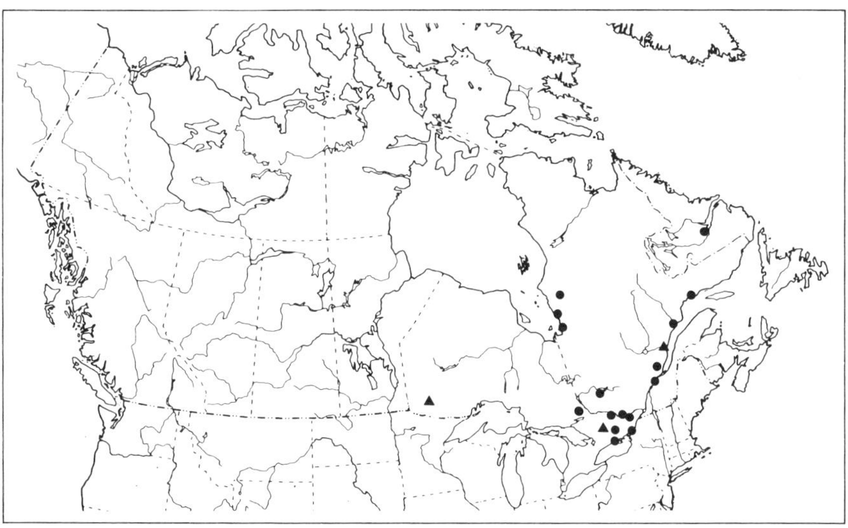

Map 12. Collection localities for *Aedes atropalpus* in Canada: • specimens we examined, ▲ literature records.

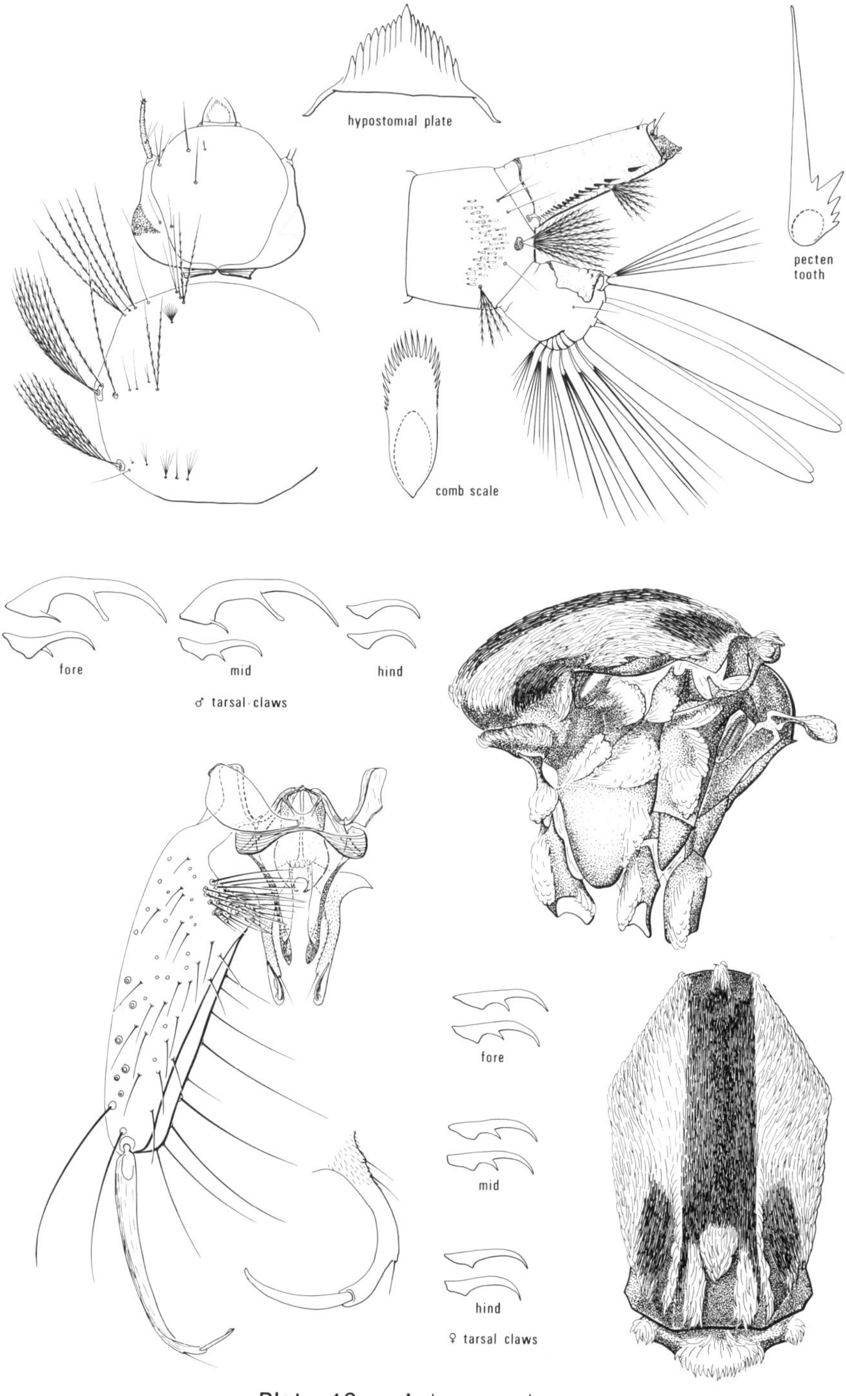

Plate 13. *Aedes atropalpus*

Male: Palpus dark-scaled, two-thirds as long as proboscis, with scattered setae, but lacking dense fringes or tufts; hind tarsal claw lacking subbasal tooth; gonocoxite without apical lobe; basal lobe with a group of setae but lacking an enlarged spine; claspette stem with a small median seta between mid length and filament; claspette filament a simple tusk-like curved unadorned rod. Coloration and arrangement of scales similar to female.

Larva. Antennal seta 1-A short, with at most three branches; mesothoracic seta 1-M as long as 1-P and usually double; siphon short and narrow, only a little longer than segment VIII; distal pecten teeth more widely and irregularly spaced than proximal teeth, the last tooth arising well beyond siphonal tuft and usually close to apex of siphon; abdominal seta 1-VIII reduced, usually double rather than multiple, and about as long as adjacent seta 2-VIII; saddle reduced, even in fourth instar, scarcely enveloping dorsal third of anal segment; saddle seta 1-X arising from membrane ventral to ventral edge of saddle; anal papillae greatly elongate. Additional characters given in Table 3.

This is the only species in Eastern Canada, along with *togoi* on the Pacific Coast, in which the saddle seta (1-X) arises outside of the reduced saddle, even in the fourth instar. In the second and third instars of other species, this seta usually arises at the edge of the saddle, the last pecten tooth arises nearer the middle than the apex of the siphon (except in *cataphylla*), the anal papillae are usually shorter, and the mesothoracic seta 1-M is usually shorter and single.

Biology. Larvae of *A. atropalpus* inhabit rock pools, that is, depressions along rocky shores or in empty riverbeds below natural or artificial dams. The species overwinters as an egg, hatching in spring "when pussy willows and red maples bloom" (Shaw and Maisey 1961). Thereafter, *atropalpus* breeds continuously, as long as water remains in the rock pool, usually until October (James 1964). Eggs are laid singly above the existing water level on the rock pool wall (Breeland et al. 1961) or on the water itself (Hedeen 1953, Kalpage and Brust 1974); hatching takes place after the eggs have dried and been reflooded. Mosaic hatching (Hedeen 1953) occurs when not all eggs hatch on the first reflooding (eggs of the tree-hole species, *triseriatus*, also hatch this way). Eggs laid in late summer enter diapause as a result of the effect of decreasing photoperiod on mature larvae and pupae. Once in diapause, eggs may be stored moist at 20°C for many months without hatching. Either a 20-day exposure to high temperatures (30°C) or a 3-month exposure to long photoperiod (light-to-dark ratio 16:8) is required to terminate diapause. Light acts directly on the egg itself, probably through the micropyle, because eggs with the micropyle covered did not respond (Kalpage and Brust 1974). After eggs have been in diapause for 60–90 days, however, they hatch in about 5 days at 30°C. The larvae are vegetatively phototropic, burrowing in the bottom detritus for long periods. Artificial containers such as concrete septic tanks, trees, and even a tree hole (Hedeen 1953) have also been utilized by *atropalpus* (Breeland et al. 1961).

Developing larvae and pupae produce a soluble substance in the surrounding water that acts as an oviposition attractant (Kalpage and Brust 1973). The attractant remains effective for several weeks.

Aedes atropalpus was previously considered (and still is by some authors) as a species with both autogenous and anautogenous populations, but the latter have been segregated as a distinct species, *A. epactius* Dyar & Knab, found in the southwestern USA, Mexico, and Central America (Zavortink 1972, Brust 1974). Hudson (1970) has shown for *atropalpus* that egg development in the first gonadotrophic cycle is solely dependent on the quality and quantity of the larval diet. Adult females from her laboratory culture did not require food, either a blood meal or a carbohydrate meal, for their first gonadotrophic cycle, and neither meal, even when imbibed, had any effect on the number of eggs developed. O'Meara and Craig (1970) and Brust (1974) also found *atropalpus* to be entirely autogenous in the first gonadotrophic cycle. Hudson (1970) found that females deprived of carbohydrate laid most of their eggs within a few days of emergence, whereas sucrose-fed females retained some of their eggs for as long as 9 months postemergence, a mechanism that under prolonged conditions of drought, could ensure the populations' survival. After oviposition, however, a blood meal can provide sufficient nutrient for a second batch of eggs. This may explain why *atropalpus* has been collected while it was blood-feeding in Ontario and Quebec and has been reported as causing annoyance in Minnesota (Owen 1937, Barr 1958). However, further studies are needed to determine if biting females taken in the field are always parous or whether some are nulliparous. If nulliparous, then do they require blood to complete the first ovarian cycle?

Distribution. Eastern North America, from Minnesota and Newfoundland (Labrador) south to Alabama.

Aedes aurifer (Coquillett)

Plate 14; Figs. 138, 175; Map 13

Culex aurifer Coquillett, 1903*a*:255.

Adult. Female: Integument medium to dark brown; proboscis and palpus entirely dark-scaled; pedicel dark-scaled on mediodorsal third; occiput yellow-scaled laterally; vertex with a narrow middorsal wedge of yellow scales flanked by a spot of dark brown scales; scutum with a very broad middorsal patch of dark brown scales occupying all of median, submedian, and postsutural sublateral stripes, as well as medial half of presutural sublateral area; lateral half of presutural sublateral area, lateral stripe, prescutellar depression, and most of postpronotum with yellowish brown scales; acrostichal and presutural dorsocentral setae few or none; postprocoxal membrane, hypostigmal area, anterodorsal corner of katepisternum, lower third of mesepimeron, and metameron without scales; lower

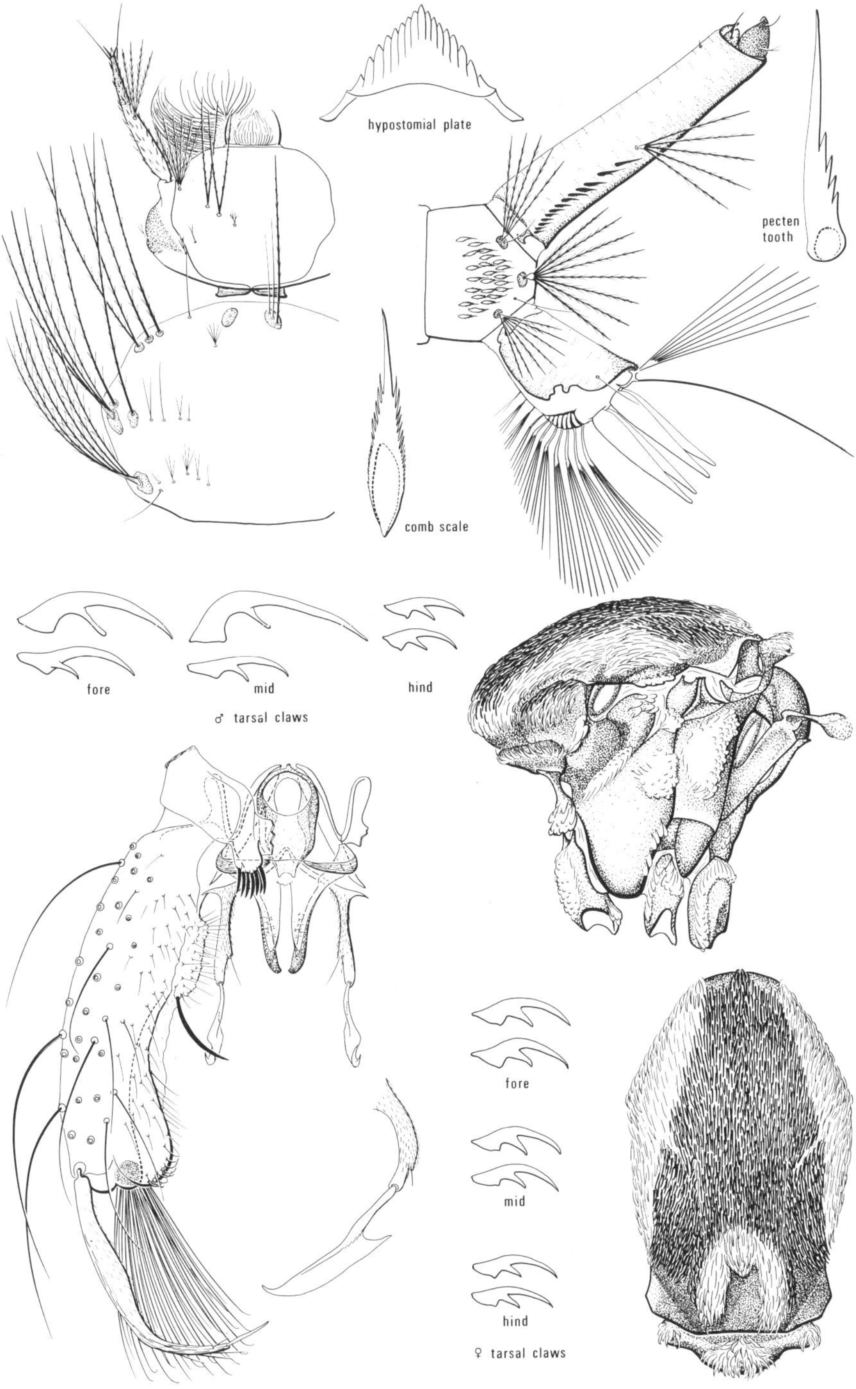

Plate 14. Aedes aurifer

mesepimeral setae absent; fore coxae usually extensively brown-scaled; tibiae dark-scaled except for narrow line of paler scales ventrally, tarsi entirely dark-scaled; wing veins including base of costa dark-scaled; abdominal tergites dark-scaled, each with lateral spot of white scales, usually visible only from below on segments I to IV, more conspicuous from above on segments V to VII; cercus long, slender, black-scaled.

Male: Palpus entirely dark; apex of third palpomere with dense tuft of setae that does not extend beyond apex of fourth segment; fourth and fifth palpomeres fringed laterally and medially with long setae; apex of gonocoxite with posteriorly directed pencil-like tuft of long setae arising from ventromedial angle; basal lobe of gonocoxite with a single long curved distally directed spine; claspette stem bearing a minute seta near its apex; claspette filament cleaver-like with long cylindrical base and well-developed crest on convex side, with pronounced retrorse angle.

Larva. Antenna peculiarly whitish translucent and swollen basally, black distal to antennal seta 1-A; base of head setae 5-C, 6-C, and 7-C aligned in a straight line; setae 5-C and 6-C with only two or three branches; siphon long and slender; distal two or three pecten teeth more widely spaced than proximal teeth; saddle enveloping three-quarters or more of anal segment, its posteroventral angle truncated; remaining characters given in Table 3.

No other species of *Aedes* has the antenna swollen basally, with translucent base and black apex.

Biology. *Aedes aurifer* is usually an uncommon species in Canada, although it is locally abundant in flooded areas along the St. Lawrence River in Quebec (Maire, personal communication). In the Ottawa area, we

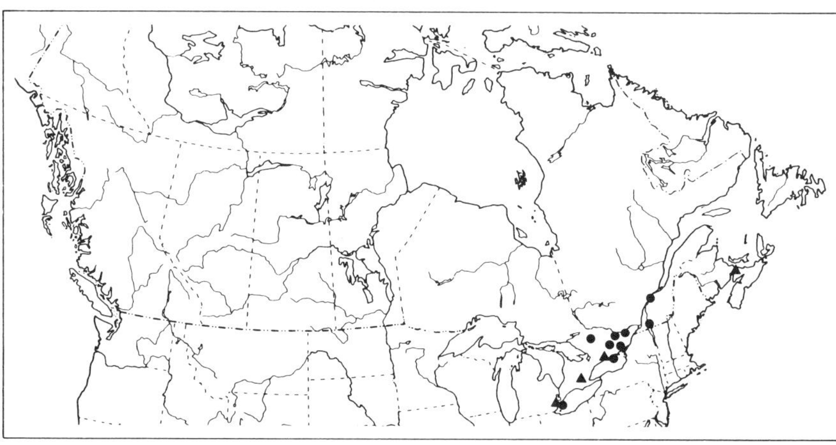

Map 13. Collection localities for *Aedes aurifer* in Canada: • specimens we examined, ▲ literature records.

have encountered *aurifer* larvae in moderate numbers only in one location, a semipermanent marsh between a roadside and woodland at Dunrobin, Ont. The marsh was an enlargement of a slow-flowing stream, and the water was choked with rotten plant material and brown algae. A few larvae of *excrucians* and *riparius* were also present. Even though *aurifer* was the most common species, no more than two or three larvae could be caught with each sweep of a dip net. In New Jersey, larvae were taken in cranberry bogs, usually associated with emergent vegetation some distance from shore (Carpenter and LaCasse 1955).

Distribution. North America, from Minnesota and Illinois east to Maine and Maryland. Twinn (1949) included Manitoba in the range of *aurifer*, but no specimens exist to support this record, and the species has not been taken there since. We have omitted Manitoba from the distribution map.

Aedes campestris Dyar & Knab

Plate 15; Figs. 77, 116, 163; Map 14

Aedes campestris Dyar and Knab, 1907*b*:213.
Aedes callithotrys Dyar, 1920*a*:16.

Adult. Female: Integument medium brown; proboscis and palpus predominantly dark-scaled, with scattered pale scales; pedicel almost completely encircled with pale scales; scales of upper half of postpronotum and of median, submedian, postsutural sublateral, and presutural lateral stripes of scutum pale reddish brown, those of dorsocentral stripe and presutural sublateral area creamy white to yellowish, resulting in five alternating bands of about equal width, three reddish brown separated by two paler yellowish bands; probasisternum extensively scaled; pleuron except for proepimeron, anterior part of postspiracular area, anteroventral part of katepisternum, mesomeron, metepisternum, and metepimeron entirely covered with creamy white scales; tibiae predominantly pale-scaled, with scattered dark scales on anterior surfaces; basal tarsomeres mostly dark-scaled except for scattered pale scales, and a fairly well defined pale apical ring; second tarsomeres dark-scaled, with distinct basal white ring, usually also with apical white ring; remaining tarsomeres of forelegs and midlegs mostly dark-scaled and of hind leg with fairly distinct basal and apical rings, except last hind tarsomere, which is predominantly white-scaled; tarsal claws, especially front claw, long and strong with relatively long subbasal tooth, each rather strongly bent ventrally beyond subbasal tooth; scales of wing veins mixed pale and dark, with the pale ones usually predominant; dorsum of abdomen with at least a well-defined median band of pale scales separating lateral brown-scaled areas, varying to almost entirely pale-scaled.

Male: Palpus relatively pale, with mixed pale and dark scales and with long, yellowish, brown-tipped setae at apex of segment III and medial and lateral fringes on remaining two segments; scutum usually colored as in female, but sometimes uniformly yellowish, without middorsal stripe; basal lobe of gonocoxite with a long enlarged seta arising near apex of lobe and three or four shorter enlarged setae arising from posterolateral corner of lobe, neither long nor short enlarged setae more strongly sclerotized than remaining fine setae, and hence not as readily discernible as in *dorsalis* and *melanimon*; apical lobe of gonocoxite prominent in comparison with the latter two species; claspette filament sickle-shaped without prominent crest or elongate cylindrical stem.

Larva. Head seta 5-C with two or three branches; seta 6-C usually single; seta 7-C with eight or more branches; mesothoracic seta 1-M two- or three-branched and as long as head seta 5-C; dorsolateral abdominal setae 1-III to 1-VII reduced, much less than half as long as lateral abdominal setae of same segments; pecten teeth occupying basal half of siphon, the distal teeth usually more widely spaced than the basal teeth, but occasionally being uniformly spaced (*campestris* larvae must, therefore, be keyed out twice, see pp. 115, 120); siphon about three times (range 2.8–3.2) as long as maximum width and slightly longer and narrower than that of *dorsalis* or *melanimon*; saddle reduced, barely covering upper half of anal segment; anal papillae bud-like, usually shorter than saddle, occasionally longer (McLintock 1944); other characters as in Table 3.

A close relative of *dorsalis*, *campestris* may be differentiated from it by its two- or three-branched head seta 5-C (which is single or rarely double on one side in *dorsalis*) and by the reduction of the dorsolateral abdominal setae 1-III to 1-VII. Apparently no single character is unique to this species.

Biology. *Aedes campestris* overwinters in the egg stage. Rempel (1950) and Barr (1958) considered this species to be univoltine, but at Churchill, Man., both diapause (75%) and nondiapause eggs (25%) were found (Brust 1968 and personal communication). However, 88 batches from

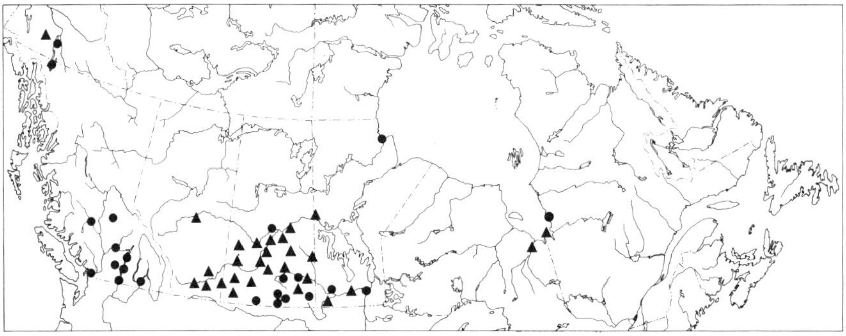

Map 14. Collection localities for *Aedes campestris* in Canada: • specimens we examined, ▲ literature records.

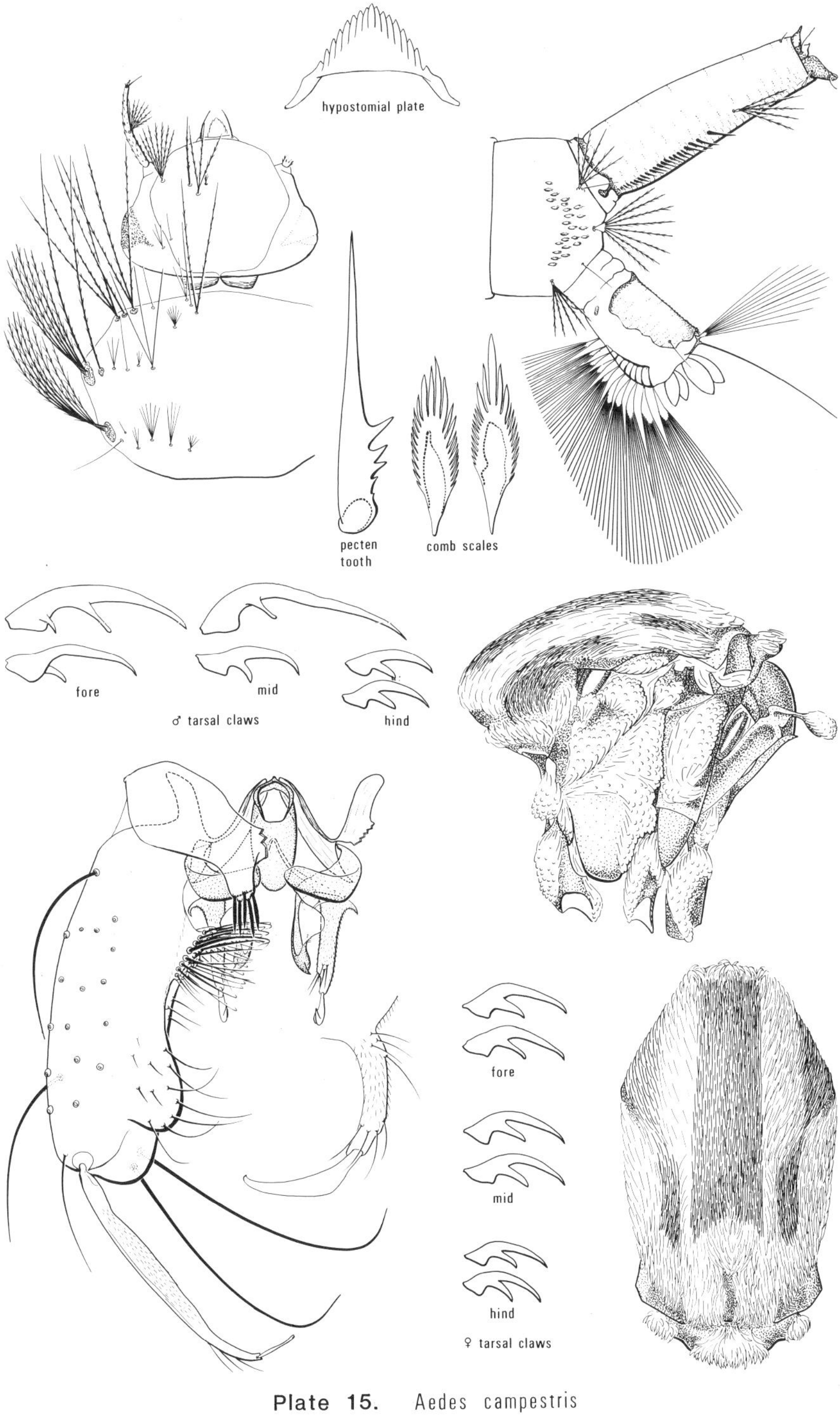

Plate 15. Aedes campestris

several prairie localities were almost entirely nondiapausing (Tauthong 1975, Tauthong and Brust 1977*a*). Thus *campestris* is essentially univoltine in the north, but multivoltine farther south.

Larvae have been collected in a variety of open situations, including prairie alkaline sloughs that have a pH of 8.4 and are rich in organic matter (Gibson 1937, Rempel 1950, 1953), temporary or semipermanent snowmelt, rain pools, roadside ditches (McLintock 1944), and subarctic pools in birch and willow scrub at tree line (Hocking et al. 1950). Larvae may develop to pupation at 15°C, but they do not live to produce adults (Tauthong and Brust 1977*b*). Optimum development and survival occurred at 23°C.

Females usually rest in grass by day and feed at night, but they will bite during the day when disturbed (Dyar 1922*b*, McLintock 1944). Beckel (1955) found that caged females preferred to oviposit on open water rather than on filter paper. The species is one of the dominant pest species at Kamloops, B.C., where it appears early in the season (Gibson 1929). Gibson (1936) estimated that *campestris* entering Kamloops came from a breeding site 10 km away.

Males swarm after sunset close to the tops of small pine trees (Dyar 1922*b*) and in forest clearings 2–4 m above ground near Churchill, Man. (Hocking et al. 1950).

Distribution. North America, from Alaska and Oregon east to Manitoba, Michigan, and Texas; also the coasts of Hudson Bay and James Bay.

Aedes canadensis (Theobald)

Plate 16; Fig. 192; Map 15

Culex canadensis Theobald, 1901*b*:3.
Culex nivitarsis Coquillett, 1904:168.

Adult. Female: Integument reddish brown; proboscis and palpus entirely dark-scaled; pedicel with at most a few small dark scales medially; scutal integument pale reddish brown, with evenly distributed narrow pale yellow scales that are slightly darker medially, lighter laterally, otherwise without pattern; scutal setae brown; postprocoxal membrane, hypostigmal area, anterodorsal corner of katepisternum, and lower third of mesepimeron bare; mesepimeral bristles absent; tibiae dark-scaled except for narrow ring of white scales at base and at apex; basal tarsomeres similar, but with more conspicuous white rings, especially on hind leg; remaining tarsomeres on fore- and mid-legs mostly dark-scaled; second tarsomere usually with a basal and apical ring or spot of white scales; third tarsomere usually with at least a basal white spot; hind tarsomeres conspicuously banded basally and apically with white scales, the basal and apical bands subequal in width; last hind tarsomere entirely white-scaled; wing veins, including base of costa, entirely dark-scaled; abdominal tergites usually dark-scaled, each with basal lateral patch of white scales, sometimes each with complete transverse basal band of white scales.

Male: Palpus mostly dark-scaled, about as long as proboscis; segments III, IV, and also usually II and V each with basal band of white scales; apex of segment III and all of IV bearing tufts of long setae; scutum and legs as in female; abdominal tergites each with rather prominent inversely U- or V-shaped white basal band of white scales connecting lateral triangular patches; basal lobe of gonocoxite prominent, rounded, with seta-bearing surface slightly convex and strongly oriented dorsally; setae shortest laterally, increasing in length to medial edge; apical lobe of gonocoxite broadly rounded, bearing short stout setae; claspette filament cylindrical, tusk-like, without blade.

Larva. Usually darkly pigmented, with reddish brown head capsule; head setae 5-C and 6-C each with five or more branches, their bases not forming a straight line with base of seta 7-C; prothoracic setae 2-P and 3-P less than half as long as and finer that seta 1-P; prothoracic seta 5-P single; mesothoracic seta 1-M minute; pecten teeth evenly spaced; anal papillae longer than anal segment; each comb scale usually with subapical spinules almost as long as median apical spine, thus rounded apically, but specimens from some parts of Quebec have shorter subapical spinules and therefore key to *sticticus* in Carpenter and LaCasse (1955); other characters as in Table 3.

A common, widespread, and rather late occurring species, often mixed with *cinereus*, which also has head setae 5-C and 6-C each with five or more branches, but from which it may be separated by the nonlinear arrangement of these setae (5-C to 7-C) and by the evenly spaced pecten teeth. Only

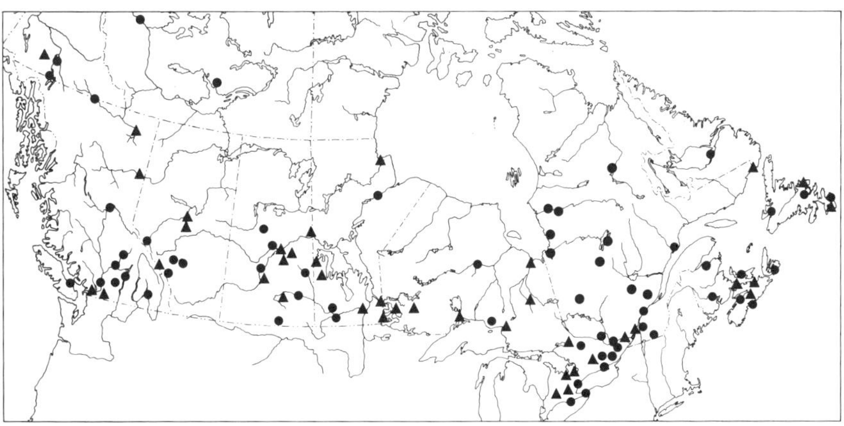

Map 15. Collection localities for *Aedes canadensis* in Canada: • specimens we examined, ▲ literature records.

three other species, *thibaulti, pionips,* and *pullatus*, have head setae 5-C and 6-C with five or more branches and evenly spaced pecten teeth; the head setae are arranged linearly (with 7-C) in *thibaulti*, and mesothoracic seta 1-M is elongate in *pionips* and *pullatus*.

Remarks. Pickavance et al. (1970) reported an unusual adult of *canadensis* from Newfoundland in which the first and second hind tarsomeres have the normal basal and apical white-scaled rings, but the third tarsomere has only a basal white ring and the last two tarsomeres are entirely dark-scaled. They assigned it to the subspecies *mathesoni* Middlekauff, which is otherwise known only from the southeastern USA. We have not seen this specimen, nor one like it from Canada. From such widely separated localities it is hardly likely to be a subspecies, but it could represent anything from a rare recessive allele to another species. Because the comb scales of some populations of *canadensis* are pointed and some are rounded and the species is univoltine in some populations and bivoltine in others, it could be a complex of sibling species.

Biology. Although overwintering eggs of *canadensis* usually hatch in spring, as do those of most *Aedes*, their hatching and subsequent development is decidedly irregular compared to other *Aedes*. Early workers suspected a staggered hatch (Dyar 1922*b*, Rudolfs 1929). In Tennessee, Breeland et al. (1961) found hatching took place from the beginning of April through May, and he postulated that hatching of some of the overwintering eggs was delayed. Bickley and Whitlan (1956) found *canadensis* larvae in mid-August in a pool that this species had vacated the previous mid-May. Belton and Galloway (1966) and Siverly and DeFoliart (1968*a*) also suspected either a staggered hatch or a second generation. Brust (1968) has shown that in Manitoba 30% of eggs laid by this species were nondiapausing, and these eggs hatched in nutrient broth at room temperature about 2 wk after oviposition.

The larval habitat is diverse. In the Ottawa area, we found *canadensis* most numerous associated with *cinereus* in small open pools in a sphagnum bog. The larvae were only in the second and third instars when *punctor*, in more shaded pools, was mature. At Radisson, Que. (53°45′N), however, *canadensis* larvae were contemporary with *punctor*, being among the first species to appear long before *cinereus* (Maire and Aubin 1976). Nevertheless, small numbers of larvae were present until the third week of August, again suggesting a staggered hatch because of the presence of some nondiapausing eggs or some exceptionally cold or shaded habitats.

In northern Minnesota, Price (1963) found larvae of *canadensis* in temporary woodland pools, roadside ditches, cattail and sedge marshes, and muskeg pools. Rudolfs and Lackey (1929) described the habitat for *canadensis* as a temporary woodland pool, with shrubs and dense cinnamon fern, that by midsummer was dry and fully shaded. McLintock (1944) usually found larvae in locations protected by trees and shrubs, but some *canadensis* were found in open prairie pools at Winnipeg.

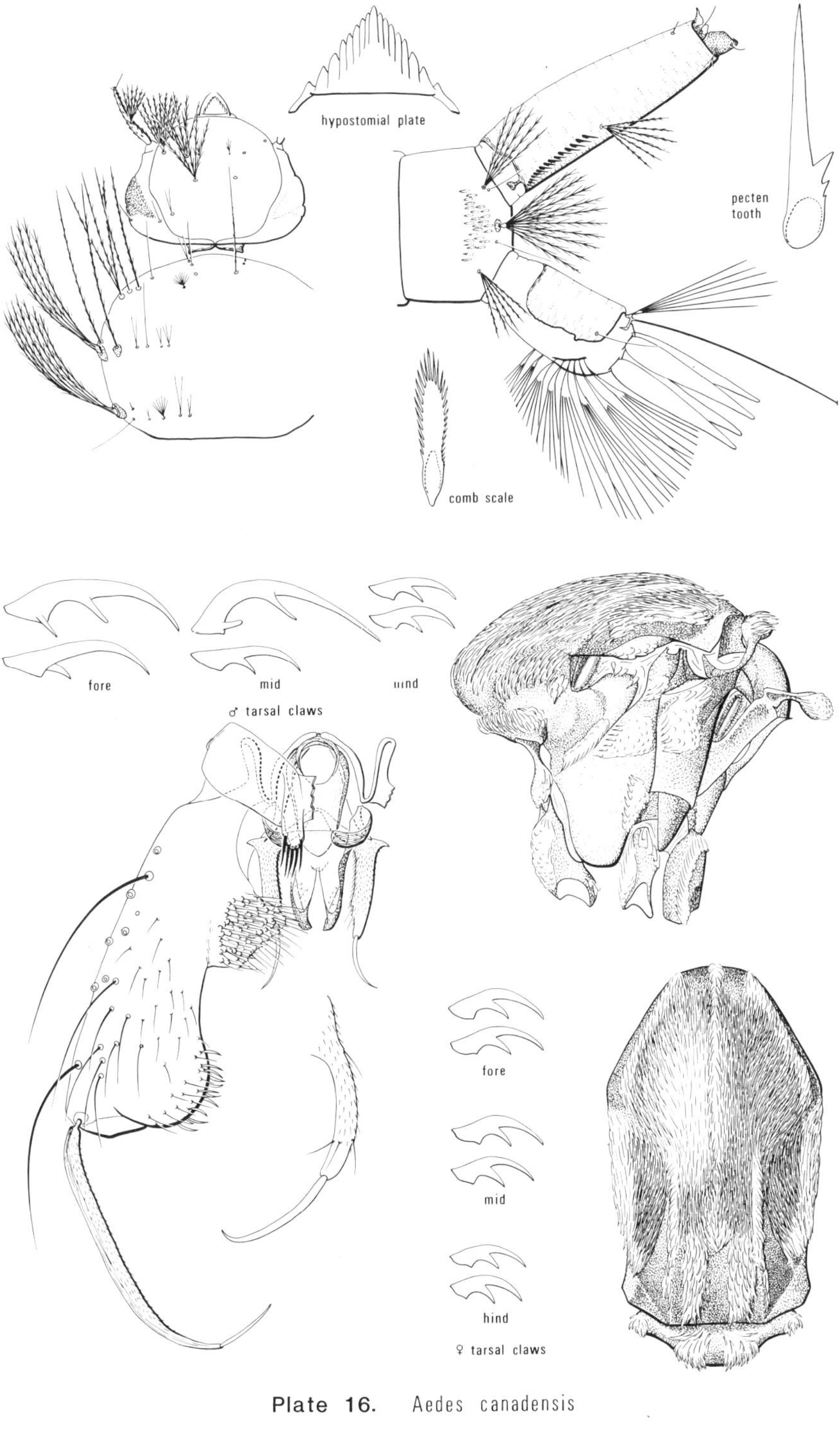

Plate 16. *Aedes canadensis*

Females were attracted to and engorged on a variety of small mammals, birds, amphibians, and reptiles (Hayes 1961), but they were more attracted to mammals than to the other hosts. In Alberta, females were readily caught in traps baited with chicken (Graham 1969*a*). The virus of California encephalitis has been isolated from this species (Parkin et al. 1972). According to Carpenter and Nielsen (1965), ovarian follicles of female *canadensis* coming to bite were in stage I or II.

Dyar (1919*b*) observed a small swarm of male *canadensis* after sunset at Prince Albert, Sask., over an overgrown road between dwarf spruces in a swamp, near another swarm of male *pionips*.

Distribution. North America, from Yukon Territory, Washington, and Texas east to Newfoundland and Florida.

Aedes cantator (Coquillett)

Plate 17; Fig. 78; Map 16

Culex cantator Coquillett, 1903*a*:255.

Adult. Female: Integument reddish brown; proboscis entirely dark-scaled; palpus mostly dark-scaled, with a few scattered white scales; pedicel dark-scaled medially, pale-scaled dorsally; scutum medium brown, uniformly covered with pale yellow scales without pattern; postprocoxal membrane, hypostigmal area, and lower third of mesepimeron without scales; mesepimeral bristles present; fore coxa mostly brown-scaled; tibiae mostly dark-scaled; tarsomeres each with narrow basal ring of pale yellow scales, the width of each ring usually no more than one-quarter the length of respective tarsomere; tarsal claws nearly straight, with slightly down curved apex; a small subbasal tooth present only on fore and mid claws, lacking on hind claws; wing veins dark-scaled; abdominal tergites each with basal transverse band of white scales; first abdominal sternite with some pale hairs and a few scales.

Male: Palpus entirely dark-scaled, about as long as proboscis, apical third of segment III and all of segment IV with long dense tufts of setae; legs, wings, and coloration of scutum as in female; hind tarsal claws usually lacking subbasal tooth, occasionally one or both may have a small tooth; basal lobe of gonocoxite rather small, conical, with long medially directed spine; apical lobe present, but not prominent; claspette filament sickle-shaped with somewhat expanded keel on convex side.

Larva. Head seta 5-C usually with four or five branches (rarely three or up to ten); seta 6-C with two to eight branches; mesothoracic seta 1-M single, as long as head seta 5-C; dorsolateral abdominal setae 1-III to 1-VI and ventrolateral abdominal setae 13-III to 13-VI double, about two-thirds as long as upper lateral setae (6-III to 6-VII) of their respective segments; pecten teeth evenly spaced; saddle reduced, occupying only upper half of anal segment; anal papillae extremely small, often encrusted with protozoa or other deposits, and easily mistaken for the stumps of normal-sized papillae that were broken off.

A nondescript species, perhaps best recognized by its habitat (coastal salt marshes), by the contrast between head setae 5-C and 6-C (the latter almost always with fewer branches), by the elongate mesothoracic seta 1-M, and by the minute anal papillae.

Biology. A species found in coastal salt marshes, *Aedes cantator* has a wider range in Canada than has *sollicitans*, the other salt marsh species. Both species are multivoltine, overwintering in the egg stage. *Aedes cantator* is the more common of the two in early spring in New Jersey, whereas *sollicitans* is more common in midsummer, possibly because of lower salinity in the marshes at that time (Headlee 1916). Larvae of *cantator* are tolerant of high salinity (Chapman 1959), although they prefer fresh or brackish water (Mitchell 1907). They occur mainly in pools periodically refilled by extra high tides and by summer rains, which because of the stable water table may not need much rain to keep them wet. Along the east coast of James Bay, Que., Maire and Mailhot (1978) found very large populations of larvae (75 000/m^3) in numerous shallow depressions along the tidal zone that had been created by large blocks of ice that had been forced on shore during the previous winter. Along the coast of New Brunswick we were unable to find many larvae in salt marshes that were connected directly to the ocean, evidently because of the presence there of fish, but larvae were

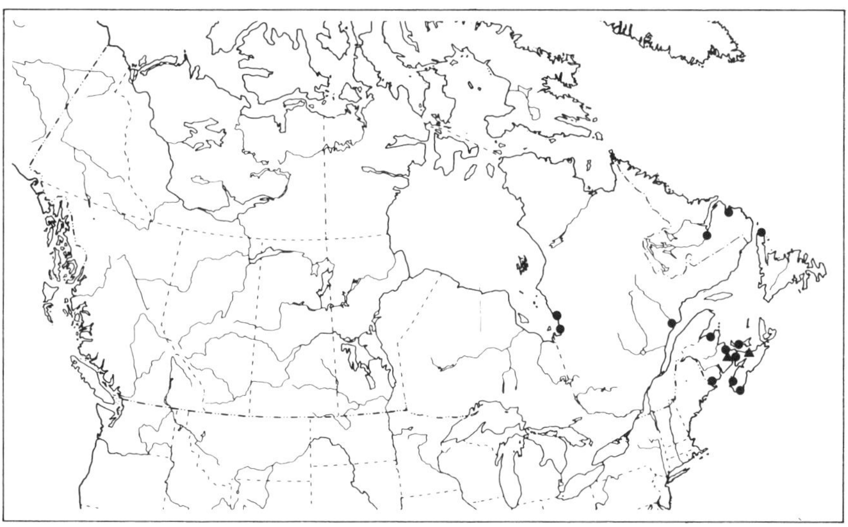

Map 16. Collection localities for *Aedes cantator* in Canada: • specimens we examined, ▲ literature records.

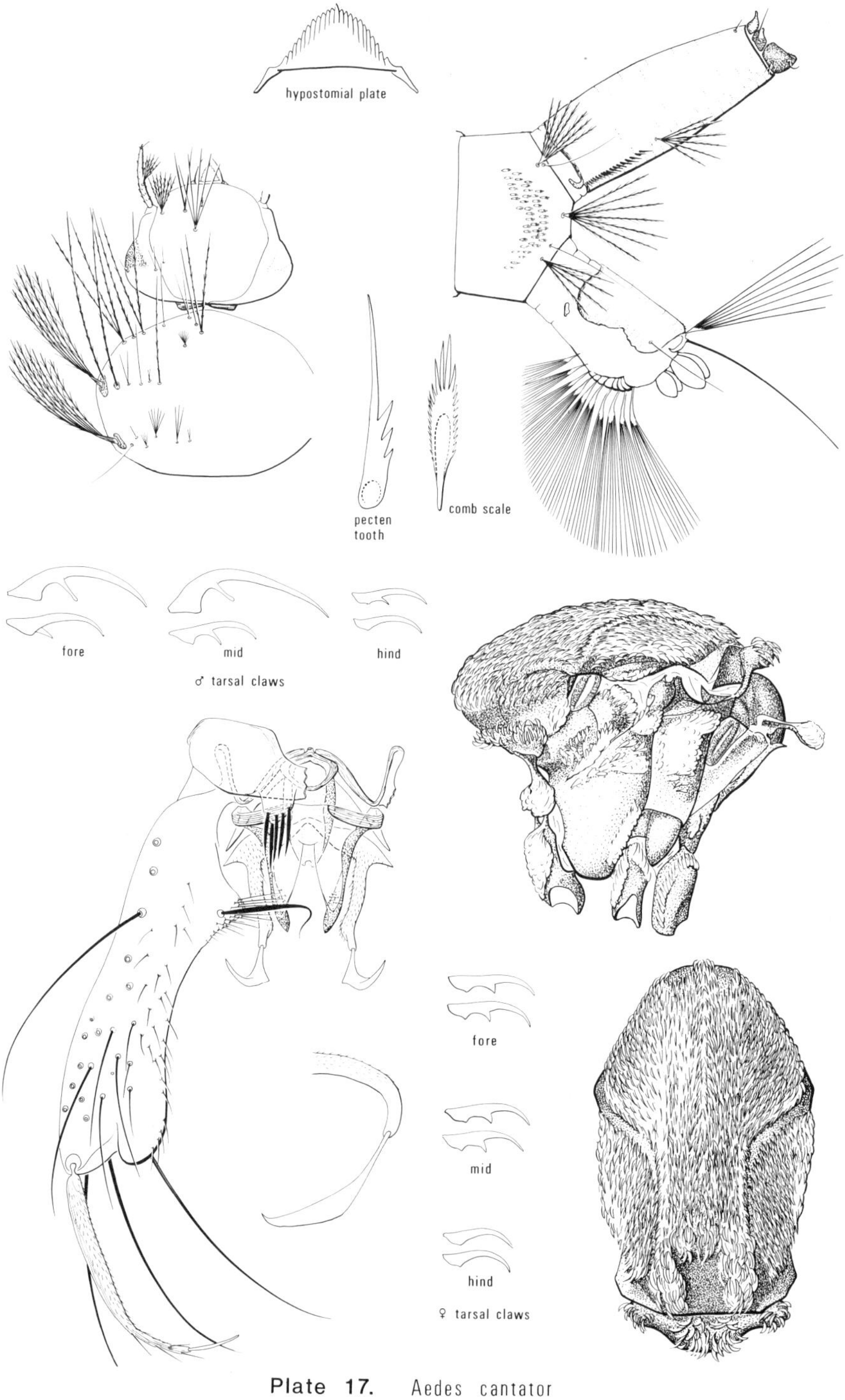

Plate 17. *Aedes cantator*

present in enormous numbers in certain small impounded pools. In one such pool in mid-August, we found *cantator* predominantly in the larval stage and *sollicitans* emerging as adults, which suggests that they may not be in direct competition.

Females of *cantator* and also *sollicitans* and a southern relative *Aedes taeniorhynchus* (Wd.) are infamous for their mass migrations inland. Mitchell (1907) observed that males accompanied the females on these mass flights. At Sackville, N.B., blood-seeking females appeared in large numbers at sunset, although they had not been in evidence during the day, even on overcast days.

Distribution. Atlantic Coast of North America, from Newfoundland south to Virginia; also in saline pools in inland New York.

Aedes cataphylla Dyar

Plate 18; Figs. 98, 156; Map 17

Aedes prodotes Dyar, 1917*c*:118.
Aedes cataphylla Dyar, 1916:86.
Aedes pacificensis Hearle, 1929*c*:101.

Adult. Female: Integument dark brown; proboscis and palpus mostly dark-scaled, but with numerous scattered white scales; pedicel and first flagellomere mostly white-scaled; scutum mostly or entirely pale brown scaled, if whitish scales present, confined to lateral area, prescutellar area, and middorsocentral spot, with sometimes a few scattered pale scales on median stripe on posterior half of scutum; sublateral area with numerous brown setae; postpronotum brown-scaled or partly brown-scaled above, pale-scaled on lower portion, usually with several setae along dorsal margin; postprocoxal membrane with scales; anepisternum almost entirely covered with scales; katepisternum extensively scaled along dorsal and posterior margins; mesepimeron scaled to lower margin; mesepimeral bristles present; tibiae and tarsi mostly dark-scaled dorsally, pale-scaled ventrally; tarsal claws small, evenly curved, each with short subbasal tooth; costa with patch of white scales basally, remainder mostly dark-scaled but with scattered white scales; subcosta and R_1 also with scattered white scales; remaining wing veins dark-scaled; abdominal tergites dark-scaled, each with broad basal transverse band of white scales.

Male: Proboscis with scattered pale scales basally; palpus about as long as proboscis; second and third palpomeres predominantly pale-scaled dorsally; fourth palpomere with as many pale as dark scales; last palpomere dark-scaled; apex of third and all of fourth palpomere with tufts of long dark setae; legs and wing as in female; scutum same color as in female; pleuron not as heavily scaled; medioventral edge of gonocoxite with numerous, long, parallel, medioventrally directed setae; basal lobe of gonocoxite conical, with single strong medially directed bristle subtending a single row of long setae; apical lobe somewhat parallel-sided, subtruncate apically; claspette filament with expanded flange on convex side arising at base of filament.

Larva. The only species in which the siphonal tuft arises proximal to middle of siphon, and in which the distal pecten teeth, though irregularly and often asymmetrically arranged, usually extend almost to the apex of the siphon. Usually two or more pecten teeth arise distal to siphonal tuft. *Aedes atropalpus*, with similarly arranged pecten teeth, occurs allopatrically in Eastern Canada and has a much shorter siphon and a reduced saddle with seta 1-X arising ventral to saddle margin. *Aedes nigripes*, which usually also has widely spaced distal pecten teeth extending beyond siphonal tuft, usually has the saddle encircling the anal segment.

Biology. This species overwinters in the egg stage and except for the overwintering species is one of the earliest species to appear in spring at Edmonton (Strickland 1938) and Kamloops (Gibson 1929). A deep snow cover favors a large hatch (Gibson 1932). The first instar was described and drawn by Bohart (1954).

Females are active both by day and by night (Dyar 1922*b*) and are among the major pest species in British Columbia and Alberta (Gibson 1928, 1929; Happold 1965*b*). Carpenter and Nielsen (1965) recorded one female of *cataphylla* that completed five ovarian cycles in 40 days.

Details of dispersal and swarming of *cataphylla* in the Edmonton area were described by Klassen and Hocking (1964). Dispersal was initiated by a downwind flight at twilight from the site of emergence. Male swarms formed about 1 m above ground, near small bushes on a golf course. Swarms also formed over animals and they moved as the animals moved. Swarming males oriented themselves by flying into the wind.

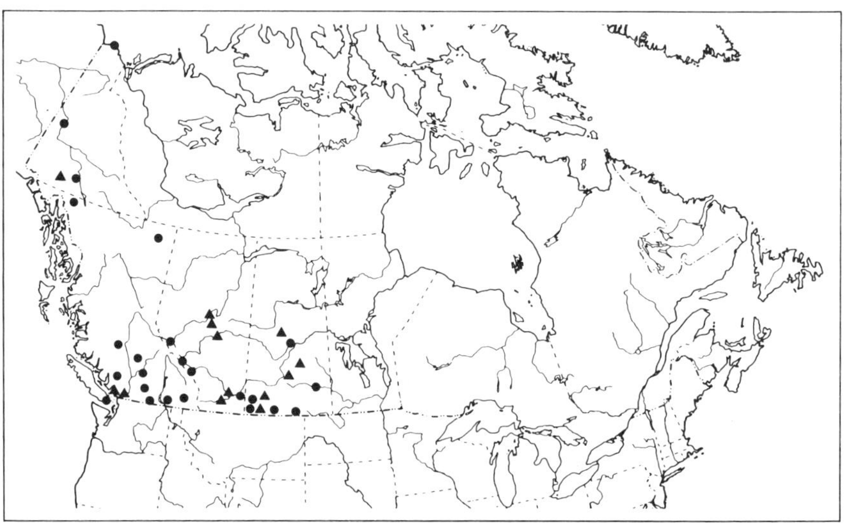

Map 17. Collection localities for *Aedes cataphylla* in Canada: • specimens we examined, ▲ literature records.

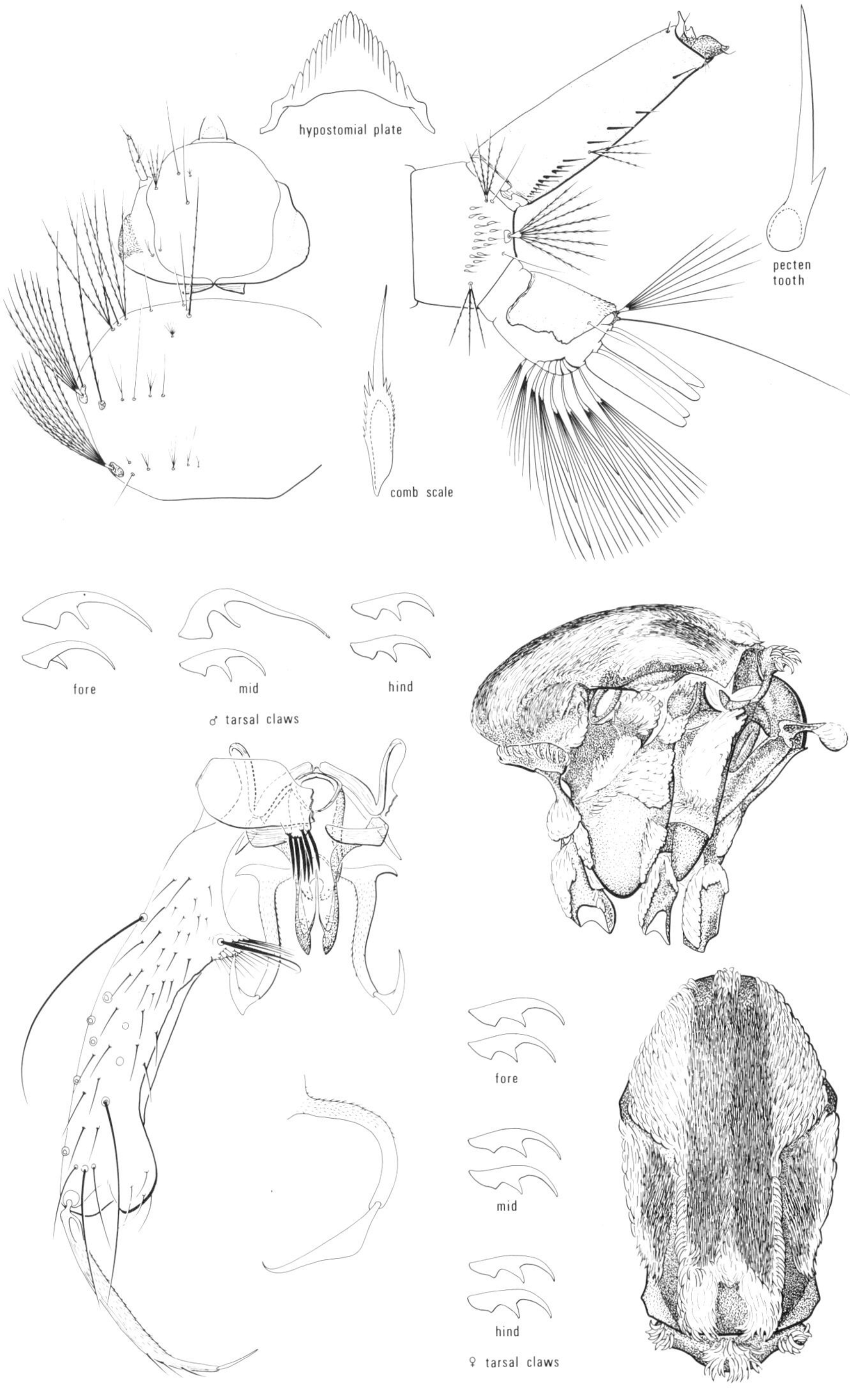

Plate 18. Aedes cataphylla

Distribution. North America and Eurasia: Alaska and California east to Saskatchewan and Colorado; Europe, Kazakhstan, Siberia, eastern USSR, north almost to the tundra, south to the Ukrainian steppe, northern Caucasus, and the mountains of Kirghizia. Stage (1943) recorded *cataphylla* from the western shore of Hudson Bay, where larvae were recovered from rock pools and forest–tundra pools. The species has not been collected in Manitoba by other workers, however, and this record should be verified.

Aedes churchillensis Ellis & Brust

Figs. 104, 154; Map 18

Aedes churchillensis Ellis and Brust, 1973:922.

Aedes communis: Hocking (1952, 1954), Beckel (1954), Beckel and Barlow (1953), Beckel (1958), Brust (1971), and other authors, in part, before 1973, not De Geer.

Adult. Female: Almost indistinguishable from *communis* externally; tarsal claws of hind leg usually more abruptly curved beyond subbasal tooth; submedian dark-scaled scutal stripe usually poorly defined, often obsolete (some *communis* lack submedian stripes as well). The salivary gland is only one-half to two-thirds as large as that of *communis*, and the median acinus is shorter and narrower with less cellular definition (Ellis and Brust 1973).

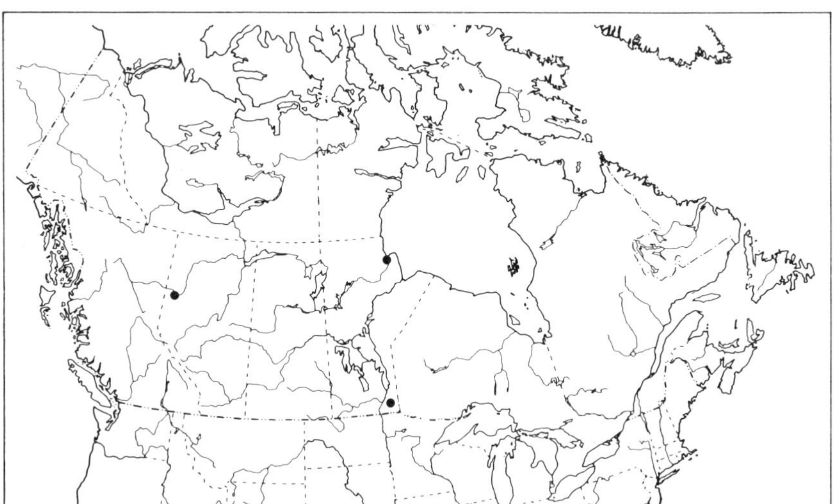

Map 18. Collection localities for *Aedes churchillensis* in Canada: • specimens we examined.

Male: Similar to that of *communis*; setae of fourth and of apex of third palpomere fewer and shorter, usually not as long as fourth palpomere and less dense (Fig. 154).

Larva. Not distinguishable from that of *communis* except by analysis of populations. Head narrower; siphon shorter and narrower (Ellis and Brust 1973).

Biology. During early work on biting flies at Churchill, Man., some populations of "*communis*" were discovered that had a smaller ratio of proboscis length to wing length and these insects produced smaller individuals than did other *communis* populations (Hocking et al. 1950). Because the larvae of this smaller form developed in large numbers in "rather special woodland pools" in the absence of other species, Hocking et al. (1950) suspected the existence of a second species. The females were subsequently found to mature their eggs autogenously (Hocking 1952). Hocking (1952, 1954) proposed that the thoracic flight muscles were partly autolyzed to provide the protein believed to be necessary for egg development. Subsequent workers, however, have not found any evidence for flight muscle autolysis (Beckel 1954, Kalpage 1970, Brust 1971, all as *communis*), but all have confirmed autogeny, which in this species is obligatory; the females do not take blood (Ellis and Brust 1973). Nevertheless, almost as many eggs are produced (in the first cycle) by females of *churchillensis* solely from food reserves carried over from the larval stage as are produced by blood-fed *communis*. Because of their autogeny, females of *churchillensis* would not be expected to mature a second batch of eggs, and Ellis and Brust (1973) found evidence of this in the relatively small size of the anterior follicles.

Males of *churchillensis* are stenogamous, that is, they will mate readily in a small cage without a preliminary swarming flight (Ellis and Brust 1973), whereas in *communis*, only an occasional mating occurred under identical laboratory conditions.

Distribution. Central North America, at present known only from Alberta and Manitoba.

Aedes cinereus Meigen

Plate 19; Figs. 102, 127, 131, 172; Map 19

Aedes cinereus Meigen, 1818:13.
Aedes fuscus Osten Sacken, 1877:191.
Culex pallidohirta Grossbeck, 1905:359.
Aedes cinereus race *hemiteleus* Dyar, 1924*c*:179.

Adult. Female: Integument medium brown; proboscis, palpus, most of vertex, tibiae, tarsi, and wing veins with flattened dark scales having a slight bluish metallic appearance; fore coxa mostly brown-scaled anteriorly;

antepronotum, postpronotum, and scutum with narrow pale brown scales; anepisternum devoid of scales except for a few white scales among post-spiracular setae; scales of katepisternum white, confined to two small areas near posterodorsal and posteroventral corners; mesepimeron bare on lower half; mesepimeral setae absent; abdominal tergites usually dark-scaled, occasionally each with narrow transverse basal band of white scales.

Male: Palpus minute, even shorter than that of female (*cinereus* is the only species of *Aedes* in Canada with this character); antenna plumose; scales of thorax, legs, wings, and abdomen arranged and colored as in female; basal lobe of gonocoxite very large, triangular, occupying basal half of medial edge of gonocoxite; apical lobe absent; gonostylus twice furcate, with the dorsal ramus from basal furcation shorter, setose, and the ventral ramus longer, furcate near apex; claspette Y-shaped, without filament.

Larva. A small species, uniformly dark brown in color; head setae 5-C and 6-C each with five or more branches, their bases forming a straight line with seta 7-C; siphonal tuft short and inconspicuous; distal pecten teeth unevenly and more widely spaced than basal teeth; other characters as in Table 3.

The only other species with five or more branched head setae 5-C and 6-C that are simultaneously arranged in a straight line with seta 7-C is *A. thibaulti*, which has evenly spaced pecten teeth, a siphonal seta 1-S of normal length, and bluish green body pigment that is not uniformly distributed but concentrated within each segment, giving the larva a beaded appearance. *A. vexans* with similarly arranged pecten teeth and reduced siphonal seta 1-S has fewer branched and nonlinearly arranged head setae.

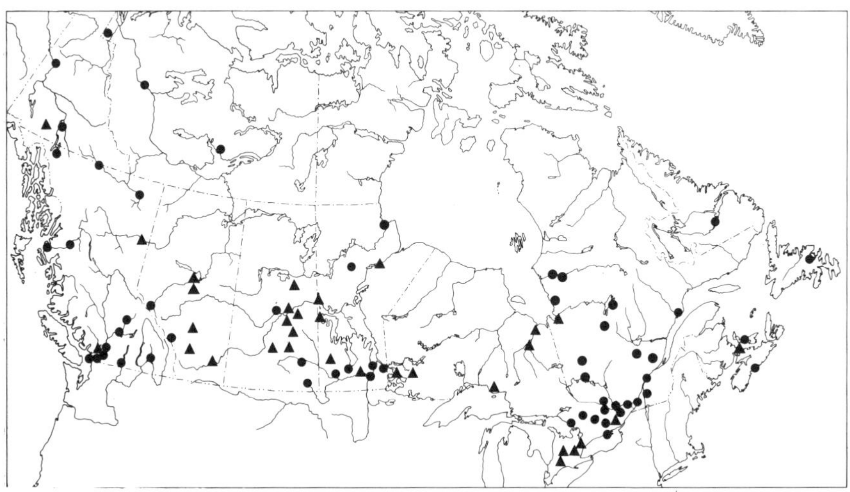

Map 19. Collection localities for *Aedes cinereus* in Canada: • specimens we examined, ▲ literature records.

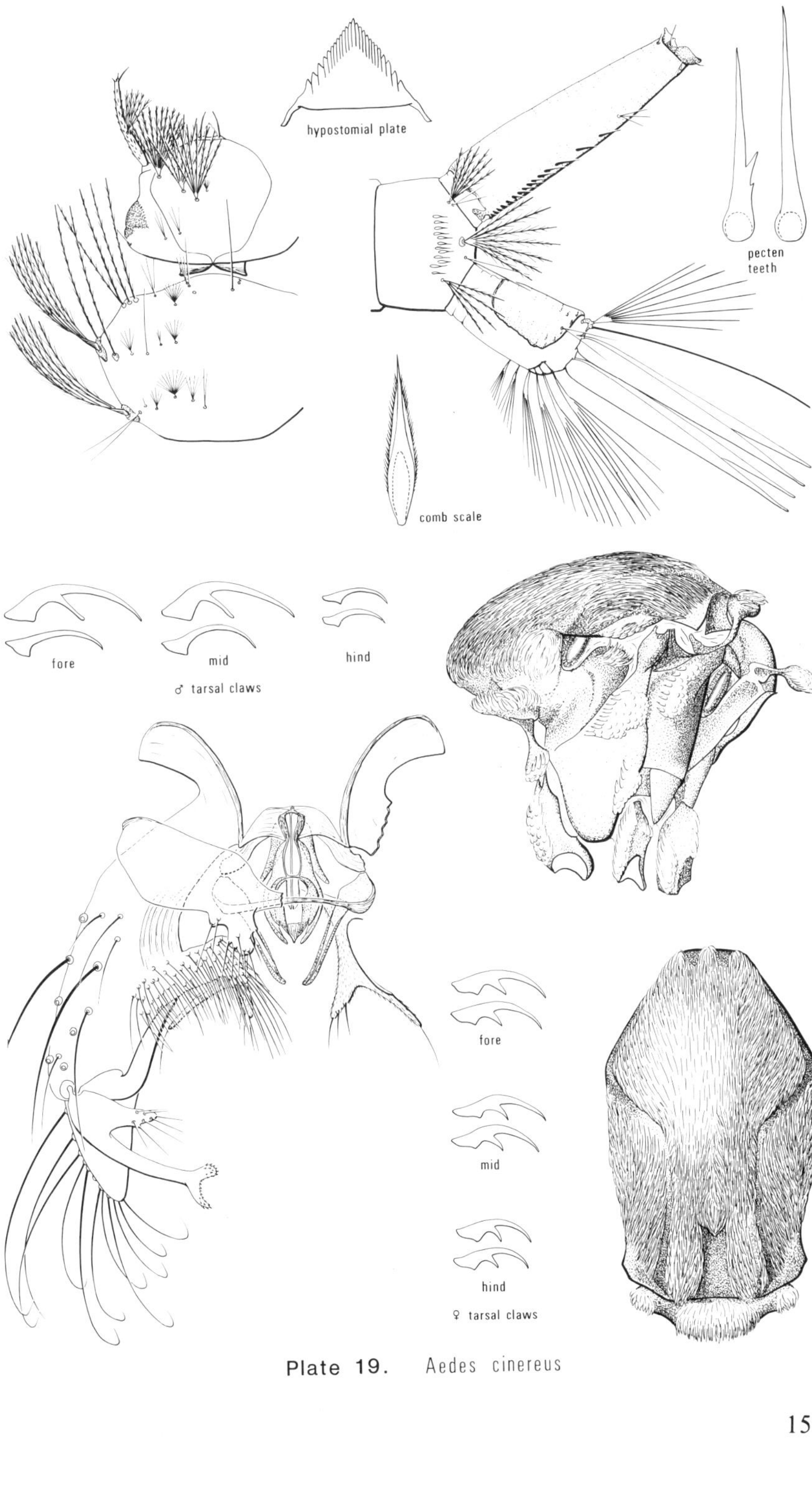

Plate 19. Aedes cinereus

Biology. This species overwinters in the egg stage. It is one of the last species of *Aedes* to hatch in spring (Happold 1965*a*, Graham 1969*a*). Brust (1968) found that about half the eggs of this species were nondiapausing and would hatch at room temperature about 2 weeks after oviposition. Thus *cinereus* is partly multivoltine as are *canadensis, campestris*, and *spencerii*, at least in southern Canada. In the USSR, *cinereus* is said to complete three generations per year (Detinova 1968). Larvae have been found in various habitats, but in the Ottawa area and in Minnesota (Price 1963) they have been most often collected at the edges of semipermanent mixed cattail and sedge marshes and in sphagnum bogs, especially in locations partly shaded by small shrubs such as alders or willows. Twinn (1931) recovered larvae from small shallow snow and rainwater pools in woodland; adults emerged in late May and June. The first instar was described and drawn by Bohart (1954).

Females of *cinereus* are most active at dusk, but they may bite anytime during the day in shaded locations. They are not particularly aggressive and usually attack man on the legs; thus they are hardly noticeable unless you are seated or insufficiently dressed. Carpenter and Nielsen (1965) found that most females completed only one ovarian cycle. Twinn (1931) assumed that the females he observed in late August were long-lived members of the spring generation, but they were probably a second or third generation.

Recently, Bohart and Washino (1978) used the name *hemiteleus* for this species. They compared male terminalia of specimens from various parts of North America with figures of terminalia given by Peus (1972) and concluded that they were indistinguishable from the European species *geminus* Peus, but different from *cinereus*. We agree with Bohart and Washino that the Nearctic populations show a wide range of variation, which suggests the existence of a complex of sibling species. However, we believe that the use of any of the names *fuscus, pallidohirta*, or *hemiteleus* for Nearctic specimens is unwarranted until a thorough study of the *cinereus* complex in North America has been completed.

Distribution. North America and Eurasia: Alaska and California east to Newfoundland and Florida; Europe; Siberia, eastern USSR, northeastern China, Korea, and Japan.

Aedes communis (De Geer)

Plate 20; Figs. 5, 103, 149, 153, 155, 198, 203; Map 20

Culex communis De Geer, 1776:316.
Culex lazarensis Felt and Young, 1904:312.
Culex borealis Ludlow, 1911:178.
Aedes tahoensis Dyar, 1916:82.
Aedes altiusculus Dyar, 1917*b*:100.
Aedes prolixus Dyar, 1922*a*:2.

Adult. Female: Integument medium brown; proboscis dark-scaled; palpus with a few scattered white scales; recumbent scales and most of erect forked scales of vertex yellow; probasisternum often with a few scattered pale scales; postprocoxal membrane bare; hypostigmal area usually bare, occasionally with three or four scales; katepisternum with scales extending to anterodorsal corner; mesepisternum scaled to lower margin; mesepimeral setae present; scutum rather variable, typically yellow-scaled except for a pair of brown-scaled submedian stripes, each about as wide as the median yellow-scaled stripe between them; submedian stripes may be weakly defined in some specimens, which then resemble *churchillensis,* or they may be absent, resembling *intrudens;* in northern specimens sublateral areas, though typically entirely yellow-scaled, may have a central triangular patch of darker brown scales concolorous with submedian stripe; postpronotum and adjacent scutum yellow-scaled, except in some northern specimens in which dark scales are present along dorsal edge; tibiae and first tarsomeres mostly dark-scaled dorsally, with mixed pale and dark scales ventrally; remaining tarsomeres dark-scaled; tarsal claws moderately to strongly curved, each with a long subbasal tooth; wing veins dark-scaled; abdominal tergites dark-scaled, each with transverse basal band of white scales, which may occasionally be reduced or lacking.

Male: Palpus about as long as proboscis, entirely dark-scaled; setae arising from apex of third and base of fourth palpomere long, numerous, and dense, as long as fourth palpomere; basal lobe of gonocoxite subcircular, with posterior surface concave, oriented more or less posteroventrally, thus not visible in dorsal view; ventral edge of subcircular setose cup-like surface of basal lobe with fringe of long, widely spaced, posteriorly directed setae; dorsal edge with the usual row of closely set, long, medially directed setae, subtended by a single enlarged bristle-like seta; convex edge of claspette filament with two parallel ridges or flanges separated by a V-shaped groove, best seen end on, as when viewed dorsally in an undistorted preparation.

Larva. Head setae 5-C and 6-C single, rarely double; maxilla ovate, with apical tuft of long setae directed distally in line with maxilla; dorsolateral abdominal setae 1-IV and 1-V usually two-thirds as long as upper lateral setae 6-IV and 6-V and as long as dorsolateral seta 1-VII on segment VII; pecten teeth evenly spaced; comb scales numerous, 35–70, with long subapical spinules, thus appearing rounded apically; spinules on posterior margin of saddle smaller than those on surface of saddle, too small to be seen under a dissecting microscope; other characters as in Table 3.

A common, widespread species, apparently without any single distinctive character. Though similar to *A. implicatus* it may readily be distinguished from that species by its larger, less darkly pigmented head capsule and by the shape of the maxilla. Also, *A. implicatus* usually has fewer than 30 comb scales. *A. stimulans* has the dorsolateral setae (1-IV and 1-V) shorter, less than two-thirds length of upper lateral setae of the same segment, and has conspicuous spinules along the posterior edge of the saddle.

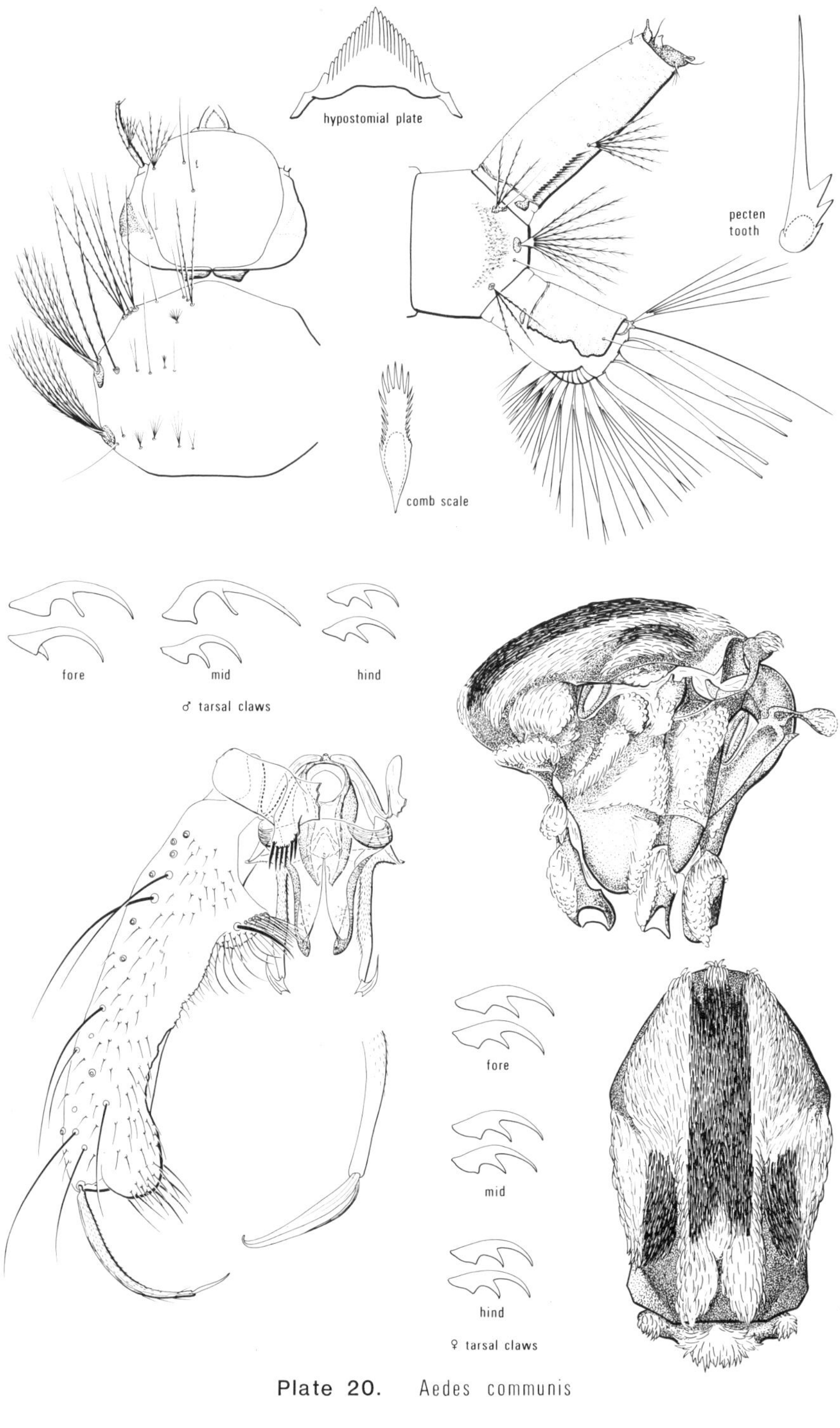

Plate 20. *Aedes communis*

Biology. One of our most abundant and widely distributed species, *communis* occurs throughout the forested part of Canada as well as northern USA and northern Eurasia. This species overwinters in the egg stage. The eggs hatch in early spring, usually before the snow has disappeared. In the Ottawa area, larvae develop in large numbers in deciduous forest pools in which there is a high tannic acid content. They have a characteristic habit of aggregating in dense masses, probably in response to certain combinations of light and temperature conditions. Larvae of *intrudens* and *diantaeus* are often associated with *communis,* but in much smaller numbers. In coniferous woodland, larvae of *punctor* often outnumber *communis.* In other regions *communis* larvae have been taken in association with a wide variety of other *Aedes* (Horsfall 1955, Ellis and Brust 1973).

Unlike *churchillensis,* females of *communis* are obligatorily anautogenous, that is, they require a blood meal to develop a batch of eggs. Ellis and Brust (1973) reviewed a few cases where some females have shown a tendency to facultative autogeny, that is, they have matured a small number of eggs without a blood meal, as have been described for *impiger* and *nigripes* by Corbet (1967). The eggs are laid from early summer to midsummer on the underside of damp leaves on the dry bed of woodland pools (Wesenberg-Lund 1920–21). In Denmark, eggs hatch anytime during the winter during a brief warm period, but larval development is greatly delayed by low temperatures. Ultimately all the larvae reach maturity at about the same time.

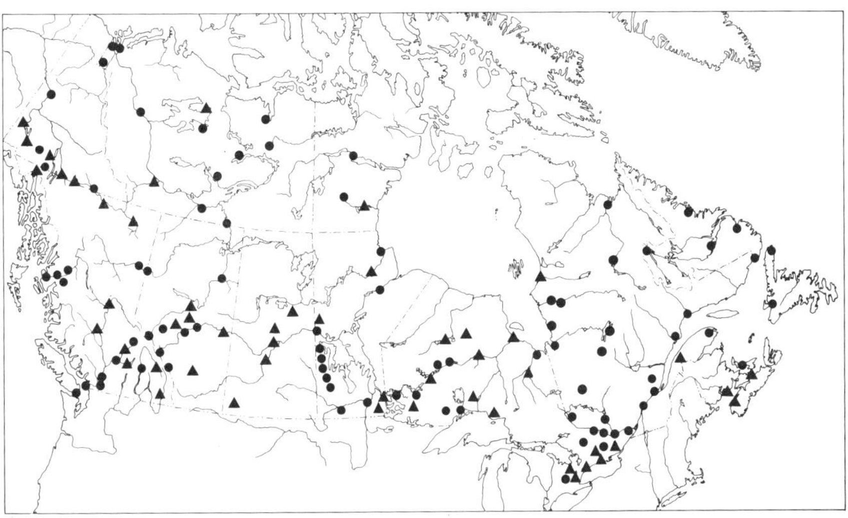

Map 20. Collection localities for *Aedes communis* in Canada: • specimens we examined, ▲ literature records.

Distribution. North America and Eurasia: Alaska and California east to Newfoundland and New Jersey; Europe; Siberia, eastern USSR, north to the Arctic Ocean, south to Bulgaria, northern Caucasus, and eastern Kazakhstan.

Aedes decticus Howard, Dyar, & Knab

Plate 21; Figs. 112, 137, 176; Map 21

Aedes decticus Howard, Dyar, and Knab, 1917:737.
Aedes pseudodiantaeus Smith, 1952:21.

Adult. Female: Integument medium to dark brown; proboscis, palpus, tibiae, tarsi, and wing veins entirely dark-scaled; pedicel usually without scales, two or three dark ones may be present medially, recumbent scales of vertex pale except for a pair of large dark spots; erect forked scales entirely dark; scutum with a pair of broad submedian dark-scaled stripes, separated by a narrow yellowish brown scaled median stripe (about half as wide as submedian stripe); postsutural sublateral stripe brown-scaled; presutural sublateral area typically yellowish brown scaled, occasionally with a triangular dark-scaled spot; lateral stripe yellowish brown scaled, concolorous with sublateral area and median stripe; postprocoxal membrane bare; katepisternum with anterodorsal corner bare; mesepimeron bare on lower third; mesepimeral bristles absent; abdominal tergites dark-scaled dorsally, each with lateral triangular spot of white scales, but lacking transverse basal white bands; tarsal claws small, more strongly bent beyond subbasal tooth than those of *diantaeus*.

Male: Palpus nearly as long as proboscis, entirely dark-scaled; third palpomere lacking setae; fourth and fifth palpomeres each with a few scattered short setae only; coloration and arrangement of scales as in

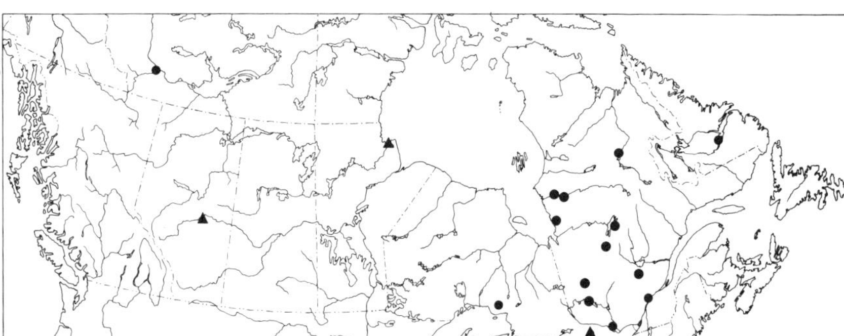

Map 21. Collection localities for *Aedes decticus* in Canada: • specimens we examined, ▲ literature records.

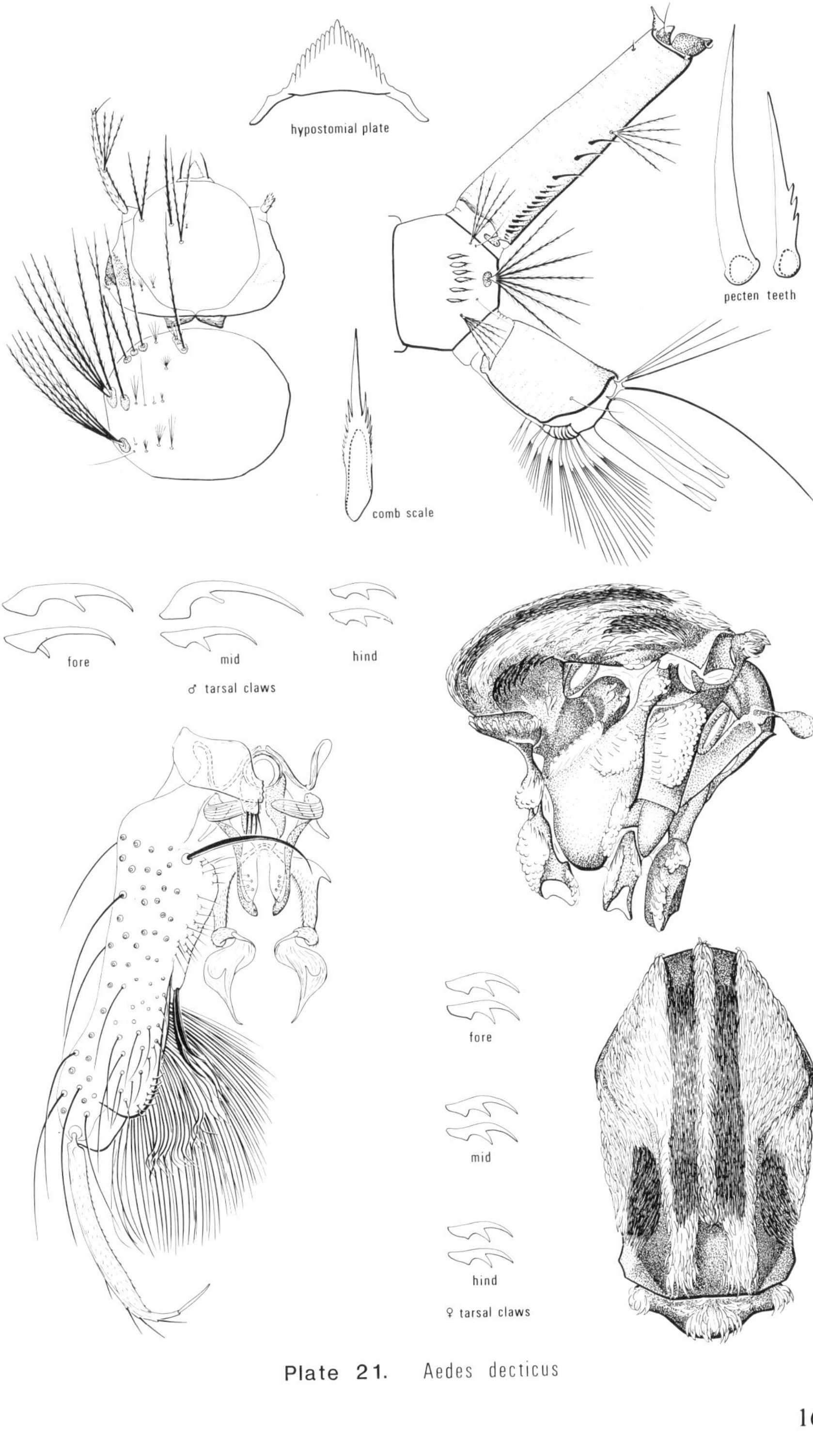

Plate 21. *Aedes decticus*

female; basal lobe of gonocoxite bottle-shaped, with the neck directed posteriorly bearing basally one long medially directed bristle and from apex of neck two long posteriorly directed curved bristles; apical lobe of gonocoxite with a dense brush of posteromedially directed setae; median ventral edge of distal half of gonocoxite with even larger denser brush of longer posteromedially directed setae; apex of claspette stem spiraled 180° dorsomedially; claspette filament ligulate, continuing the same spiral another 360°.

Larva. Antenna long, but scarcely as long as head; head and body setae thick, stiff, tapering near apex rather than from base, and sparingly branched; head seta 5-C usually single; seta 6-C usually double; setae 5-C and 6-C arising close to one another; seta 4-C arising far back, level with 6-C; all main prothoracic setae (for example, 1-P, 5-P to 7-P, 10-P) unbranched; both upper and lower lateral setae (6-I, 6-II, 7-I, and 7-II) of first two abdominal segments all single; siphon rather narrow and parallel-sided, with prominent long, unevenly spaced distal pecten teeth; comb scales 5–7; saddle almost completely encircling anal segment, with its ventral edges closely approximated and occasionally joined by narrow bridge.

Biology. Larvae of this widely distributed but relatively rare species have been collected in pools in the central unshaded quaking mats in sphagnum bogs (Smith 1952, Maire and Aubin 1976), where they may be locally common. In the Ottawa area, a few *decticus* larvae were collected in a sphagnum bog associated with mature larvae of *punctor* and *abserratus* and numerous half-grown larvae of *canadensis* and *cinereus*.

Males were observed hovering singly near females, but they were not observed swarming (Smith 1952).

Distribution. North America, from Alaska and Michigan east to Newfoundland (Labrador) and Massachusetts. Dyar (1919*b*) recorded several collections of this species from Ontario, Manitoba, and Alberta, but the specimens are not now found under this species in the U.S. National Museum. We assume they were misidentified.

Aedes diantaeus Howard, Dyar, & Knab

Plate 22; Figs. 110, 111, 133, 136, 174; Map 22

Aedes diantaeus Howard, Dyar, and Knab, 1913:pl. 24, fig. 167 (desc. 1917:758).

Adult. Female: Integument medium brown; proboscis, palpus, tibiae, tarsi, and wing veins entirely dark-scaled; pedicel with at most two or three dark scales medially; vertex entirely yellow-scaled; scutum with prominent pair of dark-scaled submedian stripes, somewhat narrower than those of *decticus*, separated by narrow yellow-scaled median stripe; sublateral and lateral areas yellow-scaled; postsutural sublateral dark stripe poorly defined or absent; postprocoxal membrane, anterodorsal corner of katepisternum, and lower third of mesepimeron without scales; mesepimeral setae absent; tarsal claw longer and straighter than that of *decticus*, evenly curved beyond subbasal tooth; abdominal tergites dark-scaled, each with small lateral patch of white scales.

Male: Palpus slightly longer than proboscis; third palpomere with a few long setae at apex forming a greatly reduced, inconspicuous tuft; fourth palpomere fringed laterally and medially with longer setae that are widely enough spaced that they could not be considered tufted; medioventral surface of gonocoxite with a dense tuft of long dorsally directed setae; basal lobe of gonocoxite flask-shaped, with one medially directed bristle at base and two posteriorly directed curved bristles at apex; claspette stem at mid length with small medially directed seta-bearing side branch; claspette filament semicircular, with curved concentrically arranged translucent "windows."

Larva. Antenna exceptionally long and straight, longer than length of head, pale yellowish brown with darker apex; bases of head setae 5-C and 6-C quite close together; head seta 4-C arising posterior to 5-C; upper lateral abdominal setae 6-I and 6-II usually single; lower lateral seta 7-I also single and as long as 6-I; lower lateral 7-II minute and multiple; other characters as in Table 3.

No other Canadian species has such a long straight antenna nor head seta 4-C arising posterior to 5-C.

Biology. In the Canadian shield area north of Ottawa, the larvae of *A. diantaeus* usually develop in temporary spring pools in hardwood forest

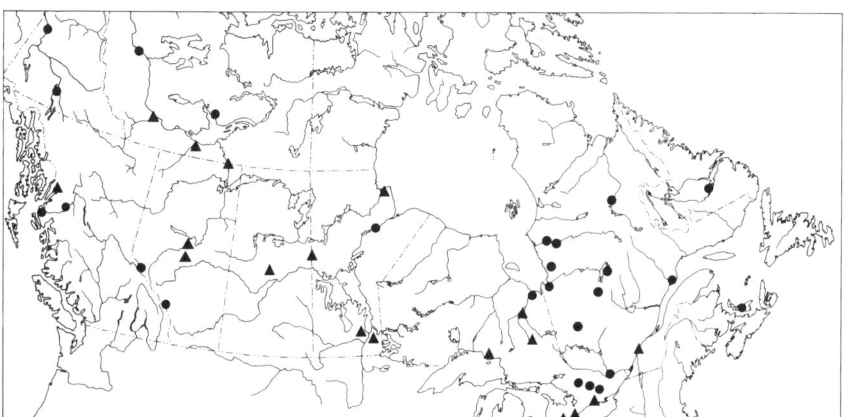

Map 22. Collection localities for *Aedes diantaeus* in Canada: • specimens we examined, ▲ literature records.

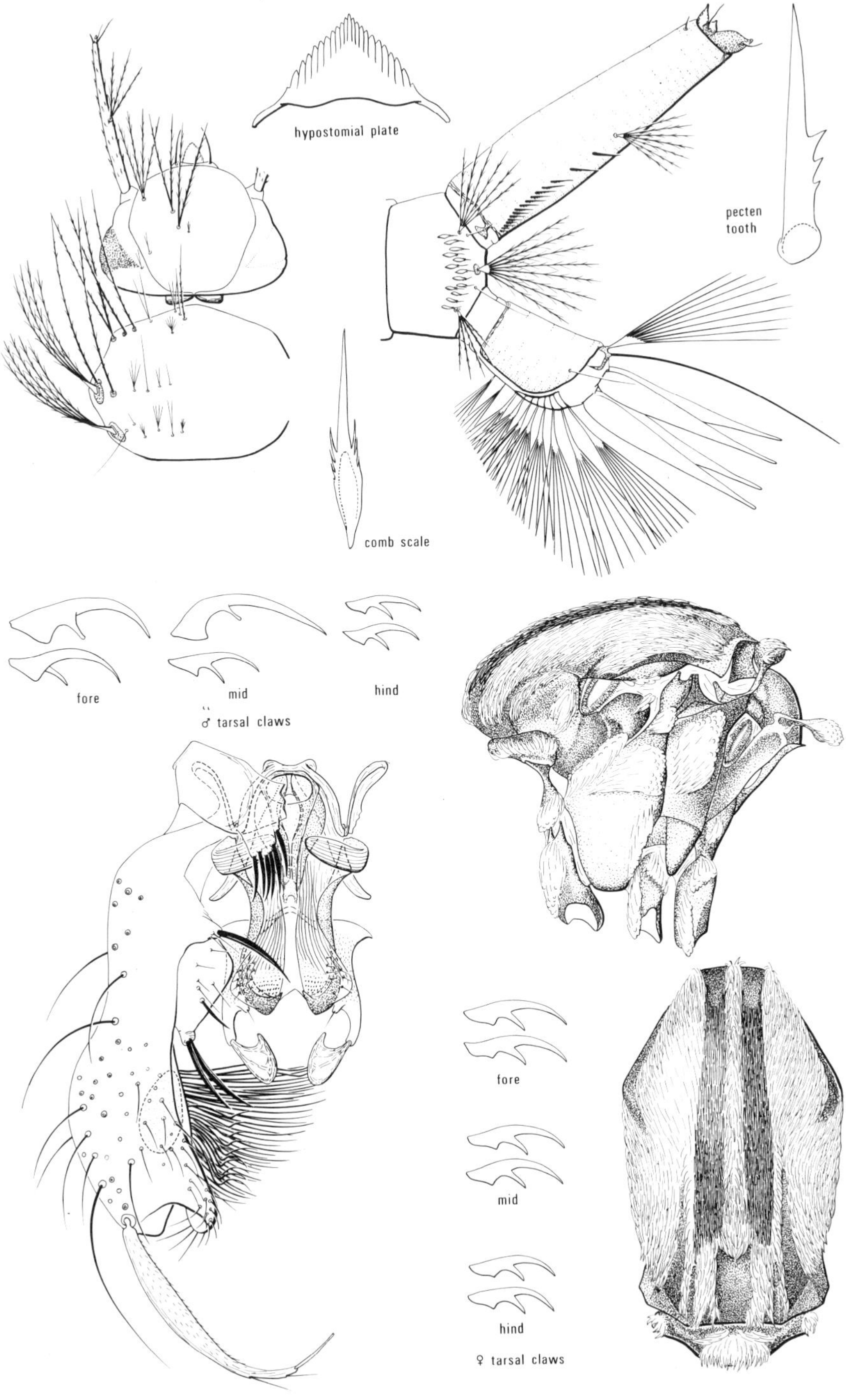

Plate 22. Aedes diantaeus

associated with, but greatly outnumbered by, larvae of *A. communis*. The water is usually stained dark brown by the abundant leaf litter. Instead of actively grazing on the bottom debris, however, as *communis* larvae do, this species adopts a peculiar feeding stance, resting vertically at the bottom apparently supported only by the tips of the antennae. In this position the labral brushes remain close to, but not in contact with, the substrate, and it is quite possible that this species gathers stray particles stirred up by *communis* larvae. We have found that perhaps as a result of this behavior, larvae of *diantaeus* that have been kept isolated from other species are more difficult to rear by standard methods.

Dyar (1922*b*), Jenkins (1948), Frohne (1957), and Price (1963) collected larvae of *diantaeus* in boggy situations. In Northern Canada most mosquito habitats are of this type, and the hardwood forest pools of the type described previously probably do not exist except in the south. Brust (personal communication) found as many larvae of *diantaeus* as of *communis* at Thompson, Man.

The males of *diantaeus* do not swarm, but are attracted to the host (as are *sierrensis*), where they intercept the female as she approaches to take a blood meal (Dyar 1920*a*). They also mate readily in small cages (Brust 1971).

Distribution. North America and Eurasia: Alaska and British Columbia east to central Quebec, Nova Scotia, and Massachusetts; Norway east to Khabarovsk Territory, eastern USSR, south to southern Ukraine and northern Caucasus.

Aedes dorsalis (Meigen)

Plate 23; Figs. 69, 76, 113, 115; Map 23

Culex dorsalis Meigen, 1830:242.
Culex curriei Coquillett, 1901:259.
Culex onondagensis Felt, 1904:304.
Culex lativittatus Coquillett, 1906:109.
Aedes quaylei Dyar and Knab, 1906*c*:202.

Adult. Female: Scutum quite variable, with middorsal reddish brown stripe either broad or narrow, sometimes having narrow streaks of whitish scales within it (as in *melanimon*); scales of lateral stripe usually concolorous with sublateral area (rather than with postpronotum as in *campestris*); scales of sublateral area chalky white to yellowish; veins R_{4+5}, M_2, M_3, M_4, and CuA_2 usually with more brown than pale scales; all other characters similar to *campestris* from which it can be distinguished with certainty only by the tarsal claw, which in *dorsalis* is straighter, evenly curved rather than bent, and with a small subbasal tooth whose length is only one-third the distance between the apices of the subbasal tooth and the claw.

Male: Hind tarsal claw rather straight, as in female, with small subbasal tooth; basal lobe of gonocoxite prominent, constricted slightly at base, with two heavily sclerotized thickened setae, one long, one short, in addition to finer setae; the longer of the two thickened setae arises near middle of lobe; the shorter of the two arises at its proximolateral corner; apical lobe of gonocoxite low, with several prominent setae; claspette filament sickle-shaped; coloration and arrangement of scales as in female.

Larva. Head seta 5-C single, rarely double on one side; seta 6-C single; seta 7-C with eight or more branches; mesothoracic seta 1-M as long as 5-C, double or triple; dorsolateral abdominal setae 1-III to 1-VII and ventrolateral setae 13-III to 13-V moderately well developed, half to two-thirds as long as upper lateral setae (6-III to 6-VI) of their respective segments; siphon shorter and wider than in *campestris*, less than three times as long as wide; pecten teeth evenly spaced, extending to mid length of siphon; saddle reduced, covering only dorsal half of anal segment; anal papillae scarcely as long as saddle; other characters as in Table 3.

The species *campestris*, *dorsalis*, and *schizopinax* are the only Canadian species with a branched mesothoracic seta 1-M; *dorsalis* usually has head seta 5-C single, rarely double on one side, whereas in *campestris* and *schizopinax* this seta is usually double or triple. The dorsolateral and ventrolateral setae are longer in *dorsalis* and *schizopinax*.

Remarks. The middorsal stripe on the scutum of *dorsalis* may be narrow, broad, or divided, called eastern, *lativittatus*, and *curriei*, respectively, though without formal nomenclatural standing, to describe these three most obvious variations (Mail 1934, McLintock 1944). Kalpage and Brust (1968) also observed two different nonoverlapping sizes of *dorsalis* eggs in southern Manitoba. Each female produced only one of these types. These differences suggest the possible presence of more than one species. This species needs further taxonomic study.

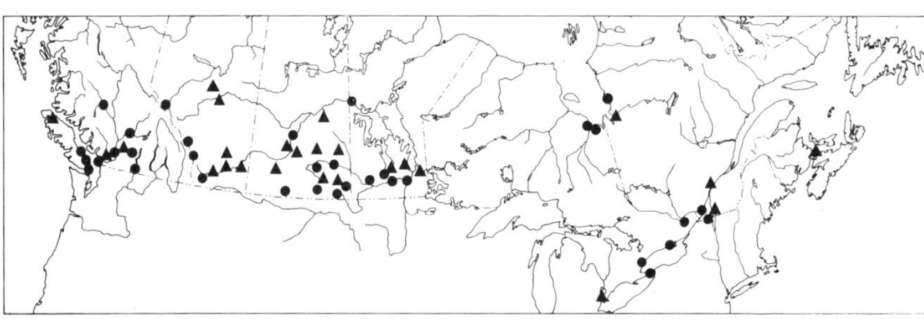

Map 23. Collection localities for *Aedes dorsalis* in Canada: • specimens we examined, ▲ literature records.

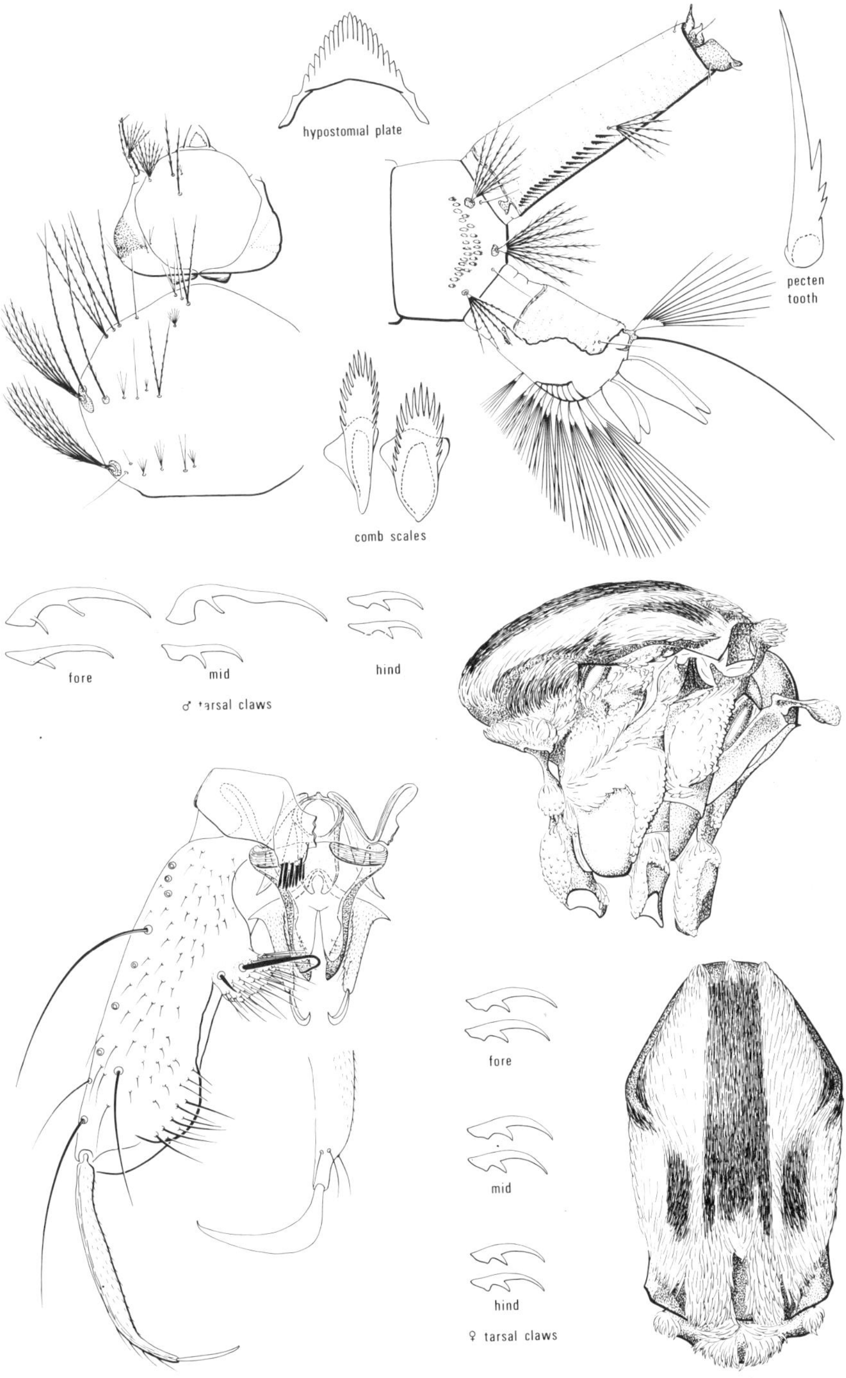

Plate 23. Aedes dorsalis

Biology. This species overwinters in the egg stage. Larvae of *dorsalis* appear in early spring in open roadside pools and shallow ponds on prairie and farmland throughout most of southern Canada. They are scarce or absent in forested country (Graham 1969*b*). They have a predilection for alkaline or saline habitats in some regions, and the length of the anal papillae varies inversely with the quantity of dissolved salts found in the water (Rempel 1950). The species has not been collected in salt marshes in the Atlantic Provinces, but occurs commonly in similar habitats of high salinity in southern British Columbia (Gibson 1932, 1935, 1936) and has even found its way across hundreds of kilometres of forest to coastal James Bay. The use of salt on highways in Eastern Canada during the winter may favor subsequent oviposition in roadside ditches.

When conditions are favorable, *dorsalis* may have one or more summer generations (as in *vexans* and *spencerii*) and the population may then build up to very large numbers. Khelevin (1958) has shown that diapause in the egg is facultative. Eggs laid by first-generation females did not go into diapause even when chilled, whereas those laid by second or third generations did go into diapause at temperatures below 18°C. Above 18°C, the percentage of diapausing eggs decreased with rising temperature and was inhibited by temperatures above 30°C. Eggs may remain viable in a dried condition "for years" (Rempel 1950). Klassen and Hocking (1964) observed considerable annual variation in numbers of *dorsalis*. Thus reports of its importance vary from low (Knab 1908) to one of the most abundant in the Prairie Provinces (Cameron 1918, McLintock and Rempel 1963). Its biology is obviously complex, further complicated by the possibility of sibling species.

Viruses of the California group have recently been isolated from larvae of *dorsalis* in Utah (Crane et al. 1977).

Distribution. North America and Eurasia: British Columbia and California east to Delaware and Louisiana; western Europe, east across the steppes of Ukraine, northern Caucasus, and mountains of central Asia to eastern USSR.

Aedes euedes Howard, Dyar, & Knab

Plates 8, 24; Figs. 91, 126, 168; Map 24

Aedes euedes Howard, Dyar, and Knab, 1913:pl. 28, fig. 191 (desc. 1917:714).

Aedes barri Rueger, 1958:34.

Adult. Female: Integument reddish brown; proboscis predominantly dark-scaled basally and apically, but the middle third extensively white-scaled, sometimes predominantly white-scaled ventrally, occasionally so concentrated as to form a distinct white ring; palpus mostly dark-scaled with

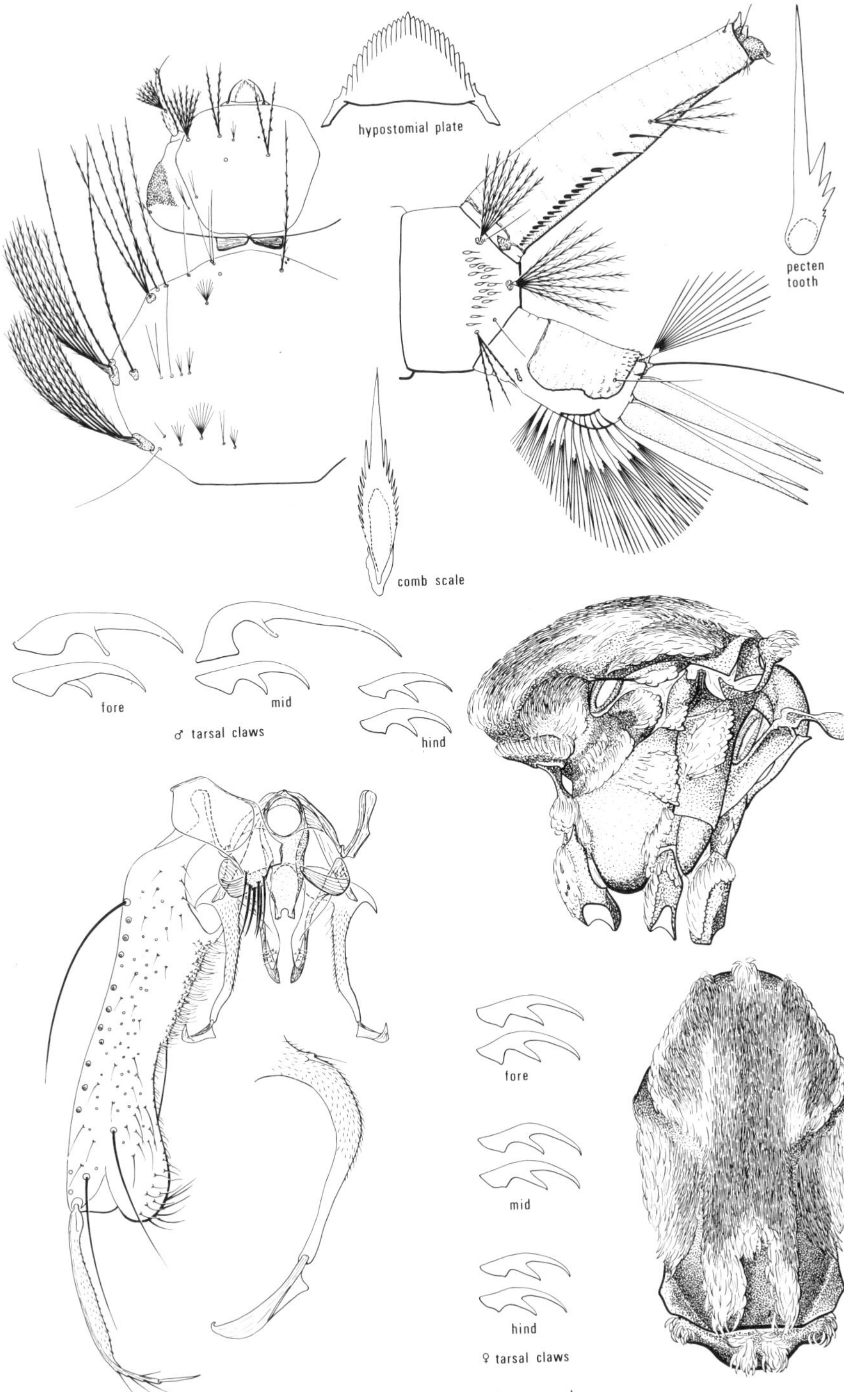

Plate 24. Aedes euedes

scattered white scales concentrated at extreme apex of palpus and at bases of second and third palpomeres; pedicel mostly white-scaled on anteromedial half; recumbent scales of vertex mixed pale and brown, usually with a pair of weakly defined dark areas; erect forked scales mostly dark; postprocoxal membrane with scales; antepronotum and fore coxa with scattered brown scales; postpronotum brown-scaled on upper two-thirds; scutum with broad middorsal stripe (median and submedian stripes combined) of reddish brown scales; sublateral area mostly brown-scaled, pale scales reduced to a dorsocentral and a lateral stripe; lateral stripe pale-scaled except for a small brown patch above paratergite; hypostigmal area, anterodorsal corner of katepisternum, and lower fourth of mesepimeron bare; lower mesepimeral setae absent; tibiae mostly white-scaled with scattered dark scales; first tarsomeres extensively white-scaled, with the basal band usually indistinct; second and third tarsomeres dark-scaled, each with a broad basal band of white scales; fourth midtarsomere and fourth and fifth hind tarsomeres with at least a lateral and a medial spot of white scales, sometimes with a complete basal band; tarsal claw rather strongly bent beyond subbasal tooth, with the subbasal tooth relatively long, enclosing an angle of 30–50°; wing veins mostly dark-scaled, with scattered yellowish scales throughout; abdominal tergites dark-scaled, each with a narrow basal band or median and a lateral patch of white scales, and numerous scattered white scales distally, often forming a narrow transverse distal band; cercus dark-scaled with scattered pale scales.

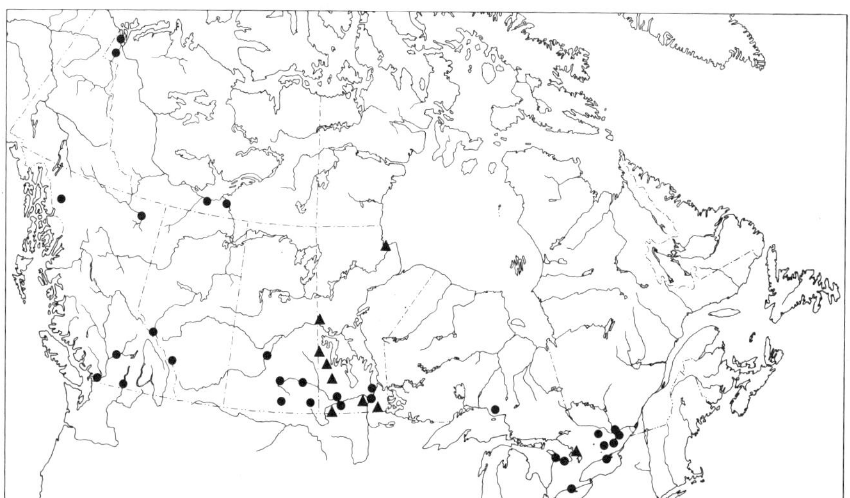

Map 24. Collection localities for *Aedes euedes* in Canada: • specimens we examined, ▲ literature records.

Male: Palpus extensively pale-scaled, longer than the proboscis by almost the length of the last palpomere; apex of third palpomere with conspicuous dense tuft of setae extending beyond base of last palpomere; fourth palpomere heavily fringed medially; coloration and scaling of thorax, legs, and wing and shape of hind tarsal claw as in female; basal lobe of gonocoxite not differentiated, without long medially directed seta; setose area extending from base of gonocoxite along mediodorsal edge almost to apical lobe; apical lobe prominent, extending beyond base of gonostylus; claspette filament with flange on convex side.

Larva. Body brown, usually darker than either *excrucians* or *flavescens*; head seta 5-C usually double; prothoracic seta 3-P long, strong, single; ventrolateral setae 13-III to 13-V of abdominal segments III to V as long and as thick as upper lateral setae 6-III to 6-V; dorsolateral abdominal setae not as well-developed, with setae 1-IV and I-V at most half as long as upper laterals of respective segment; comb scales dark brown, 11–19 (intermediate between *riparius* and *excrucians*); siphon tapering evenly from base to apex, not as abruptly narrowed as in *excrucians*; distal pecten teeth long, thick, rather widely spaced, and extending beyond middle of siphon; siphonal seta 1-S arising at two-thirds distance from base of siphon; anal papillae brown, usually more than one and one-half times longer than saddle; seta 9-S at apex of siphon not strongly hooked as in *excrucians*; other characters as in Table 3.

Mature specimens are among the largest of *Aedes* larvae, larger than *excrucians* and *riparius*, and exceeded only by *flavescens*. They are recognizable by the number of comb scales and by the widely spaced distal pecten teeth, the last of which arises beyond mid length of the siphon, with the siphonal seta 1-S arising beyond it. In life, the brown anal papillae are characteristic. Rempel's (1950) fig. 35*B*, p. 245, evidently refers to this species, not to *riparius*, as pointed out by Enfield (1977).

Biology. This species overwinters in the egg stage. It produces only one generation each year. In the Ottawa area, larvae have been collected in association with *excrucians*, *fitchii*, and *stimulans* in large open marshes where there are dense masses of decomposing sedges and cattails from previous years. It has always been the least common of the four species. Although widely distributed, it seems to be uncommon throughout Canada. Because it has only recently been recognized as a distinct species and also because of the difficulty of separating both sexes from related species, it remains little known.

Distribution. North America and Eurasia: Alaska and British Columbia east to New York, probably more widely distributed in the United States; Sweden; Finland; European USSR (Kursk, Yaroslavl, and Moscow regions); and Siberia east to the Sea of Okhotsk.

Aedes excrucians (Walker)

Plates 8, 25; Figs. 72, 85, 125, 164, 166; Map 25

Culex excrucians Walker, 1856:429.
Culex abfitchii Felt, 1904:381.
Culex siphonalis Grossbeck, 1904*a*:332.
Aedes sansoni Dyar and Knab, 1909*a*:102.

Adult. Female: Integument reddish brown, but dark brown in northern specimens; proboscis mostly dark-scaled, less extensively white-scaled than that of *euedes*; palpus dark-scaled with narrow pale-scaled ring at base of each of second and third palpomere; pedicel mostly white-scaled; postprocoxal membrane with scales; hypostigmal area, anterodorsal corner of katepisternum, and lower third of mesepimeron without scales; lower mesepimeral setae usually absent, rarely one or two; postpronotum brown-scaled on upper two-thirds; scutum predominantly reddish brown scaled, with the sublateral area usually extensively so and whitish scales restricted to an interrupted dorsocentral stripe (usually appearing as a presutural and postsutural oval spot or streak), lateral stripe, and prescutellar depression; a small patch of brown scales above paratergite; fore coxa with a few dark scales mixed among white scales; tibiae predominantly white-scaled; first tarsomeres mostly dark-scaled, with the basal white-scaled ring usually distinct except on foreleg; remaining tarsomeres dark-scaled, each with broad basal ring of white scales, except last fore tarsomere and occasionally also last mid tarsomere; tarsal claw strongly and abruptly bent beyond subbasal tooth, with the latter usually subparallel, enclosing an angle of less than 25°; claw beyond bend somewhat sinuous, with slightly hooked apex; wing veins with mixed pale and dark scales throughout; abdominal tergites dark-scaled, each with narrow transverse basal yellowish-scaled band or patch, and numerous scattered pale scales; cerci with a few scattered pale scales. Northern specimens from the vicinity of tree line are much darker; proboscis almost entirely dark-scaled; pleural scales strongly yellowish; scutal pattern darker and dark-scaled areas of abdominal terga with fewer scattered pale scales.

Male: Indistinguishable from *euedes* except that the hind claw is more strongly bent than that of *euedes*, subparallel to subbasal tooth, and the apical lobe of the gonocoxite is smaller and does not project beyond base of gonostylus.

Larva. Medium to light brown, yellowish, or greenish; head seta 5-C double or triple; prothoracic seta 1-P long, thick, single; ventrolateral setae (13-III to 13-V) on abdominal segments III to V and dorsolateral setae (1-IV and 1-V) on segments IV and V each as long and as thick as upper lateral setae of their respective segment; dorsolateral seta on segment VII usually double, but almost as long as upper lateral; comb scales brown,

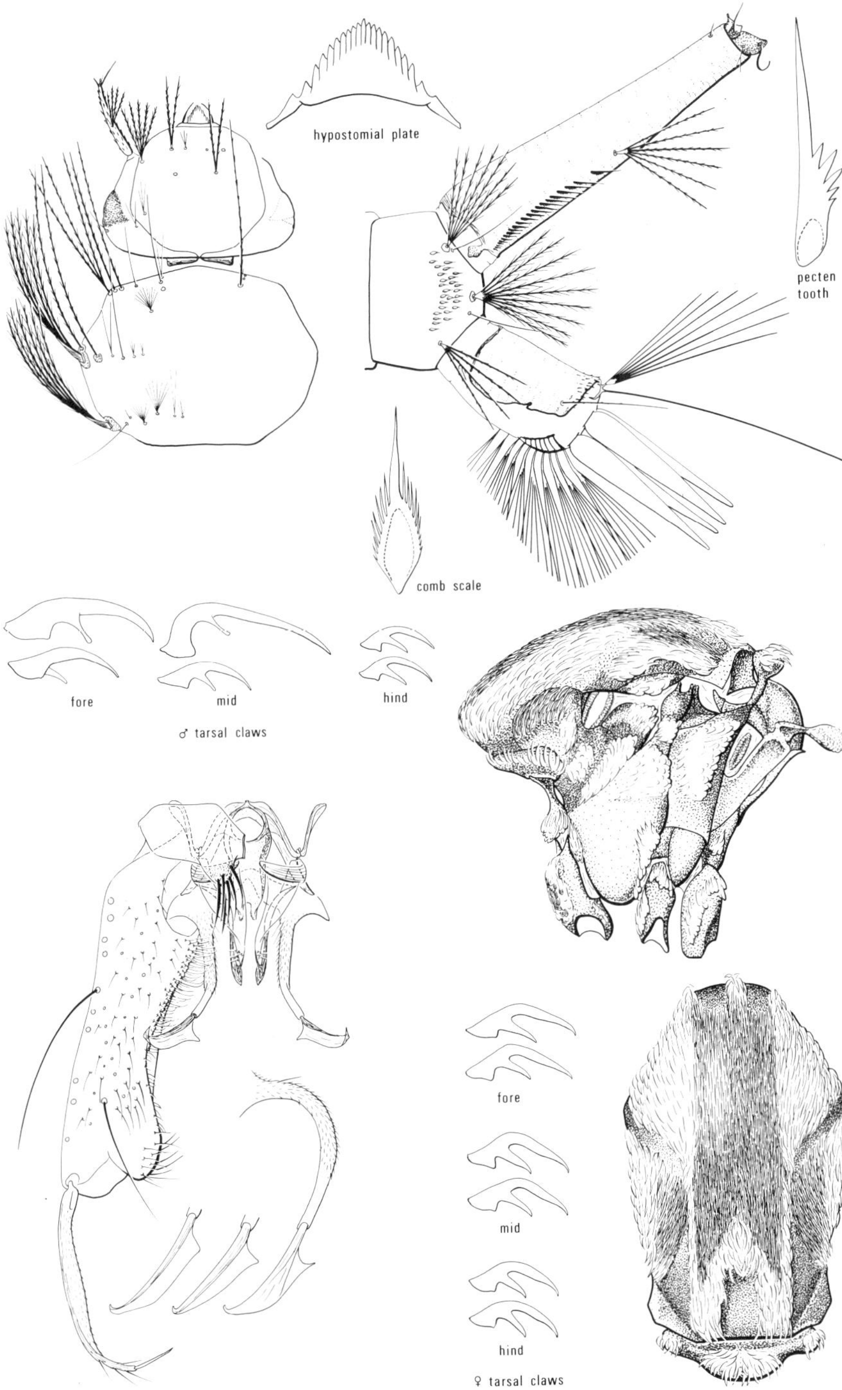

Plate 25. *Aedes excrucians*

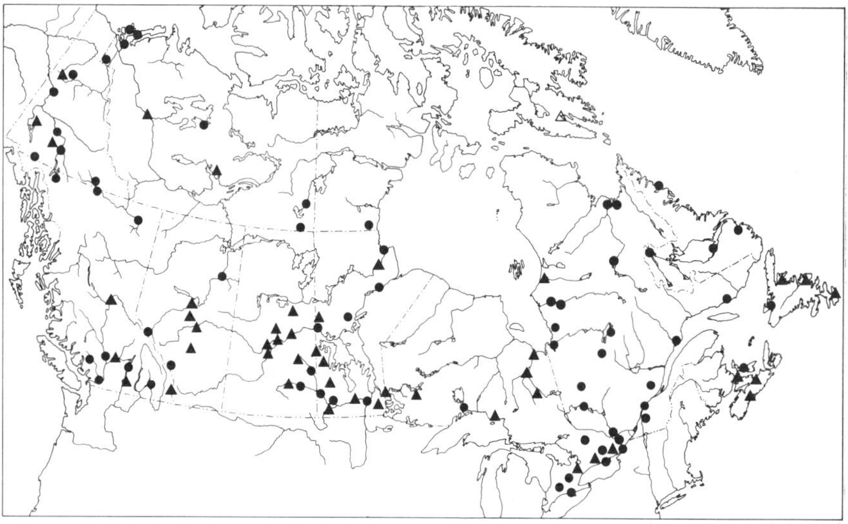

Map 25. Collection localities for *Aedes excrucians* in Canada: • specimens we examined, ▲ literature records.

17–33, in two or three irregular rows; siphon long and strongly tapered to narrow apex, five times as long as basal width and more than seven times as long as apical width; distal pecten teeth more widely spaced, but last tooth not arising beyond middle of siphon; seta 9-S on apical flap of siphon thickened and strongly hooked; anal papillae pale yellow, about one and one-half times as long as saddle; remaining characters as in Table 3.

This species and *A. fitchii* are recognizable by the long strongly tapered siphon and the well-developed strongly hooked apical siphonal seta 9-S. *A. fitchii* has an elongate mesothoracic seta 1-M, well-developed dorsolateral abdominal setae on segments III to VII inclusive, and pecten teeth evenly spaced. In southwestern British Columbia, *A. aloponotum* has a fur-like body covering of microtrichiae, which is lacking in larvae of *excrucians* and *fitchii*.

Biology. Distributed over the entire country south of the arctic tundra, *excrucians* is found in a wide range of habitats almost everywhere, but fortunately not in overwhelming numbers. It also varies considerably geographically; northern specimens are so much larger and darker than southern specimens that they appear to be different species. Similar though much less pronounced variation in northern and southern specimens is also characteristic of *euedes* and *fitchii*.

Overwintering takes place in the egg stage. The species produces only one generation each year. In the Ottawa area larvae have been found in almost every ground pool that has supported larvae, but the most common collection site was near the edges of larger semipermanent marshes and swamps. The presence or absence of shrubs and trees seems unimportant; larvae have been collected in densely forested as well as open marshy habitats, but their numbers are greater in the latter. As a direct consequence of this plasticity, they have been collected in association with nearly every other spring species. Because the ovipositing adult determines the larval habitat, we can assume that females have little or no specificity in their choice of an oviposition site. Larvae of *excrucians* and those of *euedes* and *flavescens* are the largest *Aedes* larvae in Canada and they generally do not reach maturity until after such species as *communis, intrudens, punctor,* and even *fitchii* and *stimulans* have pupated. Frohne (1953*a*) collected some *excrucians* larvae in an Alaskan coastal salt marsh in which the length of the anal papillae varied considerably.

The swarming behavior of the males of this species is unusual: they form aggregations, disperse a few minutes later, and then reform again nearby, 6–10 m away (Dyar 1919*b*, Frohne and Frohne 1954). Most swarming activity was concentrated about 3–6 m above ground in a woodland clearing and took place after sunset for about half an hour. Their flight was erratic and rapid. Frohne (1959) speculated that such rapid and unpredictable behavior might make the males of *excrucians* less subject to attack by predaceous Empididae than those of other species that remained fixed above a swarm marker.

Distribution. North America and Eurasia: Alaska and Oregon east to Newfoundland and New Jersey; Europe; Kazakhastan, Siberia, eastern USSR including Mongolia, northwest China, and Japan.

Aedes fitchii (Felt & Young)

Plates 9, 26; Figs. 16, 92, 119, 182; Map 26

Culex fitchii Felt and Young, 1904:312.

Aedes palustris Dyar, 1916:89.

Aedes palustris var. *pricei* Dyar, 1917*a*:16.

Aedes mimesis Dyar, 1917*c*:116.

Adult. Female: Integument reddish brown to medium brown; proboscis and palpus almost entirely dark-scaled; pedicel mostly white-scaled on both medial and dorsal surfaces; both erect and recumbent scales of vertex pale medially, dark laterally; antepronotum mostly pale-scaled; postpronotum reddish brown scaled on upper two-thirds; scutum with middorsal longitudinal stripe of reddish brown scales (median and submedian bands together) narrower than in *euedes* or *excrucians*; sublateral area usually entirely whitish-scaled, occasionally with a patch of reddish brown scales enclosed by whitish scales; lateral stripe whitish-scaled except for

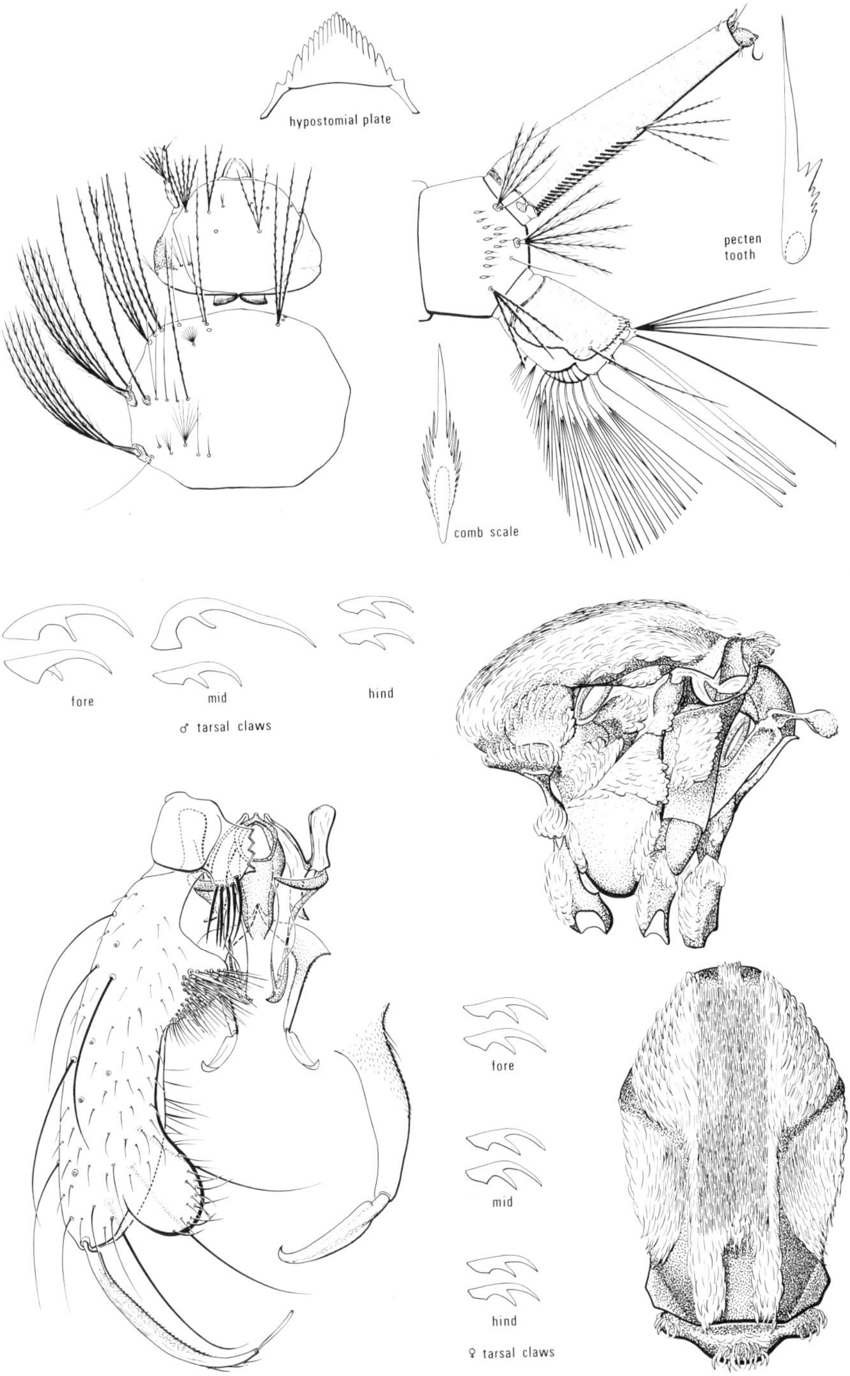

Plate 26. Aedes fitchii

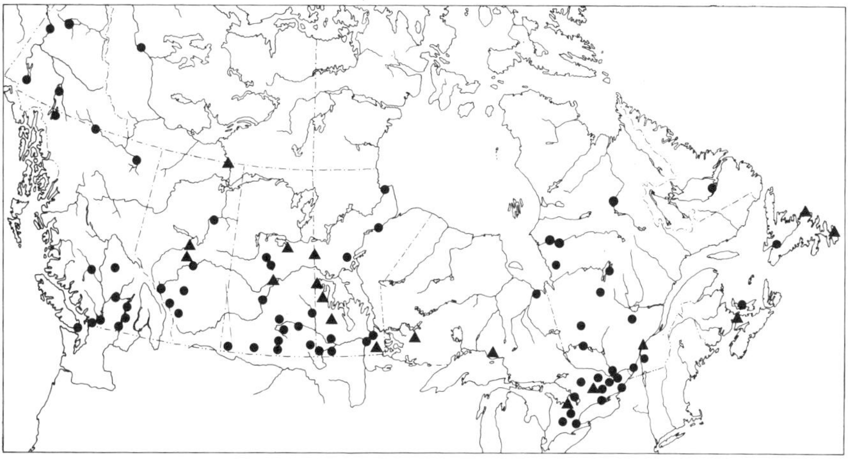

Map 26. Collection localities for *Aedes fitchii* in Canada: • specimens we examined, ▲ literature records.

small patch of brown scales above paratergite; postprocoxal membrane with scales; hypostigmal area, anterodorsal corner of katepisternum, and lower third to half of mesepimeron bare; mesepimeral bristles 1–3 or absent; mesomeron usually with a few scales in posterodorsal corner; fore coxa with brown scales mixed among white scales; tibiae with pale and dark scales; first tarsomeres predominantly dark-scaled except for basal white ring; second and third tarsomeres usually each with complete basal white ring; fourth and fifth tarsomeres entirely dark-scaled on fore- and mid-legs, with only the last tarsomere of hind leg entirely dark; wing veins usually with pale and dark scales, rarely with only dark scales. Abdominal tergites dark-scaled, each with transverse basal white-scaled band and few, if any, scattered white scales among the black.

Male: Palpus longer than proboscis by slightly more than length of last palpomere; second to fourth palpomeres each pale-scaled basally; all setae at apex of third and along lateral and medioventral sides of fourth palpomere dark, long, and dense; coloration and scaling of thorax, legs, and wings same as in female; basal lobe of gonocoxite large, conical, with relatively large clump of medially directed setae, subtended by a still larger, longer seta that is not always conspicuous because it is not always more strongly sclerotized; apical lobe large, rounded, bent medially; claspette filament flanged on both convex and concave sides, with the flanges not quite extending to base of filament.

Larva. Pale yellowish brown or greenish yellow; head seta 5-C usually three- or four-branched; prothoracic seta 3-P double or triple; mesothoracic seta 1-M exceptionally long, even longer than 3-P; dorsolateral abdominal setae 1-III to 1-VII and ventrolateral setae 13-III to 13-VII on abdominal segments III to VII as long and as thick as upper laterals of their respective segment; comb scales brown, in two or three irregular rows; siphon exceptionally long and slender, with strongly tapered apex, five times as long as basal width and seven to eight times as long as apical width; pecten teeth evenly spaced, increasingly longer distally, with the last tooth longer than apical diameter of siphon; seta 9-S at apex of siphon thickened and strongly hooked; anal papillae pale yellow, twice as long as saddle; other characters as in Table 3.

No other species has long dorsolateral abdominal setae on segments III to VII, an exceptionally long mesothoracic seta 1-M, and a long slender siphon five or more times as long as wide.

Remarks. Although the larvae of *fitchii* are easily recognized, we suspect that females have been widely and persistently misidentified. Many, perhaps most, records of *stimulans* females from Western and Northern Canada probably refer to *fitchii*, whereas the rest are probably *mercurator*. Some records of *fitchii* undoubtedly refer to *euedes*. In *fitchii* the lower mesepimeral setae may be present or absent, and we have searched in vain for a character that is peculiar to this species alone. However, our reared *fitchii* all have an ample cluster of white scales on the pedicel, and almost none on the proboscis, whereas *stimulans* has the reverse. Females of *mercurator* are most likely to be mistaken for those of *fitchii*, but they are darker, with entirely black-scaled wing veins and other characters given in the key.

Barr (1958) has shown that the number of comb scales on the larva tends to be bimodal with peaks at about 14 and 28.

Biology. Like *Aedes excrucians, fitchii* is distributed throughout Canada south of the tree line and is present nearly everywhere, though not in large numbers. Surprisingly, it does not occur in Eurasia and is one of the few northern species that is not holarctic. Its life cycle is also similar to that of *excrucians*, and larvae of the two species are usually collected together in association with many other species. Rempel (1953) found it the most abundant species in the aspen parkland of south central Saskatchewan and in the Cypress Hills of southwestern Saskatchewan. It is less common on the open prairie (Rempel 1950, McLintock and Rempel 1963).

Male swarming has been described as similar to, but less "wild and flighty" than, that of *excrucians* (Dyar 1919*b*). The swarms were loosely organized, lower, within 1 m of the ground, and 2.5 m or more in diam (Knab 1908).

Distribution. North American and eastern Asia: Alaska and California east to Newfoundland and New Jersey; Magadan region in eastern Siberia.

Aedes flavescens (Müller)

Plate 27; Figs. 79, 124, 169, 191; Map 27

Culex flavescens Müller, 1764:87.
Culex fletcheri Coquillett, 1902*c*:84.

Adult. Female: Integument reddish brown, darker on scutum; proboscis and palpus mostly dark-scaled with scattered yellowish scales throughout; vertex entirely yellow-scaled; ante- and post-pronotum yellow-scaled; scutum with broad middorsal stripe of yellowish brown scales flanked by sublateral and lateral areas of yellow scales; postprocoxal membrane, all of hypostigmal and subspiracular areas, anterodorsal corner of katepisternum, all of metepimeron, and all of metameron with pale yellowish white scales; mesepimeral setae absent; fore coxa with mixed whitish and brown scales; tibiae almost entirely yellowish-scaled except for median surfaces; basal tarsomeres mostly yellowish-scaled, except for dark-scaled apex; remaining tarsomeres (except last tarsomere on foreleg) each with broad basal ring of whitish scales; wing veins predominantly yellowish-scaled with scattered dark scales; abdominal tergites mostly yellow-scaled, with dark scales confined to lateral patches and occasionally a faint midlongitudinal band or paired spots or both.

Male: Palpus slightly longer than proboscis, predominantly yellow-scaled; apex of third palpomere with long pale brown and shorter yellow setae; fourth and fifth palpomeres fringed with dense long yellow setae; coloration and scaling as in female; gonocoxite with strongly enlarged medially directed bristle at base of basal lobe; basal lobe low and poorly defined, with setose area extending almost to base of apical lobe; apical lobe prominent, slightly constricted at base; claspette filament with cylindrical base, the keel on convex side confined to distal two-thirds.

Larva. Head seta 5-C with two to five branches; prothoracic seta 3-P long, single; prothoracic setae 8-P usually double, as long as 7-P; seta 10-P more than two-thirds as long as 8-P; ventrolateral setae 13-III to 13-V of abdominal segments III to V as long as their respective lateral setae (6-III to 6-V); dorsolateral setae 1-IV and 1-VI short, setae 1-V and 1-VII much longer, more than two-thirds as long as setae 6-V and 6-VI; lower lateral seta 7-II of abdominal segment II multiple (five or more branches) and less than one-third as long as upper lateral 6-II; comb scales brown, 20–36; pecten teeth evenly tapering from base to apex; distal pecten teeth usually evenly spaced, if not, usually only last tooth more widely spaced (two or three teeth usually more widely spaced in *excrucians*, *euedes*, and *riparius*); apical seta 9-S of siphon small, not strongly hooked; precratal setae usually more than four; anal papillae colorless, usually shorter than saddle; other characters as in Table 3.

This is the largest *Aedes* larva in Canada. It varies in such important characters as branching of head setae and arrangement of pecten teeth. Unique characters are apparently lacking.

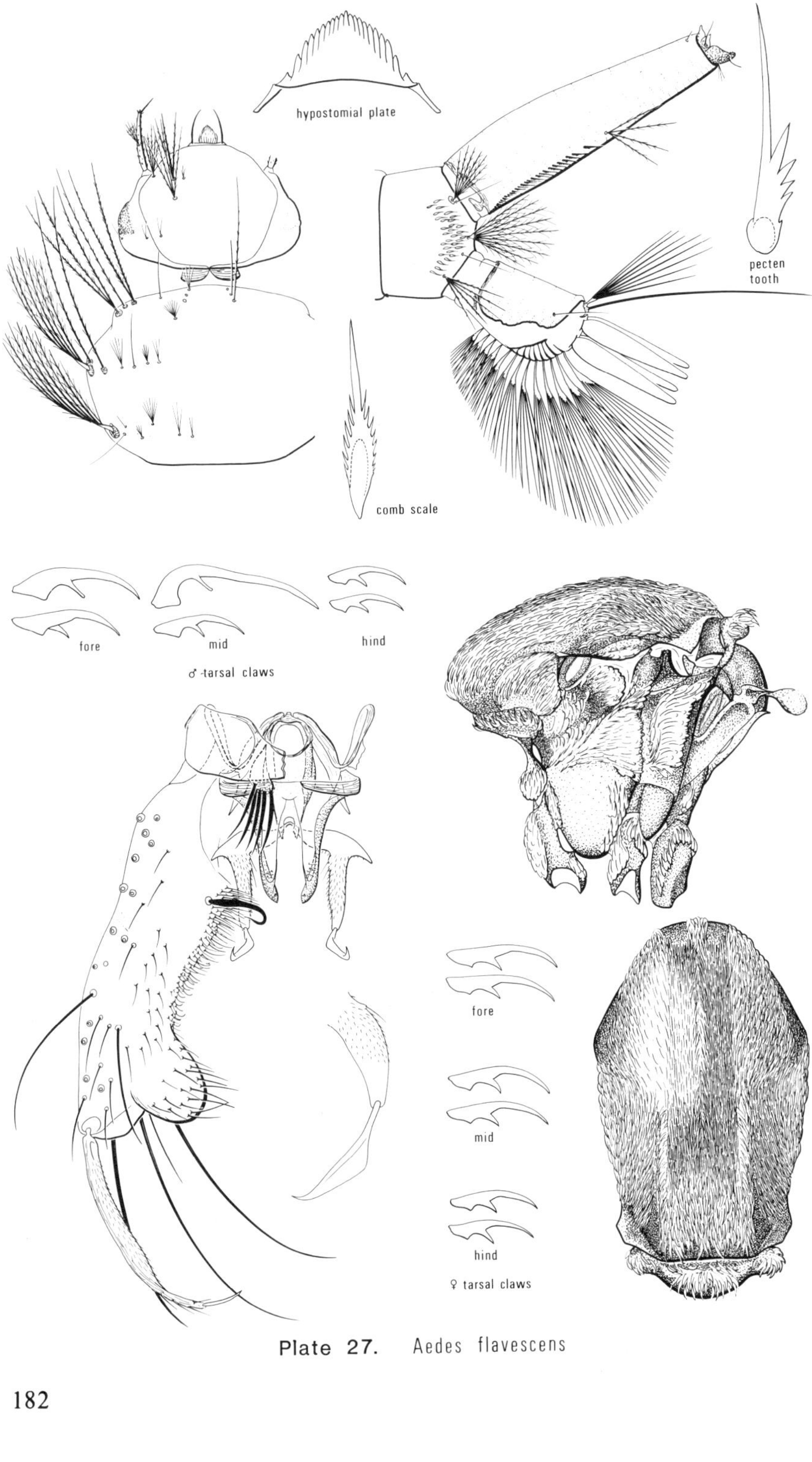

Plate 27. *Aedes flavescens*

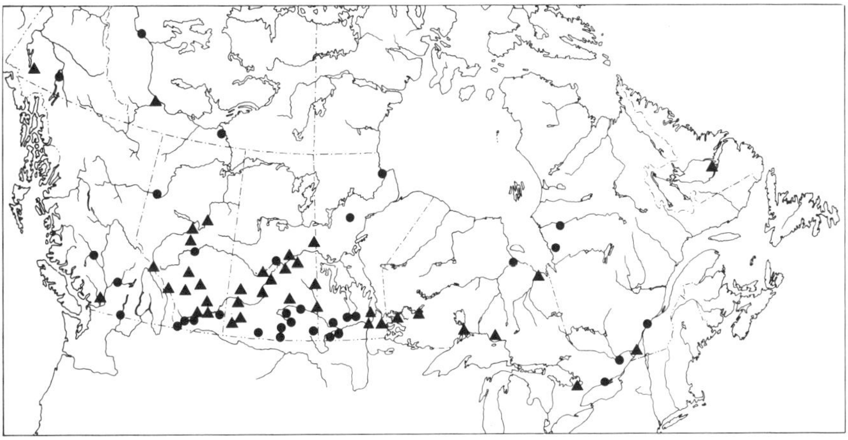

Map 27. Collection localities for *Aedes flavescens* in Canada: • specimens we examined, ▲ literature records.

Biology. Although this species, the largest *Aedes* in Canada, is widely distributed, it occurs sporadically over most of its range and is common only in the open prairies, where it breeds in large moderately deep overgrown and partly shaded semipermanent pools (Cameron 1918, Hearle 1929, Rempel 1953). Records from Eastern Canada and British Columbia (Hearle 1926) are few and scattered. Like all other members of the *excrucians* group, *flavescens* produces only one generation each year. It overwinters in the egg stage. It is one of the latest species to develop and as a result its numbers can be more affected by predators than earlier species (Hearle 1929). At Churchill, Man., it formed, with *campestris* and *excrucians*, the third and last emergence peak of the summer (Hocking et al. 1950). Hocking et al. (1950) found *flavescens* larvae in pools surrounded with birch–willow scrub not in true tundra pools. In Alaska, *flavescens* is confined to brackish coastal marshes (Frohne 1953*a*, Gjullin et al. 1961) in the area of Cook Inlet. The anal papillae of this population are small and rounded (Frohne 1953*a*), similar to those of other salt marsh species, such as *cantator* and *sollicitans*.

Although most *Aedes* can be induced to develop three or four egg batches each year, Carpenter and Nielsen (1965) obtained a maximum of only two from *flavescens*.

Male swarms were observed by Hearle (1929) over a "dry patch of typical open prairie with low and scanty vegetation" only about 1 m above ground. After coupling took place, the mated pair rose to 4.5–6 m in the air, then drifted down and flew away from the swarm. Wesenberg-Lund (1920–21) also observed swarming, near sunset, scarcely 60 cm above a patch of nettles. Earlier in the evening, just before swarming started, he was able to stimulate swarming and coupling by disturbing the adults that were resting among the nettles.

Distribution. North America and Eurasia: Alaska and California east to Newfoundland (Labrador) and New York; Europe; all of USSR, north to Karelia and Yakutsk, south to Transcaucasia, Mongolia, and northern China.

Aedes grossbecki Dyar & Knab

Plate 28; Map 28

Aedes grossbecki Dyar and Knab, 1906*c*:201.
Culex sylvicola Grossbeck, 1906:129.

Adult. Female: Integument dark brown; proboscis mostly dark-scaled, with a few scattered pale scales on basal half or less; palpus dark-scaled with scattered pale scales that are more densely aggregated at bases of second and third palpomeres; pedicel pale-scaled on anteromedial half; scales of vertex both erect and recumbent, pale medially, brown laterally with scattered pale scales; postprocoxal membrane with scales; antepronotum and fore coxa mostly pale-scaled with a few scattered dark scales; postpronotum dark-scaled except for a small patch of pale scales on lowest quarter; scutum with broad middorsal stripe of reddish to dark brown scales medially, changing to yellowish brown scales laterally, with a few scattered pale scales; presutural sublateral area pale-scaled except for a narrow longitudinal stripe of yellowish brown scales; hypostigmal area and lower quarter of mesepimeron bare; katepisternum with pale scales extending to anterodorsal corner; lower mesepimeral setae usually present; tibiae with mixed pale and dark scales; first tarsomeres with mixed pale and dark scales, but with the basal pale-scaled band usually distinct; remaining tarsomeres dark-scaled, each with a broad basal pale-scaled ring except for the last tarsomeres of fore- and mid-legs; tarsal claw evenly and strongly curved, with a rather short subbasal tooth, somewhat as in *fitchii* (Fig. 92); wing veins with large, broadly triangular scales, mostly dark, with pale scales scattered rather uniformly on all veins; abdominal tergites dark-scaled, each with a broad basal transverse band of pale scales and a few pale scales scattered among the dark ones; cercus dark-scaled.

Male: Palpus dark-scaled, with aggregations of pale scales at base of each palpomere; palpus longer than proboscis by almost the entire length of last palpomere; apex of third palpomere with conspicuous dense tuft of setae extending beyond base of last palpomere; fourth palpomere with long median fringe; middorsal longitudinal stripe of scutum narrower and paler than in female; wing scales as in female, large and broadly triangular, mixed pale and dark; basal lobe of gonocoxite conical with clump of medially directed setae accompanied by a longer, thicker, more strongly sclerotized seta; apical lobe of gonocoxite prominent, rounded, smaller than that of *fitchii*; claspette lobe relatively small, with keel-like ventral flange.

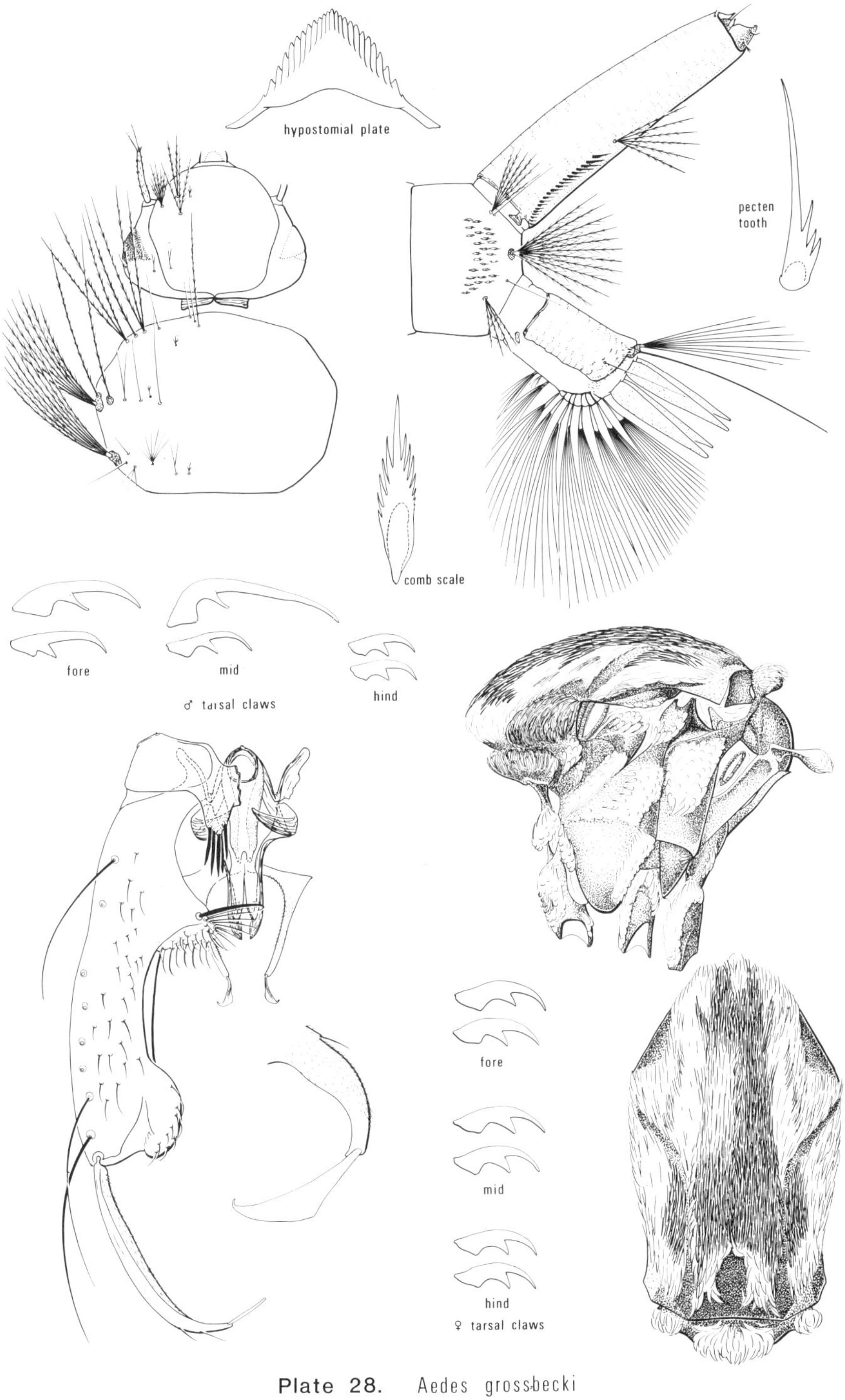

Plate 28. *Aedes grossbecki*

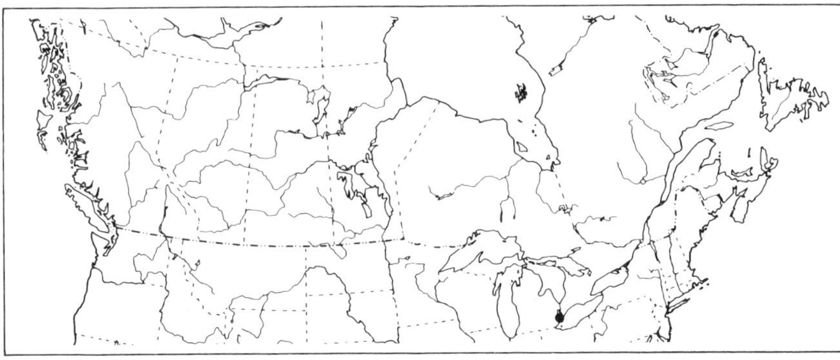

Map 28. Collection localities for *Aedes grossbecki* in Canada: • specimens we examined.

Larva. Head seta 5-C two- to four-branched; seta 6-C two- to three-branched; prothoracic seta 2-P single, more than two-thirds as long as seta 1-P; prothoracic seta 3-P much shorter, less than half as long as seta 1-P (in *flavescens* both 2-P and 3-P short); prothoracic seta 8-P and 10-P each single, each as long as seta 7-P; dorsolateral abdominal seta 1-IV elongate, two-thirds as long as lateral seta 6-IV (1-IV much shorter in *flavescens*); siphon as in *stimulans* with closely spaced pecten teeth each with several small cusps, none of which is as long as maximum width of pecten tooth.

Remarks. Adults of this species are readily recognizable by their large size, banded legs, and broad triangular scales on the wing veins. The larva would be most easily mistaken for specimens of *flavescens*, which have the pecten teeth all evenly spaced; however, seta 2-P is much longer in this species than in *flavescens*, seta 1-IV is elongate, and the pecten teeth have small basal cusps.

Biology. One female, the first record for Canada, was recently reported from Windsor, Ont. (Helson et al. 1978). Larvae of this univoltine species have been collected in woodland pools in spring (Carpenter and LaCasse 1955). In northwestern Ohio, *grossbecki* is a fairly common species (Venard and Mead 1953) and it may have been formerly overlooked at Point Pelee or Pelee Island, Ont.

Aedes hendersoni Cockerell

Plate 29; Figs. 107, 130, 180; Map 29

Aedes triseriatus var. *hendersoni* Cockerell, 1918:199.

Aedes triseriatus, authors prior to 1960, not Say, 1823.

Adult. Female: Proboscis, palpus, and pedicel dark-scaled; vertex entirely white-scaled with dark setae; ante- and post-pronotum with flat,

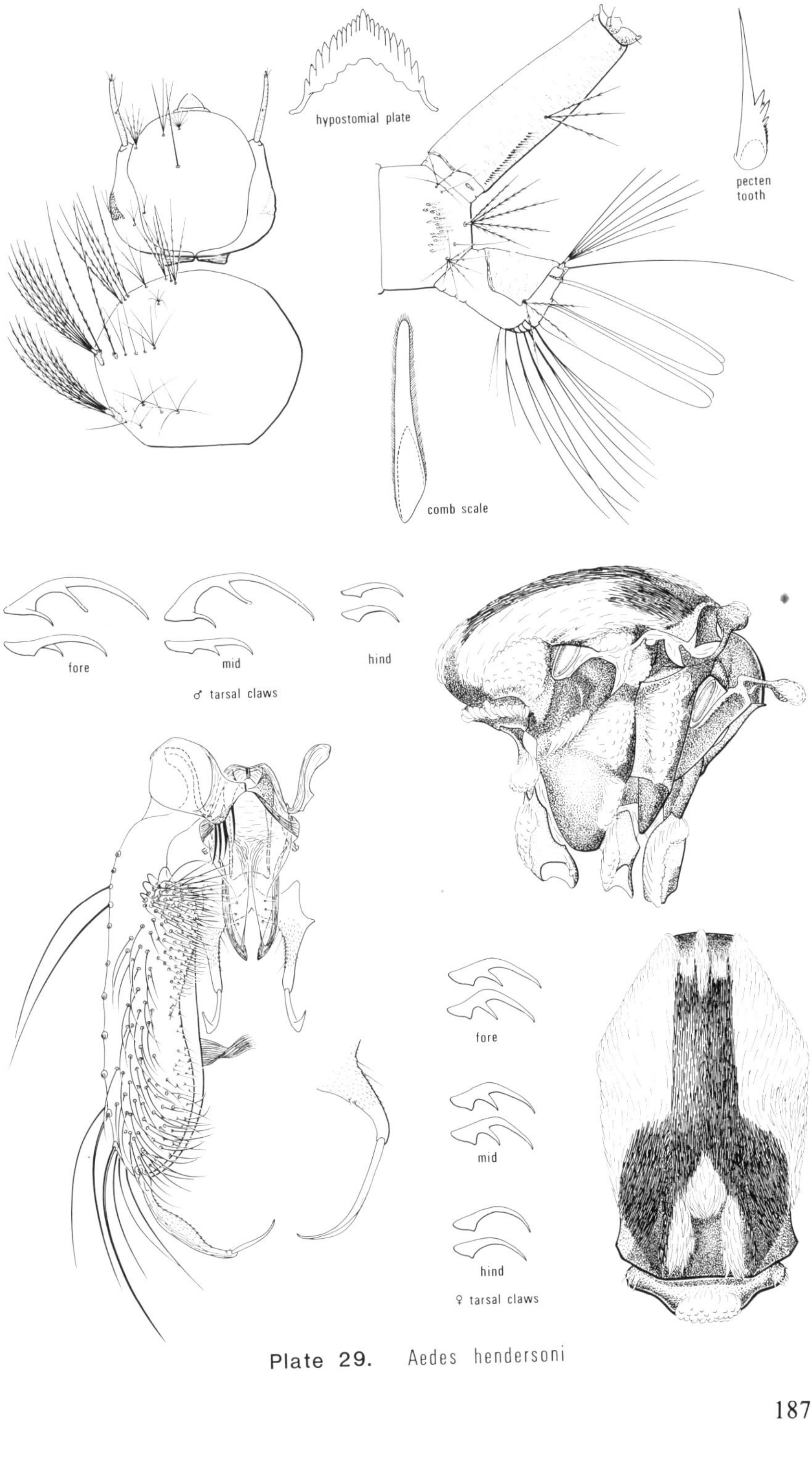

Plate 29. *Aedes hendersoni*

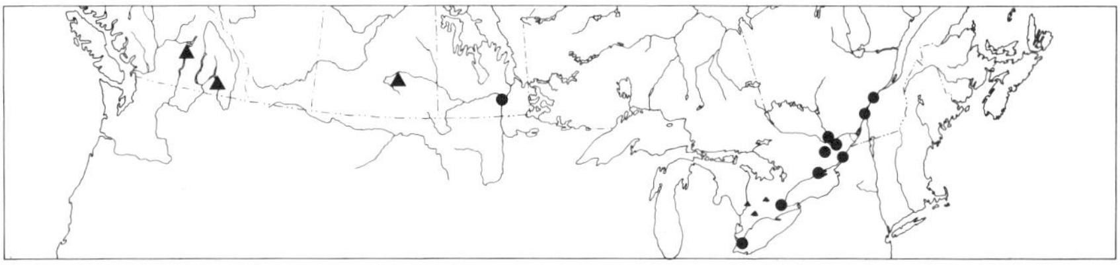

Map 29. Collection localities for *Aedes hendersoni* in Canada: • specimens we examined, ▲ literature records.

appressed, ovate, silvery white scales; scutum with middorsal longitudinal stripe of dark brown scales that, anterior to transverse suture, is parallel-sided or widening slightly posteriorly, but remains confined within dorsocentral setae at least on anterior two-thirds of presutural scutum, and that may be subtruncate or even cleft by white scales anteriorly, extending almost to anterior scutal margin, or may be tapered anteriorly, ending well behind posterior margin; presutural sublateral area entirely covered with crescent-shaped appressed white scales, which also extend across anterior margin of scutum to middorsal stripe and may even divide the anterior ends of the submedian stripes; postsutural sublateral area entirely brown-scaled; lateral stripe and periphery of prescutellar depression white-scaled as in sublateral area; lower border of anepisternum, center of postspiracular area, postero-dorsal and ventral areas of katepisternum, and upper half of metepimeron with flat ovate silvery white scales; most of tibiae and all of tarsi and wing veins dark-scaled; tarsal claws more strongly bent beyond subbasal tooth than in *triseriatus* (Harmston 1969); subbasal tooth of tarsal claw usually longer than that of *triseriatus*; abdominal tergites dark-scaled, each with large lateral white-scaled triangle.

Male: Palpus longer than proboscis, usually by at least the length of last palpomere; scales and setae entirely dark; apex of third palpomere with tuft of up to 10 long setae; fourth palpomere with moderately dense fringe of long setae on medial surface; enlarged tarsal claws of fore- and mid-legs with both subbasal and basal teeth; gonocoxites, in undistorted preparation, almost parallel to one another; dorsomedial margin slightly concave, thus accentuating a basal lobe, which is more setose than the remaining dorsomedial edge; gonostylus short, with long apical spine, which is about half as long as remainder of gonostylus; claspette filament cylindrical, slender, curved, needle-like.

Larva. Antenna smooth, without spicules; antennal seta 1-A usually a single hair, occasionally bifurcate, but not forming a tuft; head seta 4-C multiple, about half as long as seta 6-C; thoracic setae 1-P, 1-M, 1-T, and dorsolateral and ventrolateral abdominal setae on segments I to VII stellately branched, mostly three- or four-branched rather than five or more as in *triseriatus*; pecten teeth evenly spaced; acus detached from siphon; saddle subtriangular in lateral view, narrower anteriorly, deeper posteriorly; anal papillae equal in length, twice as long as anal segment, swollen sausage-like, transparent in life but readily visible against a black background.

Remarks. This species was recently resurrected from the synonymy of *triseriatus* by Breland (1960), who studied it in detail and gave a full account of its morphology. Further details were provided by Zavortink (1972). Males are readily distinguished by the curved inner edge of the gonocoxite and by the almost parallel position of the two gonocoxites. Tarsal claws of the female are more strongly bent beyond the subbasal tooth, similar to those of *impiger* and *excrucians*. The scutal pattern varies intraspecifically resulting in two color forms in both *hendersoni* and *triseriatus*, but usually the brown scales of the middorsal longitudinal stripe anterior to the transverse suture do not extend laterally to the dorsocentral setae in *hendersoni*. Larvae of *hendersoni* have less greenish brown pigment than those of *triseriatus*, though we have seen a few of the latter that have some white material in the blood, which makes them paler than usual and thus more like *hendersoni*. The main tracheal trunks are also more silvery in *hendersoni*, which increases its pale appearance when alive. In order to distinguish living larvae, the most easily visible character is the length of the anal papillae; those of *hendersoni* are equal to one another in length, consistently twice as long as the anal segment, and swollen. In *triseriatus* the ventral pair of papillae is always shorter than the dorsal pair, although their absolute length compared to the anal segment varies considerably.

The two species, *hendersoni* and *triseriatus*, can be hybridized in the laboratory (Truman and Craig 1968). Grimstad et al. (1974) have given a detailed account of the characters of naturally and artificially produced hybrids. Larvae of apparently hybrid origin formed less than 0.5% of the population. Among over 2100 Canadian specimens we have not seen any that was not readily identified as one of the two species.

Curtis (1967), Rempel (1953), and Brust and Kalpage (1967) reported *triseriatus* from British Columbia, Saskatchewan, and Manitoba, respectively. We have reidentified the material reared by Brust and Kalpage and found it to be *hendersoni*. Because Zavortink (1972) recorded *hendersoni* from Sirdar (as Sidar), Kootenay Lake, B.C., and showed that *triseriatus* does not occur west of Minnesota, we assume the records from Regina and Vernon are *hendersoni* and have plotted them accordingly on the map.

Biology. Larvae of this species have usually been collected from water-filled rot holes in trees. Zavortink (1972) recorded a series of *hendersoni* reared from a pitcherplant (*Sarracenia*). In the Ottawa area *hendersoni* was found only in tree holes, where it was usually associated with but always outnumbered by its sister species *triseriatus*. Only in one location, an elm tree hole in which the liquid was a dark tea color, did *hendersoni* occur without *triseriatus*, in successive collections made over 3 yr. This is consistent with the observations of Truman and Craig (1968), who found *hendersoni* only in tree-hole liquid darkened with organic matter. These authors suggested that the long anal papillae of *hendersoni* might be an adaptation to such high organic content. Zavortink (1972) remarked that larvae of *hendersoni* were neotenic compared to the other species of the subgenus *Protomacleaya,* and he speculated that this character may fit it for a different environment than that exploited by *triseriatus.*

Very little is known about the biology of this species, because most work on tree-hole-inhabiting larvae was done before *hendersoni* was recognized as a distinct species. In Ontario and Quebec its life cycle seems to be similar to *triseriatus*; larvae hatch in the spring from eggs that have overwintered in the tree hole. There may be several generations during the summer, depending on availability of water. No larvae of either *hendersoni* or *triseriatus* were found in the Ottawa area in collections made in October or April, and we assume that larvae cannot overwinter. Because of the difficulty of separating this species from *triseriatus*, its role as a pest or as an encephalitis vector has not been evaluated.

Distribution. North America, from southern British Columbia and south central Texas east to southern Quebec and Georgia (Zavortink 1972).

Aedes hexodontus Dyar

Plate 30; Fig. 101; Map 30

Aedes hexodontus Dyar, 1916:83.
Aedes cyclocerculus Dyar, 1920*a*:23.
Aedes leuconotips Dyar, 1920*a*:24.
Aedes masamae Dyar, 1920*c*:166.
Aedes labradorensis Dyar and Shannon, 1925:78.

Adult. Female: Integument medium to dark brown; proboscis entirely dark-scaled; palpus with an occasional pale scale; pedicel mostly white-scaled; vertex and upper three-quarters of postpronotum and scutum with yellowish brown scales that may fade to a dull yellow; scutum usually without darker markings, occasionally with a pair of indistinct submedian stripes; probasisternum almost entirely covered with yellowish white scales; postprocoxal membrane, anterodorsal corner of katepisternum, all of mesepimeron, and all of metameron with yellowish recumbent scales; tibiae and tarsi with mixed pale and dark brown scales, the latter predominating toward apices of tarsi; tarsal claws rather straight basally, curving moderately and uniformly beyond small subbasal tooth; costa usually with patch of white scales basally, proximal to humeral crossvein; remainder of costa and other wing veins dark-scaled; abdominal tergites dark-scaled, each with basal transverse band of white scales.

Male: Palpus dark-scaled, about as long as proboscis; apex of third palpomere and lateral and medial surfaces of fourth palpomere with tufts or

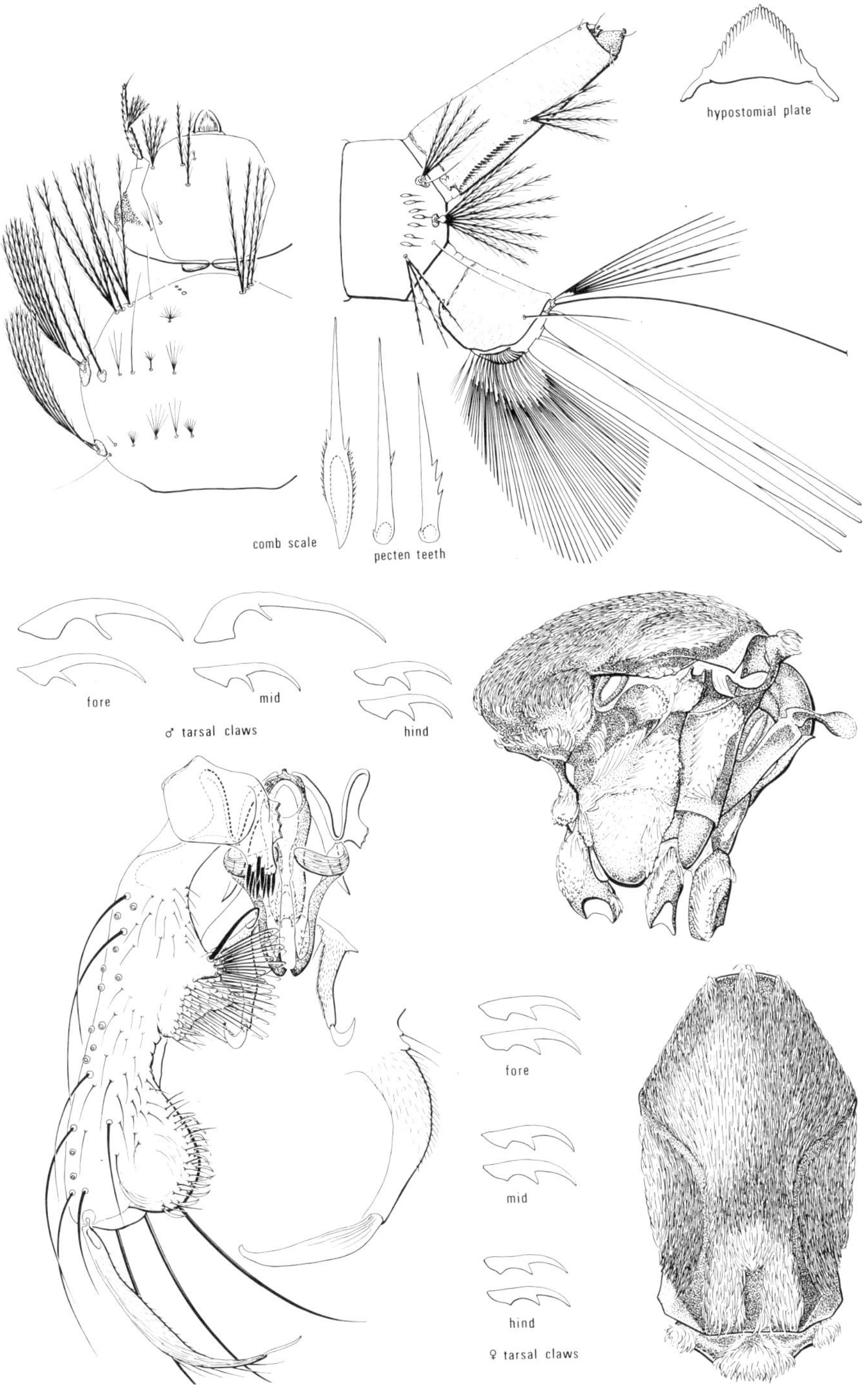

Plate 30. *Aedes hexodontus*

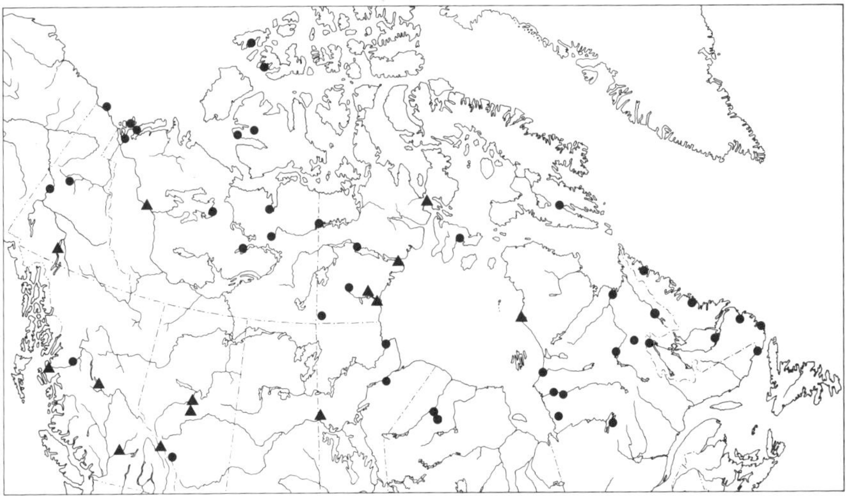

Map 30. Collection localities for *Aedes hexodontus* in Canada: • specimens we examined, ▲ literature records.

fringes of long dense setae; coloration and scaling much as in female except that probasisternum is bare; basal lobe of gonocoxite massive, with triangular seta-bearing surface inclined posterodorsally, subtended by one enlarged medially directed seta; apical lobe prominent, rounded, fringed medially with short, flattened, ventrally curved setae; claspette filament short, appearing inflated rather than compressed and keeled.

Larva. Head setae 5-C and 6-C each single; prothoracic setae 2-P and 3-P almost as long and as thick as either branch of seta 1-P (as is typical of members of the *punctor* subgroup); seta 5-P branched (usually single in *punctor*); mesothoracic seta 1-M and metathoracic setae 1-T and 3-T minute, usually multibranched (longer and single or double in *punctor*); comb scales 4–14, usually 6 or 8, each longer than 0.1 mm (Vockeroth 1954*b*, Kalpage and Brust 1968), thus usually longer than last pecten tooth, and probably always longer than second last tooth; apical spine of comb scale usually longer than oval basal portion; anal papillae exceptionally long, several times as long as anal segment.

Remarks. This species was not distinguished from *punctor* in Northern Canada before the work of Jenkins and Knight (1950) and Knight (1951). Earlier workers (Twinn 1949, Hocking et al. 1950) did not mention it. It is the dominant mosquito in the vicinity of the tree line, yet it does not extend nearly as far north as do *nigripes* and *impiger*. Records from the islands of the western arctic archipelago, Prince Patrick and Victoria islands, are based on a single female and less than 10 females respectively. Most of the specimens from inland Victoria Island were collected in June

1975, after 2 days of strong southerly wind, more than a week before *impiger* and *nigripes* began to emerge. We suspect *hexodontus* may not breed regularly in the western arctic islands, although it is resident on southern Baffin Island, because larvae and a long series of adults were taken there. *Aedes hexodontus* also ranges much farther south, to California, Nevada, and Colorado in the foothills and mountain meadows at 1920–3050 m (Carpenter and LaCasse 1955, Chapman 1959, 1961).

In Alberta, it is sympatric with, and difficult to distinguish from, *punctor*, with which it may share the same larval pools (Enfield 1977). Along the coast of British Columbia, a third closely related species, *aboriginis*, occurs with *punctor* and *hexodontus*. Though the larvae of *aboriginis* are easily distinguished, the adults of all three species are extremely similar. Females of *hexodontus* usually have a more extensively scaled probasisternum; males seem indistinguishable. The *punctor* group requires much more extensive and intensive study, especially in British Columbia, where *hexodontus* is reputed to breed in muskeg pools at sea level (Dyar 1922*b*). The records of collections made in Alberta (Graham 1969, Pucat 1964, Tawfik and Gooding 1970, Wada 1965) may have been based on larvae of *punctor* that had low numbers of comb scales; these specimens should be reexamined.

Biology. This species, probably more than any other, is responsible for the formidable reputation that the arctic has for mosquitoes. It attains extraordinarily high populations in the northern fringes of the boreal forest and adjacent tundra in Canada, Alaska, and Eurasia, where permafrost and poor drainage combine with a relatively warm summer and abundant algal growth to provide almost unlimited opportunities for mosquito breeding. Probably because farther north the tundra pools remain cold and oligotrophic and the growing season is sharply reduced, most of the arctic does not experience the enormous numbers of mosquitoes that are found near the tree line.

After a suitably long period of cold storage, eggs of *hexodontus* hatch at just above 0°C when placed in a medium that supports active decomposition (and contains low oxygen levels) (Beckel 1953). Developing larvae appeared to have a surprisingly high minimum temperature threshold, higher than the other common species at Churchill, Man., and a higher than average rate of development (Haufe and Burgess 1956). At Great Whale River, located at the tree line, larvae reached maturity in late June or early July (Jenkins and Knight 1950). In Alaska, larvae were most often found in semipermanent pools such as grass or sedge marshes (Gjullin et al. 1961).

Adult females actively continued seeking hosts in wind speeds of up to 16 km/h (Gjullin et al. 1961). Ovipositing females preferred rough dark surfaces for oviposition (Beckel 1955).

Swarming behavior of this species at Churchill was described by Downes (1958). Flying males positioned themselves 1–3.5 m above ground and either over or just upwind of a natural marker (the edge of a tundra pool) or an artificial marker (a white sheet laid on the ground). Males in the lower part of the swarm were over the marker; those in the upper part were upwind of it, the distance upwind of the marker being directly proportional to the height above ground, to a maximum of 3.5 m. At wind speeds above 0.8 km/h, each male retained his position in relation to the marker by flying forward, then drifting backward with the wind. When the wind speed dropped to 0.5 km/h, a speed insufficient to move them back fast enough, males began flying in circles over the marker, either clockwise or counter-clockwise. Females maintained a smaller swarm of their own, below that of the males, 30–60 cm above ground. Mating took place in the male swarm above, presumably after a female left her own swarm to fly upward and upwind. A swarm of male *flavescens* occurred directly over the same marker, above the swarm of *hexodontus*, at 3–4 m. Interaction with the *hexodontus* swarm was limited to an occasional *flavescens–hexodontus* coupling. Males of a third species, *Aedes excrucians*, flew at 6–9 m over the same marker, above the swarm of *flavescens*. One of us (D.M.W.) observed similar displacement of male swarms of *hexodontus* and *excrucians* over a white tent at Tuktoyaktuk, NWT.

Distribution. North America and Eurasia: Alaska south to the mountains of California and Colorado, and east through the taiga and tundra of Canada to Newfoundland; northern Scandinavia to eastern Siberia, south in the mountains. Some of the peculiarities of the North American distribution and taxonomy of *hexodontus* were reported by Wood (1977).

Aedes impiger (Walker)

Plate 31; Figs. 96, 157, 194, 196, 200; Map 31

Culex impiger Walker, 1848:6 (of authors after 1954).
Aedes nearcticus Dyar, 1919*a*:32.

Adult. Female: Integument dark brown; proboscis and palpus dark-scaled, with the palpus sometimes scattered with a few pale scales; pedicel pale-scaled; vertex, postpronotum, and scutum with abundant long, dark scattered setae and mixed yellowish and brown scales; scutal scale pattern in fresh specimens dark brown middorsally, changing through yellowish to whitish laterally, paler in faded specimens; probasisternum bare; upper half of anteprocoxal membrane and postprocoxal membrane with pale scales; anterodorsal corner of katepisternum and lower portion of mesepimeron bare; mesepimeral setae present; tibiae white- and dark-scaled; tarsi predominantly dark-scaled, with scattered pale scales basally; tarsal claw rather sharply bent beyond long subbasal tooth, with the distal portion subparallel with subbasal tooth; costa and radius pale-scaled at base, occasionally with a few pale scales scattered beyond base; remaining veins dark-scaled; abdominal tergites dark-scaled, each with transverse basal band of white scales.

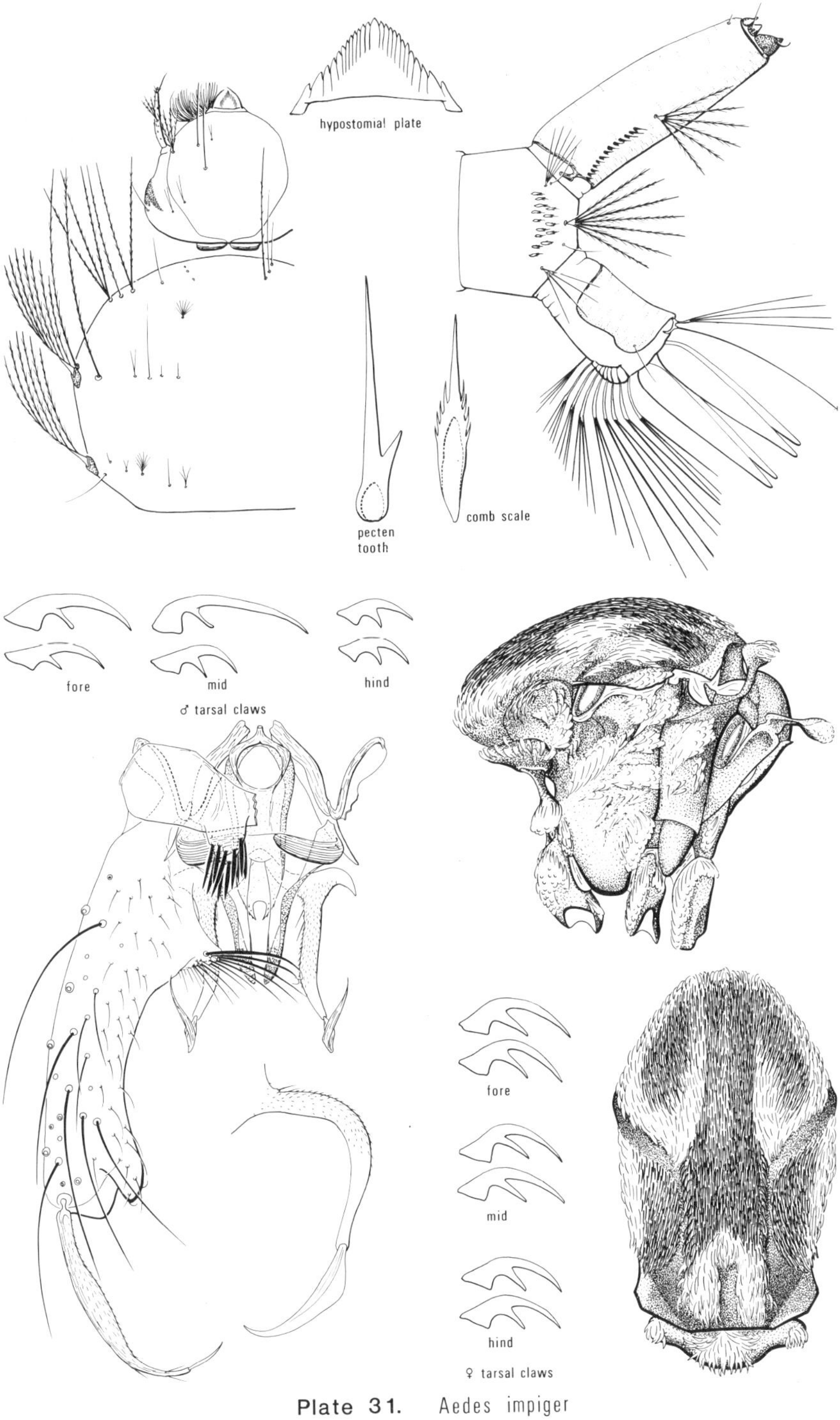

Plate 31. Aedes impiger

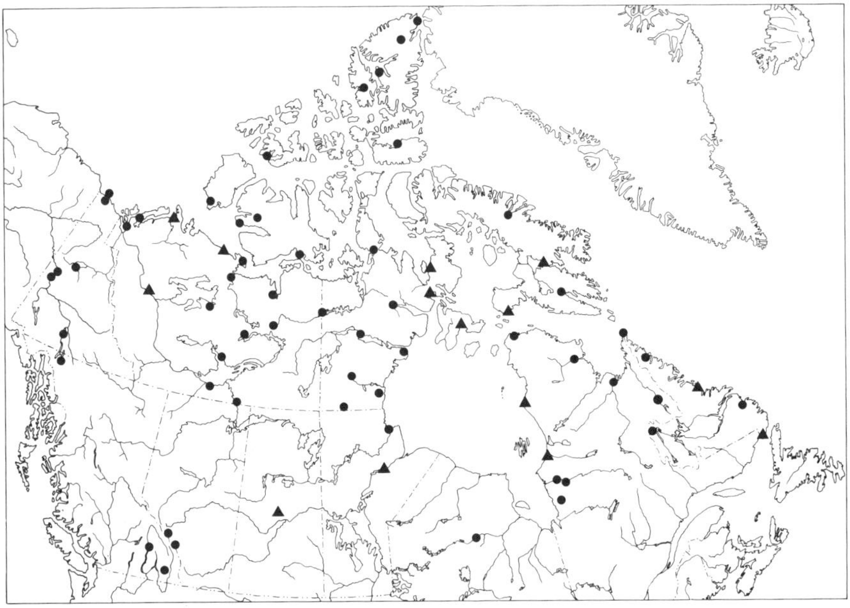

Map 31. Collection localities for *Aedes impiger* in Canada: • specimens we examined, ▲ literature records.

Male: Palpus entirely dark, slightly shorter than proboscis; apex of third palpomere and lateral and medial edges of fourth palpomeres with tufts of long setae; basal lobe of gonocoxite conical, with one long, enlarged seta and a row of finer, well-spaced, long, medially directed setae; apical lobe small, somewhat pointed; medioventral edge of gonocoxite lacking a fringe of long, medioventrally directed setae; claspette filament with flange on convex side originating at base of filament.

Larva. Head setae 5-C and 6-C each single; prothoracic setae 2-P and 3-P finer and less than two-thirds as long as seta 1-P; seta 8-P minute; seta 9-P longer than 7-P; mesothoracic seta 1-M minute; dorsolateral abdominal setae (series 1) insignificant; ventrolateral abdominal setae (13-III to 13-V) of segments III to V about as long as upper lateral setae (series 6) of their respective segments; comb scales few, 8–16, each with long apical spine and short subapical spinules; pecten teeth evenly spaced, each with single large basal tooth; saddle not completely encircling anal segment; anal papillae usually about as long as anal segment.

A rather nondescript larva, apparently without any single distinctive character. Third-instar *punctor* and *hexodontus* are often misidentified as *impiger* because of their incomplete saddle, but these two species have setae 2-P and 3-P subequal to a branch of seta 1-P, 8-P better developed and at least half as long as 7-P, and well-developed dorsolateral abdominal setae (series 1), at least on segments IV and V. Danks and Corbet (1973) presented keys to separate all four instars of *impiger* from *nigripes*.

Biology. The distribution of *A. impiger*, and of *nigripes* with which it coexists over most of the Canadian arctic, would indicate it is well adapted to cold and windy summer conditions. It has lost neither its need for a blood meal nor its mating flight. In most aspects of its life cycle it scarcely differs from other species of *Aedes* farther south (Corbet and Danks 1973), yet it succeeds in environments that are far more rigorous than any other mosquito species has been able to colonize. A few females that were denied blood but given access to sugar and nectar were able to develop some eggs by resorbing most of the oocytes, which indicates a tendency to facultative autogeny (Corbet 1964, 1965).

At Lake Hazen, Ellesmere Island, eggs were discovered in cracks in the tundra, 5–20 mm below the soil surface, on the northern (that is, south-facing) edges of tundra ponds. Oviposition occurred during the warmest part of the day, usually a few hours after solar noon (Corbet 1966). As in *nigripes*, the eggs are usually placed where the pond will first thaw the following spring.

Although adult activity at Laken Hazen was greatest around solar noon, the warmest part of the day, Corbet (1966) found small peaks of activity in the morning and in the afternoon that he interpreted as vestiges of the periods of maximum activity that usually occur at sunrise and sunset in most mosquitoes farther south.

As might be expected, larvae of this species and *nigripes* have the lowest temperature threshold of any northern species (Haufe and Burgess 1956). Probably because of this low temperature threshold and also because of its small size, it was the first species to emerge at Churchill, Man. (Hocking et al. 1950, identified as *nearcticus*), and in Alaska (Gjullin et al. 1961).

On Victoria Island, NWT, larvae of *impiger* did not share the same tundra ponds with larvae of *nigripes*, but were collected in the small interconnecting ditches and puddles between polygons that had been formed by ice wedging. In Peary Land, where such small pools were scarce, and *nigripes* was not present, larvae of *impiger* occupied the larger tundra ponds.

Distribution. North America and Eurasia: Alaska and the high mountains of Oregon and Colorado, east through the taiga and tundra of Canada to Newfoundland, north to Ellesmere Island and Greenland; northern Scandinavia east across northern Siberia to the Bering Strait, including the new Siberian Islands.

Aedes implicatus Vockeroth

Plate 32; Figs. 100, 150, 197, 201; Map 32

Aedes impiger, of authors prior to 1954, not Walker, 1848.
Aedes implicatus Vockeroth, 1954*a*:110.

Adult. Female: Integument medium brown; proboscis and palpus dark-scaled, with the palpus having a few pale scales; pedicel mostly pale-scaled; vertex mostly pale-scaled, except for a pair of dark triangles; antepronotum pale-scaled; postpronotum brown-scaled above, pale-scaled below; hypostigmal area occasionally with a few scales; median and submedian stripes and often part of sublateral stripe of scutum yellowish brown scaled, usually with a pair of narrow submedian stripes of narrower, hence apparently darker, scales; remainder of sublateral area and lateral area with grayish white scales; postprocoxal and upper half of anteprocoxal membranes with scales; anterodorsal corner of katepisternum and ventral third of mesepimeron bare; tibiae and first tarsomeres mostly dark-scaled, with scattered white scales more numerous on ventral surfaces; remaining tarsomeres dark-scaled; tarsal claws moderately and evenly curved, with relatively small subbasal tooth; wing veins dark-scaled, except for a small patch of white scales at extreme base of costa; abdominal tergites dark-scaled, each with transverse basal band of white scales.

Male: Proboscis entirely and palpus mostly dark-scaled; apex of third palpomere and lateral and medial surfaces of fourth palpomere with long dense fringes of setae; coloration and scaling as in female, except that anteprocoxal membrane is bare; medioventral edge of gonocoxite with numerous long medially directed setae along entire length; a group of 6–12 exceptionally long setae arising from dorsal surface of gonocoxite at base of apical lobe; basal lobe of gonocoxite conical, with long medially directed enlarged seta arising distal to a row of finer shorter medially directed setae; apical lobe of gonocoxite prominent, rounded, almost parallel-sided, directed posteromedially rather than posteriorly; claspette filament with flattened keel along entire length of convex side.

Larva. Head setae 5-C and 6-C usually single, occasionally double; prothoracic setae 2-P and 3-P finer and shorter than half as long as seta 1-P; mesothoracic seta 1-P shorter than head seta 5-C; dorsolateral abdominal setae 1-IV to 1-VII rather weak (stronger in *communis*); comb scales usually fewer in number than in *communis* (less than 35 in *implicatus*, usually more in *communis*); each scale broad, fringed apically with long subequal spinules; pecten teeth evenly spaced; saddle not encircling anal segment; spinules along posterior margin too small to be readily apparent under stereomagnification.

This species is similar to *communis* and *stimulans*. It is distinguishable from *stimulans* by the microscopic spinules along the posterior edge of the saddle, which are too small to be seen readily, and from *communis*,

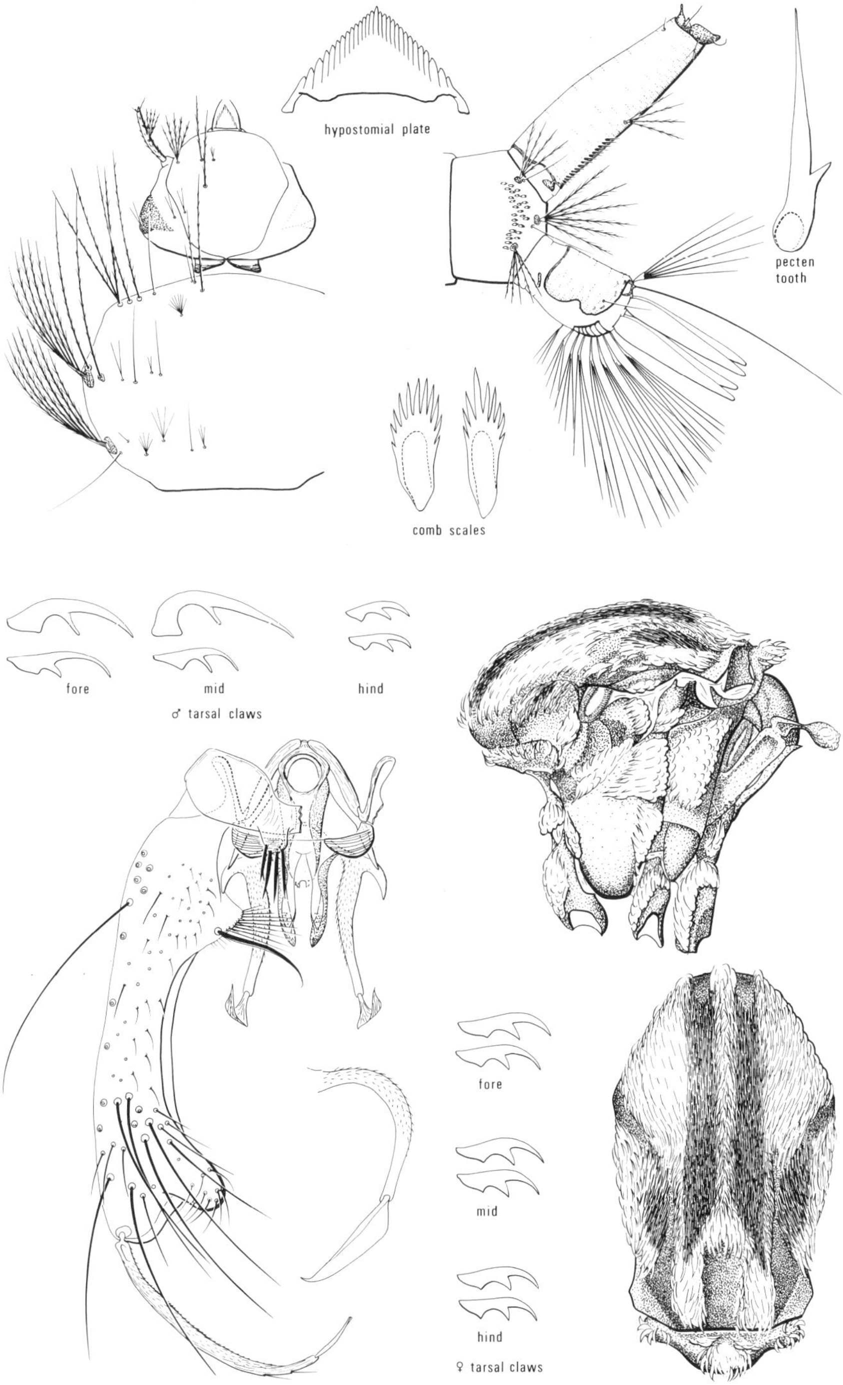

Plate 32. Aedes implicatus

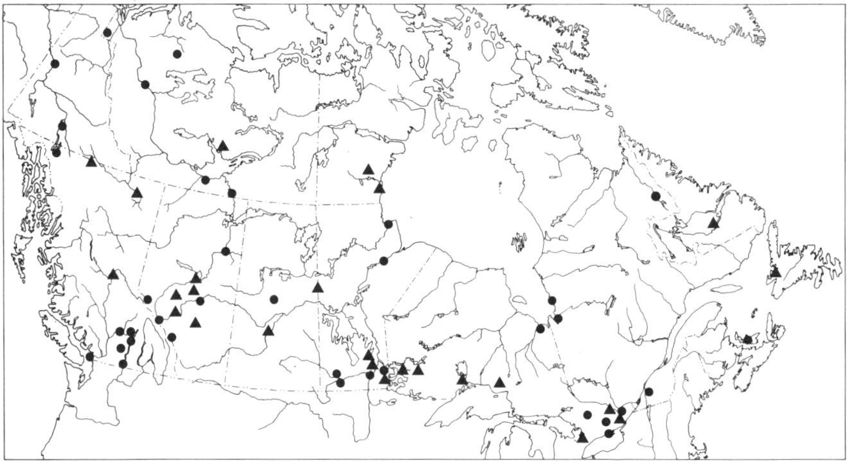

Map 32. Collection localities for *Aedes implicatus* in Canada: • specimens we examined, ▲ literature records.

stimulans, and probably most other similar species by the small rounded maxilla with short apical brush (Fig. 201). Larvae from brackish conditions along the coast of James Bay have greatly reduced anal papillae.

Remarks. Before 1954, references to *impiger* in the literature referred to this species. Vockeroth (1954*a*) realized, however, that the type material of *impiger* was the same as *nearcticus* Dyar. The latter then became a synonym of the older name *impiger*, while *impiger* of authors, not Walker, was renamed as *implicatus*. We have also examined one of Walker's specimens (designated as lectotype by Belkin 1968) and concur with Vockeroth (1954*a*).

Biology. One of the most widely distributed species in Canada south of the tree line, *A. implicatus* is seldom abundant and is not considered a serious pest. It is a forest species with one spring generation that develops very early. It is one of the earliest species to appear as an adult (Gjullin et al. 1961, Barr 1958). In Colorado, Smith (1965) found larvae of *implicatus* almost invariably in small shallow pools left by receding stream waters, shaded by willow thickets and usually fed by some running water. We have found *implicatus* larvae in large numbers in a similar habitat near Perth, Ont., associated with younger instars of *stimulans*. Adults of *implicatus* appear to be short lived. Viruses of the Snowshoe Hare strain (California group) have recently been isolated from larvae of *implicatus* in Saskatchewan (McLintock et al. 1976).

Distribution. North America, from Alaska, Washington, and Colorado east to Newfoundland and Massachusetts.

Aedes increpitus Dyar

Plates 9, 33; Figs. 83, 88, 120, 199, 202; Map 33

Grabhamia vittata Theobald, 1903*c*:313 (preoccupied in *Aedes* by *Aedes vittatus* Bigot).

Aedes increpitus Dyar, 1916:87.

Aedes mutatus Dyar, 1919*b*:24.

Aedes hewitti Hearle, 1923:5.

Adult. Female: Integument reddish brown; proboscis entirely dark-scaled; palpus dark-scaled except for narrow white-scaled basal ring on each of third and fourth palpomeres; pedicel with four or five pale scales medially among group of short setae, otherwise bare; postpronotum mostly brown-scaled; scutum mostly reddish brown scaled, usually with a narrow presutural dorsocentral streak or spot of white scales, and a more conspicuous postsutural stripe of similar scales, usually also with white scales on prescutellar area and often a narrow median line of white scales; lateral stripe of scutum white-scaled; postprocoxal membrane and anterodorsal corner of katepisternum with white scales; lower mesepimeral setae present; tibiae and first tarsomeres with numerous scattered white scales; each tarsomere except the last one with a broad basal ring of white scales; tarsal claws small, short, slightly bent beyond middle; wing veins dark-scaled, usually with a few scattered pale scales; abdominal terga dark-scaled, each with basal transverse band of white scales; first abdominal sternite usually covered with pale scales and short pale hairs.

Male: Palpus slightly longer than proboscis; second to fourth palpomeres each with conspicuous basal patch of pale scales and pale scales scattered distally; apex of third palpomere and all of fourth palpomere with fringe of long setae; scaling and coloration as in female; first abdominal sternite covered with pale scales and hairs; basal lobe of gonocoxite low, without enlarged seta; patch of medially directed setae extending along inner margin of gonocoxite one-quarter of the distance to apex; apical lobe angular; claspette filament with long cylindrical base, with the keel occupying the apical two-thirds or less of the convex side.

Larva. Head setae 5-C and 6-C single; prothoracic seta 8-P minute; seta 9-P as long as a branch of 7-P; dorsolateral abdominal setae (1-IV and 1-V) on segments IV and V short, in contrast with elongate dorsolateral seta (1-VII) on segment VII; subapical spinules of each comb scale nearly as long as apical spinule; pecten teeth evenly distributed, each with two or more basal teeth; anal papillae creamy white, almost transparent (Carpenter 1962), about as long as anal segment, with the dorsal pair longer than the ventral pair. A nondescript larva, without any distinctive characters of its own. Resembling *stimulans*, it is not known at present to overlap geographically with that species.

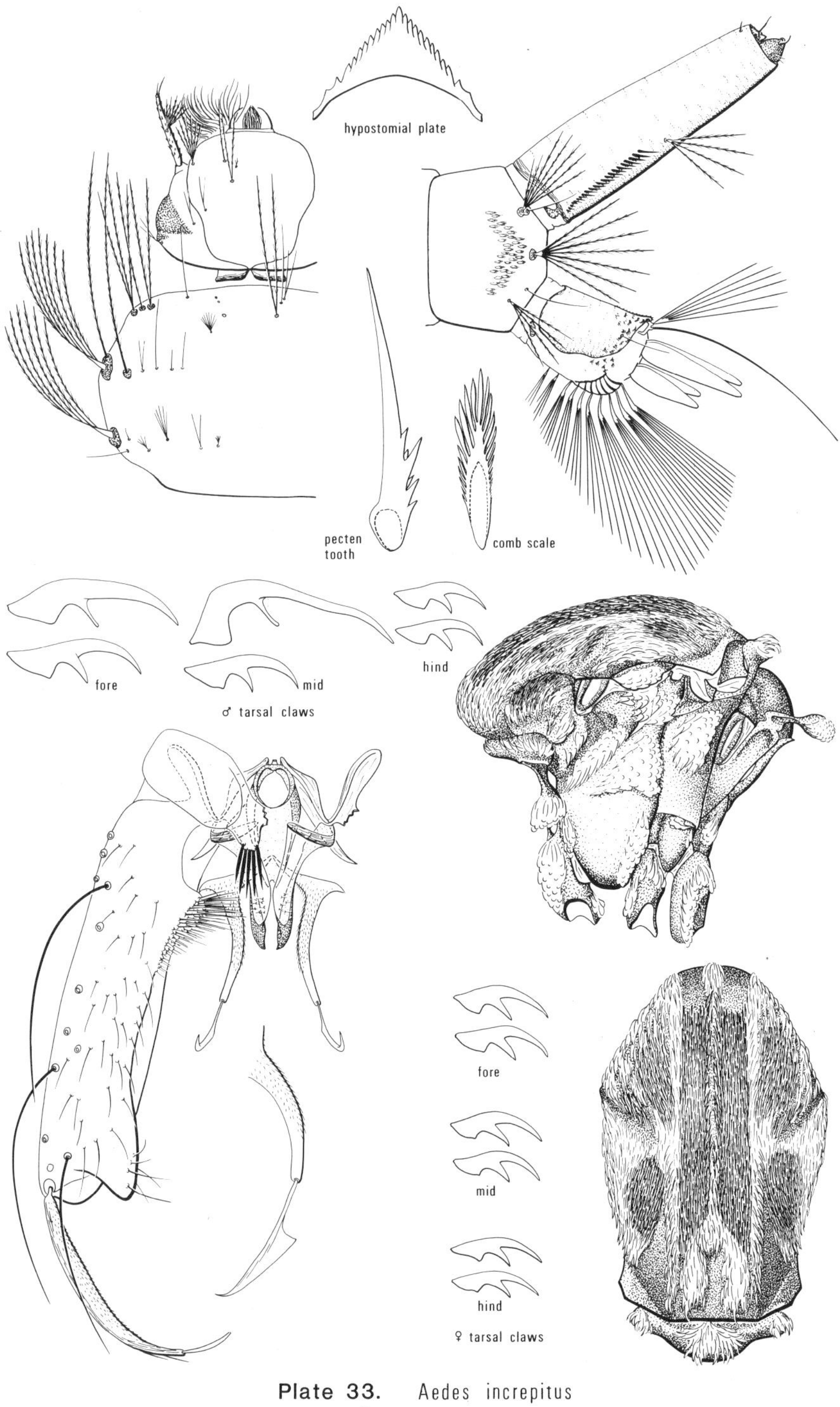

Plate 33. *Aedes increpitus*

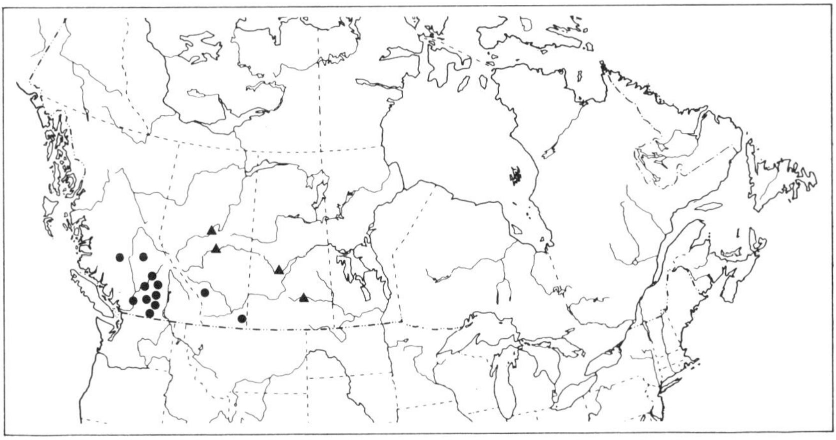

Map 33. Collection localities for *Aedes increpitus* in Canada: • specimens we examined, ▲ literature records.

Remarks. Usually the first abdominal sternite of the adult in all species except *increpitus* is bare (a few scattered scales may sometimes be present, for example, in *pionips*) in contrast with the remaining sternites, which are always covered with scales. Most adults of *increpitus* from the interior of British Columbia, where the species seems most common, have the first abdominal sternite covered with white scales and pale hairs, and in this respect the species is unique among Canadian species. Unfortunately, a series of reared specimens from Elkwater, Alta., lacks this character, and it is not always present in all specimens from British Columbia. Although the lack of a spine on a distinctly elevated basal lobe renders the male distinctive among banded-legged *Aedes* of the *excrucians* group, females lacking scales or hairs on the first abdominal sternite may be recognized by a black-scaled proboscis, nearly scaleless pedicel, mostly reddish brown scaled scutum with prominent pale dorsocentral spot, wing scales almost entirely dark, and small evenly curved tarsal claws. The larvae are best identified by elimination.

Biology. *Aedes increpitus* is uncommon and local in Western Canada (Rempel 1950). In the lower Fraser Valley, Hearle (1926, as *Aedes mutatus*) located larvae in pools left behind by high water of the Fraser River. Freeborn (1926) recorded it as the predominant mosquito in Yosemite Valley of California in circumstances similar to those of the Fraser Valley. The species overwinters as an egg, hatching in spring, but it may also be capable of a second generation, because Carpenter (1962) records the reappearance of *increpitus* larvae after a flood in August had refilled the same pools that had dried up after the spring brood of *increpitus* had left. The larvae are not confined to floodwater situations, however, but have been collected in a variety of habitats. Chapman (1961) recorded *increpitus* as the most common species in snowmelt pools in the foothills of Nevada, inhabiting open roadside ditches as well as shaded pools. Carpenter (1962) considered the typical larval habitat to be large willow-shaded pools in California. These larvae are among the last to complete development (Dyar 1916).

In Nevada and Utah females of *increpitus* are an important pest (Chapman 1961, Nielsen and Rees 1961). In Canada the females are probably too uncommon to be important as pests except locally in the foothills and in the Cypress Hills of Alberta.

Dyar (1916) observed what amounted to swarming behavior among male *increpitus* in California. Individuals drifted and circled, usually without forming a definite aggregation. Carpenter (1962) collected several male *increpitus* flying around his head. Possibly they were using him as a swarm marker, as was recorded for the males of *spencerii* (Knab 1908).

Distribution. North America, from British Columbia and California east to Saskatchewan and New Mexico.

Aedes intrudens Dyar

Plate 34; Figs. 140, 170, 179; Map 34

Aedes intrudens Dyar, 1919*b*:23.

Adult. Female: Integument medium brown; proboscis and palpus dark-scaled; pedicel yellowish, especially laterally, paler than clypeus, with at most two or three pale scales, usually without scales; vertex, upper third to half of postpronotum, and scutum with yellowish brown scales; submedian stripe of scutum sometimes with slightly narrower scales, which though concolorous with those beside them appear darker because of exposure of integument; postprocoxal membrane, anterodorsal corner of katepisternum, and lower third of mesepimeron without scales; hypostigmal area usually bare, rarely with a patch of scales; tibiae and tarsi dark-scaled; tarsal claws rather small, moderately and evenly curved, with small subbasal tooth; wing veins dark-scaled; abdominal terga dark-scaled, each with transverse basal white-scaled band.

Male: Proboscis dark-scaled; palpus slender, entirely dark, about as long as proboscis; apex of third and fourth palpomere with dark setae that are not especially long or dense; scaling and coloration as in female; gonocoxite with medially directed enlarged seta arising near base of basal lobe; basal lobe of gonocoxite bottle-shaped, with two posteriorly directed curved bristles arising from apex; medioventral apex of gonocoxite with a tuft of long posteromedially directed setae; apical lobe rounded, well-developed; claspette stem with short conical posteromedially directed offshoot at its mid length, terminating in a fine seta; claspette filament with an angular keel along its entire convex margin.

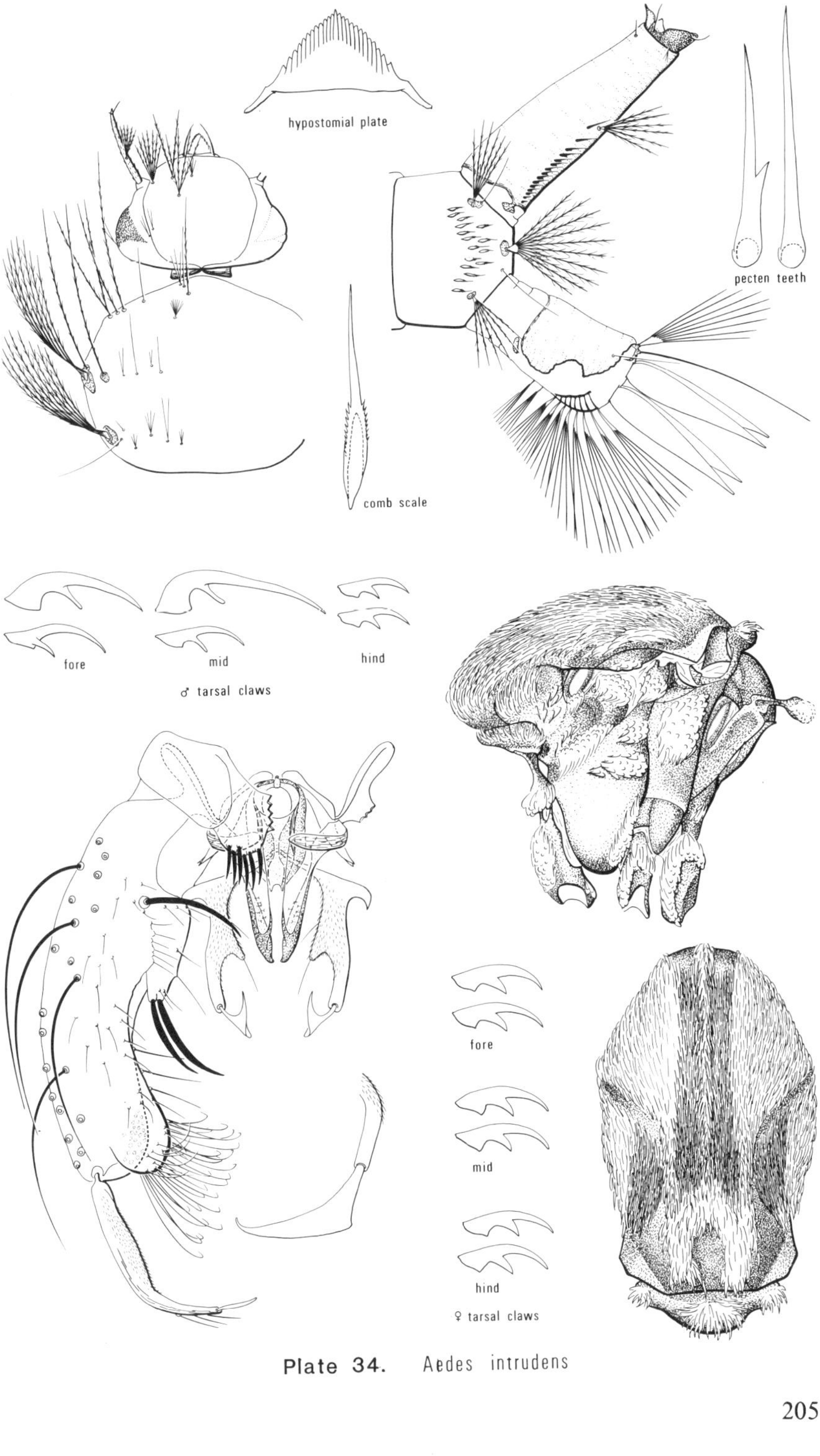

Plate 34. *Aedes intrudens*

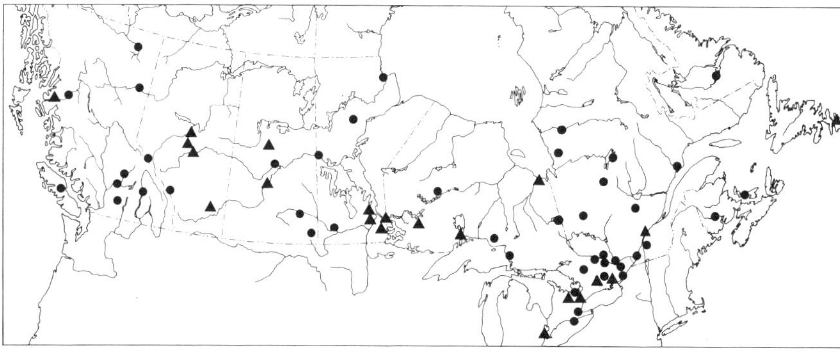

Map 34. Collection localities for *Aedes intrudens* in Canada: • specimens we examined, ▲ literature records.

Larva. Maxillary palpus oval, with long narrow pencil-like apical tuft of setae (Fig. 170); antenna slender, longer than in most *Aedes*; head seta 6-C arising close to seta 5-C, both with an average of three branches; prothoracic seta 8-P as long and as thick as a branch of 7-P; seta 10-P much shorter than 8-P; distal pecten teeth unevenly spaced, and usually also irregularly placed, even on both sides of the same larva; upper and lower lateral abdominal setae (6-I and 7-I) of segment I, and upper lateral of segment II (6-II) single; last pecten tooth usually arising close to siphonal tuft, often dorsal to it, and at or near the same level; comb scales few (12–16), each with long apical spine and short subapical spinules; saddle usually incised at posteroventral corner.

Remarks. *Aedes intrudens* is usually collected in association with *communis*. The larva and adult male of *intrudens* are readily distinguished from that species, and from any other, but the females are rather nondescript and have often been mistaken for *communis* and *sticticus*. The latter always has a broad reddish brown middorsal scutal stripe, but some *communis* have little or no evidence of a dark submedian scutal stripe, hence reliance must be placed on the absence of scales both on the anterodorsal corner of the katepisternum and on the lower fourth of the mesepimeron. *Aedes communis* has both these areas fully scaled, whereas *sticticus* has the katepisternum scaled but lacks scales on the lower border of the mesepimeron. Both *intrudens* and *sticticus* have shorter, more strongly curved tarsal claws than does *communis*, but the differences are subtle.

Biology. Like *A. implicatus*, *intrudens* is distributed throughout Canada south of the tree line, but unlike that species it is a more troublesome biter, seldom abundant, but present everywhere in late spring. Being short-lived, however, it cannot be ranked among the serious pests. It is aptly named, because blood-seeking females seem more adept than other *Aedes* in locating and entering minute openings in buildings (Dyar 1919*b*, 1928). In the Ottawa area, larvae hatch from overwintering eggs in April, in temporary woodland snowmelt pools. The species is often associated with *communis* and *diantaeus* in hardwoods, or with *punctor* and *abserratus* in coniferous woodland, emerging before any of these species. Early emergence and a tendency to being short-lived were also noted by Barr (1958).

Curtis (1967) rated *intrudens* as uncommon in British Columbia. At Waskesiu, Sask., it made up almost half of the biting population (Rempel 1953), attacking during the day as well as at night. Barr (1958) and Natvig (1948) also experienced annoyance during the day from this species. There seems to be no specific information available concerning its mating or oviposition behavior.

Distribution. North America and Eurasia: Alaska and Oregon east to Newfoundland and Massachusetts; western Europe, east to Khabarovsk and Kamchatka, north to the taiga, south to the Ukraine.

Aedes melanimon Dyar

Plate 35; Figs. 117, 189; Map 35

Grabhamia mediolineata Ludlow, 1907:129 (preoccupied in *Aedes* by *Aedes mediolineatus* (Theobald), 1901*a*).

Aedes melanimon Dyar, 1924*a*:126.

Aedes klotsi Matheson, 1933:69.

Adult. Female: Scutum with middorsal reddish brown stripe streaked with white; lateral stripe reddish brown, concolorous with upper half of postpronotum and middorsal stripe; sublateral area chalky white; veins predominantly dark-scaled, R_{4+5}, M_2, M_3, M_4, and CuA_2 usually lacking white scales; tarsal claws similar in shape to, but smaller than, those of *campestris*; abdominal tergites each with narrower, more clearly defined midlongitudinal pale band and transverse pale basal band; lateral dark areas correspondingly larger and with fewer scattered pale scales; otherwise very similar to *campestris* and *dorsalis*, but separable from *campestris* by the dark wing veins, and from *dorsalis* by the shape of the tarsal claw.

Male: Hind tarsal claw as in female, with long slender subbasal tooth, in length more than half the distance between apices of tooth and claw; scutum varying from almost entirely whitish, that is, middorsal stripe partly obliterated by whitish streaks, to extensively reddish brown scaled; basal lobe of gonocoxite with two enlarged darkened setae, one short and one long, as in *dorsalis*, except that both arise side by side on posterolateral angle of lobe; apical lobe prominent, extending distal to level of base of gonostylus; claspette filament with slender cylindrical base and low crest on convex side; coloration and scale patterns as in female.

Larva. Dorsum of head capsule with large median circular spot of dark pigment and a similarly pigmented crescent-shaped spot behind it, with concave margin directed anteriorly; head setae 5-C and 6-C each single;

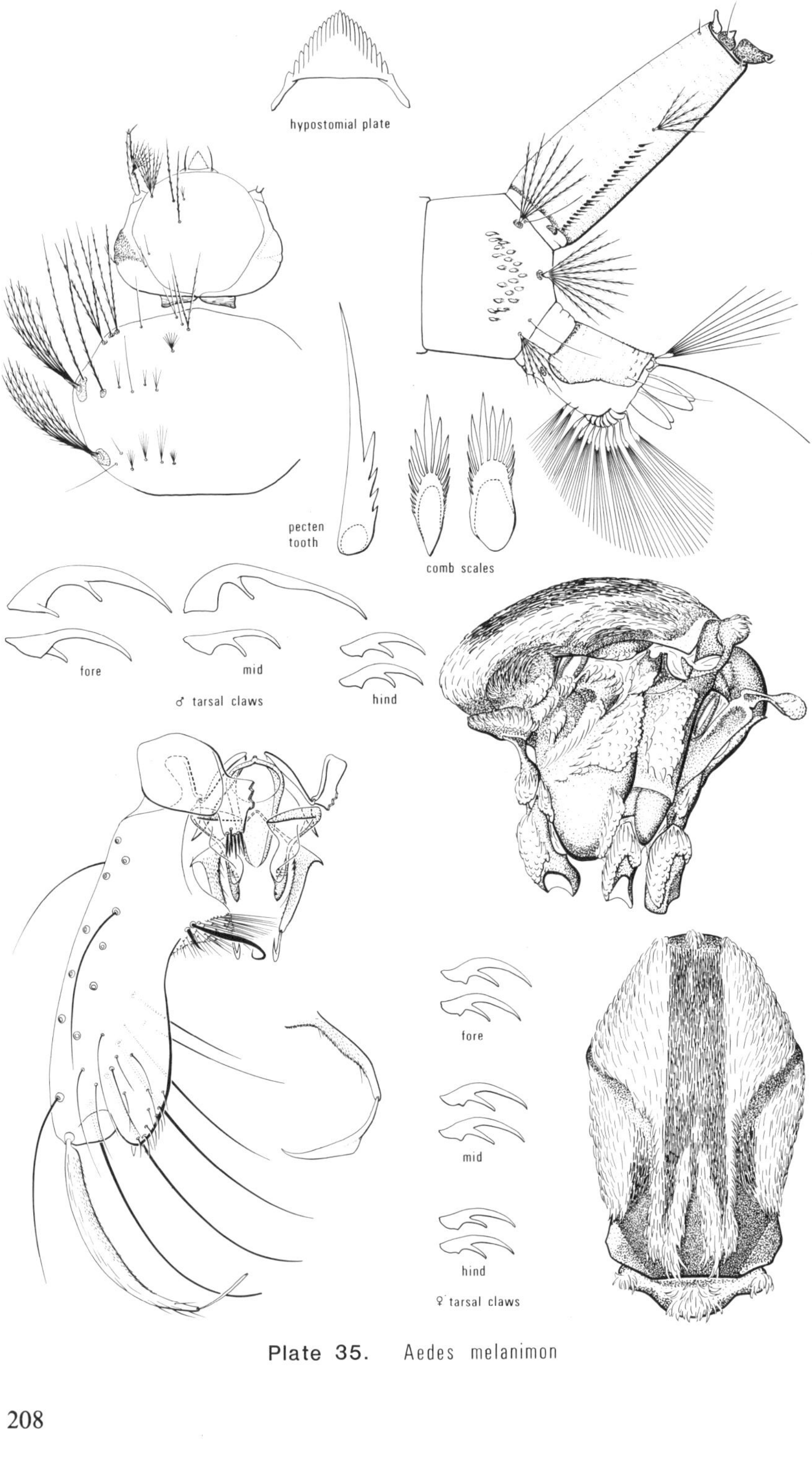

Plate 35. *Aedes melanimon*

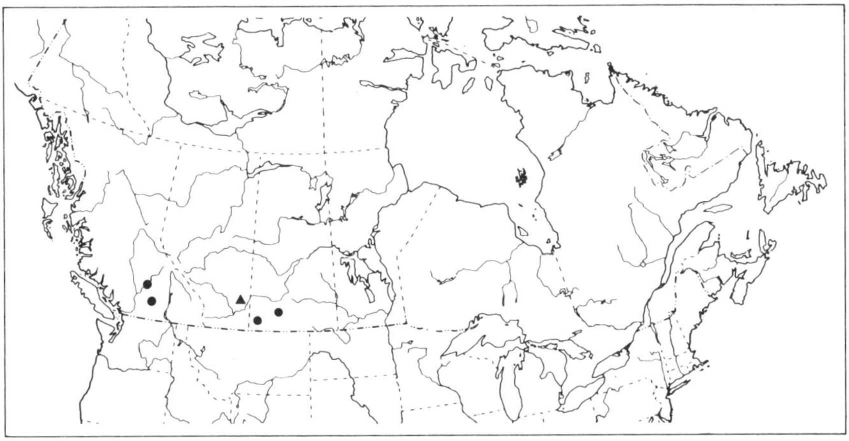

Map 35. Collection localities for *Aedes melanimon* in Canada: • specimens we examined, ▲ literature records.

prothoracic setae 8-P and 9-P subequal in length, both shorter than half length of a branch of 7-P; dorsolateral abdominal setae (1-IV and 1-V) of segments IV and V minute; ventrolateral setae (series 13) of segments III to V double, about two-thirds as long as upper laterals; comb scales 19–24, each with long subapical spinules that are two-thirds or more as long as apical spine; siphon shorter and wider than in *campestris*, less than 3 times as long as maximum width; pecten teeth evenly spaced, each with two or more small basal teeth; siphonal seta 1-S longer than basal width of siphon; saddle, in lateral view, nearly twice as deep anteriorly as posteriorly, thus appearing truncated or notched posteroventrally; anal papillae unequal, with ventral pair shorter than dorsal pair, both shorter than saddle.

Remarks. *Aedes melanimon* was first reported in Canada from Brooks, Alta. (Burgess 1957). In southwestern Saskatchewan it is locally common (Holmberg and Trofimenkoff 1968). In addition, several adults of this species from the interior of British Columbia, especially from the Kamloops area, were discovered among *dorsalis* in the Canadian National Collection. The immature stages have only recently been collected in Canada at Keremeos, B.C. (Costello, personal communication). These records appear to indicate the northern limits of its range, because it is much more common in the southwestern USA.

Biology. This species overwinters in the egg stage. It is capable of one or more summer generations, similar to its relatives *campestris* and *dorsalis*. Nielsen and Rees (1961) found larvae in overflowing streams and irrigated pastures. Larvae were collected in large numbers in Colorado in June from overflow pools along the Animas River (Harmston and Lawson 1967). Because this species was formerly mistaken for *dorsalis*, much early work on virus transmission and other information on the latter's biology is questionable.

Distribution. North America, from British Columbia and California east to Saskatchewan and Colorado.

Aedes mercurator Dyar

Plates 9, 36; Figs. 87, 123; Map 36

Aedes mercurator Dyar, 1920*a*:13.
Aedes albertae Dyar, 1920*b*:115 (as *stimulans* ssp.).
Aedes stimulans: sensu Gjullin et al. 1961, not Walker, 1848.

Adult. Female: Integument dark brown; proboscis and palpus dark-scaled, with the palpus usually having a few pale scales at base of third and fourth flagellomeres; scales on pedicel mixed pale and dark, extending over entire medial edge; vertex yellow-scaled except for a pair of dark lateral spots; antepronotum and lower two-thirds of postpronotum yellow-scaled; upper third of latter brown-scaled; scutum with broad middorsal longitudinal stripe (combined median and submedian stripes) of dark reddish brown scales, usually including within it a pair of narrow submedian stripes of narrower, hence apparently darker, scales; presutural sublateral, lateral, and prescutellar areas entirely yellow-scaled; postsutural sublateral stripe brown-scaled; postprocoxal membrane, anterodorsal corner of katepisternum, and almost all of mesepimeron with pale scales; mesepimeral bristles present; tibiae predominantly dark-scaled, but with numerous scattered white scales; first fore tarsomere almost entirely dark-scaled, with narrow indistinct basal white ring and few scattered white scales; second tarsomere with broad white-scaled basal ring; third tarsomere usually with small medial basal spot of white scales only; fourth and fifth tarsomeres usually devoid of white scales; mid tarsus similar to fore tarsus except that third tarsomere usually with complete white basal ring; fourth tarsomere with a few white scales basally; hind first tarsomere with distinct but narrow basal white-scaled ring; second to fourth tarsomeres each with broad (almost half as long as the segment) white-scaled ring; fifth tarsomere sometimes with a few pale scales basally; tarsal claws relatively large, bent near middle beyond subbasal tooth, with the apex beyond the bend tapering uniformly and not thickened; wing veins entirely dark-scaled; abdominal tergites dark-scaled, each with basal transverse band of white scales.

Male: Palpus longer than proboscis by almost the full length of the last palpomere; each palpomere dark-scaled, with pale-scaled basal ring or spot dorsally; apex of third and margins of fourth palpomere with conspicuous dense fringe of long setae; tarsi even more sparingly white-scaled than those of female, with first fore and mid tarsomeres usually without basal white rings; coloration and scaling otherwise as in female; medially directed enlarged seta on gonocoxite at base of basal lobe extremely long, extending across midline nearly to base of opposite bristle; apical lobe of gonocoxite not extending distal to base of gonostylus; claspette filament long, with cylindrical base and keel having retrorse angle confined to apical half of convex side.

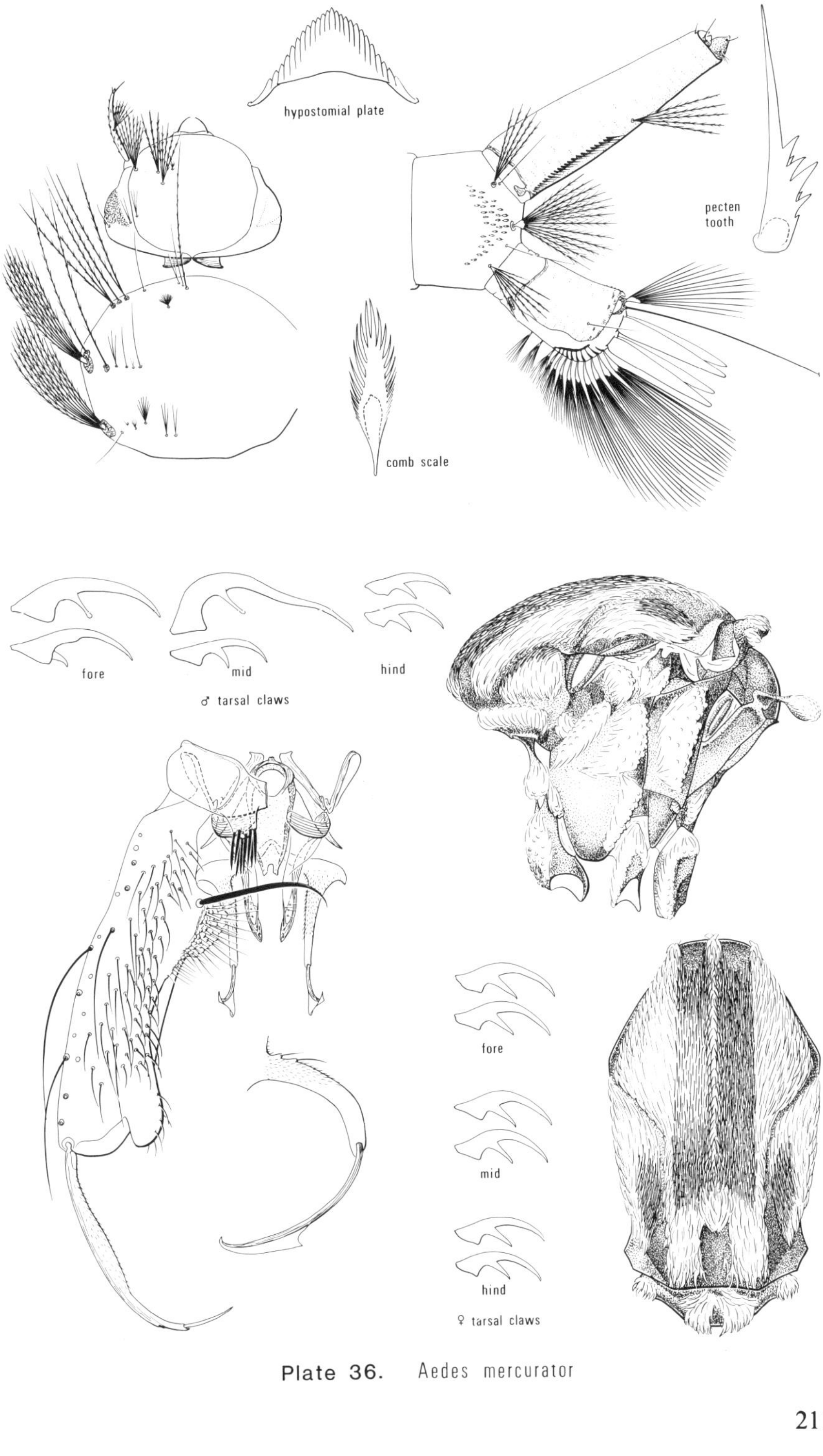

Plate 36. Aedes mercurator

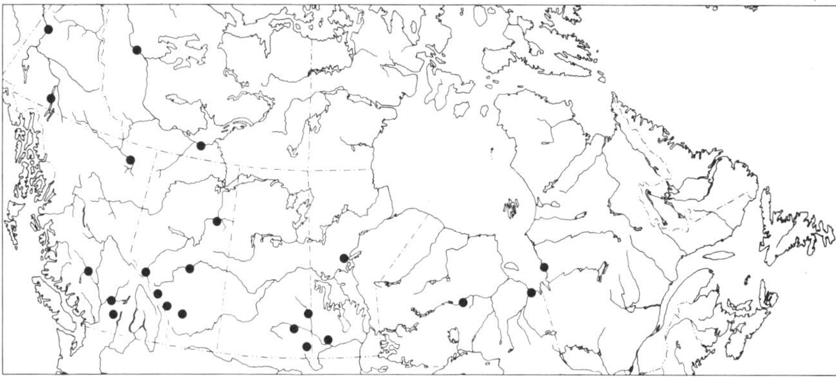

Map 36. Collection localities for *Aedes mercurator* in Canada: • specimens we examined.

Larva. Head seta 5-C with three to six branches, usually at least four or more on one side; seta 6-C with one to four branches; prothoracic setae 2-P and 3-P about half as long as 1-P; mesothoracic seta 1-P single, as long as seta 5-C; dorsolateral abdominal setae (series 1) developed only on segments IV, V, and VII, usually more than two-thirds as long as upper lateral seta (in *stimulans,* setae 1-IV is usually short, whereas seta 1-V is half as long as seta 1-VI); pecten teeth evenly spaced; saddle encircling two-thirds or more of anal segment; anal papillae longer than anal segment.

The larva is most likely to be mistaken for that of *pionips* or *aboriginis* because of the elongate 1-M and multibranched head seta 5-C. The former has 60 or more comb scales, twice as many as *mercurator,* and an elongate seta 3-M, whereas *aboriginis* has the lower lateral seta (7-II) as long as the upper lateral (6-II) above it.

Remarks. Dyar (1920*a*) remarked on the lack in Western Canada of the preferred habitat of *stimulans,* that is, of pools that were filled by high water rather than by seepage or snowmelt. He first assumed that *stimulans* had changed its habits, but later (1920*b*) concluded that *stimulans* had been replaced there by a related form that he described as *stimulans albertae.* In these two papers he noted the greater number of branches of the larval head seta 5-C of both *albertae* and *mercurator* and the longer spine on the basal lobe of the gonocoxite, but some years later (1928) he concluded that these characters were too subtle for species separation and *albertae* and *mercurator* were relegated to the synonymy of *stimulans. Aedes mercurator* was recognized as a separate species (Danilov 1974) primarily on larval differences, and *albertae* was shown to be *mercurator* rather than *stimulans* (Wood 1977).

Biology. The larvae of this species have been found earlier in the spring than most other species at the edges of open semipermanent marshes that have abundant emergent vegetation (Dyar 1920*b*, Gjullin et al. 1961, as *stimulans*). Males were observed swarming at Dawson, Y.T., over willow bushes on a hillside after sunset (Dyar 1920*b*). The species (as *stimulans)* was considered rare in British Columbia (Curtis 1967) and Saskatchewan (Rempel 1953). Both these authors commented on how much more pestiferous *stimulans* was in eastern North America than in their areas. Although Rempel (1950) did not find the larvae in Saskatchewan, he illustrated the true *stimulans* (his fig. 16) probably using a specimen from Eastern Canada.

Distribution. North America and Eurasia: northeastern part of European USSR and Crimea, Caucasus, the mountains of Turkey, east across Siberia and Alaska to Ontario. The North American distribution is still largely unknown.

Aedes nigripes (Zetterstedt)

Plate 37; Figs. 97, 142, 162; Map 37

Culex nigripes Zetterstedt, 1838:807.

Aedes innuitus Dyar and Knab, 1918:166.

Aedes alpinus, authors, not Linnaeus.

Adult. Female: Integument dark brown; proboscis and palpus dark-scaled, with the palpus sometimes having some pale brown scales; pedicel three-quarters encircled, medially, dorsally, and laterally, with mixed pale and dark scales; scales of vertex, postpronotum, and scutum dark coppery brown, paler on supraalar and prescutellar areas and fading as a result of sunlight to yellowish or silvery gray in older specimens; postpronotum and scutum except for submedian and postsutural sublateral stripes clothed with rather long, fine, erect brown setae; integument of scutum, when denuded of scales, matt blackish brown pollinose, lacking the grayish pollinose submarginal and transverse sutural markings usually present in other species; pleural scales pale yellowish brown; lower mesepimeral setae numerous; tibiae and first tarsomeres dark-scaled, with scattered paler brownish scales; remaining tarsomeres dark-scaled; tarsal claws long, moderately and evenly curved beyond subbasal tooth; wing veins mostly dark-scaled, usually with paler scales scattered at base of costa and radius, and sometimes at base of other veins as well; abdominal terga dark-scaled, each with basal transverse white-scaled band.

Male: Palpus shorter than proboscis by nearly the length of the last palpomere, entirely dark-scaled; apex of third and margins of fourth palpomere with fringes of long setae; scutum sparsely scaled, readily denuded, with the scales brown at emergence but silvery gray in most field-collected specimens; all setae of head, thorax, abdomen, and legs dark brown; hind margin of sternite VIII convex, fringed with 20 or more stiff setae; lobes of

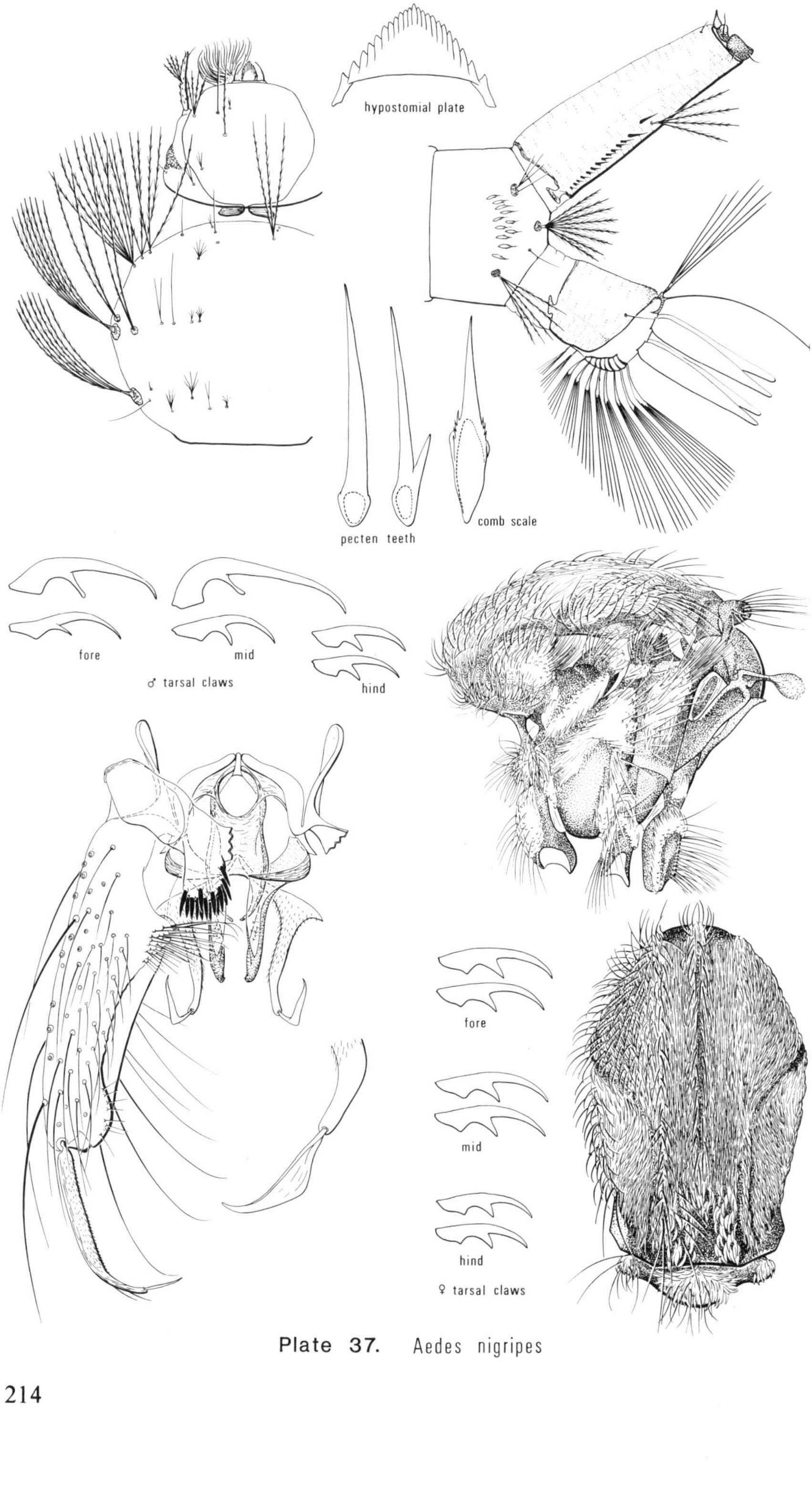

Plate 37. *Aedes nigripes*

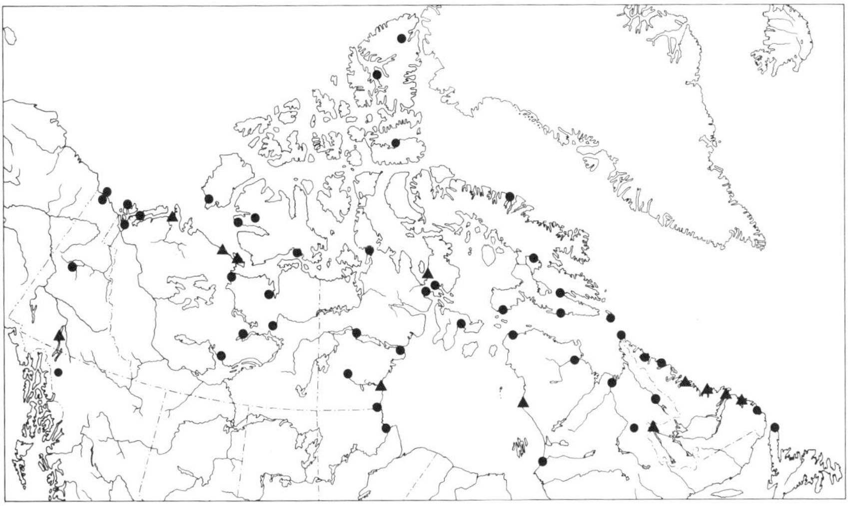

Map 37. Collection localities for *Aedes nigripes* in Canada: • specimens we examined, ▲ literature records.

sternite IX closely approximated, separated by narrow V-shaped notch, each lobe broad, rounded, studded with short stout setae; basal lobe of gonocoxite conical, covered with medially directed setae but lacking an enlarged seta; apical lobe reduced, inconspicuous; claspette filament with low keel occupying entire convex side.

Larva. Prothoracic setae highly branched; seta 1-P with three or four branches; seta 5-P with four or five branches; seta 6-P double or triple (always single in other species except occasional specimens of *atropalpus*); seta 7-P usually with 10 or more branches; dorsolateral abdominal setae (series 1) well-developed on segments III to VII inclusive, usually double and as long as upper laterals; saddle usually completely encircling anal segment (if not, leaving only a narrow gap); distal pecten teeth more widely and unevenly spaced than proximal teeth, with the last tooth usually arising beyond siphonal tuft.

No other Canadian species has such highly branched prothoracic setae. In our material, seta 7-P in most fourth instars had ten or more branches, whereas other species had only four or five. However, Carpenter and LaCasse (1955) reported five to ten as the range, which is closer to the condition in the third instar in our material. In spite of the variation in each of the above-mentioned characters, *nigripes* occurs with few other species and should be easily recognized. Danks and Corbet (1973) provided keys for the separation of early instars of this species from *impiger*.

Biology. The only arctic species that does not extend south in the mountains, *nigripes* is confined to the tundra north of the tree line and to the extensive alpine areas of Quebec and northwestern British Columbia. At Churchill, Man., it was associated with tundra and was rarely collected in birch–willow scrub and not at all in forest pools (Hocking et al. 1950). After *impiger, nigripes* was the next species to emerge at Churchill. On Victoria Island, NWT, it was more common than *impiger,* but neither species attacked in the huge numbers characteristic of *hexodontus* nearer the tree line. Larvae developed in larger tundra ponds that had grass or sedge margins; larval density was low, but was counterbalanced by the large number of ponds.

At Lake Hazen, Ellesmere Island, oviposition took place only in direct sunlight at the edges of tundra ponds, 5–10 cm above the water level, on moist surfaces that directly faced the sun (Corbet 1964, 1965, 1966, Corbet and Danks 1975). Most of the eggs were laid within an hour of solar noon and, thus, were concentrated on the south-facing slope where the ice would thaw first the following spring.

Corbet (1964) observed the autogenous development of up to 16 eggs in female *nigripes* that had been fed only nectar or sugar or both after emergence. He postulated that in bad weather or when vertebrates were few in numbers such facultative autogeny, which was also found to a lesser degree in *impiger,* might be sufficient to maintain the mosquito population. *Aedes nigripes,* like *impiger,* were observed attempting to feed on various birds and mammals and were particularly successful in feeding on incubating loons and ducks (Corbet and Downe 1966). Precipitin tests showed that at least one had probably engorged on a musk ox.

Male swarms of *nigripes* were often observed by one of us (D.M.W.), on Victoria Island, in gullies 15–20 cm deep formed by frost action between large polygons. Because such sites were below ambient ground level, they were somewhat sheltered from the wind.

Distribution. North America and Eurasia: circumpolar north of the tree line from northern Scandinavia across arctic Siberia (including Komandor Island off Kamchatka) and Alaska to Greenland and Newfoundland.

Aedes nigromaculis (Ludlow)

Plate 38; Figs. 80, 118, 173, 178; Map 38

Grabhamia nigromaculis Ludlow, 1906*c*:83 (1907:85).
Grabhamia grisea Ludlow, 1907:130.

Adult. Female: Integument medium brown; proboscis dark-scaled except for white-scaled ring, which may be obscure in some specimens, encircling middle one-fifth; palpus dark-scaled; pedicel pale-scaled except on ventral quarter; vertex pale-scaled medially, with dense group of pale erect forked scales, yellow-scaled laterally; upper half of antepronotum, upper half to two-thirds of postpronotum, and most of scutum covered with dense

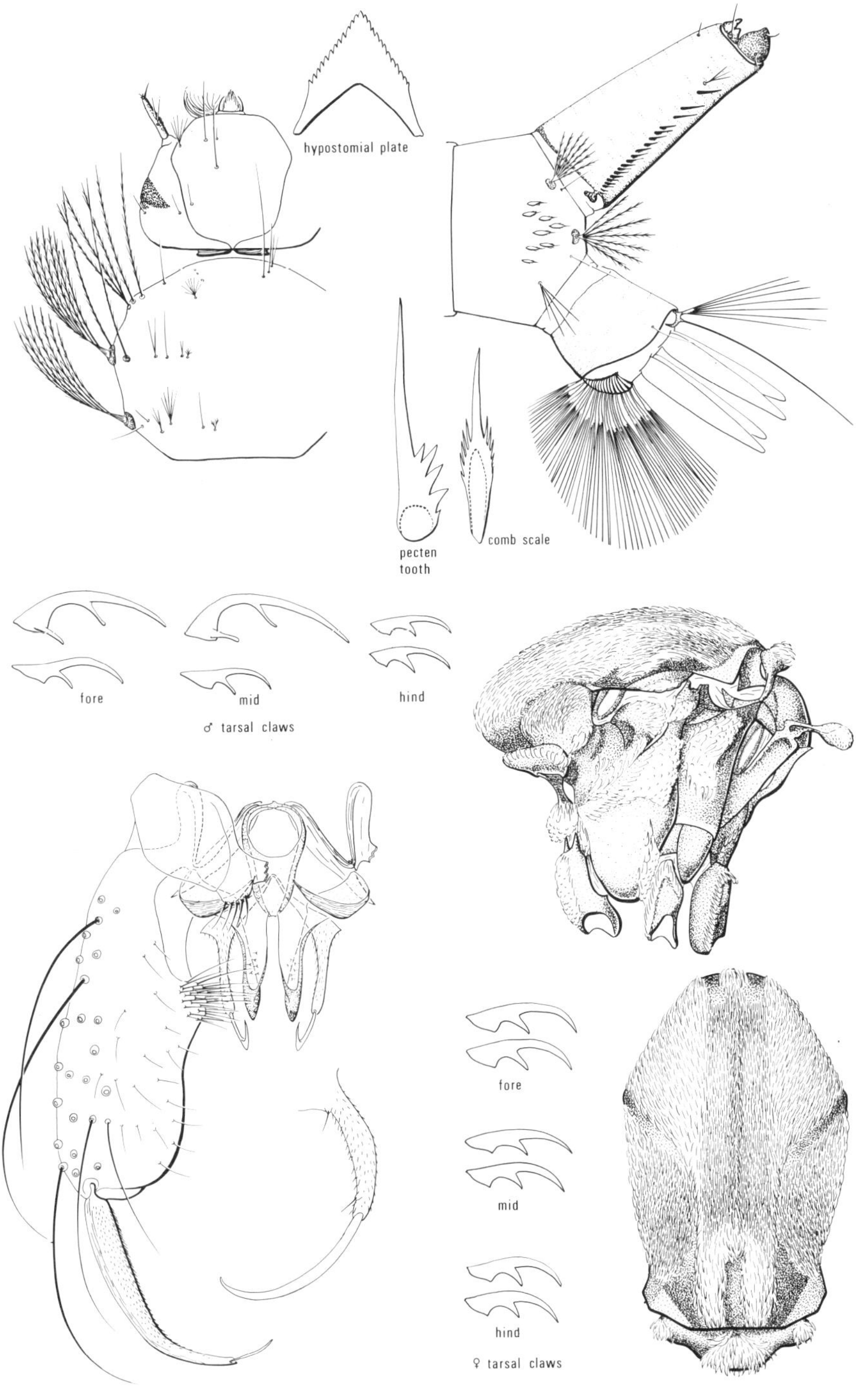

Plate 38. *Aedes nigromaculis*

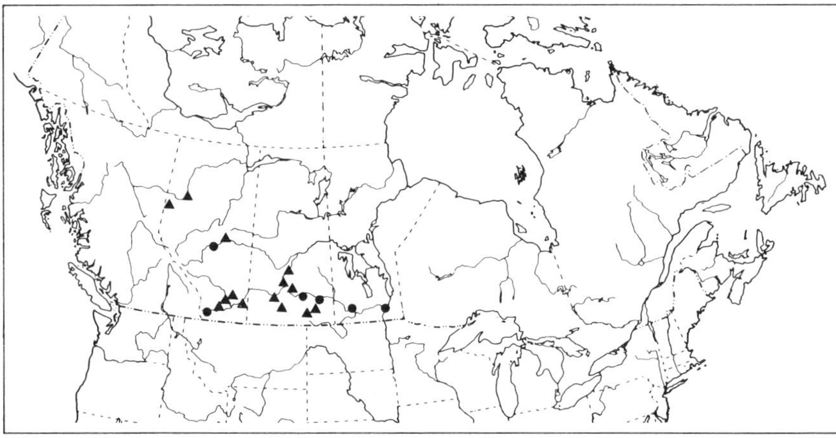

Map 38. Collection localities for *Aedes nigromaculis* in Canada: • specimens we examined, ▲ literature records.

recumbent narrow yellow scales; scutum with darker yellowish brown scaled middorsal stripe, blending to paler yellow sublateral area, then blending to darker lateral stripe; probasisternum, postprocoxal membrane, hypostigmal area, and anterodorsal corner of katepisternum with white scales; fore coxa, femora, tibiae, and first tarsomeres with mixed pale and dark scales; second and third fore and mid tarsomeres and all hind tarsomeres except the last one dark-scaled, each with a broad basal band of white scales; tarsal claws long, moderately and uniformly bent beyond middle; wing veins predominantly dark-scaled, with scattered pale scales at least on basal portions of all main veins; first and seventh abdominal terga predominantly pale yellowish scaled; second to sixth terga each with T-shaped pattern of yellowish scales formed by basal transverse band contiguous with lateral triangle and midlongitudinal band; cercus dark-scaled.

Male. Proboscis dark-scaled without pale ring; palpus slightly longer than proboscis; last three palpomeres dark-scaled, with narrow basal pale-scaled band; apex of third and margins of fourth and fifth palpomeres with long dense fringes of yellowish brown setae; coloration and scaling as in female; gonocoxite short and broad, with the basal lobe displaced distally almost to mid length; basal lobe a low rounded tubercle with a group of medially directed setae, but lacking an enlarged seta; apical lobe undifferentiated; claspette filament a simple, curved, needle-like rod without a keel.

Larva. Antenna exceptionally short; siphon short and broad, only 2–2.5 times as long as wide; pecten teeth extending to apical quarter of siphon, with the distal teeth more widely and irregularly spaced than the proximal teeth; siphonal tuft minute, scarcely longer than last pecten tooth, and much shorter than apical diameter of siphon; saddle completely encircling anal segment.

Biology. *Aedes nigromaculis* is found only on the prairies, where it may be a serious pest in irrigated areas (Strickland 1938). In southern Saskatchewan, it breeds in shallow depressions and irrigation ditches (Rempel 1953). In Manitoba (Brust, personal communication), larvae were found only in certain depressions year after year but the reason for selecting these sites was not apparent. In parts of the USA, larvae of *nigromaculis* have been associated with alkaline situations and saltgrass (*Distichlis*) meadows in Nebraska (Rapp and Harmston 1965) as well as with irrigation (Nielsen and Rees 1961). A late species, first appearing in July, it may produce a second generation under conditions of favorable summer rainfall. Adults of this second generation have been numerous at times in the vicinity of Regina, but in Manitoba this species is much less common (McLintock and Rempel 1963). Females of *nigromaculis*, similar to its relatives *sollicitans* and *taeniorhynchus*, are notoriously wide-ranging and have been recorded 50 km or more from their breeding sites (Dyar 1929). A single early record (McClure 1943) from Churchill, Man., has never been confirmed by subsequent collecting. Our conclusion is that either the species is wide-ranging or it was incorrectly identified.

In California, females of *nigromaculis* oviposited in clumps of bunchgrass in irrigated pastures, especially in moist depressions. When the temperature dropped and the hours of daylight decreased during October, the eggs went into diapause. The following spring when the temperature and the hours of daylight increased, diapause ended (Miura and Takahashi 1973).

Males swarm after sunset over prominent objects on the prairie (Dyar 1922*b*).

Distribution. Western North America, from Alberta and California east to Manitoba and Texas.

Aedes pionips Dyar

Plate 39; Figs. 184, 187; Map 39

Aedes pionips Dyar, 1919*b*:19.

Adult. Female: Integument dark brown; proboscis and palpus entirely dark-scaled; dorsomedial half of pedicel predominantly pale-scaled; vertex and presutural sublateral and lateral areas of scutum yellow-scaled; upper third of postpronotum and middorsal stripe (median and submedian stripes combined) dark brown scaled, the latter occasionally with a single row of yellow scales on the acrostichal line; postsutural sublateral stripe dark brown scaled, separated from submedian stripe and lateral edge of scutum by a narrow line of yellow scales; setae of scutum and scutellum mostly dark brown; antepronotum, lower two-thirds of postpronotum, postprocoxal membrane, anterodorsal angle of katepisternum, and all of mesepimeron with pale scales; tibiae with some scattered pale scales; tarsi and wing veins

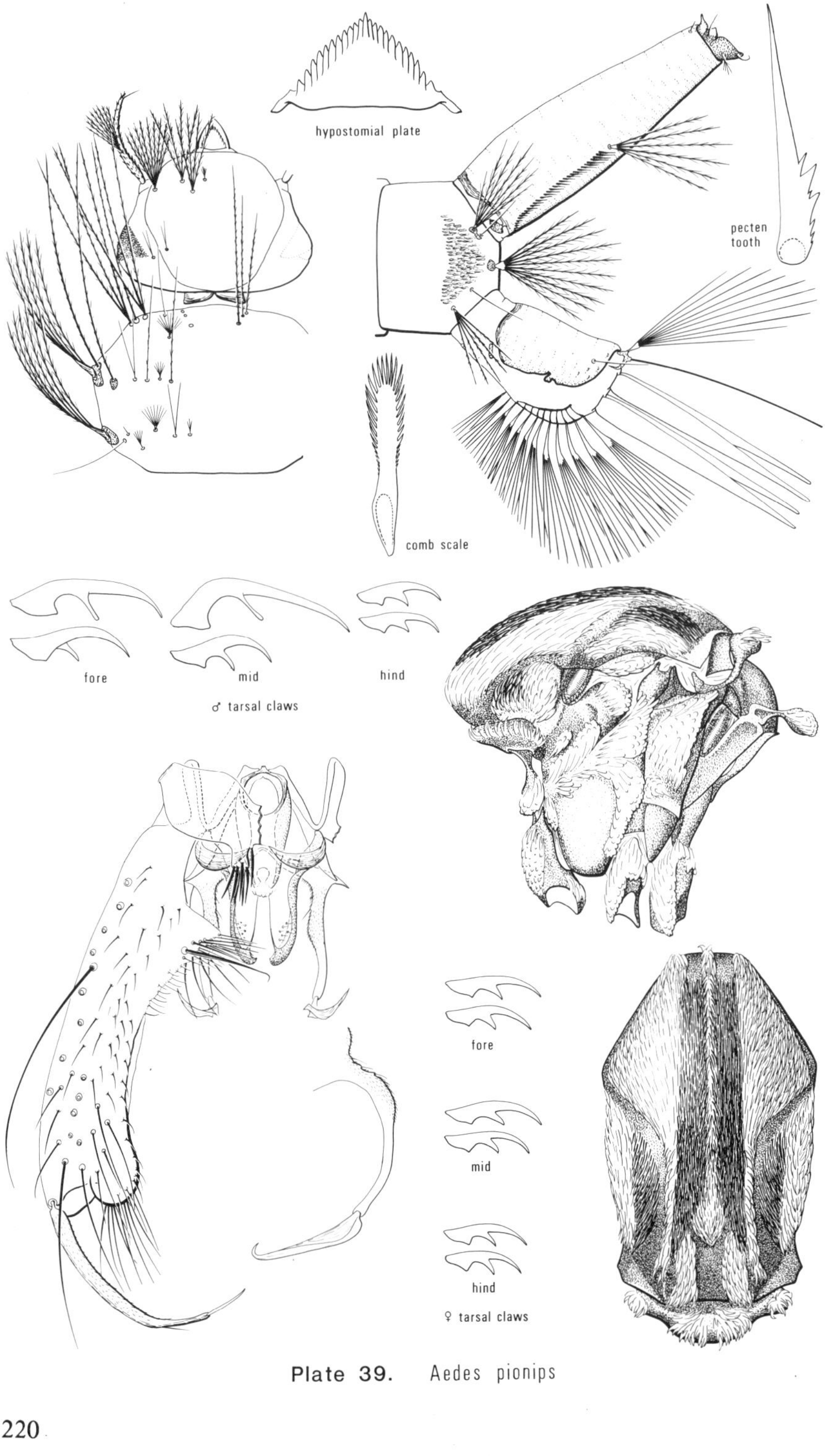

Plate 39. *Aedes pionips*

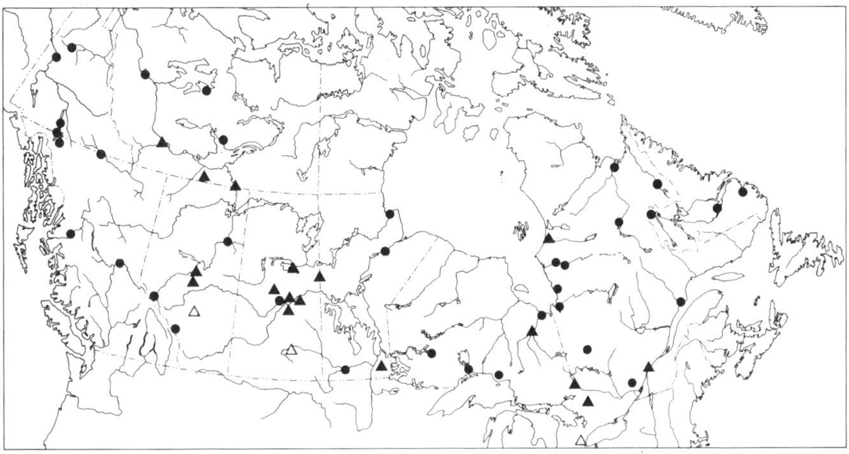

Map 39. Collection localities for *Aedes pionips* in Canada: • specimens we examined, ▲ literature records, △ literature records based on probably misidentified specimens.

entirely dark-scaled; tarsal claws moderately and uniformly curved beyond middle; abdominal terga dark-scaled, each with basal transverse pale-scaled band; first abdominal sternite with a few recumbent pale scales.

Male: Palpus entirely black, about as long as proboscis; apex of third and margins of fourth palpomere fringed with long setae; yellow scales of occiput and scutum paler than in female; submedian and postsutural sublateral brown-scaled stripes narrower, separated by correspondingly wider yellow stripes; terminalia nearly identical with those of *communis*.

An undamaged male can usually be recognized without examining the terminalia by the following characters: legs black-scaled; palpi slightly shorter than proboscis (longer in *communis*); postprocoxal scale patch present (absent in *communis*); dorsal margin of katepisternum scaled to anterodorsal corner; mesepimeron scaled almost to lower margin; scutum with broad, sharply defined submedian and supraalar stripes of dark brown scales contrasting with pale yellow scales of narrow median line, lateral areas, and postpronotum (the contrast is not so great in *punctor* and *communis*; it is more of a contrast between light and dark shades of bronzy brown); scutal bristles predominantly dark brown (bronzy yellow in *communis*), especially noticeable among supraalar group; terminalia almost identical with those of *communis*, although the ventrally projecting setae along the median edge of the apical lobe tend to be longer and the secondary crest of the claspette filament less well developed and sometimes scarcely evident.

Larva. Head and body dark greenish brown; head setae 5-C and 6-C each usually with four or five branches (rarely three or six); prothoracic setae 2-P and 3-P more than half as long as 1-P, although finer (in the *punctor* group 2-P and 3-P are the same length and thickness as 1-P); mesothoracic setae 1-M and 3-M about the same length and thickness, each about as long as head seta 5-C (in *pullatus*, 1-M is very long, almost twice as long as seta 5-C and twice as long as 3-M); comb scales 60 or more, densely crowded, and difficult to count, each with a long spatulate distal free portion fringed with numerous similar spinules; pecten teeth evenly spaced.

Biology. A transcontinental species common in the northern coniferous or boreal forest, but rare and local south of these regions (Rempel 1953). This species overwinters in the egg stage. The larvae, hatching in spring, have been found in cool clear moss-lined pools in depressions at the bases of spruce trees in northern Saskatchewan (Dyar 1919, Rempel 1953) and in snowmelt pools in southeastern Manitoba. In Alaska, the larval habitat was more varied, including roadside ditches, vehicle and animal tracks, and other small pools in disturbed ground (Jenkins 1948, Gjullin et al. 1961). Dyar (1928), Haufe (1952), Rempel (1953), and Gjullin et al. observed that *pionips* larvae were slow to develop and were still in the larval stage after *punctor*, *communis*, and most other associated spring *Aedes* had emerged.

Rempel (1953) observed that in Saskatchewan the species was not an avid feeder and rarely engorged. Although females settled on an observer, they did not probe the skin readily and were easily disturbed.

Distribution. North America and Eurasia: northern half of European USSR, east across Siberia to Alaska, south to the mountains of Idaho and Colorado, and east in forested areas of Canada to Newfoundland.

Aedes provocans (Walker)

Plate 40; Figs. 95, 99, 139, 161; Map 40

Culex provocans Walker, 1848:7.

Culex trichurus Dyar, 1904*a*:170.

Culex cinereoborealis Felt and Young, 1904:312.

Aedes pagetonotum Dyar and Knab, 1909*b*:253.

Aedes poliochros Dyar, 1919*b*:35.

Adult. Female: Integument pale brown; proboscis and palpus dark-scaled, with the palpus sometimes scattered with pale scales; vertex with pale recumbent scales medially, yellowish recumbent scales laterally, forked erect scales predominantly yellowish medially and brown laterally; antepronotum and postpronotum with oval, flat, recumbent white scales; anterolateral margin of scutum and most of lateral and presutural sublateral areas of scutum with broad crescent-shaped white scales; median, submedian, and postsutural sublateral stripes with narrow yellow and pale brown

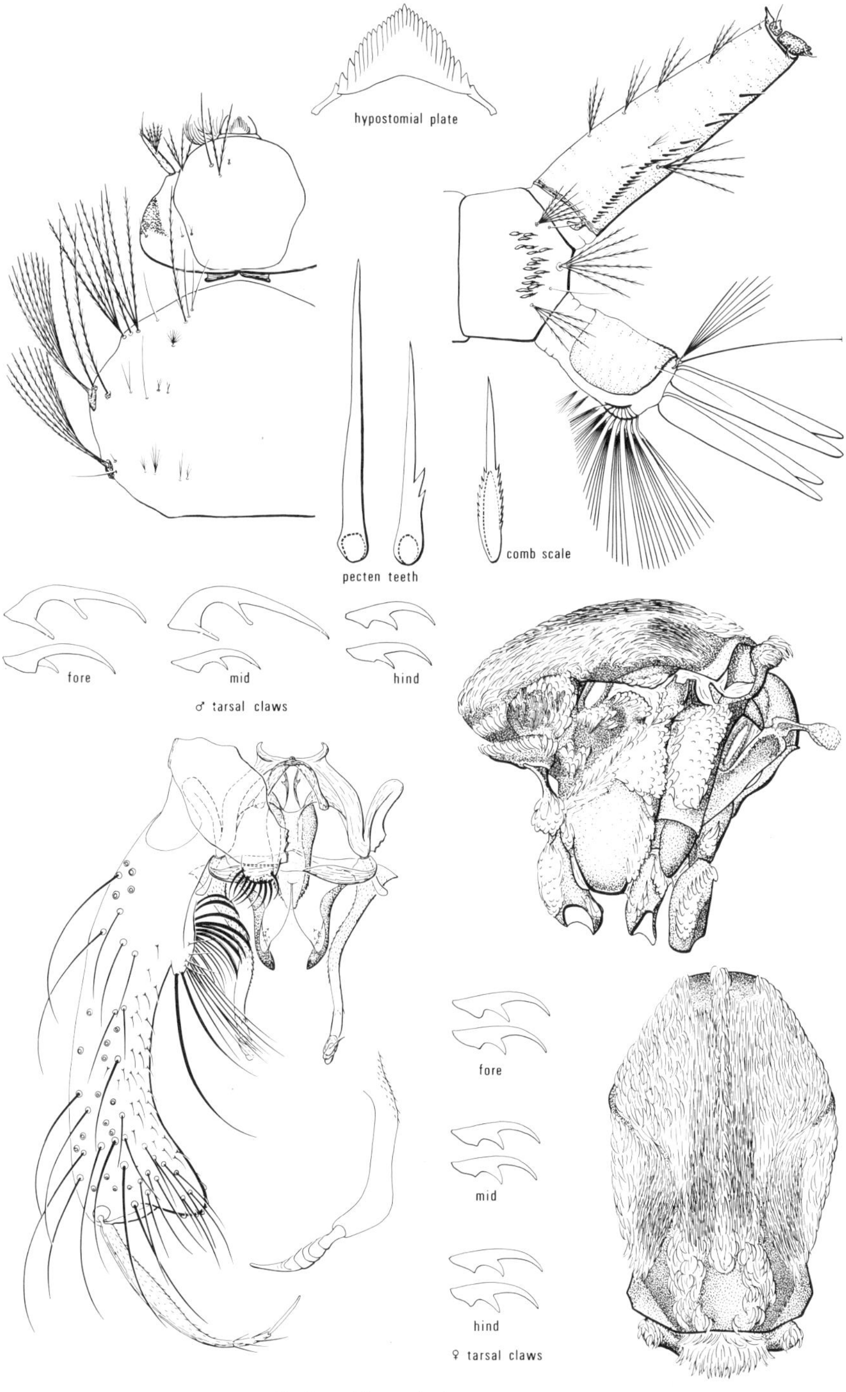

Plate 40. *Aedes provocans*

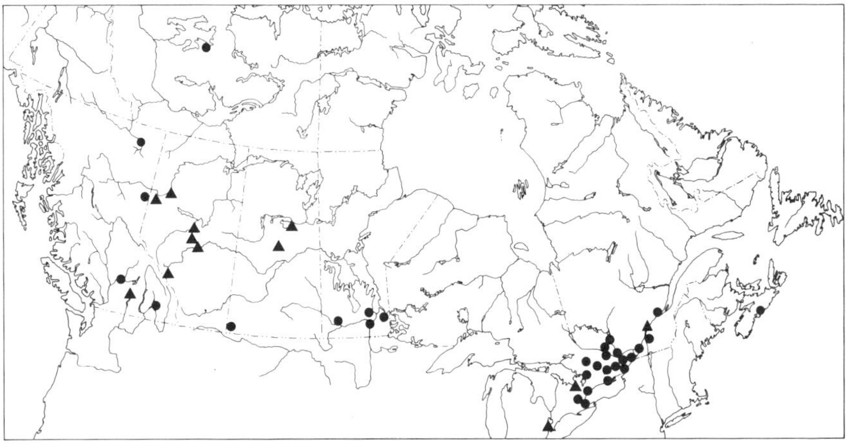

Map 40. Collection localities for *Aedes provocans* in Canada: • specimens we examined, ▲ literature records.

scales intermixed; a triangular patch of creamy yellow scales occupying posteromedial corner of presutural sublateral area bordered medially by brown scales of submedian band and posteriorly by white scales along transverse suture; hypostigmal area, probasisternum, and anteprocoxal membrane adjacent to proepisternum; all of proepisternum, postprocoxal membrane, anterodorsal corner of katepisternum, and all of mesepimeron with broad oval white scales; lower mesepimeral setae numerous; tibiae and first tarsomeres extensively pale-scaled; remaining tarsomeres dark-scaled; wing veins dark-scaled except for small patch of white scales at base of costa; abdominal terga dark-scaled, each with transverse basal band of white scales; intersegmental membrane between first tergum and sternum with a patch of flat oval white scales continuous with those on lateral margin of tergum.

Male: Palpus longer than proboscis by the length of the last palpomere; integument of palpus yellowish, sparsely clothed with dark scales; apex of third and margins of fourth palpomere densely fringed with long yellowish dark-tipped setae; anteprocoxal and postprocoxal membranes, proepisternum, hypostigmal area, and membrane between first tergum and sternum with white scales as in female; other coloration and scaling as in female; basal lobe of gonocoxite small, with longitudinal row of long, curved, medially directed setae, and with two long thick distally directed setae and several finer posteriorly directed bristles; apical lobe prominent; claspette filament short, conical, with four or five concentric ridges.

A medium to large black-legged species, with postprocoxal scale patch and an extensively scaled pleuron, especially the anepisternum, which is almost completely covered with broad waxy white scales that are characteristic of this species. It is also the only species with an extensive patch of scales on the membrane between the tergite and sternite of the first abdominal segment. This membrane, in dried material, is folded under the lateral edge of the tergum and can only be seen from below. If the abdomen is also folded down touching the hind coxae, the entire region is hidden. Scales of the central area of the scutum are yellowish, with a pair of ill-defined brown-scaled submedian stripes infiltrated with yellowish scales; periphery of scutum white-scaled, thus resembling *cataphylla* (which has white scales scattered along the costa, subcosta, and R_1) and *implicatus* (in which the anepisternum is mostly bare), which both have narrower pleural scales.

Larva. The only species of *Aedes* with extra hair tufts on the siphon in addition to the siphonal seta 1-S. Arranged in a dorsolateral and a lateral row on each side, these tufts create a superficial resemblance to a larva of *Culex*, but in *provocans* most of the tufts are dorsolateral rather than ventrolateral. In *Culex*, most of the tufts are ventrolateral, and the dorsolateral tufts are lacking (in the Canadian species at least). Also, species of *Culex* do not have unevenly spaced distal pecten teeth.

Biology. In the Ottawa area, adults of *provocans* are among the first species of *Aedes* to emerge in the spring, after *spencerii* and usually before *implicatus*. Adults seem to be short-lived, because they are seldom seen after the other species of *Aedes* emerge. Larvae usually develop along the edges of semipermanent marshes (Barr 1958), but they have also been found in woodland snowmelt pools and in roadside ditches. It is a forest species, and except for a single female collected at Great Slave Lake, all the records are from southern Canada.

We have observed males swarming after sunset in clearings in the forest, at about 5 m above the ground. Dyar (1923) observed swarm after swarm of males drifting across a meadow, also at the same height above ground.

Distribution. North America, from the Northwest Territories (Great Slave Lake), British Columbia, and Washington State east to Nova Scotia and Georgia.

Aedes pullatus (Coquillett)

Plate 41; Figs. 105, 141, 185, 188; Map 41

Culex pullatus Coquillett, 1904:168.

Aedes acrophilus Dyar, 1917*d*:127.

Aedes pearyi Dyar and Shannon, 1925:78.

Adult. Female: Integument dark brown; proboscis and palpus dark-scaled, with the palpus usually having a few scattered pale scales; pedicel dark brown, concolorous with clypeus, with pale scales dorsomedially; upper margin of postpronotum usually with a few dark setae; scutum pale brown

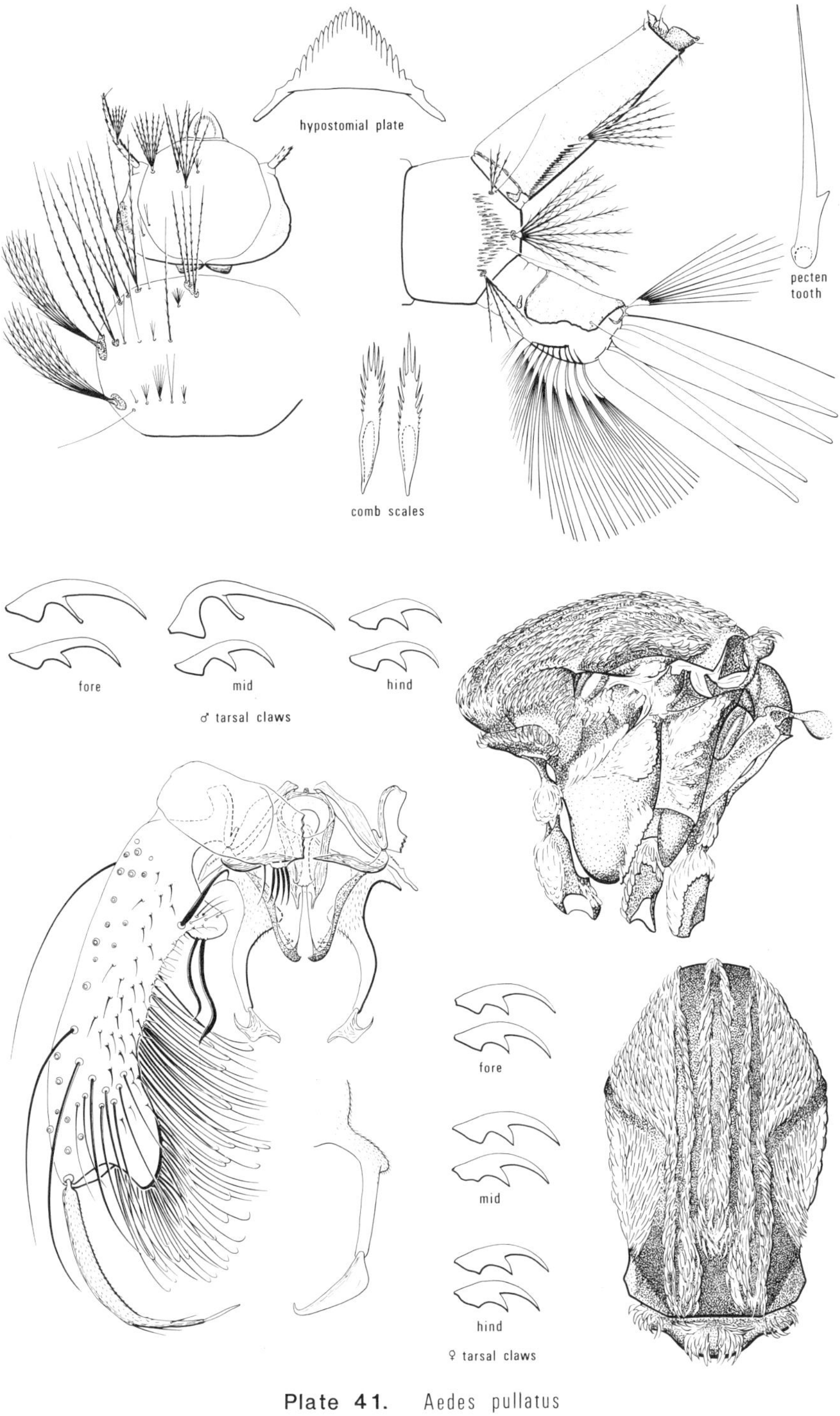

Plate 41. *Aedes pullatus*

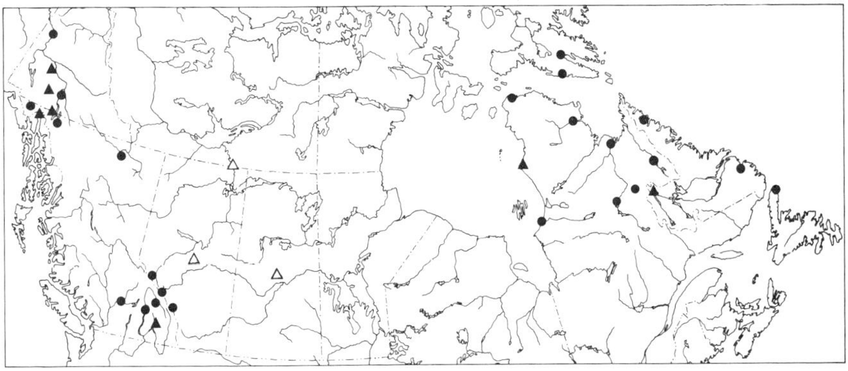

Map 41. Collection localities for *Aedes pullatus* in Canada: • specimens we examined, ▲ literature records, △ literature records based on probably misidentified specimens.

scaled medially, grading to pale-scaled laterally, with a pair of dark submedian stripes that are conspicuous because of the scarcity of scales and the resulting exposure of integument rather than by darker colored scales; setae of scutum numerous, though fewer than in *impiger* or *nigripes*; postprocoxal membrane and anterodorsal corner of katepisternum bare; hypostigmal area and lower portion of mesepimeron usually scaled; lower mesepimeral setae present; tibiae and tarsi mostly dark-scaled; tarsal claws curving evenly beyond middle; costa with patch of pale scales dorsally between base and humeral crossvein; remaining wing veins dark-scaled; abdominal tergites dark-scaled, with each tergite having a basal transverse band of white scales.

Male: Palpus dark-scaled; setae on apex of third and margins of fourth palpomere long but relatively sparse, with those on apex of third palpomere forming no more than a slender pencil-like tuft; hypostigmal area with scales, sometimes with only two or three; scutum more sparsely scaled than in female but coloration and scaling otherwise similar; medioventral edge of gonocoxite with dense fringe of long medioventrally projecting setae; basal lobe of gonocoxite lobulate, flattened front to back, with two long sinuate posteriorly directed bristles arising from its posterior surface; a long stout dorsomedially directed bristle arising from gonocoxite at base of basal lobe; apical lobe well-developed; claspette stem geniculate near base, with rounded medially directed protuberance at bend; claspette filament with prominent keel on convex side.

Larva. Head and body medium brown; head setae 5-C and 6-C each with four to eight branches, most often with five to seven; prothoracic seta 1-P two-branched; setae 2-P and 3-P subequal in length to seta 1-P; mesothoracic seta 1-M unusually long, equal in length, or nearly so, to a branch of 1-P, and twice as long as head seta 5-C; seta 3-M only half as long as and finer than seta 1-M; comb scales about 40–60, with the distal free portion more pointed than in *pionips*, each scale no longer than the attached basal portion, the subapical comb scale teeth fewer and longer; dorsolateral abdominal seta 1-VIII shorter than setae 2-VIII or 5-VIII; pecten teeth evenly spaced; saddle not encircling anal segment.

In chaetotaxy, the larva of *pullatus* is similar to that of *pionips*, with which it coexists in the Yukon Territory, the Rocky Mountains, and in northern Quebec, but is smaller, has lighter brown pigment, and the mesothoracic seta 1-M is twice as long as 3-M. Larvae of *punctor* and *hexodontus*, which also have the prothoracic setae 1-P, 2-P, and 3-P subequal in length have a complete saddle in the last instar, minute mesothoracic setae 1-M and 3-M, and far fewer comb scales.

Remarks. This species is unusual in having a disjunct range (Map 41), not only in North America but in Eurasia as well (Gutsevitch et al. 1974). Although no other mosquitoes in North America appear to show such a disjunct distribution, some other animals and plants are similarly distributed. We suppose this to be a result of two populations that were widely separated during glacial times that have not yet rejoined and not an artifact of insufficient collecting (the west coast of Hudson Bay has been well collected for mosquitoes, more so than the boreal forest farther south).

We have seen a few specimens of *communis*, *implicatus*, and *intrudens* that have a patch of hypostigmal scales; any one of these could be easily misidentified as *pullatus*. Rempel (1953) described the female scutum as having a narrow longitudinal middorsal bare line, but this is not characteristic of our material. We doubt that *pullatus* occurs in Saskatchewan.

Biology. Gjullin et al. (1961) usually found larvae of *pullatus* in small, clear, unshaded snowmelt pools in *Carex* meadows, mostly in tundra. In central Alaska, larvae of *pullatus* were found at all elevations up to the alpine zone (915 m), but were found more often at higher elevations in association with *communis, impiger, punctor,* and *pionips* (Jenkins 1948, Frohne 1957). It was a late emerging species, similar to *pionips*; it was still in the larval stage when *communis* and *impiger* had already emerged (Frohne 1957). As in all northern species of *Aedes, pullatus* overwinters in the egg stage. It produces only one generation each year.

Details of male swarming behavior at Great Whale River, Que., were described by Jenkins and Knight (1950). Some individuals patrolled only 15 cm above the tops of large boulders, whereas others formed aggregations of about 30 males above the hilltops. The latter aggregations remained compact, occupying only a cubic metre, but stationed themselves much higher, about 3.5 m above ground. Swarms also formed over the observer's head. This behavior took place only at dusk, with the males hiding during the day, in contrast to those of *impiger* and *nigripes*, which swarm at midday.

Distribution. North America and Eurasia: in mountains and on arctic tundra, in several disjunct areas; mountains of central Europe, including the Pyrenees, Alps, Tatras, and Carpathians; the mountains of central Asia, including the Altai and the Pamirs of eastern Kazakhastan; Alaska to the mountains of California, east to Alberta and Colorado; also northern Quebec, Newfoundland, and southern Baffin Island. Records from Saskatchewan (Rempel 1950, 1953) and Michigan (Irwin 1941) are probably based on misidentified specimens.

Aedes punctor (Kirby)

Plate 42; Figs. 11, 148, 152, 186; Map 42

Culex punctor Kirby, 1837:309.
Culex implacabilis Walker, 1848:7.
Culicelsa auroides Felt, 1905*b*:449.

Adult. Female: Scutum usually with a middorsal dark brown scaled stripe (median and submedian stripes combined), sometimes with a pair of dark brown scaled submedian stripes separated by an indistinct, slightly paler median stripe; presutural sublateral and lateral areas yellowish brown scaled; postsutural sublateral stripe dark brown scaled, separated from submedian stripe by a narrow line of paler scales; probasisternum usually bare, sometimes with scattered pale scales, but not as extensively covered with scales as in *hexodontus*; costa usually dark-scaled at base; otherwise scarcely distinguishable from *aboriginis, abserratus,* or *hexodontus*, with all of which *punctor* is sympatric in some part of its Canadian range.

Male: Apparently indistinguishable from *aboriginis* and *hexodontus*.

Larva. Head setae 5-C and 6-C each usually single; seta 1-P usually double; setae 2-P and 3-P each single and as long and thick as either branch of 1-P; comb scales 5–25, usually more than 10, each averaging only 0.08 mm long, shorter than last three or four pecten teeth; terminal spine of each comb scale usually no longer than oval basal portion; distal pecten teeth uniformly spaced; saddle completely encircling anal segment; anal papillae usually less than twice as long as anal segment; upper caudal seta 2-X branched (unbranched in *abserratus*) and shorter than seta 3-X.

Remarks. Although most *abserratus* females have the scutum unstriped or with a faint middorsal stripe or sublateral stripes, and most *punctor* females have a pronounced middorsal stripe, some specimens cannot be identified with certainty without reference to the larval skin. Female *hexodontus*, at least those in the subarctic, typically have an unstriped scutum and a patch of white scales at the base of the costa. Those from the mountains are even more like *punctor*, having a similar dark-scaled midlongitudinal stripe on the scutum. If such females have a heavily scaled probasisternum, they may tentatively be considered *hexodontus*, or a bare probasisternum, *punctor*, but many specimens are doubtful.

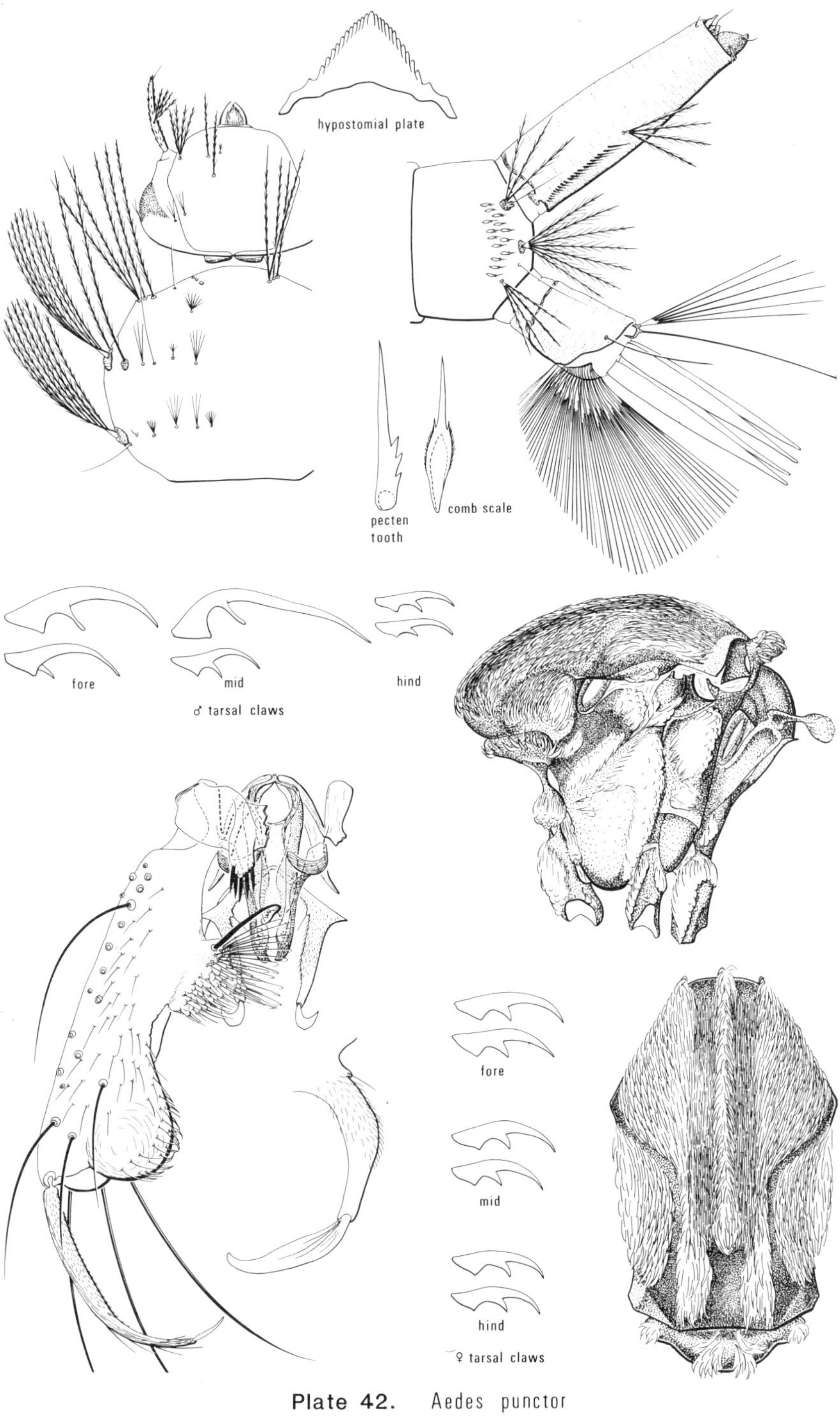

Plate 42. Aedes punctor

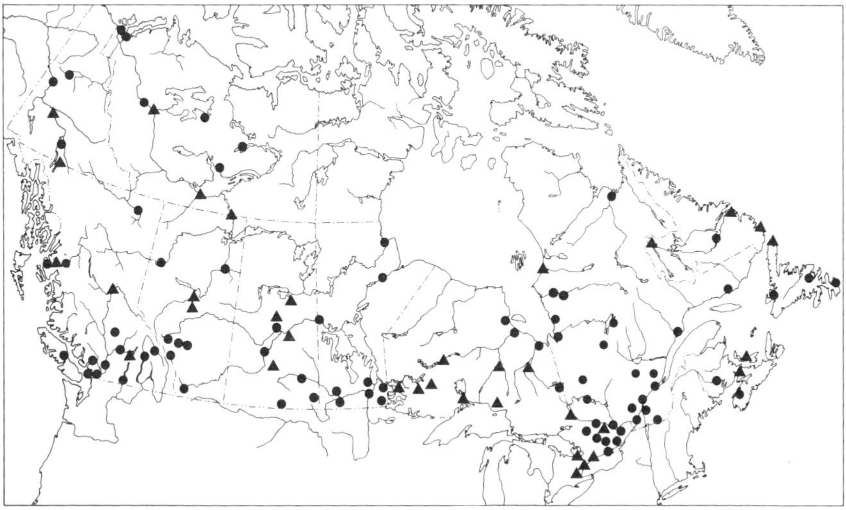

Map 42. Collection localities for *Aedes punctor* in Canada: • specimens we examined, ▲ literature records.

Larvae of *punctor* and *hexodontus* cannot be reliably distinguished by the number of comb scales, but only by the length of the comb scales (Vockeroth 1954, Kalpage and Brust 1968). An earlier reliance on the number of comb scales apparently led to the belief that there were two forms of *punctor*, the normal type and a "tundra" variety (Knight 1951). The latter has since been shown to be *hexodontus* (Wood 1977). The two species are apparently distinct in northern Canada, but their relationship in British Columbia needs further study. In Alaska the identity of *punctor* is further complicated by the presence of *punctodes* Dyar, a salt-marsh species said to be distinguishable from *punctor* only in the larval stage. *Aedes punctodes* has not been collected in Canada but it can be expected in British Columbia. *Aedes aboriginis* is also present in coastal British Columbia, and the separation of *punctor* from these other three species in British Columbia seems hardly possible at present. Curiously, *punctor* is present only in northern Washington and Idaho (Gjullin and Eddy 1972), and earlier records of the species from Utah have been changed to *hexodontus* (Nielsen and Rees 1959).

Biology. Hatching in early spring before the ice has disappeared, the larvae of *punctor* develop rapidly and are among the first species to pupate. *Aedes communis* and *punctor* are the two most common mosquitoes of the boreal forest (Rempel 1953, Gjullin et al. 1961, Pickavance et al. 1970, Maire and Aubin 1976), occurring from coast to coast. Unlike *communis*, however, the distribution of *punctor* is practically confined to the boreal forest, and it seldom occurs south of it, nor does it range out onto the tundra (Jenkins and Knight 1950). In the Ottawa area, *punctor* is much less common than *communis* and its larvae have been collected consistently only in coniferous forest pools, usually surrounding sphagnum bogs, where they were associated with *abserratus* and early instars of *cinereus*. Several authors have remarked on its association with acidic water (Barr 1958, Jenkins and Knight 1950). At Goose Bay, Nfld., Haufe (1952) found *punctor* larvae most often in open grassy pools in clearings surrounded by willows. In Alaska, both sphagnum bogs and grassy marshes were habitats for *punctor* (Gjullin et al. 1961); this species accounted for at least 39% of all mosquitoes captured there (Sommerman 1977).

Hocking et al. (1950) described a male swarm of *punctor*, associated with *communis*, that appeared in the evening as a dark cloud over a railway track at Churchill, Man., and extended as far as the eye could see, even following the track around a curve. At Moose Factory, Ont., aggregations of male *punctor* formed over a road bordered by spruce trees. The swarms moved back and forth and became more compact as the wind increased (Jenkins and Knight 1952). Between 25 and 150 matings/min were observed in several immense swarms of several thousand male *punctor* in a forest glade in Alaska in mid-June (Frohne and Frohne 1952). Mating was most often observed at the beginning of the swarming season and less often as the summer passed.

Distribution. North America and Eurasia: northern Europe and Crimea east across Siberia to Alaska, south to British Columbia and the mountains of Colorado, then east to Newfoundland and New York.

Aedes rempeli Vockeroth

Plate 43; Figs. 106, 143; Map 43

Aedes rempeli Vockeroth, 1954*a*:112.

Adult. Female: Integument of head and body dark brown; proboscis and palpus dark-scaled, with the palpus usually having scattered pale scales; pedicel dark brown, concolorous with clypeus; scales of pedicel mostly pale, only on ventromedial surface, but sometimes lacking; recumbent scales of vertex mostly pale, erect forked scales dark; thorax sparsely scaled, with indefinite pattern of two rather prominent middorsocentral spots of white scales and a line of white scales connecting this spot to edge of scutum above spiracle; median band yellowish; submedian band with sparse dark brown scales; lateral area basically yellowish scaled, enclosing an area of brown scales on scutal fossa; postprocoxal membrane and hypostigmal area bare; scales of katepisternum reduced to posterodorsal and posteroventral patches; mesepimeron without scales on basal third; lower mesepimeral bristles present; tibiae, tarsi, and wing veins dark-scaled, except for small pale-scaled patch at extreme base of costa; abdominal tergites dark-scaled, each with narrow basal transverse pale-scaled band.

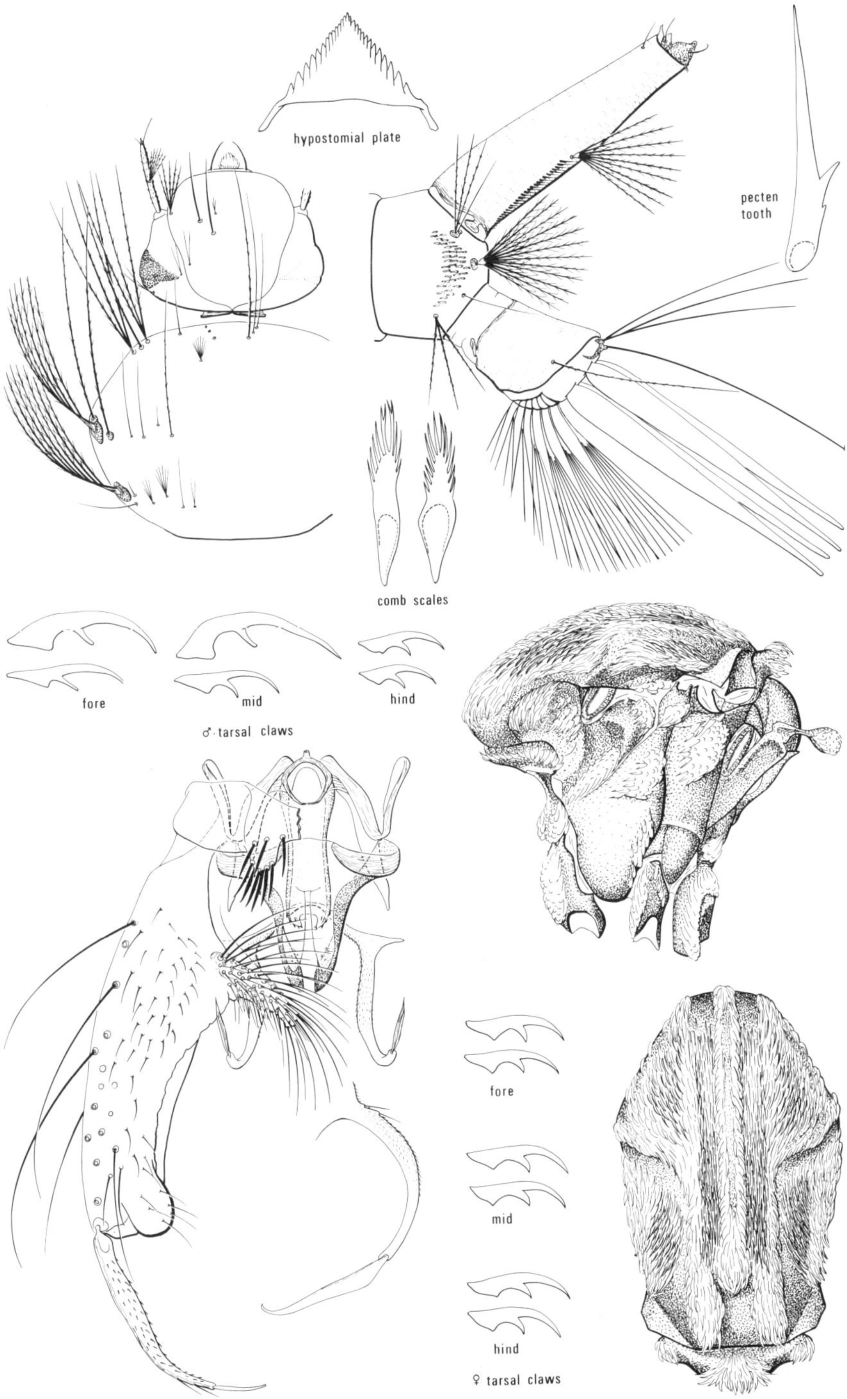

Plate 43. Aedes rempeli

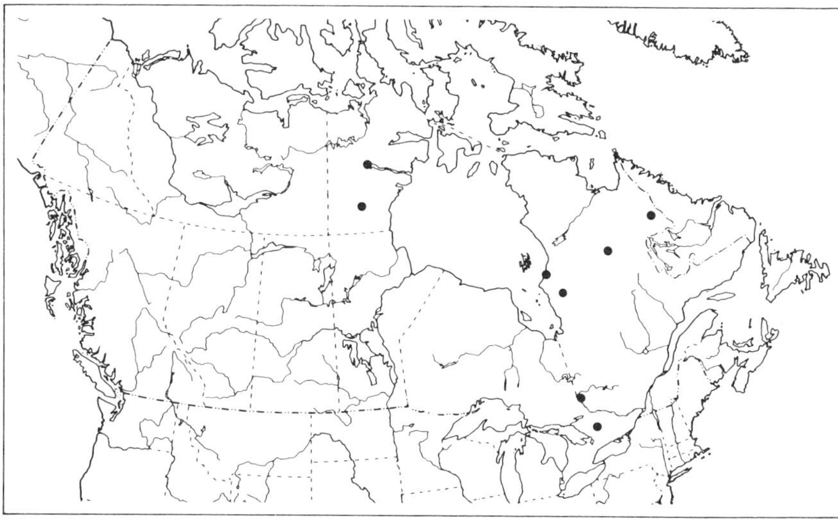

Map 43. Collection localities for *Aedes rempeli* in Canada: • specimens we examined.

Male: Palpus dark-scaled; apex of third and median and lateral edges of fourth palpomere with usual fringe of long setae, but with few setae; gonocoxite with long finger-like setose basal lobe lacking an enlarged basal bristle; apex of gonocoxite tapering, with the apical lobe small and arising close to base of gonostylus; claspette filament rather straight, with abruptly curved apex and low keel on posterior edge.

Females of *rempeli* could be easily mistaken for those of *pullatus*, which usually have a distinct group of hypostigmal scales and a more strongly curved tarsal claw. These two species occur sympatrically across north central Quebec and possibly also in parts of northern Saskatchewan, Alberta, and adjacent Northwest Territories.

Larva. Head setae 5-C and 6-C each single; prothoracic seta 1-P usually single, twice as long as seta 5-C; setae 2-P and 3-P one-half to two-thirds as long as seta 1-P but finer; mesothoracic seta 1-M as long as head seta 5-C; dorsolateral setae (series 1) present on abdominal segments II to VII, but lacking on VI; comb scales 30–45, each fringed with subequal spinules; pecten teeth evenly spaced; saddle completely encircling anal segment.

No other species has elongate dorsolateral setae on segments III to VII except for VI. Larvae of *rempeli* might be mistaken for *punctor* because of the encircling saddle, but setae 2-P and 3-P are not as long or as thick as 1-P, seta 1-M is elongate, and the comb scales are more numerous and lack an elongate apical spine.

Biology. The larva of *rempeli* was first collected from an unshaded rock crevice pool, 300 × 90 × 30 cm deep, at Great Whale River, Que., in association with *communis* and *impiger*. Two *rempeli* larvae were also collected at Padlei, NWT, in a much larger semipermanent pool, again associated with these two species plus *nigripes*. The same association of species, including larger numbers of *rempeli* larvae, was found in similar deep rock crevice pools at Baker Lake, NWT (Smith and Brust 1970). These pools were almost without emergent or aquatic vegetation except for an encrusting alga that grew on the underwater rock surfaces. Fissures in the bottom were about 1 m deep or slightly deeper; therefore water in the pool remained in contact with the underlying permafrost. Hatching of *rempeli* eggs was delayed until mid-June because of so much ice and snow among the boulders. Larval development was slow, occurring during most of July 1968, when the water temperature never rose above 6.1°C (mean 3.7°C). Of the four species of larvae present, *rempeli* was the last to develop. In late summer, after *rempeli* had emerged, the water in these habitats disappeared quickly as a result of thawing in the deeper layers.

At Baker Lake, all 70 reared adult females of *rempeli* developed their ovaries without a blood meal, that is, they were autogenous (Smith and Brust 1970). They emerged with the ovarian follicles in an undifferentiated state, but with an extensive nutrient reserve in the form of larval abdominal musculature. At 15–20°C, they completed ovarian development. Dissections of 37 adult females showed 15–39 eggs (mean 25.8). When offered moist paper as an oviposition substrate, eggs were freely deposited. Some females were even capable of completing ovarian development when denied a carbohydrate source.

Aedes rempeli may be widely but sparsely distributed in the boreal forest of northern Manitoba, Ontario, and Quebec. Some evidence suggests that females in this region may seek blood, perhaps only in the second or third gonadotrophic cycles, but because they are outnumbered by other mosquitoes, are rarely noticed. The single female that was collected by one of us (D.M.W.) in Algonquin Park, Ont., in a cedar swamp containing many deep, cold, water-filled crevices was among other mosquitoes coming to bite. The material from Indian House Lake, Que., and Laniel, Que., consists of a series of 14 females and 3 females, respectively, without associated males or any other evidence that they had been reared. However, the specimens are undamaged, and are not likely to have been collected by sweeping. The dates of collection are late, 27.VII.1954 and 10.VIII.1954 for Indian House Lake and 30.VIII.1932 for Laniel. These last specimens, which were recently found among unidentified material, are undoubtedly the first examples of the species ever collected.

Distribution. North America and Asia: Northwest Territories and north central Quebec south to central Ontario; eastern Kazakhastan (Altai Mountains) and eastern Siberia (Yakutia).

Aedes riparius Dyar & Knab

Plates 9, 44; Figs. 82, 90, 121; Map 44

Aedes riparius Dyar and Knab, 1907*b*:213.

Adult. Female: Integument pale reddish brown; proboscis predominantly dark-scaled at base and apex, mostly pale-scaled on middle third or more; palpus dark-scaled with spot or band of pale scales at base of each palpomere; scales of pedicel few, mixed pale and dark; upper half of postpronotum, median and submedian stripes of scutum, a spot in presutural sublateral area, a short postsutural sublateral stripe, and a spot above paratergite orange-scaled; occiput, antepronotum, lower third of postpronotum, and most of presutural sublateral and lateral areas yellowish-scaled; hypostigmal area often with scales; postprocoxal membrane and anterodorsal corner of katepisternum with pale scales; lower third of mesepimeron and mesomeron bare; lower mesepimeral setae absent; tibiae and first and usually second tarsomeres with scattered pale and dark scales; all tarsomeres except last on fore- and mid-legs each with basal pale-scaled ring; tarsal claw long, moderately and evenly curved, that of foreleg half again as long as claw of hind leg; subbasal tooth short, directed ventrally, enclosing an angle of 45° or more with remainder of claw (Fig. 90); wing veins predominantly dark-scaled, with scattered pale scales; abdominal tergites each with well-defined but narrow basal transverse band of pale scales and a slightly broader apical transverse band containing many pale scales in addition to dark scales.

Male: Proboscis and palpus extensively pale-scaled; base of third palpomere especially pale, somewhat translucent; apex of third palpomere and fourth and fifth palpomeres densely fringed with long setae that are yellowish basally and darker apically; middorsal stripe of scutum reddish brown scaled rather than orange scaled; gonocoxite with a long, pointed, finger-like basal lobe and an enlarged medially directed seta arising at its base; claspette filament short, with broad keel on convex side extending from base to apex.

Larva. Head setae 5-C and 6-C double or sometimes triple; lower lateral seta 7-II more than half as long as 6-II (much shorter in *flavescens*); ventrolateral setae (series 13) on abdominal segments III to V single, as long as corresponding upper lateral setae; dorsolateral setae (series 1) insignificant on all segments; comb scales fewer than 10, each rather large and darkly pigmented, with long apical spine; distal pecten teeth more widely and unevenly spaced than proximal teeth; last pecten tooth arising before midlength of siphon; siphonal seta 1-S arising just beyond middle, but before apical third of siphon; seta 9-S on ventrolateral apical flap of siphon not enlarged or hook-like; four or more precratal setae; anal papillae pigmented, about as long as saddle.

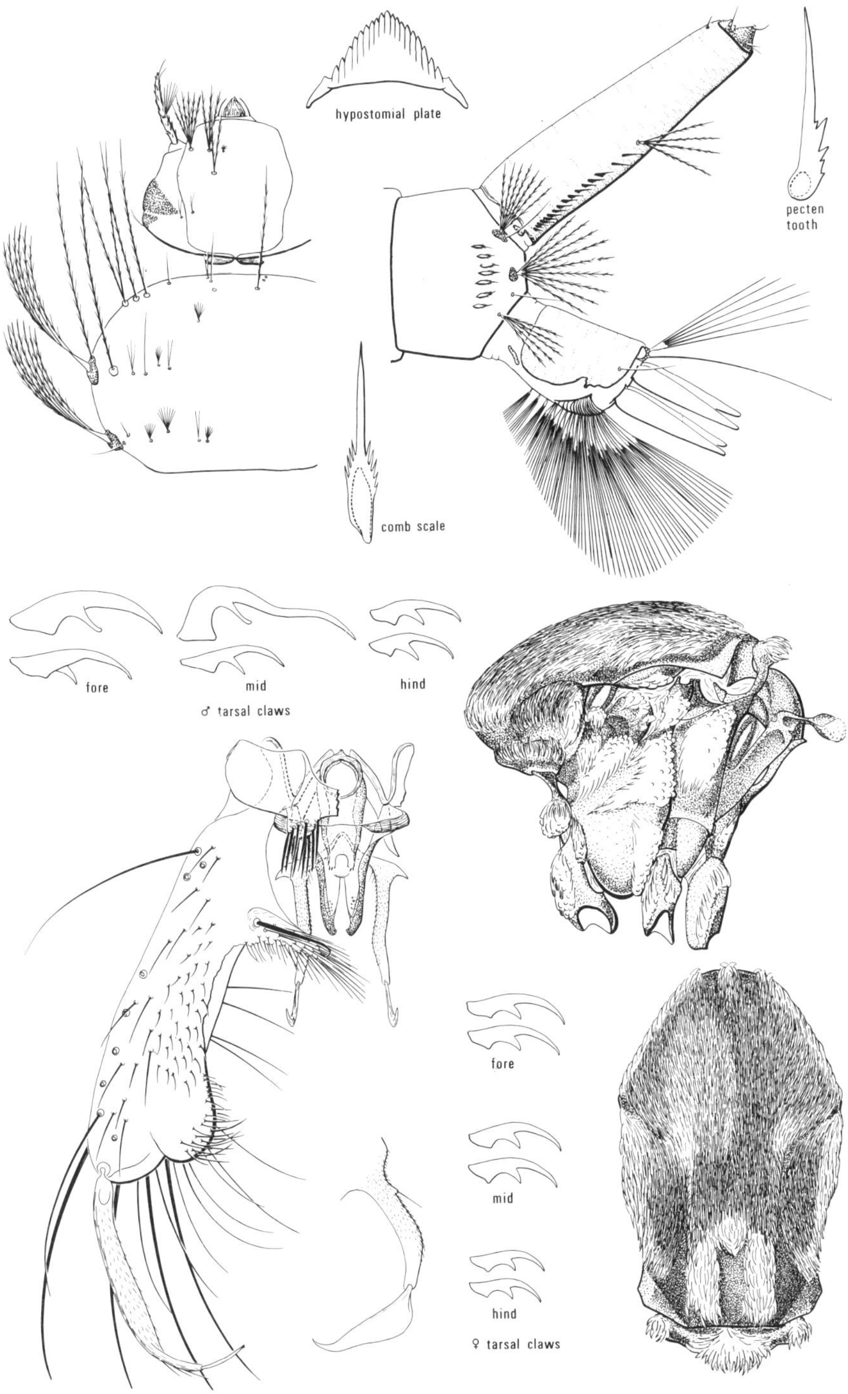

Plate 44. *Aedes riparius*

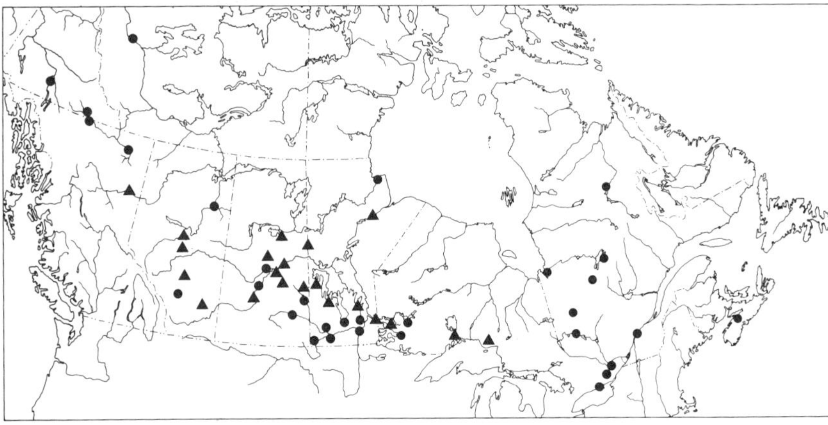

Map 44. Collection localities for *Aedes riparius* in Canada: • specimens we examined, ▲ literature records.

Biology. This species is uncommon to rare over most of its range. It has been reported as common only in a few localities in Saskatchewan and Alberta (Rempel 1950, Happold 1965*a*, Graham 1969*b*). Unfortunately these and other references to *riparius* probably included specimens of *euedes*, which was not recognized as a separate widespread species until recently. Its life cycle is similar to that of *euedes*, *excrucians*, and *fitchii*, and the larvae of *riparius* have been occasionally collected with one or more of these species in the Ottawa area.

Distribution. North America and Eurasia: Europe and Siberia east to Alaska, south to Colorado, and east to Nova Scotia and New York.

Aedes schizopinax Dyar

Plate 45; Map 45

Aedes schizopinax Dyar, 1929:1.

Adult. Female: Integument dark brown; dorsal surface of proboscis dark-scaled, occasionally with a few scattered pale scales; ventral surface of proboscis mostly pale-scaled except for some dark scales near apex; palpus dark-scaled usually with some scattered pale scales; pedicel and first flagellomere mostly pale-scaled; vertex with pale brownish scales, paler middorsally and on ventral half; antepronotum and postpronotum mostly pale-scaled, with a few darker sand-colored scales dorsally; lower edge of cervical sclerite and membrane below it with a few pale scales; probasisternum, proepisternum, procoxa, and anteprocoxal and postprocoxal membranes almost entirely covered with pale whitish scales; scales of scutum uniformly sand-colored, paling to whitish laterally and around prescutellar depression; scales of narrow submedian stripe concolorous with, though narrower than, adjacent scales (as in many *abserratus* and *intrudens*) and thus appearing

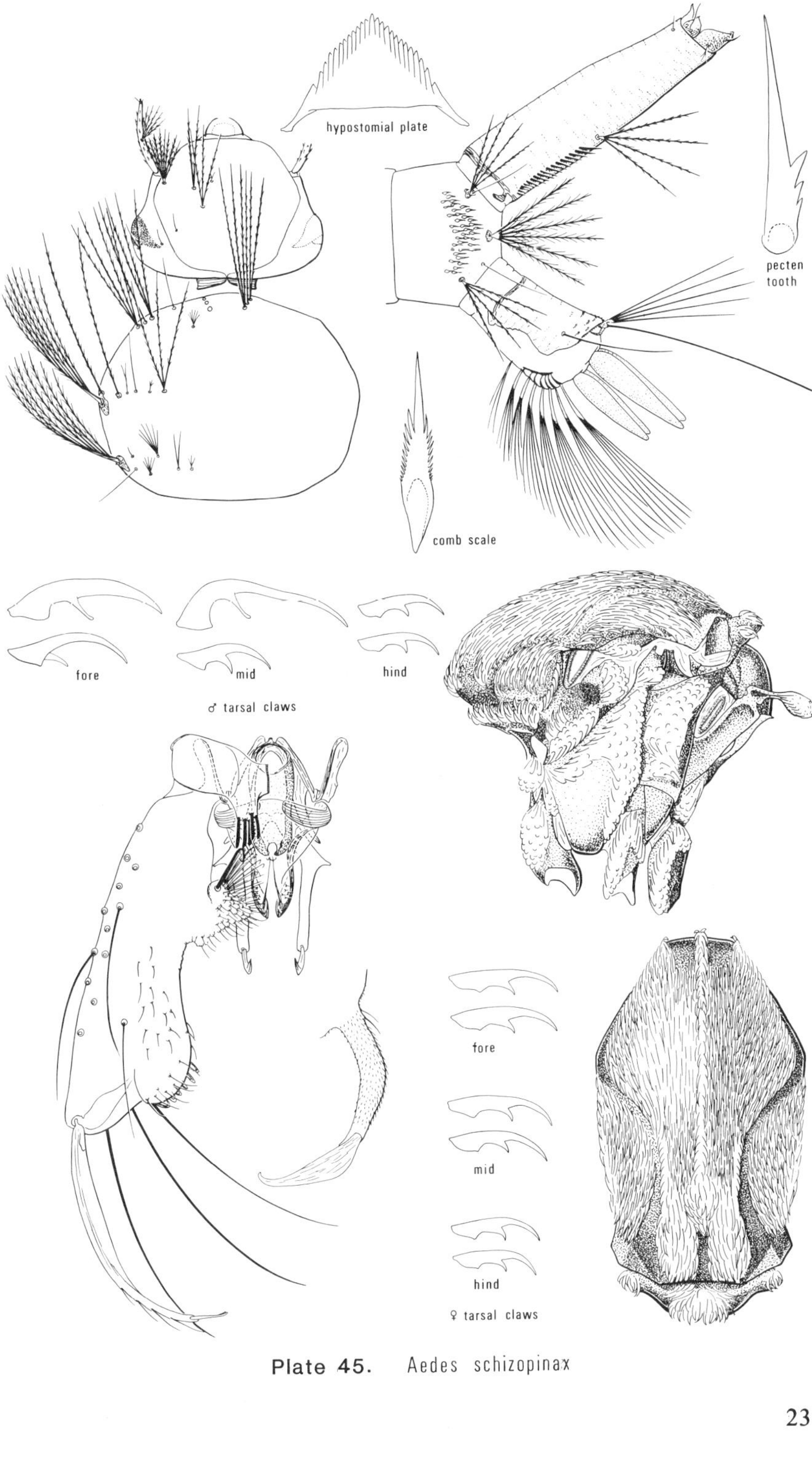

Plate 45. Aedes schizopinax

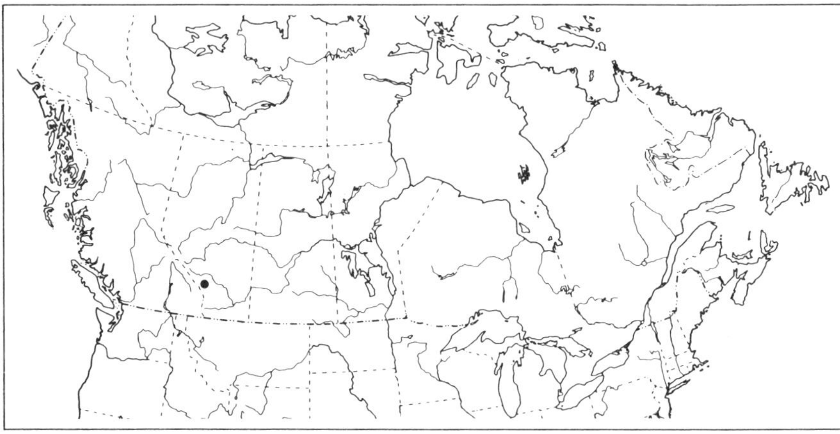

Map 45. Collection localities for *Aedes schizopinax* in Canada: • specimens we examined.

darker because of greater exposure of dark brown integument; subspiracular and postspiracular areas pale-scaled; katepisternum scaled dorsally to anterodorsal corner, even with a few scattered pale scales on normally bare anteroventral area; mesepimeron scaled to lower margin; mesepimeral setae present; tibiae mostly dark-scaled dorsally with scattered pale scales, mostly pale-scaled ventrally; tibiae dark-scaled; tarsal claws moderately and evenly curved, with small subbasal tooth; wing veins dark-scaled except for a small patch of pale scales and a few scattered scales beyond this patch at base of each major vein except Cu; abdominal tergites each dark-scaled, with transverse basal band of white scales.

Male: Proboscis mostly dark-scaled, with a few scattered pale scales both above and below; palpus about as long as proboscis; second and third palpomeres extensively pale-scaled; apex of third palpomere and fourth and fifth palpomeres dark-scaled, with medioventral patches of dense, long, black setae; precoxal bridge and anteprocoxal membrane below it with pale scales, but probasisternum bare; mesothorax with coloration and arrangement of scaling as in female; terminalia scarcely distinguishable from *aboriginis*, *hexodontus*, or *punctor*; claspette stem said to be pilose to near apex, unlike that in the three other species (Carpenter and LaCasse 1955).

Larva. Head seta 5-C two- to four-branched; prothoracic setae 2-P and 3-P as long and as thick as any branch of seta 1-P (typical of all species in the *punctor* subgroup); mesothoracic seta 1-M with three or four branches, as long and as thick as head seta 5-C; dorsolateral abdominal setae 1-IV to 1-VII of segments IV to VII usually double or triple, each branch as long and thick as a branch of upper lateral seta of its respective segment; pecten teeth uniformly spaced; saddle not completely encircling anal segment.

No other species in Canada, except *dorsalis* and *campestris*, has the mesothoracic seta 1-M multiple; *dorsalis* has the head seta 5-C single (rarely one side double) and both *dorsalis* and *campestris* have shorter dorsolateral abdominal setae and much shorter and finer prothoracic setae 2-P and 3-P. *Aedes aboriginis*, which is much like *schizopinax*, has an unbranched mesothoracic seta 1-M.

Biology. *Aedes schizopinax* has only recently been discovered in Canada (Enfield 1977). The species has seldom been collected and little is known of its habits.

Distribution. Western North America, from Alberta to California and Wyoming.

Aedes sierrensis (Ludlow)

Plate 46; Figs. 70, 114; Map 46

Taeniorhynchus sierrensis Ludlow, 1905*a*:231.
Aedes varipalpus, authors before 1957, not Coquillett.

Adult. Female: Integument medium brown; proboscis and palpus dark-scaled; all palpomeres except the first one with an apical ring of white scales; scales of scutum forming complex pattern of pale yellow and dark brown lines and spots; median and sublateral stripes anterior to suture pale, separated by narrow dark line; lateral stripes, transverse suture, postsutural dorsocentral stripe, and perimeter of prescutellar depression pale-scaled, with the remainder dark-scaled; postprocoxal membrane and hypostigmal area bare; scales of katepisternum and mesepimeron reduced to two patches each; lower mesepimeral setae absent; legs dark-scaled except for white-scaled bases and apices of each segment; basal and apical white rings of first tarsomeres subequal, those of remaining tarsomeres four or more times wider apically than basally; last hind tarsomere entirely white-scaled; hind tarsal claws lacking subbasal tooth; wing veins entirely dark-scaled except for white patch at base of costa; abdominal tergites dark-scaled, each with lateral and middorsal basal triangle of white scales; cercus pointed apically, but shorter and broader than in most *Aedes* except *atropalpus* and *togoi*.

Male: Palpus about as long as proboscis, dark-scaled except as follows: first palpomere entirely white-scaled, second entirely dark-scaled, third with apical ring of white scales, fourth with basal and apical white rings, fifth with basal white ring only; long setae reduced to a few on lateral and medial surfaces of fourth palpomere; scutum predominantly pale-scaled; fore- and mid-legs dark-scaled except for a few white scales at base and apex of tibia and first tarsomere; hind leg as in female except that basal white rings on second to fourth tarsomeres greatly reduced, appearing absent; hind tarsal claws, as in female, lacking subbasal tooth; basal lobe of gonocoxite with dense tuft of long, thick, pale brown setae; apical lobe lacking; claspette filament narrow, unkeeled, tusk-like.

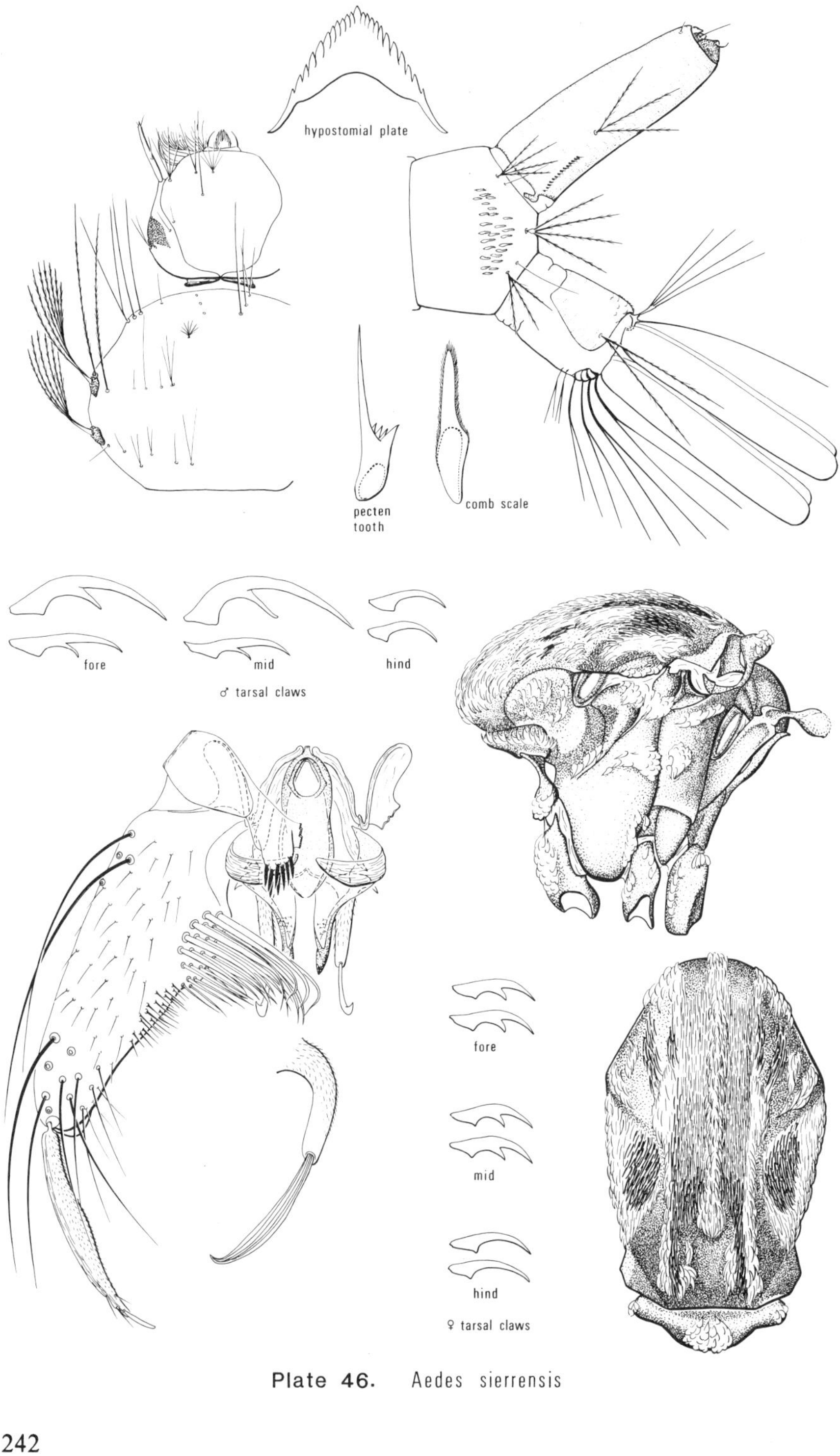

Plate 46. *Aedes sierrensis*

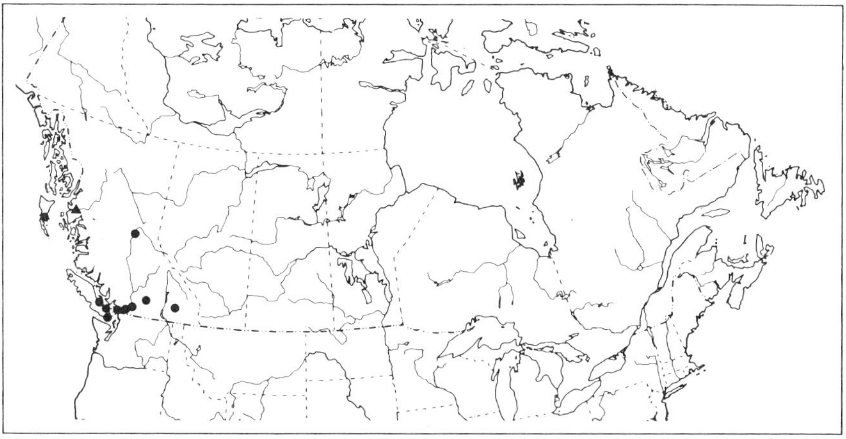

Map 46. Collection localities for *Aedes sierrensis* in Canada: • specimens we examined, ▲ literature records.

Larva. Antenna smooth; antennal seta 1-A unbranched; head seta 4-C rather prominent, as in *hendersoni* and *triseriatus*; seta 5-C single; seta 6-C usually branched; prothoracic setae all rather short and fine; dorsolateral abdominal setae (series 1) typically double; abdominal setae of series 4 usually stellately three-branched; upper lateral seta usually two- or three-branched; comb scales 12–28, each with long finger-like process fringed with minute spinules; acus not separate from siphon (separate in *hendersoni*); pecten teeth few, small, confined to basal third or less of siphon; saddle in profile narrower anteriorly than posteriorly, though not as strongly truncated anteroventrally as in *hendersoni*; saddle seta 1-S branched; anal papillae subequal in length, each twice as long or longer than anal segment, broadly rounded apically as in *hendersoni*.

Remarks. References to this species before 1956 are under the name *varipalpus* (Coquillett) (Hearle 1927*a*, Rempel 1950). Belkin and McDonald (1956, 1957) showed that the name *varipalpus* properly belonged to a species restricted to a small part of the southwestern USA and that the widespread coastal species was *sierrensis*.

The subgeneric placement of *sierrensis* is unclear at present. Carpenter and LaCasse (1955) placed *sierrensis* in the subgenus *Finlaya*. Belkin and McDonald (1957) transferred it to *Ochlerotatus* (under the name *varipalpus*). Rohlf (1963) and Cupp and Horsfall (1969) restored it to *Finlaya* on the basis of shared morphological, behavioral, and cytochemical characters, as well as parasitic affinities. Finally Zavortink (1972) transferred it back to *Ochlerotatus*. We suspect *sierrensis* may belong in *Protomacleaya*, a possibility not explored by the authors mentioned above. The larva of *sierrensis* lacks most of the stellate body setae that are characteristic of *Protomacleaya*, but like them it has a smooth antenna; a single (occasionally double) antennal seta 1-A; head seta 4-C rather prominent; head seta 6-C with more branches than 5-C; the comb scales having a long flat finger-like apical portion fringed with closely set, minute, equal spinules; and the saddle hair 1-S branched. The saddle is triangular, and the anal papillae are long and sausage-like as in *hendersoni*, but the acus is fused to the siphon, as in *triseriatus*. The comb scales of all three species are similar and unlike those of other species of *Ochlerotatus*. The adult female is also peculiar in having untoothed hind claws and short rounded cerci.

Biology. *Aedes sierrensis* is the only treehole mosquito known from western British Columbia. In central and eastern British Columbia, it may be accompanied by *hendersoni*. According to Hearle (1926) it overwinters in the egg stage in British Columbia, but larvae may be found during the winter in Oregon (Gjullin and Eddy 1972) and possibly also in southern British Columbia. Although typically an inhabitant of tree holes, larvae of *sierrensis* have also been collected in hollow stumps and various artificial containers, including tires, in wooded locations.

The males do not swarm; they are attracted to the host, as in the male of *diantaeus*. When a female approaches, the male may intercept her and either mate with her in the air (Carpenter and LaCasse 1955) or while she is engorging (Hearle 1926).

The species is usually uncommon and of little importance as a pest or vector.

Distribution. Western North America, from the Queen Charlotte Islands, B.C., south to California.

Aedes sollicitans (Walker)

Plate 47; Figs. 73, 81; Map 47

Culex sollicitans Walker, 1856:427.

Adult. Female: Proboscis always with well-defined white-scaled ring encircling as much as middle third; palpus dark-scaled except for pale-scaled apex; postpronotum bare except for narrow band of yellowish brown scales along posterodorsal margin concolorous with those on adjacent scutum and a small patch of white scales on ventral corner; scutum with weakly defined darker middorsal stripe or none, in which case the scales of scutum palest medially, gradually becoming darker laterally; first tarsomeres each with basal band of white scales and well-defined middle band of yellowish scales, alternating with bands of predominantly dark scales (Fig. 73 in key to females); abdominal tergites each with lateral patch of white scales, contrasting with transverse basal and midlongitudinal bands of yellowish scales; otherwise as in *nigromaculis*.

Male: Postpronotum bare except for narrow band of yellowish brown scales along dorsal margin and patch of white scales in ventral corner; first

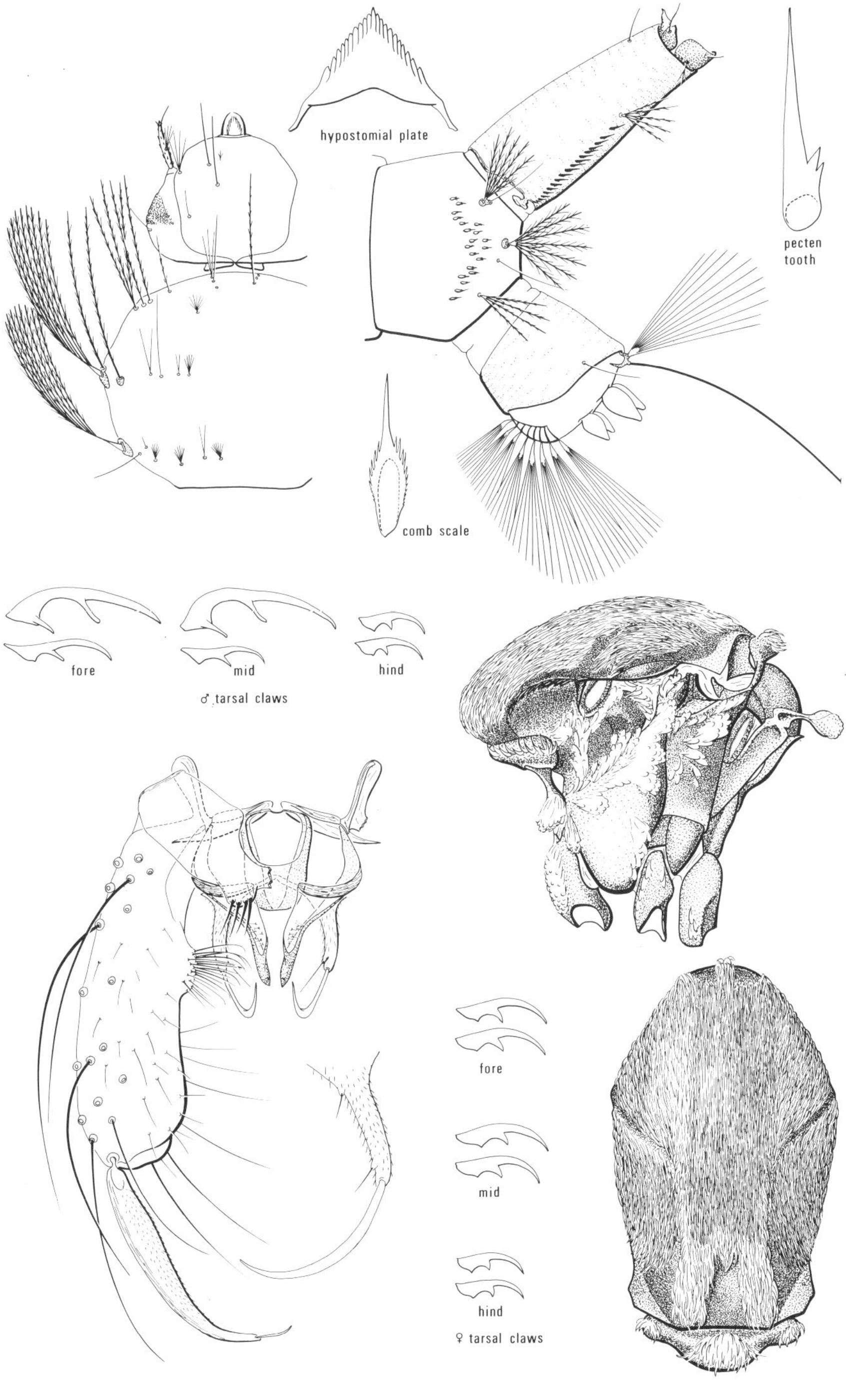

Plate 47. Aedes sollicitans

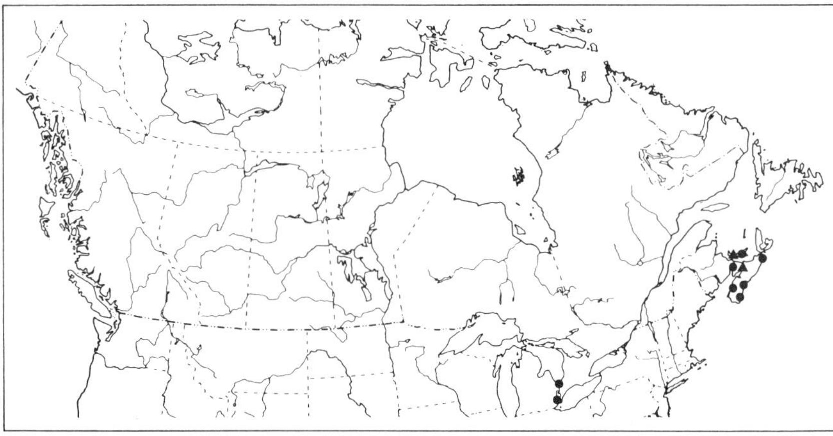

Map 47. Collection localities for *Aedes sollicitans* in Canada: • specimens we examined, ▲ literature records.

hind tarsomere with well-defined middle yellow-scaled ring; otherwise as in *nigromaculis*.

Larva. Head setae 5-C and 6-C each single; prothoracic seta 1-P single; setae 2-P and 3-P weak; mesothoracic seta 1-M minute; pecten teeth evenly spaced, each tooth with only one subbasal tooth, occasionally with a second smaller tooth; siphonal seta 1-S about as long as apical diameter of siphon; saddle completely encircling anal segment in last instar (though poorly sclerotized and difficult to discern); anal papillae minute, much shorter than saddle in our material (although some authors have reported longer papillae).

Biology. Until recently discovered in a brine dump at Sarnia, in southwestern Ont. (Helson et al. 1978), *sollicitans* was known in Canada only from the Maritime Provinces, where it breeds in coastal salt marshes. In the USA it ranges widely inland, wherever saline conditions occur. It is especially numerous and troublesome in coastal areas, and as a result is one of the most infamous species in North America.

Aedes sollicitans overwinters in the egg stage and hatches in early summer when it has been flooded with suitably warm water. There are probably several generations each summer in Canada; 5–8 broods have been recorded in New Jersey (Headlee 1945). Eggs are laid on the moist mud of salt-marsh pools, between the prevailing waterline and the level of maximum inundation (Knight and Baker 1962). Oviposition sites have 1644 ppm or more of soluble salt content compared to 1000 ppm or less in freshwater (Knight 1965). Eggs hatch within minutes of being flooded, and larvae mature within a week of hatching (Britton 1912, Headlee 1945). Larvae are tolerant of salinity and pollution (Chapman 1959) and often develop in the discharge from coal mines or gas wells (Peterson and Smith

1945). They have also been collected in freshwater in association with *Aedes vexans* and *Psorophora* spp. (Breeland et al. 1961). However, larvae are not found in seawater that has a salinity of 19 000 ppm of chloride or that contains predatory fish, but they are found in tide pools that form above the usual high tide line from flooding by spring tides and rain. Because of the perpetually high water table, even moderate summer rains may be enough to flood the eggs.

Females of *sollicitans* are capable of flying great distances. Carpenter and LaCasse (1955) mention the collection of females in light traps 160 km from their breeding sites. This capacity for dispersal has undoubtedly aided their exploitation of man-made saline habits away from the coast, such as the brine dump at Sarnia, Ont., from which the nearest known breeding place is in Ohio, much farther south. They are aggressive biters, attacking in full sun and in strong wind. They rest in the grass during the day and approach directly, attacking the legs. They do not enter houses (Headlee 1945). In Kansas, their blood meals were entirely of mammalian origin (Edman and Downe 1964).

Distribution. Atlantic Coast of North America, from New Brunswick to Texas and the West Indies, and in isolated inland stations west to North Dakota and Arizona.

Aedes spencerii (Theobald)

Plate 48; Figs. 146, 165, 167, 171; Map 48

Culex spencerii Theobald 1901*b*:99.
Grabhamia spencerii var. *idahoensis* Theobald 1903*a*:250.

Adult. Female: Integument medium brown; proboscis dark-scaled; palpus usually with scattered pale scales; pedicel almost completely encircled with pale scales; upper half of antepronotum and of postpronotum and middorsal stripe of scutum reddish brown scaled; occiput and sublateral and lateral stripes of scutum pale yellowish; postprocoxal membrane, anterodorsal corner of katepisternum, and all of mesepimeron with pale scales; lower mesepimeral setae absent; legs predominantly pale-scaled, even on basal tarsomeres, with scattered dark scales; distal tarsomeres mostly dark-scaled on dorsal surfaces; tarsal claws moderately and uniformly curved with small subbasal tooth; wing veins mostly pale-scaled, except for apical portions of costa, R_1, R_{4+5}, and all of Cu; abdominal tergites extensively and sometimes almost entirely pale-scaled, but when dark scales present, forming paired lateral spots only. Some specimens from southern Alberta and central British Columbia have the tergites dark-scaled, each with typical basal transverse white-scaled band. This variant has been named *idahoensis*. Some females, perhaps bleached by the sun, are entirely pale-scaled, lacking the usual brown-scaled areas.

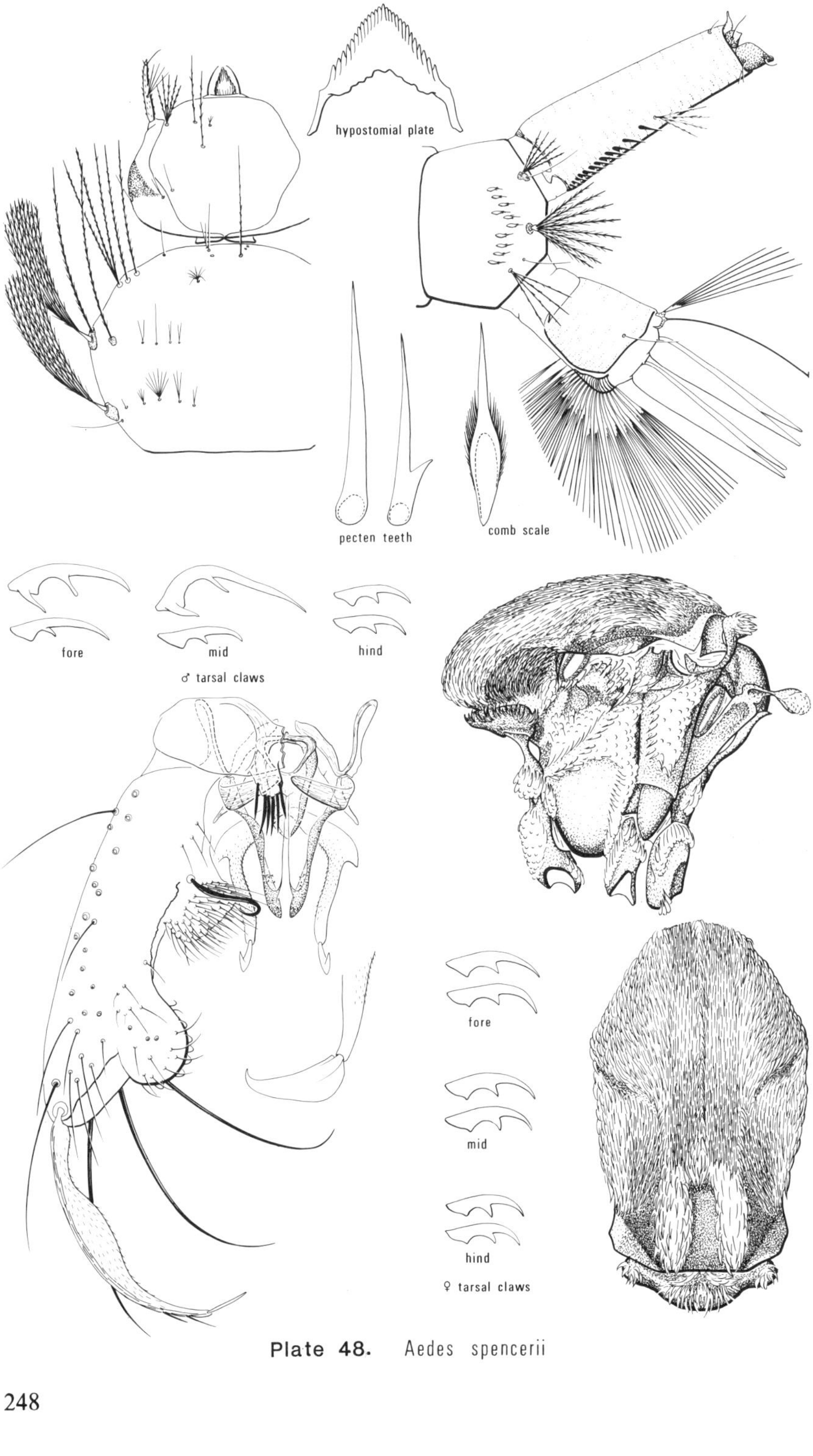

Plate 48. *Aedes spencerii*

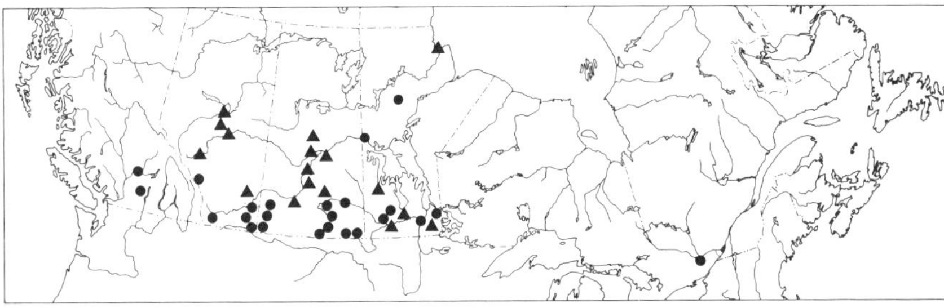

Map 48. Collection localities for *Aedes spencerii* in Canada: • specimens we examined, ▲ literature records.

Male: Much darker and more sparsely scaled than female; palpus about as long as proboscis, dark-scaled except for a few scattered pale scales; apex of third and fourth palpomeres fringed with long dark setae; scutum pale-scaled, with the middorsal stripe indistinct (and usually denuded); legs and abdomen predominantly dark-scaled; gonocoxite with strongly enlarged medially directed seta arising dorsal to basal lobe; basal lobe well developed, with strongly constricted base and large quadrate setose surface; apical lobe prominent, arising from near mid length of gonocoxite, subcircular with slightly constricted base; claspette filament with prominent keel on convex side and low keel on concave side, which does not extend to base of filament.

Larva. A pale medium-sized rather nondescript species, quite similar to *sticticus* except for unevenly spaced pecten teeth; head setae 5-C and 6-C single (rarely double); prothoracic setae 1-P and 5-P single; setae 8-P and 10-P extremely short, about one-fifth as long as seta 7-P; mesothoracic setae 1-M and 3-M minute; thorax and abdomen clothed with minute, dark, blunt spicules that extend perpendicular to the integument and can only be observed from the side, such as along the edge of the thorax or along a fold in the integument, with the most conspicuous setulae in a transverse band across the middle of the mesothorax, which though visible at 40× magnification under a dissecting microscope may easily be overlooked; distal pecten teeth more widely and unevenly spaced than basal teeth; each pecten tooth with only one basal tooth; siphonal seta 1-S about as long as apical diameter of siphon; in last instar saddle extending ventrally, almost to bases of ventral tuft (4-X).

Remarks. The status of *idahoensis* has not been satisfactorily resolved. A few females in the Canadian National Collection have the characteristic abdominal pattern of *idahoensis,* but there are no males or larvae in the collection. According to Carpenter and LaCasse (1955), larvae of *idahoensis* have more comb scales (13–29 instead of 7–13 as in *spencerii*), and each scale has a less pronounced apical spine. Males of the two species were not separated. Nielsen and Rees (1959) found "almost every degree of variation" between the typical female abdominal pattern of *idahoensis* and that of *spencerii* in Utah, Montana, and Wyoming. Furthermore, they cited three other authors who described similar variations in *spencerii.* They concluded that *idahoensis* was a more southerly subspecies of *spencerii* and that a zone of integration occurred in the three states previously mentioned. Until some detailed biological or genetical studies can resolve the status of *idahoensis,* we have no choice but to consider it a synonym of *spencerii.*

Biology. This species overwinters in the egg stage. It is the earliest species to appear in spring. Larvae are found in ditches and snowmelt pools throughout the Prairie Provinces, even when the water is standing over ground that is still frozen (Barr 1958). The species is unusual in that, in spite of its early appearance, it is not univoltine. More than half (58%) of the eggs laid by this species were nondiapausing, hatching after 2 wk at room temperature (Brust 1968). Thus, the species may reappear as a smaller second or even third brood during the summer (Rempel 1953). Because of its ubiquitous occurrence, it is considered the worst of the pests in spring in the Prairie Provinces (Knab 1908, McLintock 1944, Rempel 1953). The larva is capable of developing rapidly at low temperatures. In the Ottawa area, *spencerii* was collected as a pupa in mid-April, when other species of *Aedes* in the same pool were only half-grown larvae. For this reason, it has probably been overlooked in other parts of southern Ontario and Quebec. Hocking et al. (1950) reported a similar finding at Churchill, Man., where three adults were collected at the end of May before any other species had reached the pupal stage. Because strong winds had been blowing continuously from the south, he concluded that the three adults may have been blown from farther south. Larvae of *spencerii* have not been collected at Churchill. The species is uncommon in predominantly forested areas (Happold 1965*a*, Graham 1969*b*, Rempel 1953).

Females are most active for several hours before sunset and after sunrise (Burgess and Haufe 1960). However, they can be annoying throughout the day, especially on hot sunny afternoons (Rempel 1953). They readily invade cities and towns, often traveling several miles to do so. Owen (1937) believed that the species was diurnal only.

Males swarm at midday (Knab 1908), which is unusual for a nonarctic species. Swarms formed over the highest point on the prairie, which at times was the observer's head. Swarming males faced into the wind and then dropped down into the grass when the wind velocity became too great. Knab (1908) found that the swarms accompanied him over the prairie. He was able to observe and describe pairing within the swarm. He also found males probing willow catkins. Brust (personal communication) also observed swarming in the evening, just after sunset, at Winnipeg, Man.

Distribution. North America, from southern British Columbia and Washington State east to eastern Ontario and Nebraska.

Aedes sticticus (Meigen)

Plate 49; Figs. 145, 147, 190; Map 49

Culex sticticus Meigen, 1838:1.
Culex hirsuteron Theobald, 1901*b*:98.
Culex pretans Grossbeck, 1904*b*:332.
Culex aestivalis Dyar, 1904*b*:245.
Aedes aldrichi Dyar and Knab, 1908:57.
Aedes gonimus Dyar and Knab, 1918:165.
Aedes vinnipegensis Dyar, 1919*b*:34.
Aedes lateralis, authors, not Meigen.

Adult. Female: Integument medium brown; proboscis and palpus dark-scaled; pedicel pale-scaled on median half; upper two-thirds of postpronotum and middorsal stripe (median and submedian stripes combined) of scutum reddish brown; vertex, antepronotum, lower third of postpronotum, and sublateral and lateral stripes of scutum mostly pale yellowish scaled; presutural sublateral area often with patch of reddish brown scales; postsutural sublateral stripe reddish brown; postprocoxal membrane, hypostigmal area, and lower quarter of mesepimeron bare; anterodorsal corner of katepisternum with pale scales; mesepimeral setae absent; tibiae and tarsi mostly pale-scaled ventrally, dark-scaled dorsally, with last tarsomeres mostly dark; tarsal claws small, moderately and evenly curved, each with small subbasal tooth; wing veins dark-scaled, except for some pale scales on posterior edge of upper surface of costa; abdominal tergites dark-scaled, each with basal transverse band of white scales.

Male: Integument darker and scales sparser than in female; palpus longer than proboscis, dark, with a few scattered pale scales dorsally; apex of third palpomere and fourth and fifth palpomeres with the usual fringes of long dark setae; terminalia similar to those of *spencerii* except that apical lobe wider at base than its distance from base of gonostylus.

Larva. Similar to *spencerii* except that head setae 5-C and 6-C are usually branched, the integumental spicules on the thorax and abdomen are lacking, and the pecten teeth are uniformly spaced. In our material, the anal papillae are twice as long as the anal segment, and the comb scales are smaller and more numerous than in *spencerii,* their subapical spinules being virtually absent.

Biology: *Aedes sticticus* is primarily a floodwater mosquito, associated with bottomlands and floodplains of large rivers. Eggs remain viable for at least 5 years (Cook et al. 1974) and perhaps much longer. Consequently a season or two may pass when the species does not appear at all (Gibson 1933). Eggs may hatch anytime during late spring or summer, but usually they hatch in late spring immediately after spring flooding (Gibson 1926, Hearle 1926, Cook et al. 1974). Low water temperatures do not elicit hatching (Siverly and DeFoliart 1968*a*). Summer generations may result

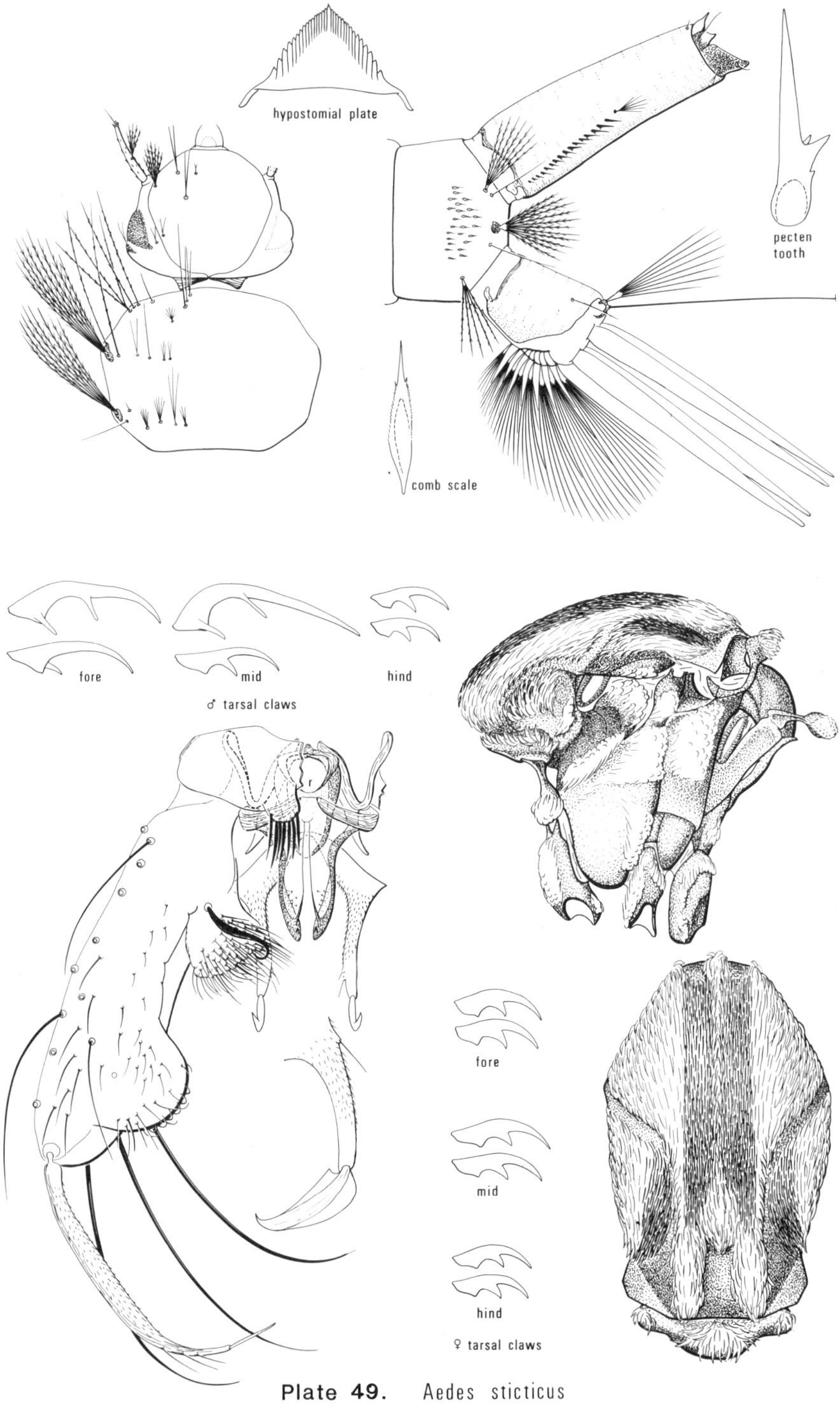

Plate 49. *Aedes sticticus*

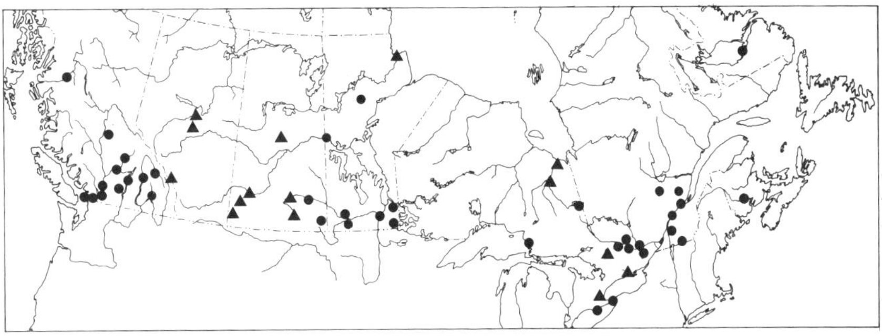

Map 49. Collection localities for *Aedes sticticus* in Canada: • specimens we examined, ▲ literature records.

from heavy rains, which may flood a large area and create many temporary pools or raise the river level sufficiently to flood the eggs. However, excessive flooding may wash the larvae downstream (Twinn 1931). Brust (personal communication) has recently found that more than 25% of the eggs laid in midsummer were nondiapausing and would hatch without cold treatment.

Although females can bite during the day, they are most troublesome in the evening (Hearle 1926, McLintock 1944). In some years, large numbers have appeared suddenly, especially in the lower Fraser Valley, B.C. (Hearle 1926, under the name *A. aldrichi*), and in the Ottawa area (Twinn 1929, as *A. hirsuteron*). Populations of *sticticus* are almost always associated with *A. vexans*. However, the converse is seldom true because *vexans* usually develops in summer rainpools after local flooding, whereas *sticticus* requires extensive flooding, which only follows widespread excessive precipitation.

Distribution. North America and Eurasia: western Europe, north to Finland and Karelia, south to Yugoslavia, Ukraine, and northern Caucasus, east across Siberia to Khabarovsk and Japan; British Columbia and California, east to Newfoundland (Labrador) and Florida.

Aedes stimulans (Walker)

Plates 9, 50; Figs. 86, 89, 122, 193, 195; Map 50

Culex stimulans Walker, 1848:4.
Culicada subcantans Felt, 1905*b*:474.
Aedes stimulans ssp. *classicus* Dyar, 1920*b*:113.
Aedes stimulans ssp. *mississippi* Dyar, 1920*b*:113.

Adult. Female: Integument medium brown or reddish brown; proboscis dark-scaled apically, with pale scales increasingly numerous toward base; palpus with mixed dark and pale scales, the pale scales concentrated at bases of third and fourth palpomeres; scales of pedicel few,

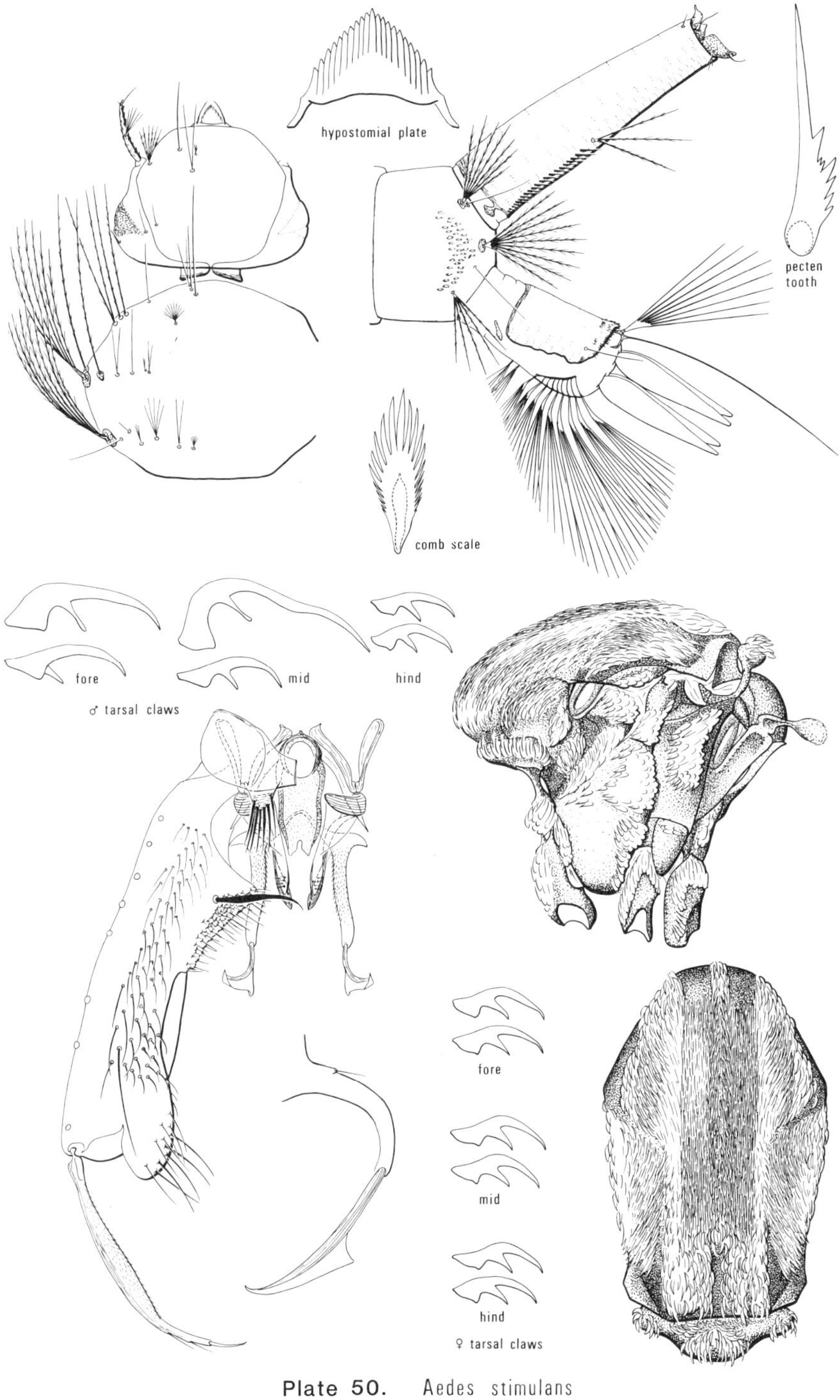

Plate 50. Aedes stimulans

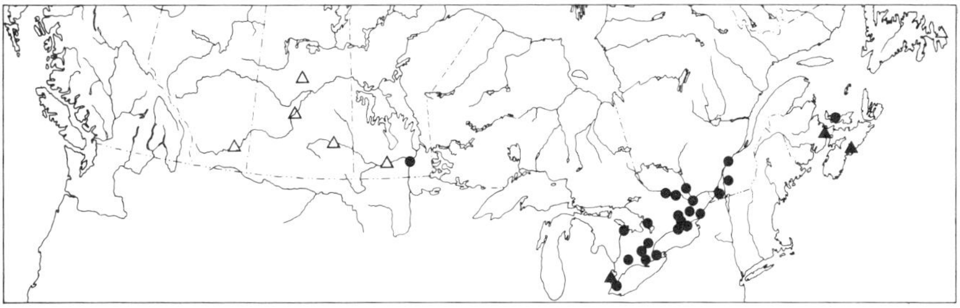

Map 50. Collection localities for *Aedes stimulans* in Canada: • specimens we examined, ▲ literature records, △ literature records based on probably misidentified specimens.

mostly dark, confined to median surface, usually absent from dorsomedian area; upper two-thirds of postpronotum and all of scutum with reddish brown scales except for pale-scaled dorsocentral line, lateral stripe, and prescutellar area; postprocoxal membrane and anterodorsal corner of katepisternum pale-scaled; hypostigmal area and lower third of mesepimeron bare; lower mesepimeral setae present; mesomeron usually with a small group of scales in posterodorsal corner; tibiae mostly pale-scaled except for scattered dark scales dorsally; tarsomeres each with broad basal ring of white scales except last tarsomere on fore- and mid-legs; tarsal claw somewhat bent beyond middle as in *euedes*, but not so strongly bent as in *excrucians*, with the portion beyond the bend slightly swollen and the apex more abruptly tapered and slightly curved ventrally as in *excrucians*; subbasal tooth moderately long; wing veins mostly dark, but with many scattered pale scales; abdominal tergites dark-scaled, each with transverse basal white-scaled band; first abdominal sternite bare or occasionally with a few pale hairs laterally; cercus dark-scaled.

Male: Integument medium brown, darker than that of female; palpus longer than proboscis by about half length of last palpomere; second to fourth palpomeres extensively pale-scaled above, especially basally; apex of third palpomere and medial and lateral edges of fourth palpomere with long dark setae, of which a few have paler yellowish bases; scutum with distinct middorsal stripe of reddish brown scales; scales of sublateral area pale yellowish; coloration and scaling darker and sparser, but otherwise as in female; basal lobe of gonocoxite with prominent, enlarged, medially directed bristle overlapping midline but not extending to base of opposite bristle, thus not as long as that of *mercurator*; apical lobe of gonocoxite long, extending distally to base of gonostylus; claspette filament long, with cylindrical base, the keel on concave surface arising near middle of claspette.

Larva. Head capsule of last instar yellowish brown (dark brown in *implicatus*); apotome with scarcely any pattern; head seta 5-C usually two-branched, often single, rarely triple; seta 6-C single; prothoracic setae 2-P and 3-P less than two-thirds as long as seta 1-P; seta 8-P minute; seta

10-P about two-thirds as long as a branch of 7-P; mesothoracic seta 1-M minute; dorsolateral abdominal setae 1-IV and 1-V shorter than 1-VII and less than two-thirds as long as upper lateral setae 6-IV and 6-V; each comb scale fringed with subapical setae; pecten teeth evenly spaced, each with several basal teeth decreasing in size basally; saddle not encircling anal segment in last instar, with the spinules along its posterior edge conspicuously projecting beyond the edge, visible under 40× stereomagnification; anal papillae unpigmented, equal in length, and almost as long as anal segment.

Remarks. Wood (1977) has shown that past records of *stimulans* from western and northern Canada refer to a different species, *mercurator*, and that *stimulans* is restricted to southern Manitoba and southeastern Canada. The western and northern limits of *stimulans* are unknown, although it could coexist with *mercurator* in western Manitoba and southern Saskatchewan. Unlike *mercurator*, which is generally uncommon, *stimulans* is one of the most abundant species of *Aedes*, especially in southern Ontario.

Biology. This species overwinters in the egg stage. The larvae of *stimulans* develop, often in huge numbers, on wooded floodplains inundated by snowmelt and spring runoff, especially near the edges of lakes and large swamps where silver maple, *Acer saccharinum*, is the dominant tree species. Such habitats, because they are at lake level, drain slowly as water levels recede, thus providing a reliable water supply. Eventually the water disappears, but the ground stays damp and becomes choked with sedges and ferns by midsummer. The maple canopy provides heavy shade by the time the females are ready to oviposit. In spring in the Ottawa area and to the south, *stimulans* is the most common and troublesome species of *Aedes*, but on the Precambrian Shield to the north it scarcely occurs. Perhaps the acidity of the water there may be an important limiting factor. Throughout southern Ontario in the vicinity of wooded areas it probably is the worst pest species, especially in early summer before *vexans* appears.

Female *stimulans* seem to be exceptionally long-lived. The black-legged *Aedes*, such as *communis*, *punctor*, and *intrudens*, usually disappear from the Ottawa area in early July. However, *stimulans* is often still unpleasantly common in August. Females have been collected in early September. Yet there is no evidence that *stimulans* is anything but univoltine.

Distribution. Eastern North America, from southern Manitoba, southern Ontario, and Nova Scotia south to Mississippi. Previous records from the west and north have been referred to *mercurator*.

Aedes thibaulti Dyar & Knab

Plate 51; Figs. 109, 128, 135; Map 51

Aedes thibaulti Dyar and Knab, 1910:174.

Adult. Female: Integument dark brown; proboscis, palpus, tibiae, tarsi, and wing veins dark-scaled, sometimes a few white scales at base of costa and radius; scales of vertex, postpronotum, and a narrow V-shaped wedge at anterior edge of median stripe, presutural sublateral area, and lateral stripe of scutum (including supraalar area) yellow; remaining scutal scales dark brown; postsutural sublateral stripe contiguous with submedian stripe, not separated from it by a line of yellow scales; middorsal dark stripe sometimes partly divided by a few yellow scales along acrostichal line; postprocoxal membrane, hypostigmal area, anterodorsal corner of katepisternum, and lower third of mesepimeron without scales; lower third of mesepimeron without scales; lower mesepimeral setae absent; abdominal tergites dark-scaled, each with lateral spot or triangle of white scales.

Male: Palpus slender, dark-scaled, slightly longer than proboscis, with a few scattered setae apically, but lacking tufts or fringes of long setae; coloration and scaling as in female; basal lobe of gonocoxite small, with a group of medially directed setae, but lacking an enlarged bristle; apical lobe small; claspette stem bifurcate, with a conical median seta-bearing branch and a shorter lateral branch bearing claspette filament; filament flattened, transparent, with medially directed thumb-like lobe and contorted apex.

Larva. Antenna slender, evenly tapering, nearly as long as head; clypeal setae short, pale, wider apart than their own length; head seta 5-C with five or more branches; seta 6-C usually with four or five branches; bases of setae 5-C, 6-C, and 7-C almost in a straight line; body with greenish brown pigment concentrated intersegmentally, giving the larva a jointed appearance; siphon rather straight-sided, 4.5 or more times as long as basal diameter; pecten teeth evenly spaced.

Biology. *Aedes thibaulti* has been collected twice in southern Ontario. The first specimen was a single male collected in a light trap near Belleville, Ont. (Belton and French 1967). A series of larvae was collected in early May at Rondeau Provincial Park from a hollow stump (about 60 cm in diam) that was standing in an upright position in a semipermanent woodland swamp. The interior of the stump, including the core of each major root, had rotted away, leaving only the outer shell with a slot in one side through which larvae of other species of *Aedes* entered the interior of the stump. Larvae of *thibaulti* were not collected outside of the stump. Most of the larvae were concealed in the dark recesses within the hollow roots. One larva was collected from inside the end of a floating hollow log in the same swamp. Many other apparently suitable stumps and logs were examined, but they did not yield *thibaulti*.

In the United States, larvae of *thibaulti* are usually associated with flooded basal cavities in various species of swamp-inhabiting trees such as

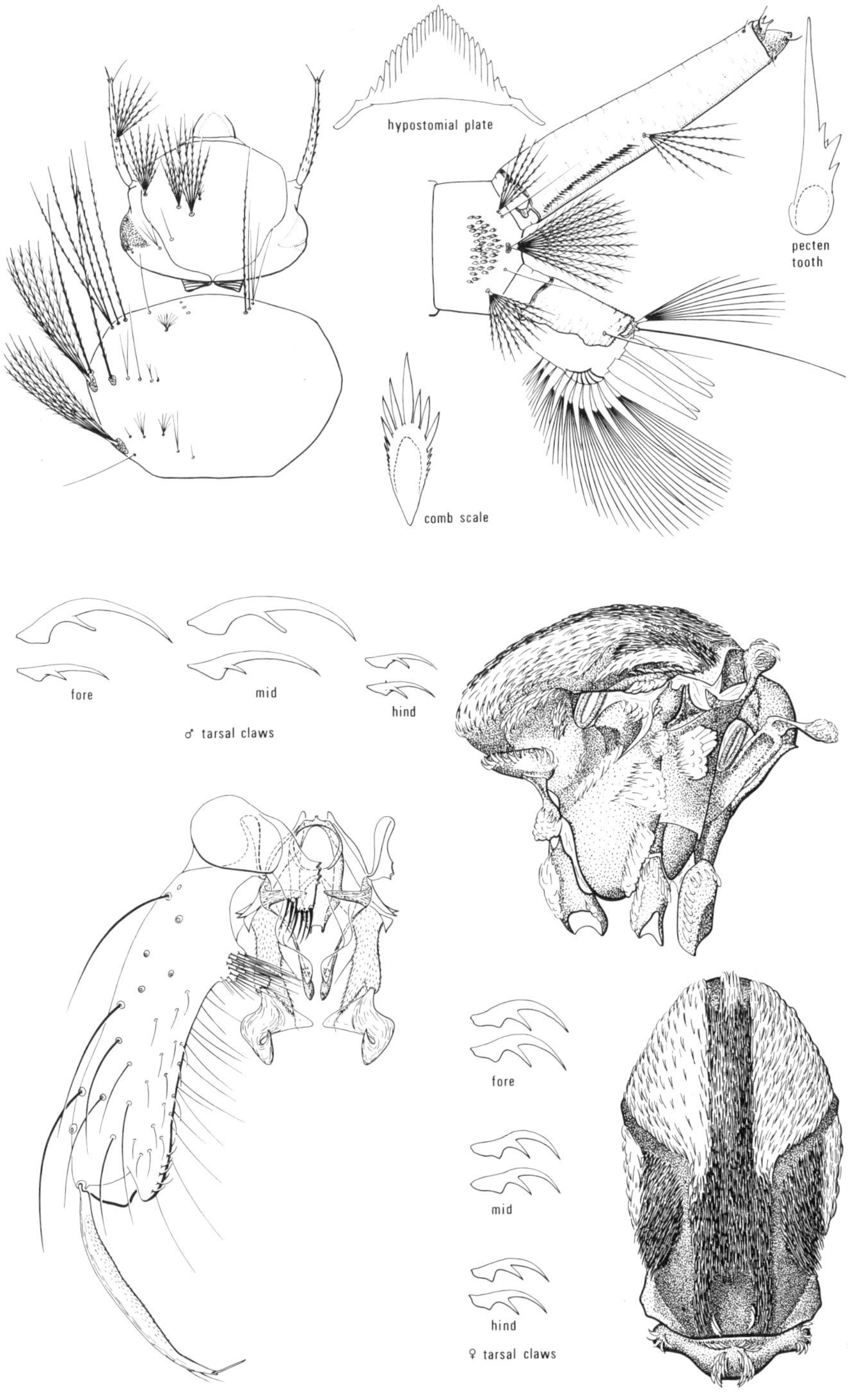

Plate 51. *Aedes thibaulti*

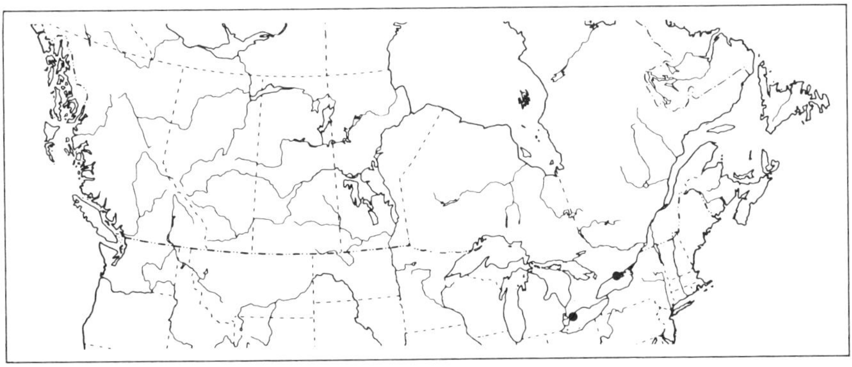

Map 51. Collection localities for *Aedes thibaulti* in Canada: • specimens we examined.

gums and bald cypress (Breeland et al. 1961). Except for an occasional small specimen of one species of gum, *Nyssa sylvatica*, these trees do not occur in Canada, and hollow bases in living trees may not be found here.

This species overwinters in Canada and the northern USA in the egg stage. However, Snow (1949) reported that larvae overwinter in the southern USA. The eggs are attached to the inner basal wall of the tree cavity. If not flooded the next season, eggs may remain viable for 2 years (Horsfall 1939). Larval feeding and duration of larval and pupal development were described by Shields and Lackey (1938).

Distribution. Southeastern North America, from Illinois and Texas east to southern Ontario, New York, and Florida.

Aedes togoi (Theobald)

Plate 52; Fig. 159; Map 52

Culiselsa togoi Theobald, 1907:379.

Adult. Female: Proboscis dark-scaled; palpus predominantly dark-scaled, with white scales at apices of segments III and IV; postprocoxal membrane, hypostigmal area, metameron, and lower third of mesepimeron without scales; scales of katepisternum separated into a posterodorsal and a posteroventral patch; scutum predominantly yellow-scaled, but with a subtle complex pattern of darker lines and spots, as follows: median stripe yellow, unusually wide, overlapping half of what is usually the submedian stripe; submedian stripe ill-defined, with yellow and darker narrower scales; dorsocentral stripe more prominent, with yellow scales; transverse suture emphasized with a curving line of yellow scales; scutal fossa and postsutural sublateral area posterior to suture each with a spot of darker scales; tibiae dark-scaled, each with a spot of white scales at apex; first and second tarsomeres each dark-scaled except for narrow basal and apical rings of white scales; apical ring only half as wide as basal ring; third tarsomeres with basal white ring, usually also with a few white scales at apex; last hind tarsomere brown-scaled; wing veins brown-scaled; abdominal tergites dark-

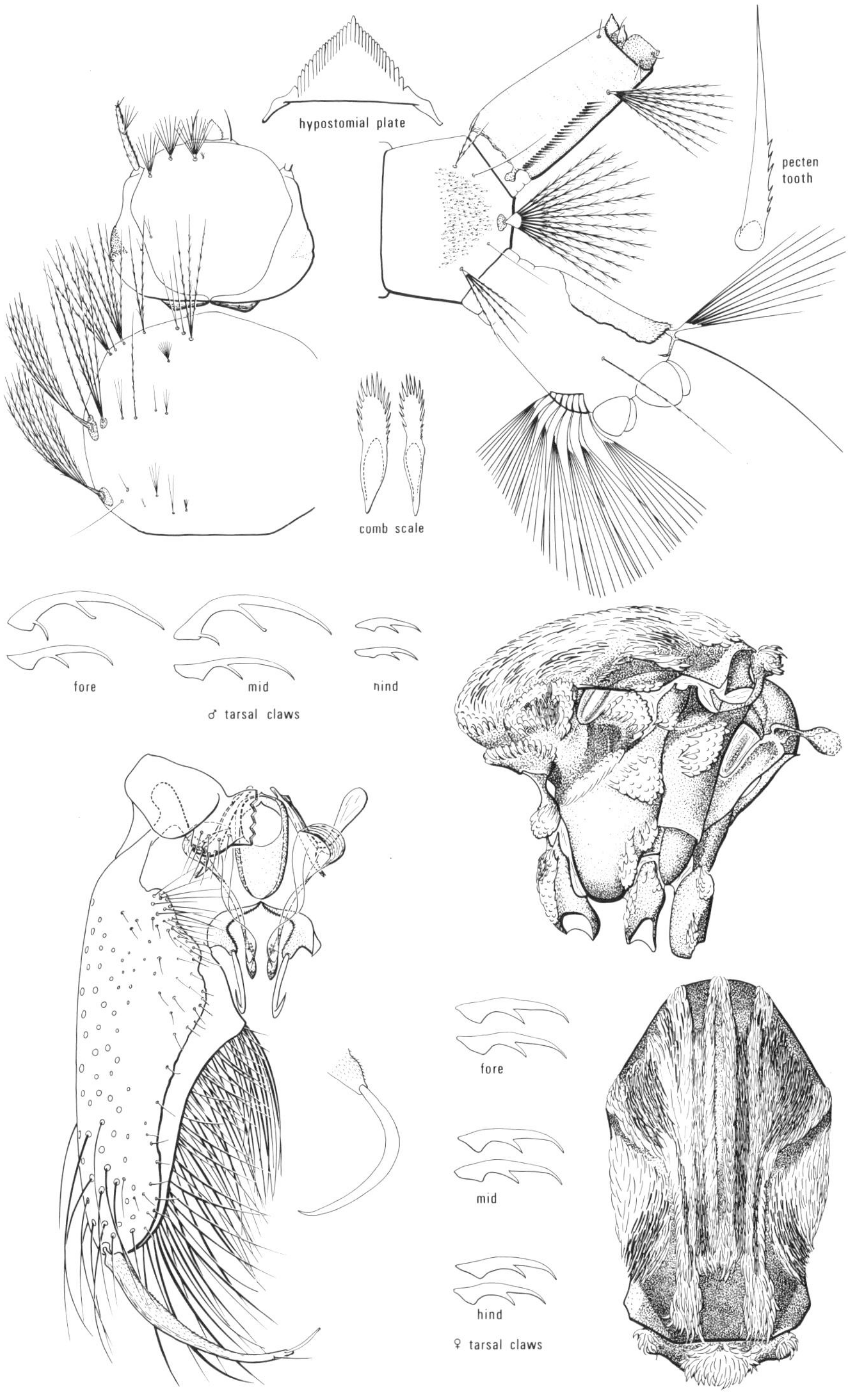

Plate 52. *Aedes togoi*

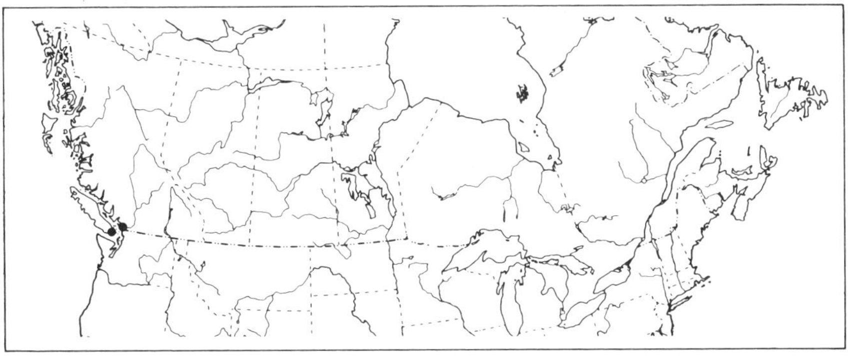

Map 52. Collection localities for *Aedes togoi* in Canada: • specimens we examined.

scaled, each with transverse basal band of white scales; cerci small, rounded, inconspicuous, rather larger than those of *A. atropalpus*.

Male: Palpus reduced, only about three-quarters as long as proboscis, dark-scaled, with narrow ring of white scales at base of fourth and fifth palpomeres (and a few white scales at apices of third and fourth palpomeres as well); palpal setae reduced in length and density; apex of third palpomere with only three or four setae; medioventral edge of gonocoxite with dense brush of long setae intersecting the setae on opposite gonocoxite; basal lobe of gonocoxite scarcely differentiated; apical lobe of gonocoxite absent; claspette filament a sickel-shaped rod; coloration of scutum and legs as in female.

Larva. Head setae 5-C and 6-C each with many branches, arising side by side far forward on apotome, anterior to seta 7-C; prothoracic seta 4-P double, as long as 1-P or 5-P; mesothoracic seta 3-M almost as long as 4-P; comb scales 80 or more, small, and densely crowded; seta 1-VIII shorter than 2-VIII or 5-VIII; siphon shorter than anal segment; pecten teeth evenly spaced, occupying basal two-thirds of siphon; saddle scarcely covering one-quarter of anal segment even in last instar; saddle seta 1-X isolated in the membrane ventral to saddle in all instars; anal papillae minute.

Biology. *Aedes togoi* was probably introduced into Western Canada from Japan. At present, the species is known only from the vicinities of Vancouver and Victoria, B.C., where larvae occur in rock pools along the coast. Our material came from several rock pools at the southern end of Lighthouse Park, North Vancouver, where the rocky shore sloped steeply down to the sea. Pools varied considerably in depth and extent. Only those above high tide level contained *togoi* larvae. The pools contained suspended green algae and harbored larvae of *Culiseta incidens* as well. Although these pools would undoubtedly receive salt spray during storms, rain was probably the most important source of water. When it was tested in September 1976, the salinity was no greater than that of freshwater (Brust, personal communication). The larvae of *togoi*, though numerous in several of the pools, were not observed at the surface, like those of *incidens*, but had to be sieved from the bottom detritus.

In Japan, larvae of *togoi* were collected from artificial containers, bilge water, and brackish coastal rock pools (Hsaio and Bohart 1946). Weathersby (1962) transferred larvae from seawater to freshwater, apparently without ill effects. Sasa et al. (1952) found that *togoi* larvae can tolerate a 5% concentration of salt and that overwintering takes place in the larval stage.

In September, at Vancouver, adult females began coming to feed in the vicinity of their breeding pools, about an hour before sunset. They were rather timid in their approach and were easily frightened before their mouthparts were imbedded. Reports on their habits in Japan are conflicting. Sasa et al. (1952) described *togoi* as a vigorous human biter, readily invading human dwellings, but LaCasse and Yamaguti (1955) collected only a few female specimens in biting collections and suggested that there may be two indistinguishable strains: one that did not bite man, which bred in artificial containers in villages, and the other that fed readily on humans, which bred in brackish saline coastal rock pools. Omori (1962) found only a small number of females in horse or cattle sheds despite the presence of many larvae in the vicinity and suggested that the strain that bred in villages might be autogenous. If two such strains exist, it is evidently the coastal form that has been introduced into Canada.

Females of *togoi* were experimentally infected with the virus of Japanese "B" encephalitis in Japan (McLintock and Iversen 1975). They were also infected with the larvae of *Wuchereria bancrofti*, the causative agent of Bancroftian filariasis (Yamada 1927). However LaCasse and Yamaguti (1955) stated that *Culex pipiens pallens* Coquillett is a much more important vector of filariasis than is *togoi*.

Distribution. Pacific Coast of Asia from Sakhalin Island and the southern part of the Maritime Territory, USSR, south through Japan, Korea, and China; introduced along the coast of southern British Columbia.

Aedes triseriatus (Say)

Plate 53; Figs. 108, 129, 132, 134, 177, 181; Map 53

Culex triseriatus Say, 1823:12.
Finlaya nigra Ludlow, 1905*b*:387.

Adult. Female: Similar to *hendersoni* except that the brown-scaled middorsal stripe of the scutum is wider, bulging rather than parallel-sided, thus usually extending lateral to the dorsocentral setae midway between the anterior margin and the transverse suture, and the tarsal claws are more uniformly curved, not bent beyond subapical tooth.

Male: Palpus slightly shorter than proboscis, its scales and setae entirely dark; third palpomere with at most two or three long apical setae;

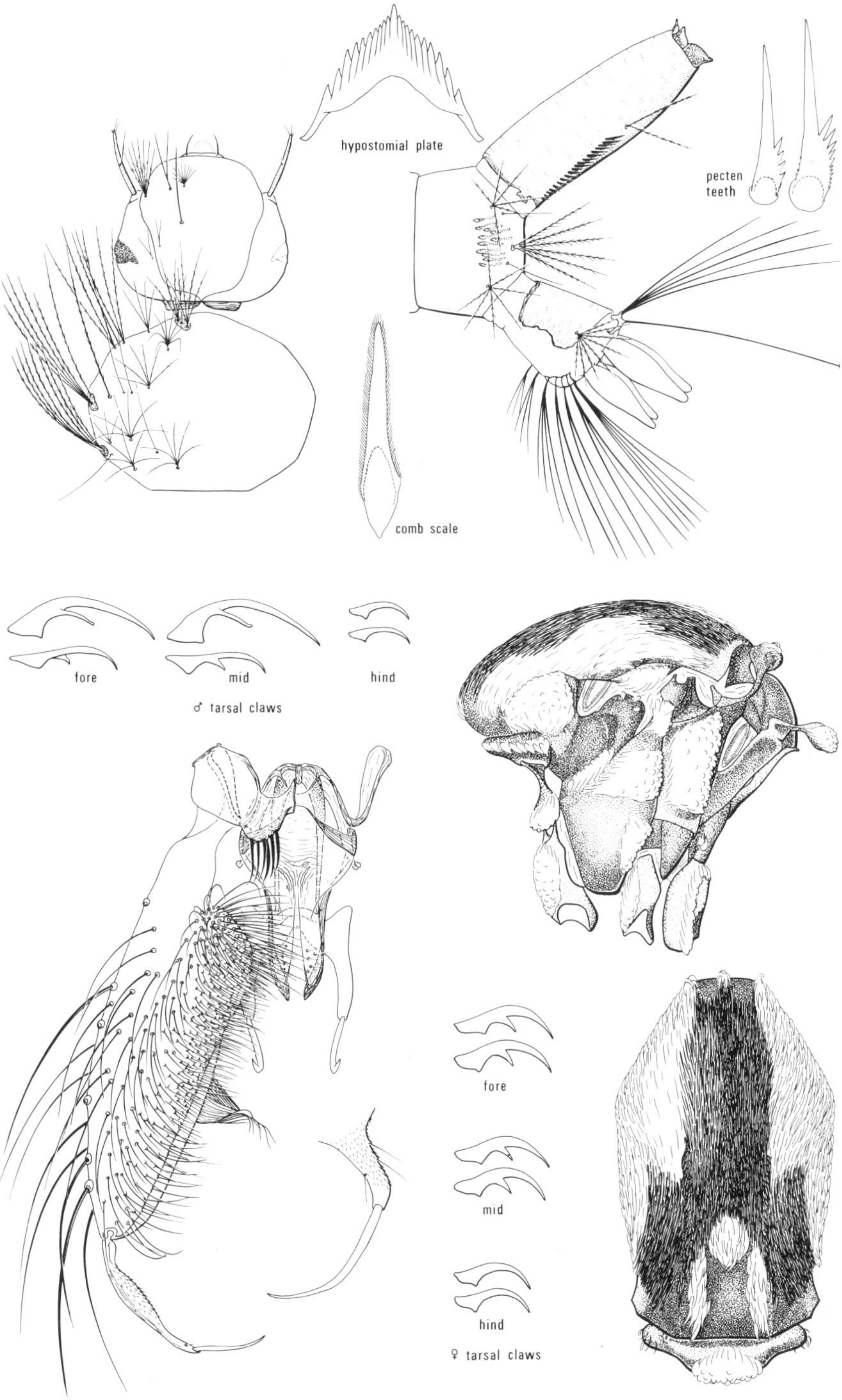

Plate 53. Aedes triseriatus

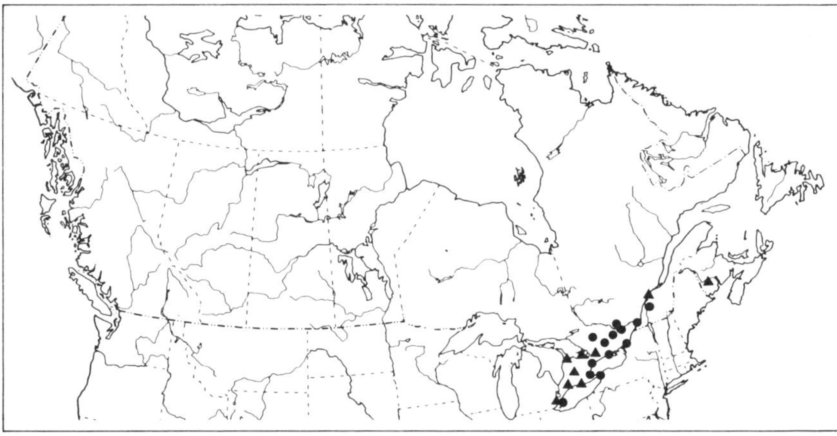

Map 53. Collection localities for *Aedes triseriatus* in Canada: • specimens we examined, ▲ literature records.

fourth palpomere usually less densely fringed than that of *hendersoni*; lateral enlarged claw of fore- and mid-legs each with a subbasal tooth, but lacking a basal tooth (which is present in *hendersoni*); gonocoxites diverging distally from one another rather than parallel as in *hendersoni*; dorsomedial margin of gonocoxite straight, not convex, with the setae on the basal lobe extending in diminishing density almost to apex of gonocoxite; gonostylus and claspette as in *hendersoni*.

Larva. Antenna smooth; antennal seta 1-A usually single; head seta 4-C multiple, about half as long as seta 6-C; thoracic setae 1-P, 1-M, and 1-T and dorsolateral and ventrolateral abdominal setae on first seven segments stellately branched, usually with five or more branches; pecten teeth evenly spaced; acus joined to siphon; saddle rectangular in lateral view, its anteroventral corner not truncated; ventral pair of anal papillae shorter than dorsal pair; all papillae more slender and more pointed (tapering rather than rounded apically) than in *hendersoni*, and usually not more than one and a half times as long as anal segment. In life, *triseriatus* larvae are more darkly pigmented than those of *hendersoni*, although an occasional specimen has been found that has white hemolymph, which gives it a paler appearance. Loor and DeFoliart (1970) described some exceptional larvae of *triseriatus* that had anal papillae three or more times as long as the anal segment.

Biology. Larvae of this species are more tolerant of habitat than are those of *hendersoni*, and they may be found in almost any kind of tree cavity, including water-filled cavities in the crotch of a tree, even near the ground, and in artificial containers. Water with a high organic content is evidently not as important for *triseriatus* as for *hendersoni*. In Ontario, *triseriatus* has been collected more often than *hendersoni*, perhaps because of its greater tolerance for various larval habitats.

Much of what has been written concerning *triseriatus* may also apply to *hendersoni*, because the two species were considered as one before 1960. Nevertheless, we have summarized some observations from the literature on *triseriatus*, because the two species are closely related and probably have similar biologies.

In Canada the species probably overwinters only as an egg. An intricate oviposition sequence was described by Snow (1949). Standing with the head down, the female inserted an egg into a crevice in the wood inside a suitable tree cavity. After five eggs were laid in this fashion, one every 2 sec, the female moved sideways 6 mm and similarly deposited five more eggs. After four repetitions, the female moved up 6 mm and, reversing direction, laid four more batches of eggs in a parallel but higher row. Several parallel rows were thus laid, each 6 mm higher than the last, before the female flew to another part of the tree hole and repeated the procedure, laying over 100 eggs in less than 10 min.

Eggs laid in summer that had become fully embryonated by September did not hatch in the laboratory unless they were exposed to continuous artificial light (Baker 1935). Eggs collected in winter (in Tennessee) hatched within a few days if held at 21°C (Breeland et al. 1961). Larvae also required long days to complete their development. In Georgia, larvae hatching from ova that developed during decreasing photoperiod were photosensitive and would not pupate unless the duration of daylight was more than 12 hr (Love and Goodwin 1961). Baker (1935) found that larvae required 16 hr of light to complete their development.

Most authors have reported *triseriatus* found in cavities in hardwoods, but Snow (1949) found larvae of this species in white pine in Wisconsin. Larvae have also been recovered from various artifical containers, such as barrels, tires, bottles, urns, cans, cisterns, and gutters, provided they were filled with organic debris, such as leaves, and were shaded (Breeland et al. 1961, Michener 1947, Lake 1954). Larvae are rarely found in rock pools associated with *Aedes atropalpus* (Zavortink 1972). Larvae seek the darkest parts of the container and they burrow in the bottom debris if they are disturbed. They can survive for a short time in the wet debris without a layer of water.

Females of *triseriatus* have often been implicated in the transmission of arboviruses. The species is a vector of the La Cross strain of the California encephalitis virus and can also transmit the viruses of Eastern, Western, and Venezuelan equine encephalitides and yellow fever under laboratory conditions (see Zavortink 1972 for references). The La Cross virus was first shown to be transmitted transovarially, from one infected female to her progeny, by the discovery of the virus in a larval population (Watts et al. 1973, Pantuwatana 1974), and then in both larvae and in newly emerged males (Watts et al. 1974). Transovarial transmission is evidently the mode of overwintering of this virus, and probably also of other arboviruses. The species is also receptive to the larvae of dog heartworm (Intermill 1973).

Loor and DeFoliart (1970) reported that males swarmed over the head of an observer in bright sunlight, long before sunset. Hayes and Morlan (1957) observed mating of *triseriatus* in laboratory cages during periods of semidarkness.

Distribution. Eastern North America, from Minnesota and Texas east to Maine and Florida (Zavortink 1972).

Aedes trivittatus (Coquillett)

Plate 54; Figs. 6, 144; Map 54

Culex trivittatus Coquillett, 1902*b*:193.
Culex inconspicuus Grossbeck, 1904*b*:332.

Adult. Female: Integument medium brown; proboscis and palpus dark-scaled; pedicel with at most a few small dark scales; scales on anterior half of postpronotum and median, submedian, and lateral stripes of scutum reddish brown, with the median stripe sometimes slightly paler; scales of posterior half of postpronotum and all of sublateral stripe of scutum pale silvery gray or yellowish; postprocoxal membrane, hypostigmal area, anterodorsal corner of katepisternum, paratergite, and lower third or more of mesepimeron without scales; lower mesepimeral setae absent; tibiae and first tarsomeres dark-scaled anterodorsally, pale-scaled posteroventrally; remaining tarsomeres dark-scaled; wing veins dark-scaled; abdominal tergites dark-scaled, each with lateral triangular spot of white scales, sometimes with complete transverse basal band of white scales.

Male: Palpus dark-scaled, longer than proboscis by almost the length of last palpomere; apex of third palpomere and fourth palpomere with moderately dense fringe of long dark setae; coloration and scaling similar to that of female; basal lobe of gonocoxite small, conical, with medially directed tuft of setae, one of which is greatly enlarged; apical lobe reduced in size, strongly angulate; keel on convex side of claspette filament with retrorse angle.

Larva. Head setae 5-C and 6-C each single, relatively short and fine; body integument covered with rather long, colorless microtrichiae that can only be seen by transmitted light through the transparent cuticle; prothoracic setae 2-P and 3-P less than two-thirds as long as seta 1-P; mesothoracic seta 1-M shorter than head seta 5-C; dorsolateral setae (1-IV to 1-VI) of abdominal segments IV to VI each about half as long as corresponding lateral setae (6-IV to 6-VI); abdominal setae (probably series 4) midway between dorsolateral setae (series 1) and upper lateral setae (series 6) also elongate, almost as long as the dorsolateral setae; ventrolateral setae (series 13) of segments III to VI elongate, two-thirds or more as long as upper lateral setae of their corresponding segments (6-III to 6-VI); abdominal setae (probably series 10) about midway between upper lateral setae (series 6) and ventrolateral setae, also elongate, almost as long as ventrolateral

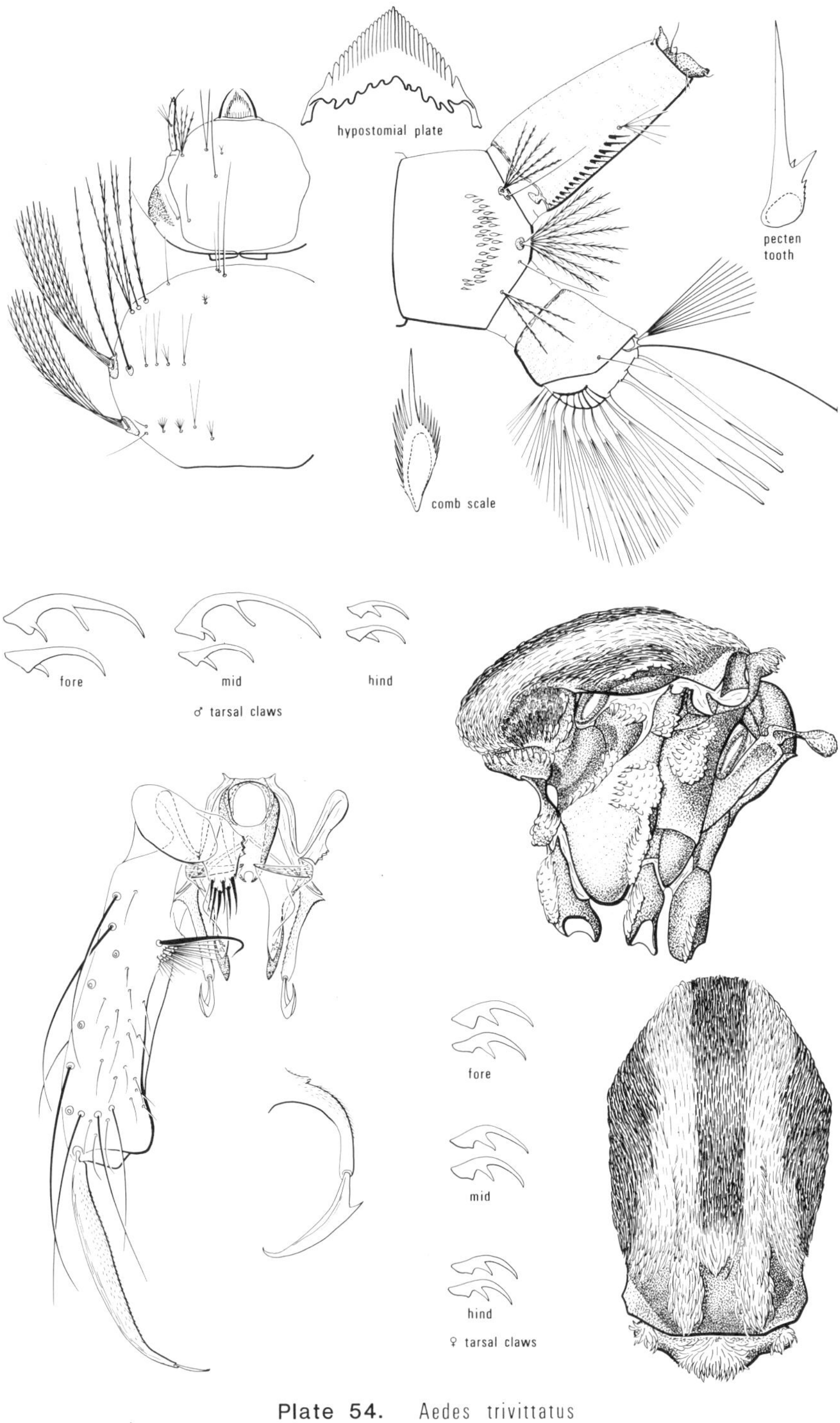

Plate 54. *Aedes trivittatus*

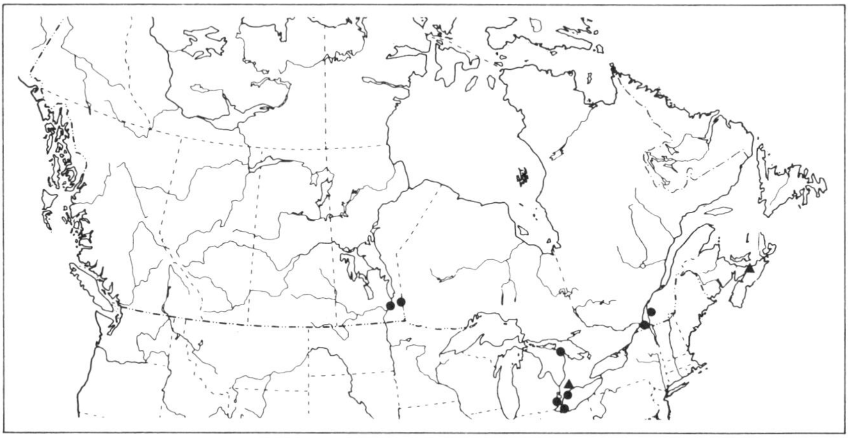

Map 54. Collection localities for *Aedes trivittatus* in Canada: • specimens we examined, ▲ literature records.

setae; siphon short and stout, slightly more than twice as long as basal width; pecten teeth evenly spaced, occupying basal half of siphon, each tooth with one large (and sometimes one small) basal cusp; saddle completely encircling anal segment; anal papillae longer than anal segment.

Biology. Overwintering as an egg in the soil, *trivittatus* is one of the last species to appear in spring. Judd (1954) collected larvae in the first week of June at London, Ont. Larvae occur in shallow, temporary, rain-filled grassy depressions, swamps, or bottomland forests. Last-instar larvae are rarely seen at the surface because they spend their time concealed in vegetation at the bottom of the pool (Breeland et al. 1961). In northern Minnesota, larvae were collected in a temporary woodland pool (Price 1963). Larvae have rarely been collected in Canada. Although adults can be fairly common in southwestern Ontario, they are only rarely collected in southern Manitoba. In the USA, they are associated with *Aedes vexans* and *Psorophora* spp.

In Wisconsin, adults appeared in the second half of June, were most abundant in early July, and disappeared in September (Loor and DeFoliart 1970). Females attack in woods, especially river bottomlands (Dyar 1922*b*), and are active day and night (Amin and Hageman 1974). This species is a vector of Trivittatus virus, a member of the California encephalitis (CE) virus group (Watts et al. 1976).

Distribution. North America, from Idaho and New Mexico east to Nova Scotia and Georgia.

Aedes vexans (Meigen)

Plate 55; Fig. 71; Map 55

Culex vexans Meigen, 1830:241.

Culex sylvestris Theobald, 1901*a*:406 (preoccupied by *sylvestris* Ross).

Culex montcalmi Blanchard, 1905:307 (new name for *sylvestris* Theobald).

Aedes euochrus Howard, Dyar, and Knab, 1917:716.

Adult. Female: Integument pale to medium reddish brown; proboscis and palpus dark-scaled; palpus with some pale scales at apex; pedicel pale-scaled on median half; upper half of postpronotum and scutum reddish brown to medium brown scaled except for a few pale scales along anterior border of scutum, on supraalar area, and around prescutellar depression; postprocoxal membrane, hypostigmal area, anterior half of katepisternum, and lower half of mesepimeron without scales; lower mesepimeral setae absent; tibiae dark-scaled dorsally, pale-scaled ventrally; first, fourth, and fifth fore tarsomeres usually entirely dark-scaled; second and third tarsomeres each with narrow basal white-scaled ring, less than one-quarter the length of the respective tarsomere; last mid tarsomere usually without basal white-scaled ring; first to fourth mid tarsomeres and all five hind tarsomeres each with narrow basal white-scaled ring; tarsal claws moderately and evenly curved, each with short subbasal tooth; wing veins dark-scaled; abdominal tergites dark-scaled, each with transverse basal white-scaled band.

Male: Palpus dark-scaled, with patch of white scales at base of each palpomere; palpus longer than proboscis by about half the length of the last palpomere; apex of third palpomere and edges of fourth and fifth palpomeres fringed with long, dark, dense setae; coloration and scaling as in female; gonocoxite without basal or apical lobes; seta of gonostylus arising subapically on a small tubercle; claspette stem broad, truncate, terminating in a group of ventrally curved setae; claspette filament absent.

Larva. Head seta 5-C with three to five branches (single in *spencerii*); seta 6-C two- or three-branched, arising anterior to 5-C, their bases not forming a straight line with 7-C (the bases of 5-C and 6-C in *cinereus*, with five or more branches, forming a straight line with 7-C); head and body darkly pigmented; body integument without minute spicules, as in *spencerii*; siphon at least three times as long as basal width; distal pecten teeth more widely and unevenly spaced than basal teeth; siphonal seta 1-S no longer than apical diameter of siphon; saddle seta 1-X branched; saddle not encircling anal segment.

Biology. During exceptionally wet summers, the population of *vexans* can increase dramatically, and as a result the species is usually considered to be the worst mosquito pest in Canada. Larval populations of about 80 million/ha of water surface were recorded at Winnipeg (Dixon and Brust

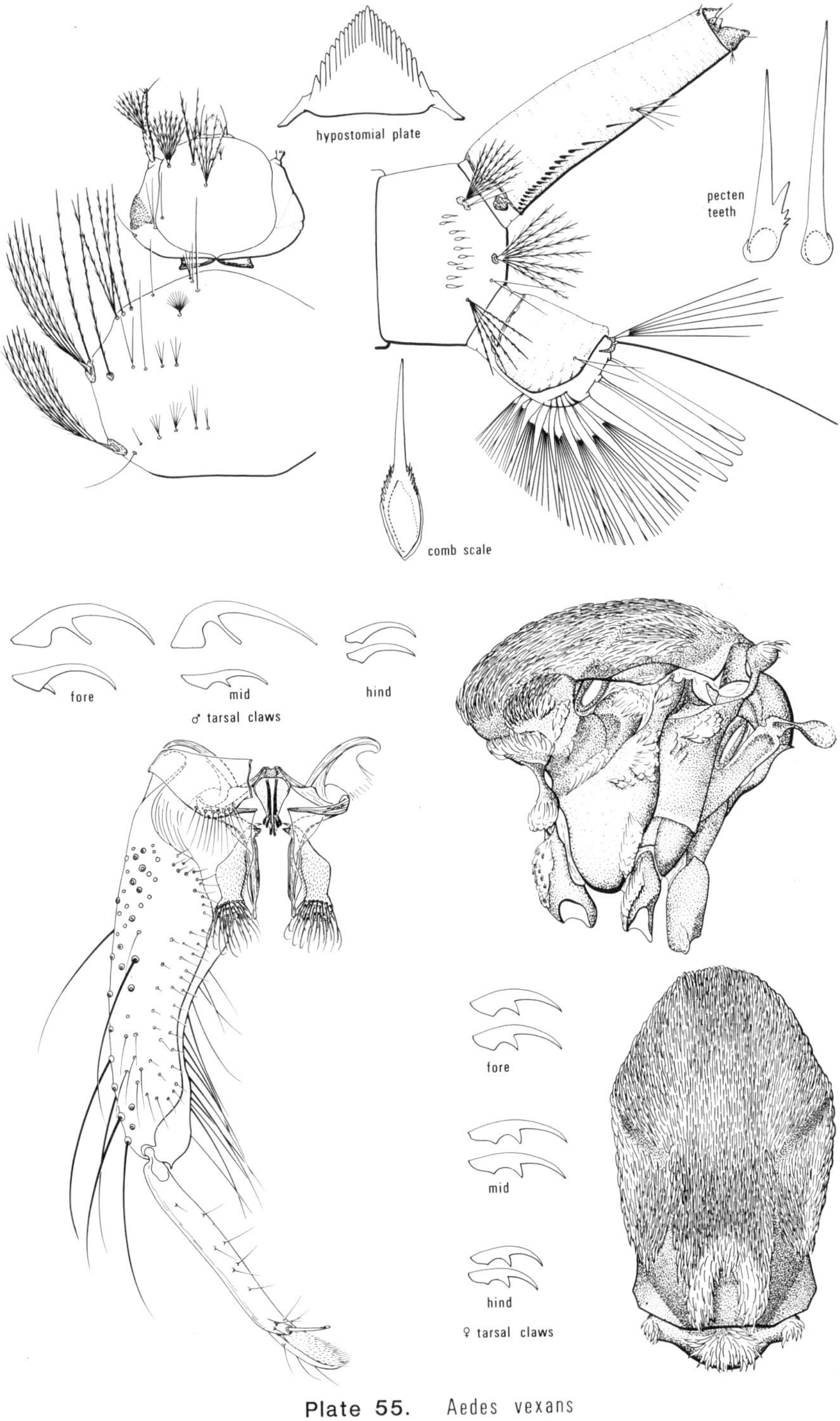

Plate 55. *Aedes vexans*

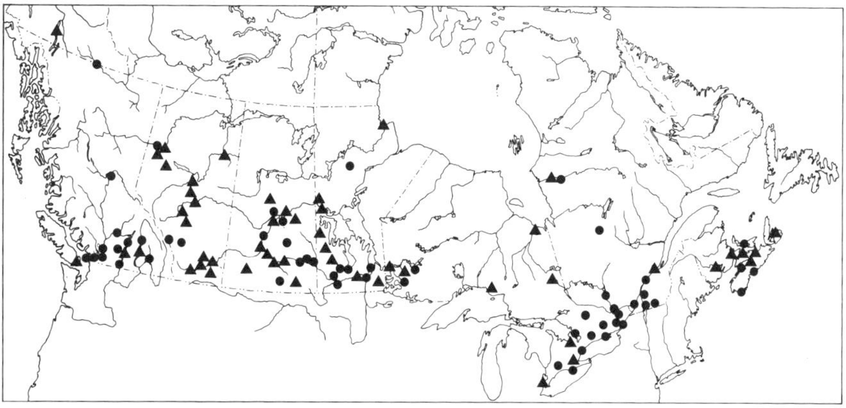

Map 55. Collection localities for *Aedes vexans* in Canada: • specimens we examined, ▲ literature records.

1972). The eggs overwinter; unlike most *Aedes* they do not hatch in spring as a result of low oxygen levels in the water but in response to temperatures above 8–10°C (Breeland et al. 1961, Brust and Costello 1969). Consequently, *vexans* larvae are among the last to appear in May. Thereafter, eggs laid by females from this first generation can hatch after only 3–5 days of dry weather (Breeland et al. 1961) anytime during the summer that enough rainwater runoff collects to inundate them. In the heat of summer, the life cycle is extremely rapid: hatching in 3 hr, maturing in 90 hr, and the adult emerging in 120 hr and engorging after 142 hr (Breeland and Pickard 1963). However, even with the capabilities of such rapid development, larvae must often be stranded and killed because such temporary pools can disappear quickly (James 1960). The well-known example of over 1000 larvae in 280 mL of water (Matheson 1944) was probably the result of a pool that was on the verge of drying up. If rainfall is insufficient, the eggs lie dormant, remaining viable for several years (Gjullin et al. 1950). Not all eggs hatch when inundated, and therefore periodic flooding and drying is needed for a complete hatch.

Larval habitats vary, but open, shallow grass-filled depressions in pastures or along roadsides are favored locations. Temporary woodland pools can also support large numbers of *vexans* larvae, but larvae are not usually found in permanent and semipermanent habitats.

Adult females are most active between sunset and sunrise (Burgess and Haufe 1960) and are usually not troublesome during the day. They may take up to eight blood meals and mature as many as 546 eggs during their lifetime (Breeland and Pickard 1964), which is very much less than the number produced by *Anopheles quadrimaculatus*. Females naturally infected with western equine encephalitis (WEE) virus and experimentally infected with St. Louis encephalitis (SLE) virus were reported by Stage et al. (1952).

Gjullin et al. (1950) reported that *vexans* mated in 1 m^3 cages. However, Taylor and Brust (1974) were unable to obtain inseminated females of *vexans* in a larger cage (120 × 120 × 210 cm high) unless males and females of *dorsalis* were also present in the cage. The swarming activities of *dorsalis* in the cage evidently stimulated *vexans* to swarm, and 10% mated successfully. Coupling took place most often during periods of changing light intensity, at simulated dawn and dusk.

Distribution. North America, Eurasia, and Africa: Europe, north to latitude 60–62° in river valleys, east across Siberia to Japan, south to South Africa, Turkey, Iran, India, China, and Indochina; Yukon and California east to Nova Scotia and Florida.

Genus *Culex* Linnaeus

Adult. Integument medium brown; submedian stripe of scutum with narrow bare line; pleural scaling and hairing greatly reduced; spiracular and postspiracular setae absent; anepisternum entirely bare except for a line of scales along upper edge of postspiracular area in *C. tarsalis*; postprocoxal membrane, paratergite, anterior half of katepisternum, and lower half of mesepimeron without scales; lower mesepimeral setae present, usually only one; tarsal claws of female all without subbasal tooth; abdomen parallel-sided, rounded apically; cercus small, rounded, inconspicuous; gonocoxite of male with prominent dorsomedially directed subapical lobe (in much the same position as the "apical" lobe of *Culiseta* but directed more dorsally), bearing several highly modified tusk-like spatulate or hooked setae; basal lobe of gonocoxite absent, its place occupied by an extensive membranous area; lobe of tenth sternite, or paraproct, with a finger-like projection, straight or curved, pointed or blunt apically, arising laterally about one-quarter or one-third the distance from its base; apex of paraproct crowned with a row or patch of denticles; aedeagus usually complex, adorned with various processes; aedeagus and paraprocts superimposed over one another, their respective outlines difficult to disassociate; claspette absent.

Larva. Bases of head setae 5-C, 6-C, and 7-C usually oriented in a straight line; comb scales arranged in a triangular patch of three almost regular rows, with the scales of the distal row longer than those of the proximal; siphon lacking basal pair of setae 1-S, but with four or more pairs of setae arising beyond pecten teeth; pecten teeth small, slender, and not as conspicuous as those of *Aedes*; saddle always encircling anal segment in final instar.

Key to the species of *Culex* of Canada—adults

1. Hind tarsomeres each with basal and apical bands of white scales, the remainder dark-scaled; proboscis encircled with band of white scales near mid length; apices of third and fourth palpomeres with white scales; subapical lobe of

gonocoxite of male with five tusk- or spike-like appendages (specialized setae), none of which has recurved apices (Fig. 204) .... ***tarsalis***

Hind tarsomeres dark-scaled, with at most a few scattered white scales; proboscis lacking an encircling white ring; palpus entirely dark-scaled; subapical lobe of gonocoxite with six or eight appendages, most of which have recurved apices (Figs. 205–207) 2

2. Abdominal tergites each with transverse apical band of white scales; the three most apically placed setae of subapical lobe of gonocoxite of male flattened, blade-like, and serrate on one side, with recurved apices; paraproct without finger-like lateral projection; apex of paraproct of male with comb-like row of about 10 short blunt teeth (Fig. 205) .... ***territans***

Abdominal tergites each with transverse basal band of pale scales (sometimes reduced to a few isolated scales in *pipiens*); setae of subapical lobe of gonocoxite with simple, unserrated margins, with one of the two most apically placed setae flattened, triangular, and paddle-like; paraproct with finger-like lateral projection; apex of paraproct with dense tuft of numerous pointed spines 3

3. Scutum uniformly brown-scaled; subapical lobe of gonocoxite with eight appendages, the basal one thickened, straight, and blade-like without recurved apex; lateral projection of paraproct a short, straight pointed protuberance (Fig. 206) .... ***pipiens***

Scutum with a pair of pale-scaled spots near middle; subapical lobe of gonocoxite with six appendages, the basal one with recurved apex; lateral projection of paraproct long, curved, blunt at apex (Fig. 207) .... ***restuans***

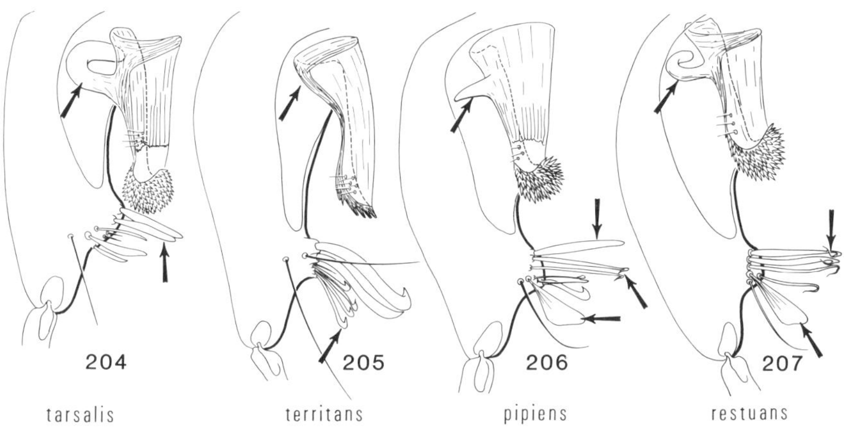

Figs. 204–207. Dorsal view of gonocoxite and paraproct of *Culex* species.

Key to the species of *Culex* of Canada—fourth-instar larvae

1. Antenna tapering uniformly to apex; antennal seta 1-A arising near middle antenna; siphon with three or four long, usually unbranched setae on each side, one dorsolateral, one lateral, and two ventrolateral .... ***restuans***

Apical third of antenna abruptly constricted; antennal seta 1-A arising at base of constriction; siphon with four or more pairs of three- to six-branched setae, all arising laterally or ventrolaterally 2

2. Head setae 5-C and 6-C usually single or double (5-C rarely three-branched); two or three precratal setae .... ***territans***

   Head setae 5-C with four or more branches and 6-C with three or more branches; no precratal setae 3

3. Siphon with five tufts on each side, all arising dorsoventrally at the same level .... ***tarsalis***

   Siphon with four tufts on each side, the second last displaced laterally out of line .... ***pipiens***

Culex pipiens Linnaeus

Plate 56; Fig. 206; Map 56

Culex pipiens Linnaeus, 1758:602.
Culex consobrinus Robineau-Desvoidy, 1827:408.

Adult. Female: Proboscis and palpus brown-scaled; apex of palpus usually with a few pale scales; erect forked scales of vertex mostly dark laterally, paler yellowish middorsally; postpronotum and scutum with uniformly pale brown scales; the scales on supraalar area and on both sides of prescutellar depression usually slightly paler; tibiae and tarsi dark-scaled dorsally, paler below; abdominal tergites dark-scaled, each usually with transverse basal band of paler scales, which may, however, be obscure or absent.

Male: Palpus entirely brown-scaled, slightly longer than proboscis, usually exceeding it by less than length of last palpomere; long fine setae of third palpomere arising ventrolaterally from distal half of palpomere; subapical lobe of gonocoxite with eight appendages; the most proximal appendage thickened, straight, and blunt apically, without recurved apex as in the two adjacent appendages; paraproct with a short, straight, sharply pointed lateral projection.

Larva. Antenna almost as long as head, with apical third to half abruptly narrowed; antennal seta 1-A arising beyond middle at point of abrupt narrowing; head setae 5-C and 6-C each with four or more branches; siphon four to almost six times as long as maximum width, widest not at base but about one-quarter of the distance from base, with four pairs of setae, all branched and almost equal in length; basal two pairs and apical pair all arising ventrolaterally, with the third pair arising more laterally and out of line with the other three pairs; upper caudal seta 2-X double, about three-quarters as long as seta 3-X; precratal setae absent.

Remarks. A report of *salinarius* from Nova Scotia (Twinn 1945) was probably based on *pipiens*. The larval siphon of *pipiens* can be almost six times as long as its greatest width and about ten times as long as its apical width. Such specimens could easily be misidentified as *salinarius* with the use of the keys of Matheson (1944) and Carpenter and LaCasse (1955).

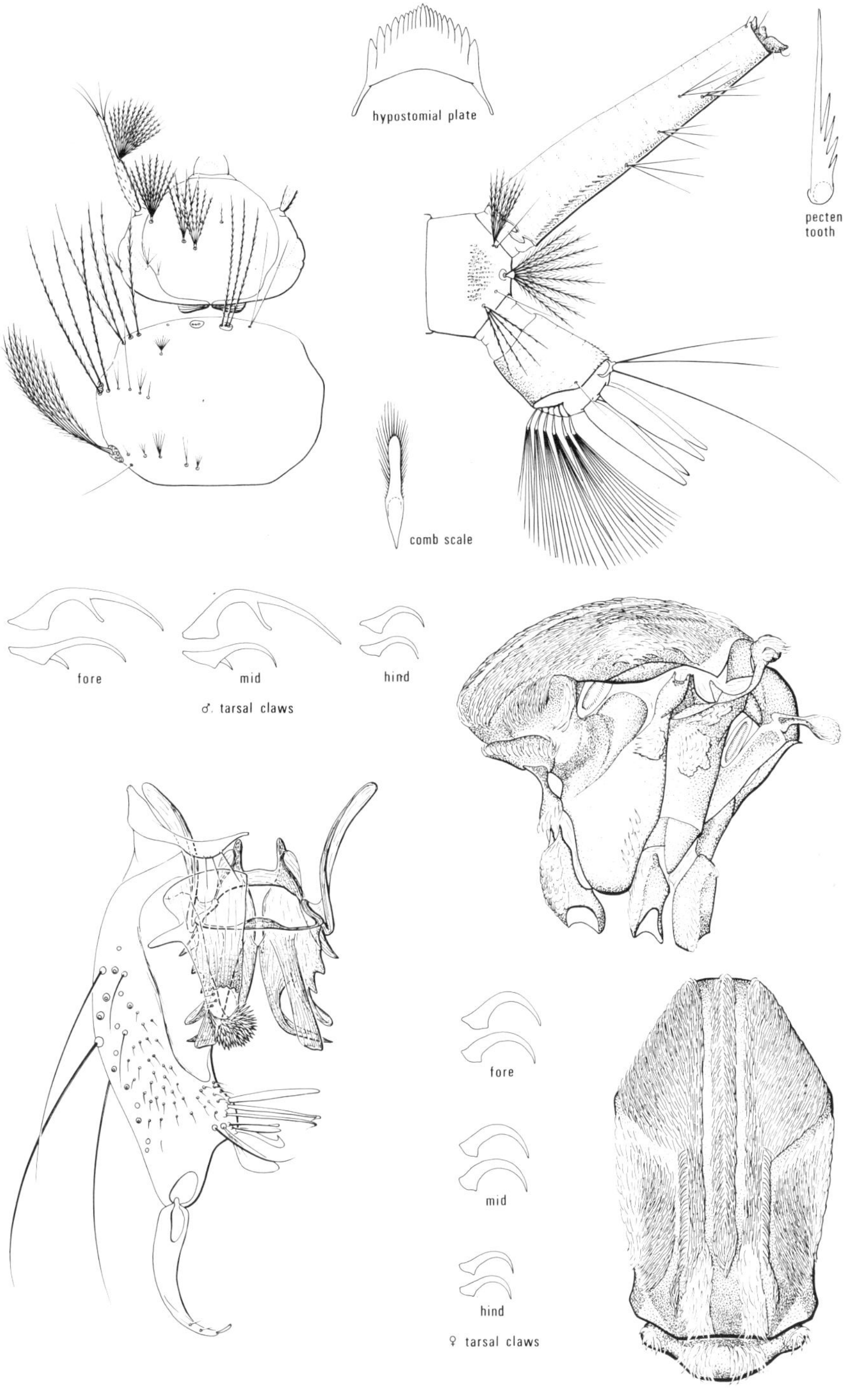

Plate 56. *Culex pipiens*

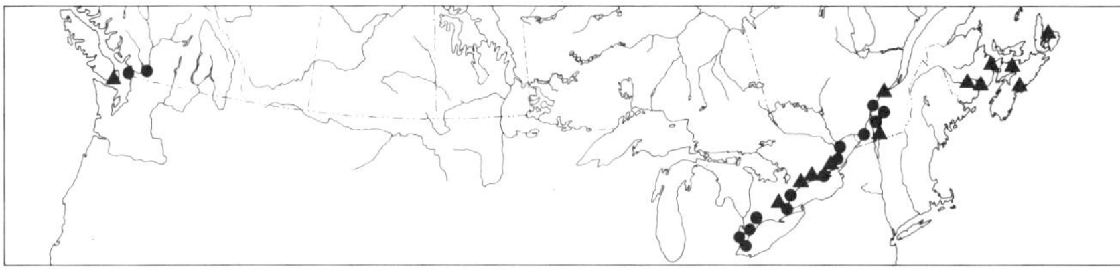

Map 56. Collection localities for *Culex pipiens* in Canada: • specimens we examined, ▲ literature records.

However, the siphon of *pipiens* is broadest beyond its base, so that its dorsal and ventral margins appear slightly bowed outward. The siphon of *salinarius* (in the few specimens we have examined) is broadest at its base and in profile its margins are nearly straight; therefore, it tapers evenly along its entire length. The siphon of *salinarius* is also longer and narrower, as long as the last four or five body segments, seven or more times as long as its basal width, and nearly fifteen times as long as its apical width. The antenna of *salinarius* larvae is also longer than that of *pipiens*. Females of *pipiens* occasionally lack transverse pale-scaled bands on the abdominal tergites, a character of *salinarius*.

Biology. As its common name implies, the rain-barrel or northern house mosquito, *Culex pipiens*, has a close association with man. Larvae can be found in almost every type of water container (listed by Headlee 1945 and Niebanck 1957), especially if the water is polluted with sewage or pulp mill or barnyard wastes (Headlee 1916, Gibson 1928, Twinn 1931). Larval development can be completed in as few as 8 days (Headlee 1945) and because there is no interruption by diapause in early summer several generations can be completed in a season. In southern Ontario the *pipiens* population builds up to a peak in late summer, especially in wet seasons, and then larvae may be found in almost every roadside ditch, rain puddle, or artificial container. Breeding continues until it is curtailed by cold weather in October.

Like other *Culex*, the unfed adult female overwinters in basements, caves, and similar shelters. The time of their emergence from hibernation in the spring is not known, but larvae begin to appear in June (Judd 1950). The females are generally believed to be predominantly bird feeders and to take human blood reluctantly (Mattingly et al. 1951). Hayes (1961) showed that they prefer birds to mammals, although various small mammals, snakes, and turtles have been bitten. They feed mainly after dark and enter houses freely, probably feeding on man when their numbers are high and other hosts less available.

Culex pipiens has been implicated in the transmission of western equine encephalitis (WEE) and St. Louis encephalitis (SLE) (Hammon et al. 1945, Stage et al. 1952). The species has been widely assumed to be the principal vector of SLE virus in eastern Northern America. The first recorded epidemic of this disease in Canada occurred in southern Ontario in 1974.

The eggs are laid in rafts, each consisting of about 100–400 eggs cemented side to side. The raft floats so that one end of each egg is immersed in water. Females avoid ovipositing on water that has an NaCl content of 1% (Wallis 1954*a*). In Illinois, a few female *pipiens* were induced to deposit egg rafts autogenously, that is, without a blood meal (Wray 1946). From their progeny six generations were reared autogenously, but their egg rafts contained less than half the usual number of eggs.

Culex restuans Theobald

Plate 57; Fig. 207; Map 57

Culex restuans Theobald, 1901*b*:142.
Culex brehmei Knab, 1916:161.

Adult. Female: Not always distinguishable from *pipiens*; palpus entirely dark-scaled; erect forked scales on vertex entirely dark; scutum usually with pair of middorsocentral spots of paler scales in addition to pale-scaled areas on both sides of prescutellar depression and on supraalar area.

Male: Palpus brown-scaled, longer than proboscis by almost the combined length of last two palpomeres; long, fine setae of third palpomere arising along its entire length, those arising ventrolaterally on apical half nearly twice as long as the setae arising ventrally and medioventrally from basal half; subapical lobe of gonocoxite with six appendages, the most proximal appendage with recurved apex as in the second, third, and fourth appendages (numbered from base); paraproct laterally with blunt, hook-like projection.

Larva. Antenna shorter than head, tapering uniformly to apex; antennal seta 1-A arising near middle of antenna; head setae 5-C and 6-C each with four or more branches; siphon with four pairs of setae, unequal in length, all except the short apical pair single, the basal and apical pairs arising ventrolaterally, the second seta arising laterally to somewhat ventrolaterally, and the third and longest seta arising dorsolaterally; upper caudal seta 2-X single and almost as long as seta 3-X (as in *Aedes abserratus*); precratal setae absent.

Remarks. Michener (1947) found that adults reared from larvae in acid water full of decaying vegetation were generally smaller and darker than those from other sources, and more than half of these smaller specimens lacked the characteristic pale middorsocentral scutal spots.

Biology. The life cycle of *restuans* is similar to that of *pipiens*. Larvae inhabit various artificial containers, rock pools, and tree cavities as well as ditches and temporary puddles when decaying vegetation is present.

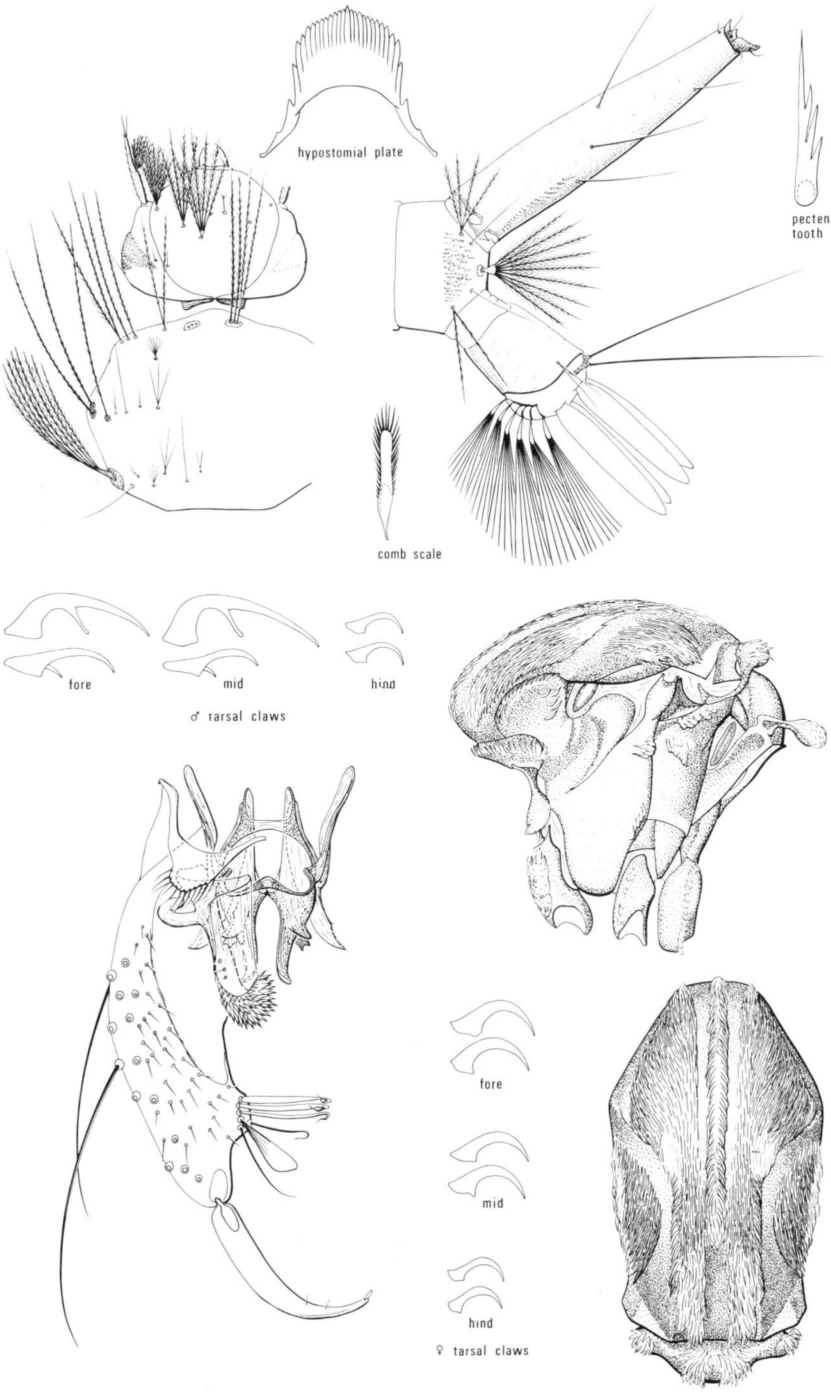

Plate 57. *Culex restuans*

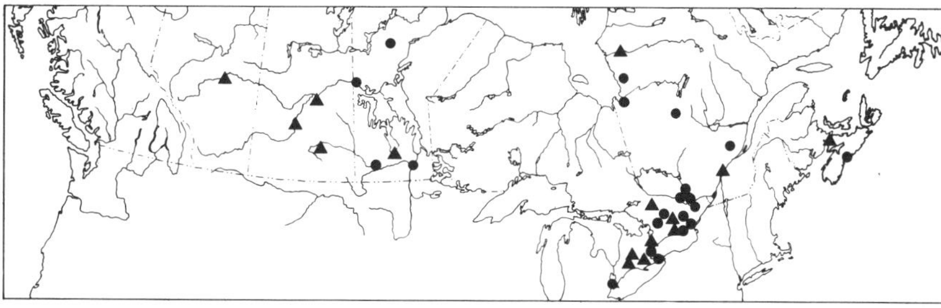

Map 57. Collection localities for *Culex restuans* in Canada: • specimens we examined, ▲ literature records.

However, the population reaches its peak much earlier in the season than does *pipiens* (Carpenter and LaCasse 1955). The earlier emergence and more northerly distribution of *restuans* reflect a greater cold tolerance than in *pipiens*. In Ontario *restuans* larvae have been associated with *Anopheles punctipennis*, *Culex territans*, and *Aedes atropalpus* in rock pools (James 1964), whereas in polluted locations they commonly occur with *Culex pipiens*.

The females hibernate in basements, caves, and hollow trees, probably in company with those of *pipiens*. This species, like *pipiens*, prefers feeding on birds, although mammals, snakes, and turtles are also hosts (Hayes 1961). They occasionally feed on man outdoors, beginning to feed at dusk and during the day in shaded areas (Breeland et al. 1961). However, even when they are abundant, the females are rarely seen biting (Michener 1947).

Although *restuans* is not usually suspected of transmitting viruses of the encephalitides, Hayes et al. (1960) isolated that of eastern equine encephalitis (EEE) in New Jersey. Apparently the distribution of EEE coincides with that of *restuans*. Western encephalitis virus was isolated from *restuans* in Manitoba (Norris 1946).

Distribution. North America, from central Alberta south to Mexico, east to Nova Scotia and Florida. Twinn (1945) recorded *restuans* from British Columbia, but recent records suggest those of Twinn were based on *pipiens*.

Culex tarsalis Coquillett

Plate 58; Fig. 204; Map 58

Culex tarsalis Coquillett, 1896:43.
Culex willistoni Giles, 1900:281.
Culex kelloggii Theobald, 1903*b*:211.

Adult. Female: Proboscis dark-scaled except for white-scaled ring encircling mid length and occupying about one-fifth of proboscis; palpus

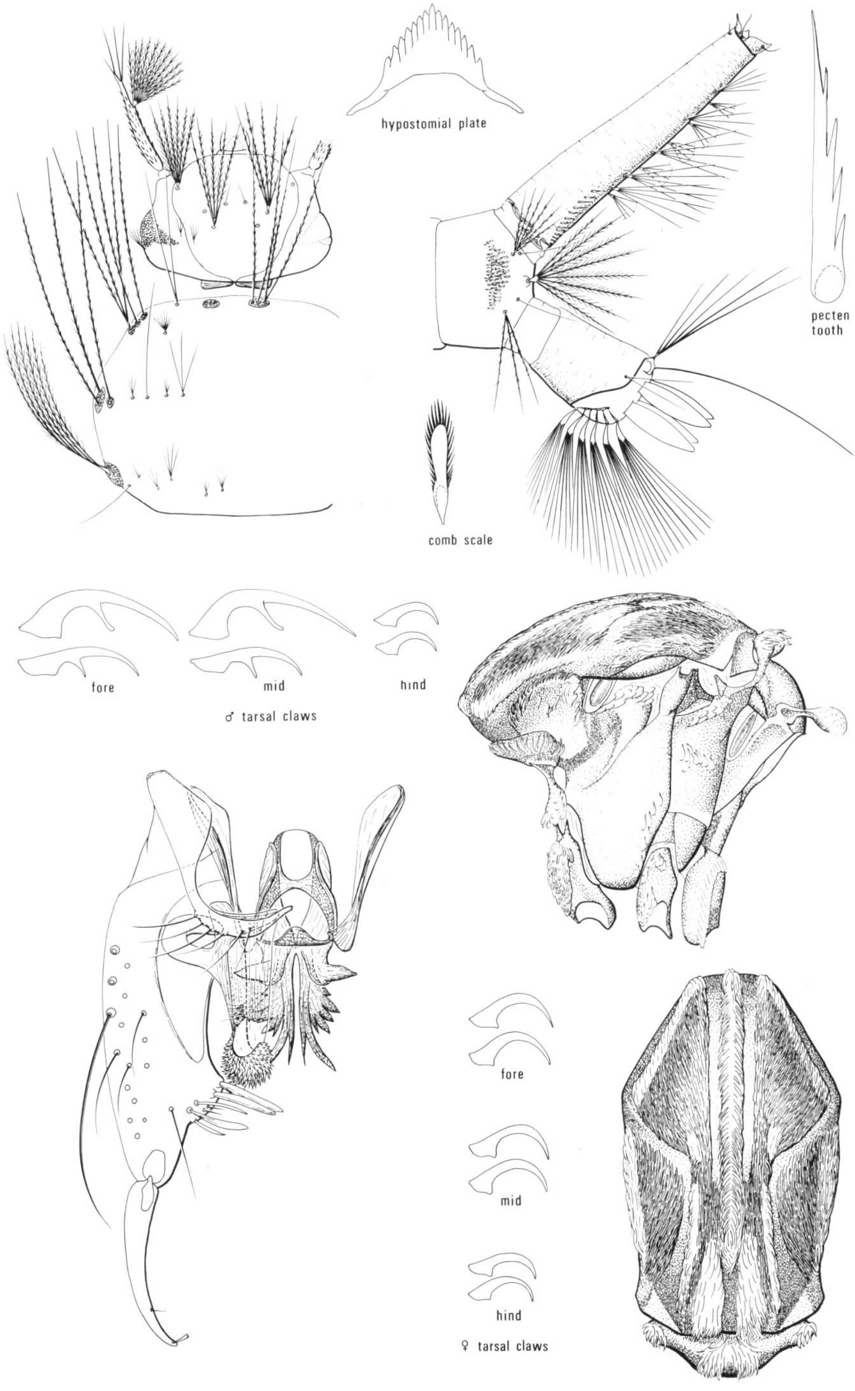

Plate 58. *Culex tarsalis*

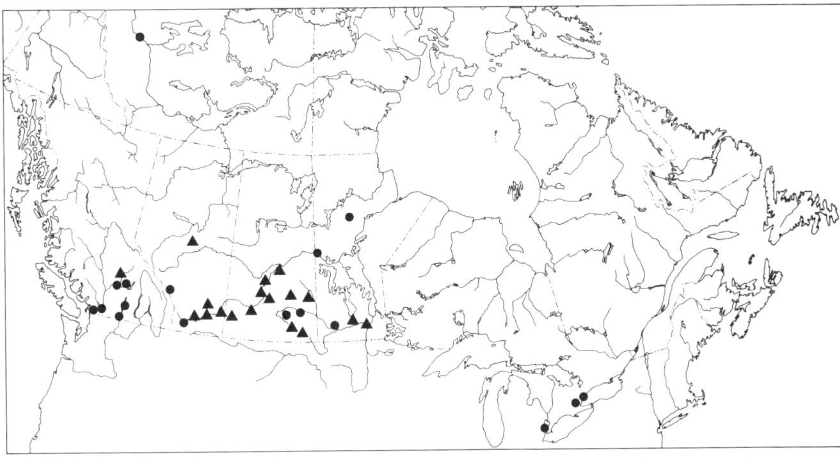

Map 58. Collection localities for *Culex tarsalis* in Canada: • specimens we examined, ▲ literature records.

mostly dark-scaled, with white-scaled apex; vertex brown-scaled, with line of white scales along eye margin and patch of white scales medially; scutum reddish brown scaled, with white scales along lateral margin anterior to transverse suture, on middorsocentral spot, on both sides of prescutellar depression and supraalar area, and as a narrow postsutural dorsocentral line; upper margin of postspiracular area with row of white scales; anterior surface of femora dark-scaled with narrow longitudinal white-scaled line or row of spots; tibiae with white-scaled line on both anterior and posterior surfaces, otherwise dark-scaled; first tarsomeres dark-scaled except for faint white-scaled line on both anterior and posterior surfaces and white-scaled basal and apical ring; second fore tarsomere with white spot of scales at base and at apex; remaining fore tarsomeres dark-scaled; second and third mid tarsomeres with clearly defined basal and apical bands; fourth mid tarsomere with patch of pale scales basally and apically; last mid tarsomere dark-scaled; all hind tarsomeres, including the last one, with basal and apical white-scaled bands; wing veins dark-scaled except for a few white scales usually at base of costa and some scattered pale scales along subcosta; abdominal tergites dark-scaled, each with basal transverse band of pale scales.

Male: Palpus longer than proboscis by the length of last palpomere and about half the length of second last palpomere; third, fourth, and fifth palpomeres brown-scaled dorsally except for narrow basal white-scaled transverse band; fourth and fifth palpomeres ventrally with a longitudinal band of white scales; long, fine setae of third palpomere arising, as in *restuans*, along entire length of the palpomere, with the setae on apical half more than twice as long as the more ventrally directed setae arising on basal half; proboscis brown-scaled, often with numerous scattered white scales and, as in female, with an encircling band of white scales; legs and scutum patterned as in female; subapical lobe of gonocoxite with eight appendages, all without recurved apices, with the two basal ones straight, somewhat flattened, and blade-like; paraproct with a long, thick, recurved, blunt lateral projection; abdominal tergites each with basal transverse pale-scaled band.

Larva. Head capsule more darkly pigmented, especially on posterior half, than in *territans*; antenna more than two-thirds as long as head, arched ventrally, and strongly narrowed on apical third; antennal seta arising at point of constriction; head seta 5-C with four or more branches; seta 6-C with three or more branches; siphon moderately stout, about five times as long as basal width, broadest slightly distal to base and tapering evenly to apex without constriction between middle and apex as in *territans*; siphonal tufts 8–10 arising in two parallel rows on both sides of midventral line as in *territans*, seldom paired opposite one another; upper caudal seta 2-X with two or three branches; precratal setae absent.

Biology. Larvae of *tarsalis* are usually found in permanent and semipermanent ponds and irrigation and roadside ditches that have emergent vegetation, but as populations build up during the summer, they may spread to small temporary pools and artificial containers (Beck 1961). Davis and Stage (1959) reported larvae found in seepage-filled hoofprints in the mud of a creek bed.

Females overwinter in caves, abandoned mines, rock piles, talus slopes, and rodent burrows (Chapman 1961, Beck 1961, Shemanchuk 1965). They emerge in spring (late March – early April in Colorado) following the spring soil temperature inversion, that is, when the temperature of the surface soil becomes warmer, rather than colder, than the deeper layers (Bennington et al. 1958*a*). There appears to be no migratory phase between emergence and host-seeking (Bailey et al. 1965). They are most active between sunset and sunrise (Burgess and Haufe 1960). Breeland et al. (1961) found females in Tennessee resting during the day, in and under various man-made structures including buildings. In Texas, Eads (1965) did not find many females in such sites, even during population peaks. In California, artificial resting boxes occupied by female *tarsalis* were vacated 17–30 min after sunset, their arrival and departure being correlated with sunrise and sunset. Peak activity occurs within 2 hr after sunset (Bailey et al. 1965).

Though females of *tarsalis* feed on birds like other *Culex* (Reeves and Hammon 1944, Beck 1961), they seem to prefer mammalian blood. From precipitin tests, Edman and Downe (1964) found more *tarsalis* containing beef blood than bird blood, although this may have been a result of what was most available. These authors recorded the blood of various mammals in a sample of *tarsalis* that included multiple feedings.

Egg rafts of this species may contain between 100 and 250 eggs (Edmunds 1955). Chao (1958) and Bellamy and Kardos (1958) discovered autogenous females among the *tarsalis* population. As in autogenous *pipiens*, the number of eggs produced was lower and was affected by larval diet. Some females oviposited without having had food of any kind, even sugar.

Because of its involvement in the transmission of western equine encephalitis (WEE), *tarsalis* has been studied intensively. Jenkins (1950) showed that the range of *tarsalis* and that of WEE roughly coincide. WEE is assumed to be primarily a disease of wild birds, among which epizootics are undoubtedly frequent, if sporadic, events, and *tarsalis* is assumed to be responsible for spreading the virus. Prevalence of the virus in an area is monitored by regular testing of individual birds in chicken flocks (called sentinel flocks) located in strategic areas. It is only during a build-up of both the virus and *tarsalis* that the disease is likely to appear in horses and humans. Horses are particularly susceptible to WEE, hence the name equine encephalitis. Fatalities among horses have always preceded symptoms among humans, and have thus served as a warning of a potential epidemic.

The mode of overwintering of the virus has evaded investigators. Overwintering in female *tarsalis* was naturally suspected, but several studies have shown this to be unlikely, because a female's first blood meal is taken in spring after she emerges from hibernation (Bennington et al. 1958*b*). WEE virus has been isolated from snakes caught in early spring and from the young of infected female snakes (Gebhardt et al. 1964). Another virus, the La Cross strain (see under *A. triseriatus*), has recently been recovered from larvae, which suggests that transovarial transmission may be the mode of overwintering of WEE virus as well.

Although *tarsalis* has a wide distribution in Western Canada, it is most prevalent in the Prairie Provinces and uncommon in the aspen parkland belt (Graham 1969*b*). Only one specimen, a reared male, is known from the Northwest Territories (Map 57), which is far beyond its main center of distribution. This is only one example of many species that reach their northern limits of distribution in the Mackenzie Valley.

Distribution. Western North America, from the central Mackenzie Valley south to Mexico, east to southwestern Ontario and Florida.

Culex territans Walker

Plate 59; Fig. 205; Map 59

Culex territans Walker, 1856:428.
Culex saxatilis Grossbeck, 1905:360.
Culex frickii Ludlow, 1906*a*:132.
Culex apicalis, authors before 1948, not Adams.

Adult. Female: An unmarked brown species, similar to *pipiens* and *restuans*, except that the transverse pale-scaled band on each abdominal tergite is apical rather than basal in position.

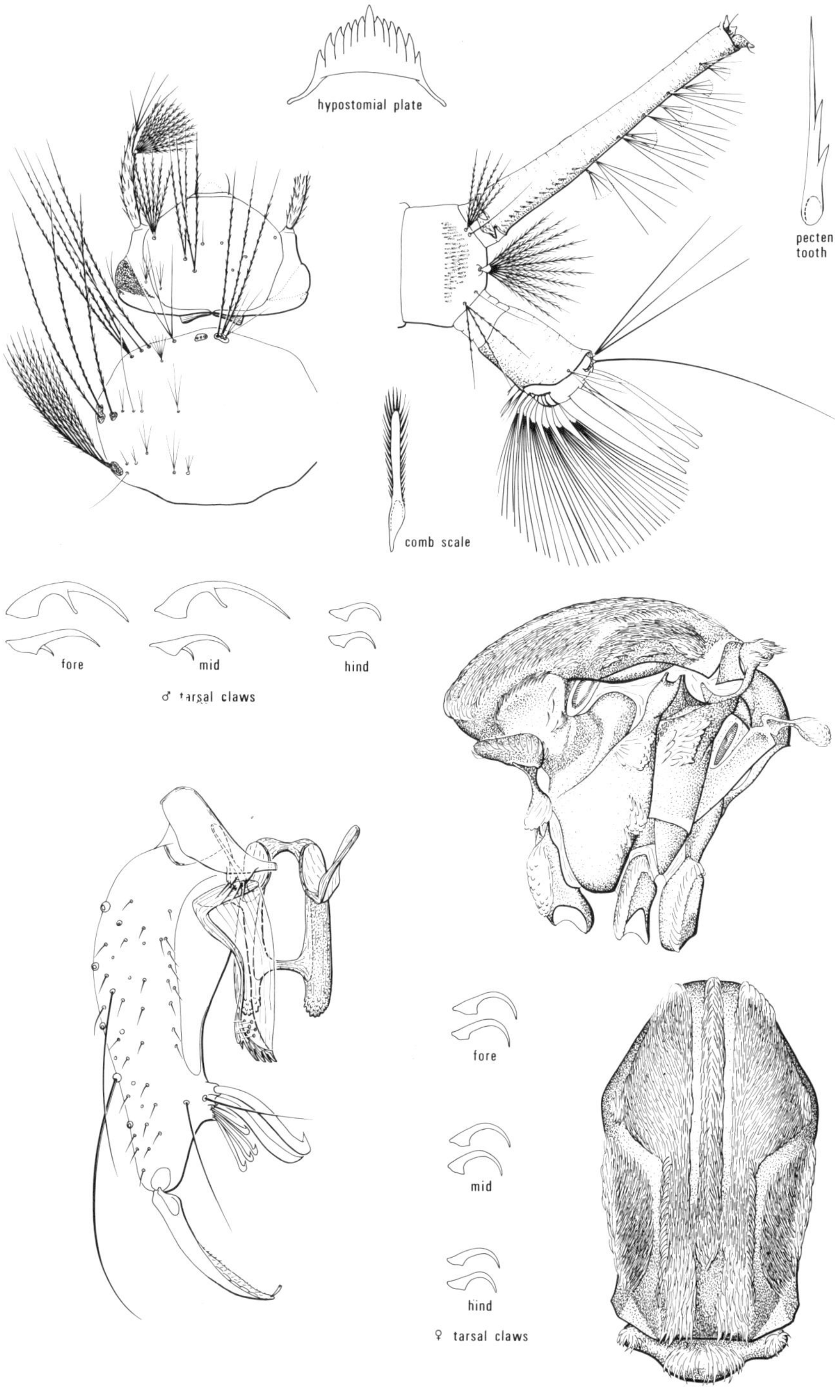

Plate 59. *Culex territans*

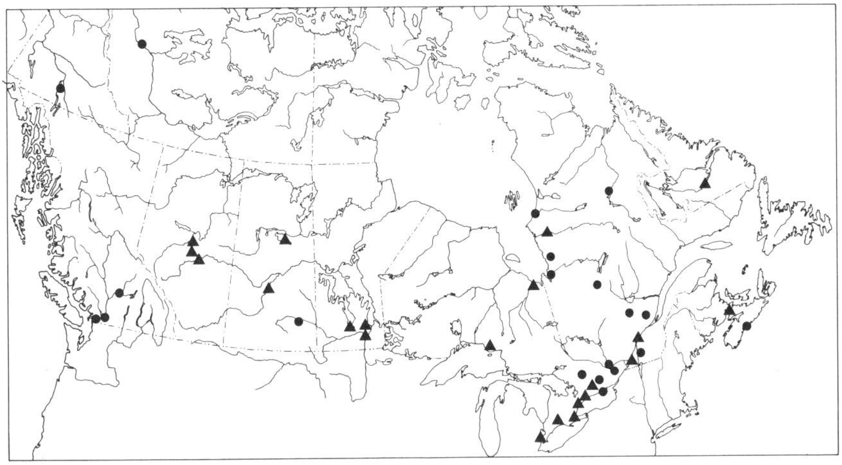

Map 59. Collection localities for *Culex territans* in Canada: • specimens we examined, ▲ literature records.

Male: Palpus longer than proboscis by combined length of last two palpomeres; long fine setae of third palpomere few, confined to apical one-quarter of segment; pale-scaled transverse bands at apices of each abdominal segment irregular, not always clearly defined; subapical lobe of gonocoxite with eight appendages, the two most basally placed ones long, flattened, slightly sinuous, each with recurved apex, the four apical ones similar but much smaller, and the remaining two long, setiform; paraproct without lateral projection and few apical denticles.

Larva. Antenna nearly as long as head and strongly curved ventrally, apical third or fourth abruptly narrowed; antennal seta arising at point of constriction; setae at apex of maxilla as long as maxilla itself; head setae 5-C and 6-C each single or double, rarely triple; siphon especially slender, five or six times as long as maximum width and seven or eight times apical width, broadest at base and tapering from base to about two-thirds to three-quarters distance from base, then widening again to apex; actual length of siphon no more than last three abdominal segments, thus not as long as siphon of *pipiens*, but apparently longer because of its slenderness; siphonal tufts 8–10, arising in two parallel rows, one on each side of midventral line, but usually not regularly paired opposite each other, all tufts branched, the last shorter and arising slightly out of line laterally; upper caudal seta 2-X with two or three branches; two or more small precratal setae.

Biology. Larvae of *territans*, like *tarsalis* and *restuans*, can be found in artificial containers, tree cavities, and other small bodies of water, but their usual habitat seems to be large vegetation-choked permanent marshes, especially those covered with duckweed (*Lemna* spp.). They have not been found in grossly polluted locations (Headlee 1945). Larvae appear in late

May and early June (Twinn 1931) and are usually common throughout Canada south of the tree line and perhaps on the prairies.

Females emerge from hibernation early in spring (Twinn 1931, Haufe 1952) and are believed to feed most commonly on frogs (Means 1965, Crans 1970). Feeding has been observed also on snakes (Dyar 1928), birds (Crans 1970), cattle (Edman and Downe 1964), and humans (West and Hudson 1960). However, some authors (Rempel 1950, Stage et al. 1952) have denied that *territans* takes mammalian blood. Crans (1969) and Benach and Crans (1973) reported that *territans* transmits the causative agent of frog filariasis, *Foleyella flexicauda,* to bullfrogs, *Rana catesbeiana*, in New Jersey.

Frohne and Frohne (1954) described males swarming in midafternoon, 2–5 m over clumps of bushes on the bluff of a river in Alaska; the weather was calm, cloudy, and humid.

In the north, there may be only one generation each year (Happold 1965*a*), but in southern Canada there are several, as in other *Culex.*

Distribution. North America, Eurasia, and North Africa: Europe and North Africa east across the USSR and the Middle East to Japan and Alaska, south to California, east to Newfoundland and Florida.

Genus *Culiseta* Felt

Adult. Integument medium brown; spiracular setae present; postspiracular setae absent; recumbent scales of vertex all pale, contrasting with dark, erect, forked scales and postocular setae; prealar area with setae, but without scales; scales of scutum mixed pale and brown (all brown in *C. melanura*), narrow, not especially dense, usually arranged in an obscure pattern of narrow longitudinal stripes; tarsal claws of female all without subbasal tooth; base of subcosta ventrally with a group of hairs; margin of calypter fringed; abdomen rounded apically, not strongly tapered; cercus of female short, rounded apically, inconspicuous; distal margin of eighth tergite of male usually lobulate and armed with a row or group of modified setae; gonocoxite of male tapering toward apex, with flattened triangular basal lobe usually bearing two enlarged apically directed bristles; apical lobe of gonocoxite a low tubercle or absent; gonostylus tapering, unbranched, its apex with short, bifurcate, peg-like seta; apex of arm of tenth sternite with four or five dorsally recurved teeth; claspette absent.

Larva. Prothoracic seta 4-P multiple, usually rather long; mesothoracic seta 7-M single, its base separated from that of 6-M; comb scales of eighth abdominal segment arranged in a triangular patch (one transverse row in *melanura*); siphon with row of pecten teeth; siphonal seta 1-S arising at base of siphon, from thickened ventral rim; saddle completely encircling anal segment, pierced ventrally by the anteriormost precratal setae of ventral brush (4-X).

Remarks. In the previous three genera, *Toxorhynchites, Aedes*, and *Culex*, it was not advantageous for us to use the subgenera (of which, among the Canadian species, there are five in *Aedes* and two in *Culex*) for identification, so almost no mention was made of them. However, in *Culiseta*, the subgenus *Culiseta* differs in adult and larval structures from the other two, *Culicella* and *Climacura*, and use of them seems advantageous.

Key to the subgenera and species of *Culiseta* of Canada—adults

1. Ventral margin of anepisternum (*a*), postspiracular area (*pa*), anterodorsal angle of katepisternum (*k*), and underside of paratergite (*pt*) each with a patch of scales (Fig. 208); gonocoxite with a subapical setose lobe (Figs. 210, 211); lateral halves of aedeagus separate from one another distally (Figs. 213*a*–216*a*) (subgenus *Culiseta*) 2

   These four pleural areas bare (Fig. 209) except for *minnesotae*, which usually has a few scales on postspiracular area, and *melanura*, with setae on underside of paratergite; gonocoxite lacking a setose subapical lobe (Fig. 212); lateral halves of aedeagus fused medially to apex (Figs. 217*a*–219*a*) (subgenera *Culicella* and *Climacura*) 5

2. Second and third tarsomeres each with basal ring of pale scales; last three palpomeres of male with conspicuous dense fringes of long setae, those of third at least as long as fourth palpomere (Fig. 220); eighth tergite of male (usually in a ventral position along with remainder of the terminalia) with a median apical cluster of up to 10 short stout spines (Figs. 213*b*, 214*b*) 3

   Tarsomeres dark-scaled; male palpus with scattered short setae (Figs. 221, 222); eighth tergite of male with row of 20 or more short closely set setae (Fig. 216*b*) or none (Fig. 215*b*) 4

3. Basal white rings of second and third tarsomeres broad, each occupying at least one-quarter of its respective segment; hypostigmal area usually with a patch of scales (Fig. 208, *h*); wing spots on R_s formed by aggregations of dark scales confined to its base and bifurcation of R_{2+3} and R_{4+5} (Fig. 225); costa, subcosta, and radius with scattered white scales along their entire length; crossveins bearing scales; setae of subapical lobe of gonocoxite uniform in size; each lateral half of aedeagus with reflexed apex .... ***alaskaensis***

   Basal white rings of second and third tarsomeres narrow, each occupying one-tenth of the segment or less; hypostigmal area bare; wing spots especially prominent, occupying all of R_s between its base and bifurcation (Fig. 226); white scales of wing, if present, confined to apex of costa (in males, especially, most dark wing scales appear rather pale and iridescent and may be mistaken for white scales); crossveins without scales; most distal setae of subapical lobe of gonocoxite longer and thicker than remaining setae; lateral half of aedeagus with pointed but distally directed apex .... ***incidens***

4. Costa, subcosta, and radius with scattered yellow scales; wing scales not aggregated to form spots (Fig. 227); tarsi with many pale scales; scutal scales all yellow; antenna of male with fewer than 20 setae per segment; male palpus with a few short coarse setae, but without fine setae, its terminal segment inflated (Fig. 222); eighth abdominal tergite of male without spines (Fig. 215*b*); lobe of ninth tergite with two rows of short stout peg-like setae .... ***inornata***

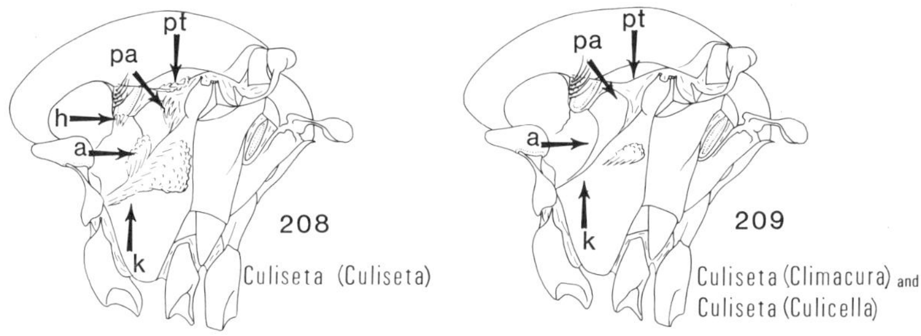

Figs. 208, 209. Left lateral view of thorax.

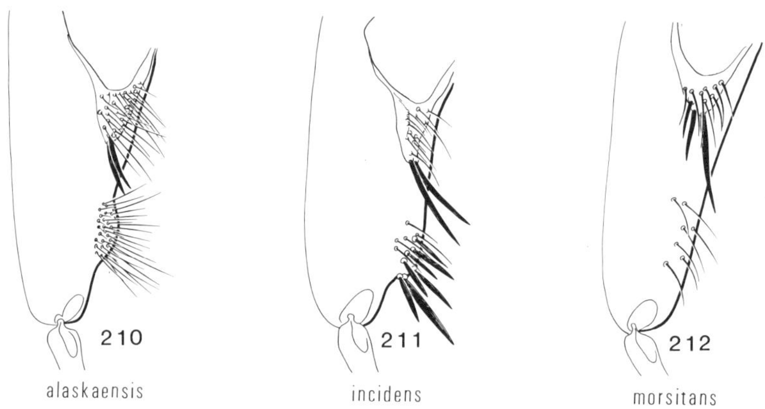

Figs. 210–212. Dorsal view of gonocoxite of species of subgenus *Culiseta*.

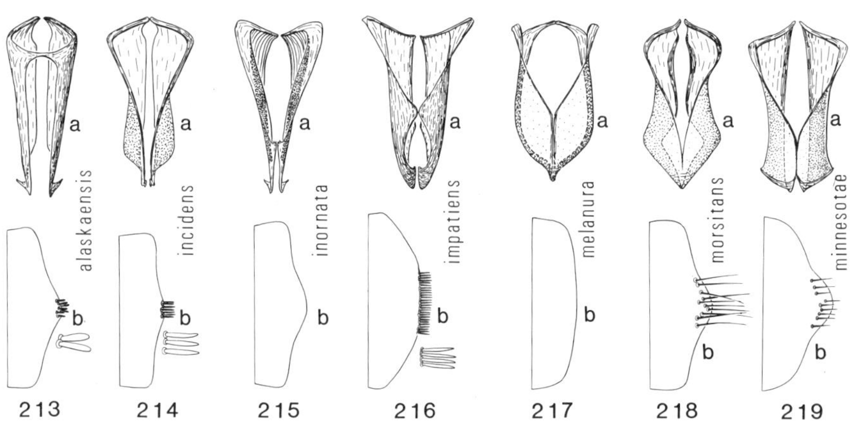

Figs. 213–219. Male of *Culiseta* species: *a*, dorsal view of aedeagus; *b*, eighth abdominal tergite.

Wing veins lacking pale scales (in some lights the scales have a bluish iridescence making them appear somewhat pale); wing scales weakly aggregated at bases of R_s, R_2, R_3, and R_{4+5} forming spots (Fig. 228); tarsi almost entirely brown-scaled; scales of acrostichal, dorsocentral, and sublateral stripes of scutum yellow, the remaining scutal scales reddish brown; male antenna with 30 or more setae per segment; fourth palpomere of male with sparse fringes of fine setae (Fig. 221); terminal palpomere cylindrical; distal margin of eighth tergite of male with row of 20 or more short spines (Fig. 216*b*); lobe of ninth tergite with one row of long curved setae .... ***impatiens***

5. Spiracular bristles few, usually two or three, more or less concolorous with adjacent bristles on postpronotum; paratergite with one to five delicate yellow setae on underside; abdominal tergites entirely dark-scaled; posterior margin of eighth tergite of male truncate, without patch of setae (Fig. 217*b*) (subgenus *Climacura*) .... ***melanura***

Spiracular bristles more than five, pale yellow, contrasting with darker bristles on postpronotum; paratergite bare; abdominal tergites each with transverse pale bands of yellowish brown scales; posterior margin of eighth tergite of male lobulate and setose medially (Figs. 218*b*, 219*b*) (subgenus *Culicella*) 6

6. Postspiracular area bare; abdominal tergites each with transverse band of paler scales only at base; tarsal claw thicker at base, shorter, and more strongly curved (Fig. 223); integument of scutum dark brown; aedeagus diamond-shaped (Fig. 218*a*), with apex pointed; setae on median lobe of posterior margin of eighth tergite of male as long as width of lobe (Fig. 218*b*) .... ***morsitans***

Postspiracular area usually with a small patch of scales; abdominal tergites each with basal and apical bands or patches of paler scales; tarsal claw slender, longer, and straighter (Fig. 224); integument of scutum reddish brown; aedeagus subparallel-sided and broadly rounded or truncate apically (Fig. 219*a*); setae on median lobe of posterior margin of eighth tergite of male much shorter than width of lobe (Fig. 219*b*) .... ***minnesotae***

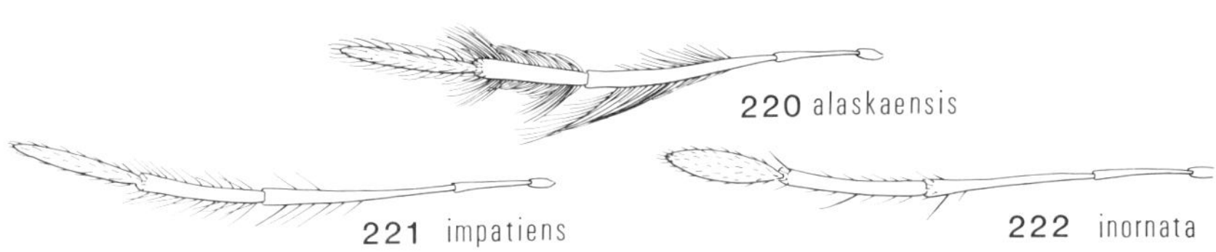

Figs. 220–222. Palpus of male of *Culiseta* species.

Figs. 223, 224. Tarsal claw of female of *Culiseta* species.

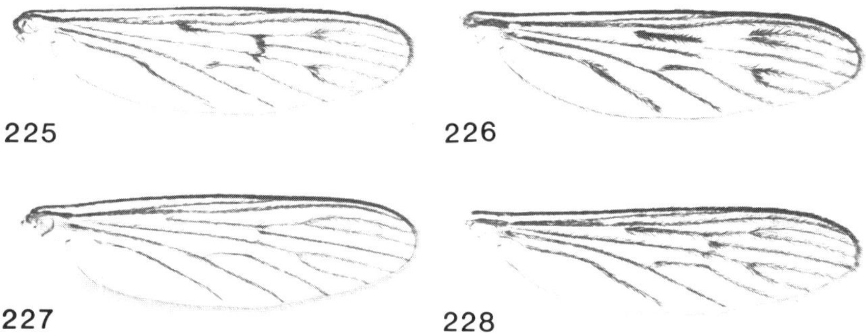

Figs. 225–228. Dorsal view of wing of *Culiseta* species: 225, *alaskaensis*; 226, *incidens*; 227, *inornata*; 228, *impatiens*.

Key to the subgenera and species of *Culiseta* of Canada—fourth-instar larvae

1. Antenna about half as long as head capsule; antennal seta 1-A inserted near middle of antenna; siphon rather short and stout, about three times as long as greatest width or less; pecten of two distinctly different lengths, a basal row of short teeth, followed by a row of longer setae extending to apical third or quarter of siphon (Fig. 40) (subgenus *Culiseta*) 2
   Antenna as long as head capsule; antennal seta 1-A inserted near apical third of antenna; siphon six or more times as long as greatest width; pecten teeth few, small, confined to basal third of siphon (Figs. 41, 42) (subgenera *Culicella* and *Climacura*) 5
2. Dorsolateral abdominal setae (1-VI and 1-VII) of segments VI and VII multiple, much shorter than seta 1-V, and less than one-tenth as long as upper lateral setae of their respective segments (Fig. 229); head setae 5-C and 6-C similar to each other in length and number of branches .... ***impatiens***
   Dorsolateral abdominal setae (1-VI and 1-VII) of segments VI and VII double or triple, as long as those on preceding three segments, and nearly as long as upper lateral setae of their respective segments (Fig. 230); head seta 6-C longer and with fewer branches than seta 5-C 3
3. Saddle seta 1-X as long as saddle or longer (Fig. 230) .... ***inornata***
   Saddle seta 1-X less than three-quarters as long as saddle 4
4. Prothoracic seta 1-P single; antenna coarsely spiculate, the spicules readily visible under a dissecting microscope .... ***alaskaensis***
   Prothoracic seta 1-P three- to five-branched; antenna appearing smooth, the spicules minute under a dissecting microscope .... ***incidens***
5. Comb scales arranged in a single vertical regular row; siphon with a single midventral row of unevenly spaced branched setae beyond pecten (Fig. 41) (subgenus *Climacura*) .... ***melanura***
   Comb scales arranged in a triangular patch; siphon without branched setae except for paired basal tuft 1-S (Fig. 42) (subgenus *Culicella*) 6

6. Head setae 5-C and 7-C usually with fewer than six and nine branches, respectively; ventral brush of anal segment (seta 4-X) usually with 20 or more setae .... ***morsitans***

Head setae 5-C and 7-C usually with more than six and eight branches, respectively; ventral brush of anal segment (seta 4-X) usually with 18 or fewer setae .... ***minnesotae***

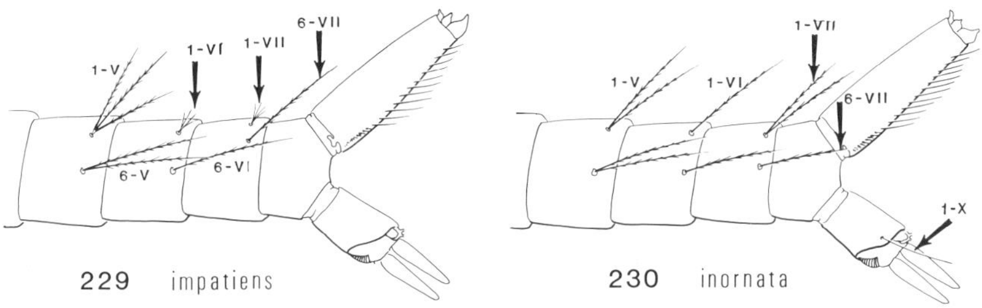

Figs. 229, 230. Left lateral view of terminal abdominal segments of larva of *Culiseta* species.

Subgenus *Culiseta* Felt

Adult. Scutum evenly though rather sparsely scaled, with narrow poorly defined median, dorsocentral and supraalar stripes, and periphery of prescutellar depression pale yellow scaled; middorsocentral spot of yellow scales prominent; lower margin of anepisternum, postspiracular area, and underside of paratergite each with small patch of pale scales; anterodorsal corner of katepisternum and most of mesepimeron scaled; scales of wing veins (except *inornata*) denser in some areas than in others, particularly at base of R_s, R_2, R_3, M_1, and M_2 appearing as dark spots; abdominal tergites dark-scaled, each with basal transverse band of pale scales; apical lobe of gonocoxite (except *inornata*) moderately developed, bearing a dense group of differentiated setae.

Larva. Head rather small in relation to body, as in *Aedes*; antenna much shorter than length of head, seta 1-A arising near or before its mid length; innermost setae of labral brush with comb on inner edge at apex, as in *Aedes* and *Psorophora*; siphon less than four times as long as maximum width; pecten consisting of a proximal row of short teeth, each with a small subbasal cusp, followed by a longer distal row of longer hair-like setae; base of siphon ventrally with quadrate indentation, the large basal siphonal seta 1-S inserted on its thickened, sclerotized margin at or near corner of indentation, and thus arising near middle of proximal row of short pecten teeth; siphonal seta 1-S conspicuous, longer and with more branches than abdominal seta 3-VIII.

Biology. All four species overwinter as mated but nulliparous, non-blood-fed females (Knab 1908, Dyar 1922*b*, Hearle 1926, Rempel 1953, Graham 1969*b*) emerging to seek blood in early spring. Shemanchuk (1965) found hibernating females of *inornata* in animal burrows, and Rempel (1953) stated that females enter houses in the fall as do *Culex*. Hibernating *alaskaensis* have been found in hollow logs (Dyar 1922*b*), cellars, and sheds (Yates 1953).

Eggs are laid in rafts on semipermanent bodies of water. Larvae of all four species of the subgenus occur during the summer. Whereas *impatiens* and *alaskaensis* can be uncommon and difficult to collect as larvae, *inornata* and *incidens* are often very abundant.

Culiseta alaskaensis (Ludlow)

Plate 60; Figs. 210, 213, 220, 225; Map 60

Theobaldia alaskaensis Ludlow, 1906*b*:326.

Adult. Female: Proboscis and palpus mostly dark-scaled, usually with scattered pale scales; scutum with mixed dark brown and whitish scales; hypostigmal area usually with small patch of scales; second, third, and on hind leg, fourth tarsomeres each with basal ring of white scales occupying about one-quarter of their respective tarsomere; costa, subcosta, and R_1 with mixed dark and pale scales throughout their length; wing spots resulting from aggregations of dark scales small but distinct, confined to the bases of veins R_s, R_2, R_3, R_{4+5}, M_1, M_2, and CuA_1, and a small spot on M at junction of crossvein r-m.

Male: Palpus longer than proboscis by almost entire length of last segment; distal two-thirds of ventrolateral surface of third palpomere with dense fringe of long setae, some of which are more than half as long as third palpomere; fourth and fifth palpomeres with median and lateral fringes of shorter setae; scales of scutum predominantly yellowish rather than brown as in female; leg bands and wing pattern much as in female but less distinct; posterior margin of eighth tergite with prominent lobe bearing a small group of short stout flattened clavate setae; lobe of ninth tergite with row of long curved setae; apical lobe of gonocoxite scarcely elevated, but distinguishable by dense group of medially directed setae; aedeagus conical, straight-sided, tapering apically, each lateral half terminating in strong laterally reflexed point.

Larva. Antenna with coarse spicules easily visible under a dissecting microscope; head seta 5-C five- to seven-branched, with the longest branches shorter than those of head seta 6-C, which is only two- to four-branched; prothoracic seta 1-P single; seta 3-P usually two- to four-branched; metathoracic ventral seta 13-M rather short, its branches subequal; dorsolateral abdominal setae (1-III to 1-VII) of segments III to VII well-developed, each usually double, more than two-thirds as long as its respective upper lateral (6-III to 6-VII); saddle pierced by at least two or three precratal setae; saddle seta minute, inconspicuous, much shorter than saddle.

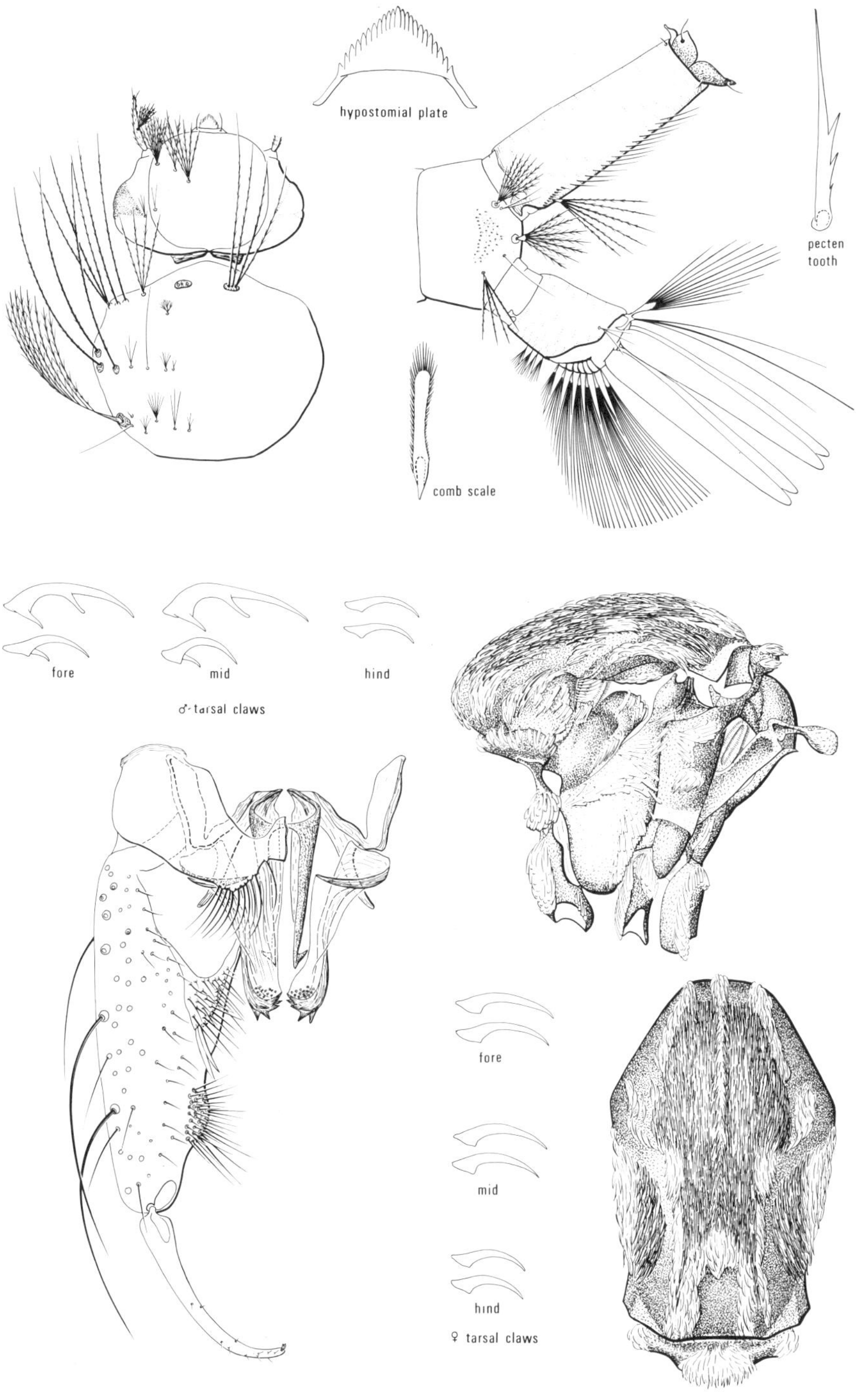

Plate 60. *Culiseta alaskaensis*

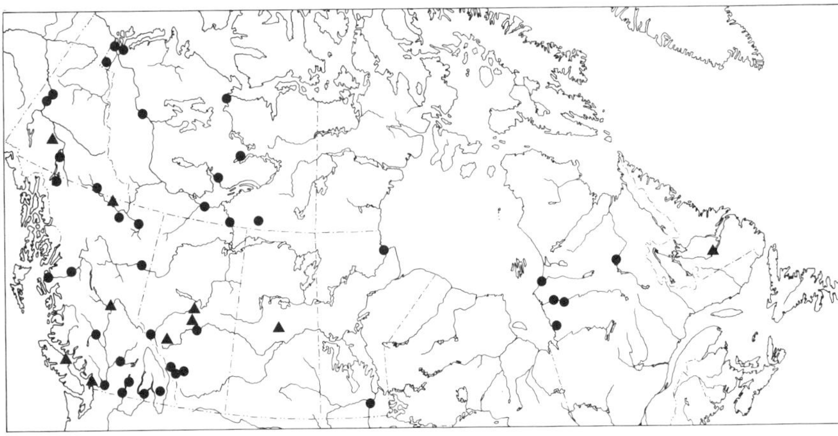

Map 60. Collection localities for *Culiseta alaskaensis* in Canada: • specimens we examined, ▲ literature records.

Biology. *Culiseta alaskaensis* is widely distributed in the boreal forest across northern Europe, the USSR, and northern Canada. In Western Canada it occurs also at low elevations in river valleys in southern Alberta (Shemanchuk, personal communication) and in the lower Fraser Valley of British Columbia (Hearle 1926), whereas in Eastern Canada it is a more northern species. Most collection records for unreared females are in May and June, and for reared adults, July, but there is one specimen in the Canadian National Collection of a female that was collected while biting on 3 August at Banff, Alta. In Alaska, the last biting record was 2 July (Frohne 1951), and no fall biting was observed; at Flatbush, Alta., Happold (1965) saw no more females after 27 May.

Various larval habitats for this species have been described by Jenkins (1948) and Frohne (1957), although this species has most often been collected among *Carex* near the edges of small deep semipermanent pools. We have encountered larvae associated with *impatiens* in central Yukon Territory, in similar pools where water was retained by permafrost. When the permafrost melted, the water disappeared rapidly, stranding many larvae that had not completed development. Jenkins (1948) found egg rafts as late as 3 July in Alaska. All investigators agree that there is only one annual generation. Sommerman (1964) described male swarming behavior in front of the entrance to a small man-made cave on the side of a cliff in Alaska.

Distribution. North America and Eurasia: Norway east across Siberia and Alaska to Newfoundland, south to the mountains of Iran, Pakistan, northern India, and Colorado.

Culiseta impatiens (Walker)

Plate 61; Figs. 216, 221, 228, 229; Map 61

Culex impatiens Walker, 1848:5.
Culex pinguis Walker *in* Lord, 1866:337.
Culex adsobrinus Felt, 1904:318.

Adult. Female: Proboscis and palpus dark-scaled; scutum brown-scaled, with especially prominent pair of yellow-scaled middorsocentral spots; scale patches of ventral edge of anepisternum and of postspiracular area smaller, more compact than in the other species; apices of femora and tibiae pale-scaled; remainder of tibiae and tarsi dark-scaled; wing veins entirely densely dark-scaled, with four weakly defined spots or aggregations of scales at bases of R_s, R_2, R_3, R_{4+5}, M_1, and M_2.

Male: Palpus about as long as proboscis; setae of third and fourth palpomeres short, sparse, and delicate, none as much as half as long as fourth palpomere; last palpomere slightly inflated; coloration and arrangement of scaling as in female except paler; posterior margin of eighth tergite with a row of 20 or more closely and evenly spaced short thick setae; lobe of ninth tergite with row of long curved setae; apical lobe of gonocoxite scarcely elevated, bearing a small group of setae; aedeagus tapering, rounded apically, each lateral half terminating in an anteromedially reflexed point.

Larva. Antenna with coarse spicules, visible under a dissecting microscope; head setae 5-C and 6-C almost equal in length and in number of branches, each with more than five branches; prothoracic seta 1-P single; seta 3-P with three or four branches; metathoracic ventral seta 13-M fan-shaped, with the longest branches in middle; dorsolateral setae 1-IV and 1-V of abdominal segments IV and V weak, between half and two-thirds as long as their respective upper lateral setae 6-IV and 6-V; remaining dorsolateral setae 1-III, 1-VI, and 1-VII much smaller, multiple, shorter than half as long as dorsolaterals 1-IV and 1-V; saddle seta fine, shorter than saddle; saddle pierced by two or three precratal setae.

Biology. Overwintering females of *impatiens* are possibly the earliest species of mosquito to seek blood in the spring. Frohne (1951) recorded females biting on warm days in March in southeastern Alaska. The females are exceptionally long lived. Some females that had already overwintered were caught in mid-May and fed sugar, water, and blood. These females lived until mid-September, probably more than 12 months from the time of their emergence. Frohne (1954*b*) believed that this explained the presence of biting females throughout the summer and early fall in Alaska, because overwintering females presumably do not feed before going into hibernation, and there is only one annual generation. Egg rafts containing about 100 eggs each were described by Dyar (1922*b*) and Frohne (1954*a*).

Larvae were collected in semipermanent ponds and bog pools (Frohne 1954*a*) and in dark, permanent, shaded, forest pools (Rempel 1950, Stage et

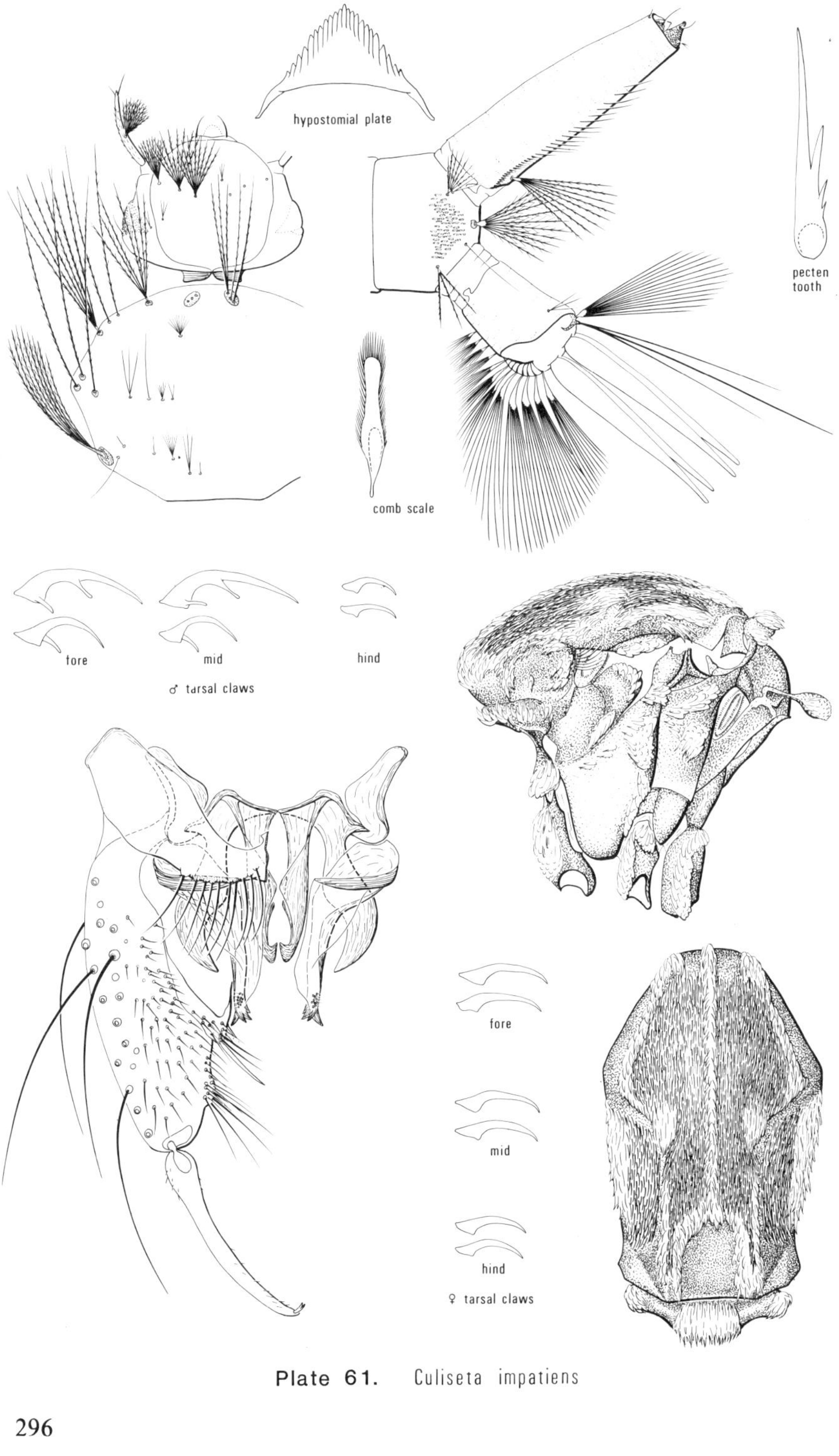

Plate 61. Culiseta impatiens

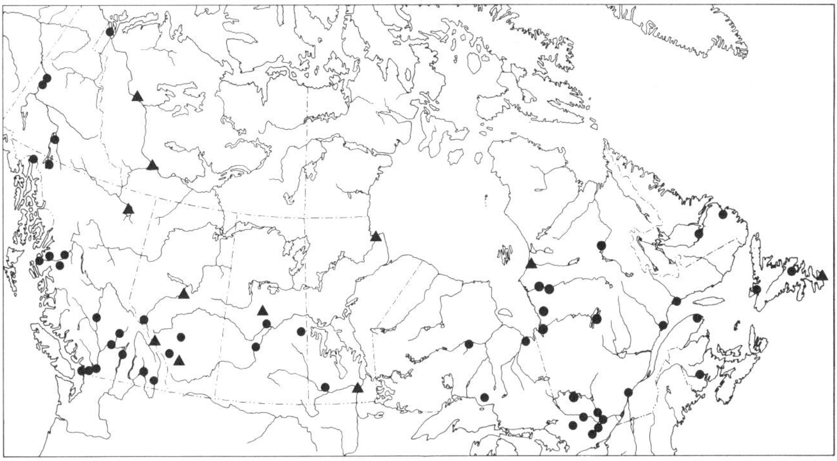

Map 61. Collection localities for *Culiseta impatiens* in Canada; • specimens we examined, ▲ literature records.

al. 1952). They have been found in association with *pullatus*, *punctor*, *excrucians*, *hexodontus*, and *alaskaensis* (Frohne 1952) and with *alaskaensis* and *morsitans* (Frohne 1953*b*).

Mating behavior was described by Frohne (1953*b*) and Sommerman (1964). Males do not swarm. Sommerman collected several coupled pairs from the ceiling of a cave in Alaska.

Distribution. North America, from Alaska and California east to Newfoundland and Massachusetts.

Culiseta incidens (Thomson)

Plate 62; Figs. 211, 214, 226; Map 62

Culex incidens Thomson, 1869:443.

Adult. Female: Proboscis and palpus mostly dark-scaled, with scattered pale scales; hypostigmal area bare; second and third tarsomeres with narrow indistinct basal and apical rings of pale scales, each ring no more than one-sixth as long as its respective tarsomere; pale wing scales only on upper distal half of costa; wing spots formed by aggregations of dark scales more extensive than in *alaskaensis*, occupying all of R_s between its base and bifurcation of R_{4+5}, an equally long adjacent section of M, basal third of each of R_2, R_3, M_1, and M_2, and most of CuA_1 between its base and crossvein m-cu; spots absent from base of R_{4+5} and from M at junction of crossvein r-m.

Male: Palpus about as long as proboscis; setae of third palpomere shorter and less dense than in *alaskaensis*, about as long as fourth palpomere; fourth palpomere fringed laterally and medially; coloration and scaling as in female, except that wing pattern is much less distinct and

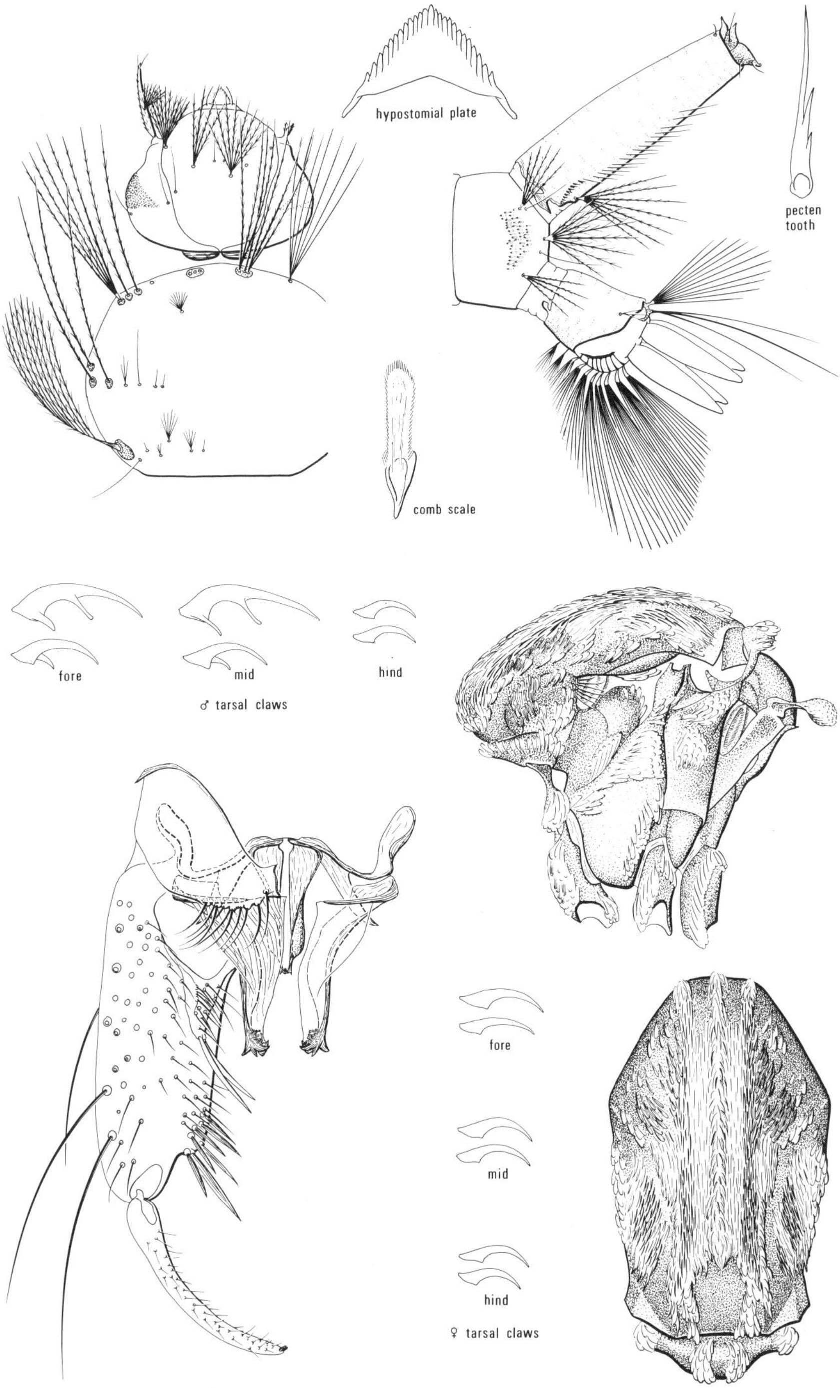

Plate 62. *Culiseta incidens*

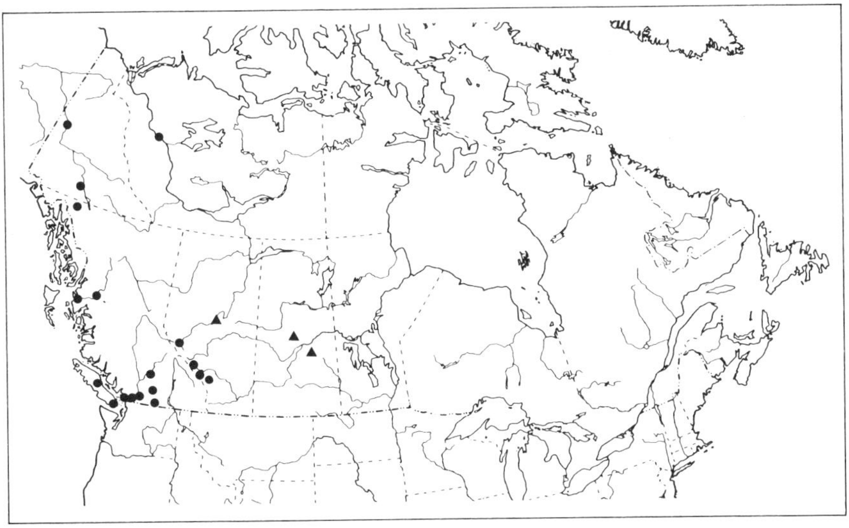

Map 62. Collection localities for *Culiseta incidens* in Canada: • specimens we examined, ▲ literature records.

tarsomeres have scarcely any trace of basal white-scaled rings; posterior margin of eighth tergite with scarcely any median lobe, bearing a small group of short, pointed, flattened setae similar to those of *alaskaensis*; apical lobe of gonocoxite better developed than in other *Culiseta*, armed with four or five long thick bristles; each lateral half of aedeagus with lateral flange and short, pointed apex.

Larva. Antenna rather smooth, with a few small, low spicules visible only under compound magnification; head seta 5-C with five or more branches, all shorter than those of seta 6-C, which is two- to four-branched; prothoracic seta 1-P multiple; seta 3-P usually four-branched; metathoracic ventral seta 13-M asymmetric, with its branches increasing in length and thickness ventrally; dorsolateral abdominal setae (1-III to 1-VII) of segments III to VII all well developed, each usually double and more than two-thirds as long as its respective upper lateral setae (6-III to 6-VII); saddle pierced by two or more precratal setae; saddle seta 1-X fine, shorter than saddle.

Biology. Unlike *alaskaensis* and *impatiens, incidens* is multivoltine, at least in British Columbia, having several generations during the summer in the lower Fraser Valley where it is the most common and widespread species of mosquito (Rempel 1950). The larval habitats are diverse: open or shaded pools and permanent or semipermanent streams and ditches (Chapman 1961, Curtis 1967). It is tolerant of polluted water (Dyar 1922*b*), including warm sulfur pools (Gibson 1926), artificial containers, and coastal rock pools.

Females overwinter in rock slides (Rush 1962) and also probably the other usual overwintering sites used by *Culex* and *Culiseta*. Hubert (1953) found that egg rafts were deposited at night. Hubert achieved male swarming in a cage 2.5 m high.

Distribution. Western North America from Alaska to California, east to the Prairie Provinces. Records east of the Prairie Provinces (cited by Carpenter and LaCasse 1955) need verification, and the record from Newfoundland (Freeman 1952) is undoubtedly a misidentification.

Culiseta inornata (Williston)

Plate 63; Figs. 215, 222, 227, 230; Map 63

Culex inornatus Williston, 1893:253.
Culex magnipennis Felt, 1904:322.

Adult. Female: Proboscis and palpus dark-scaled, with scattered pale scales; hypostigmal area bare; scale patches of lower edge of anepisternum and of postspiracular area usually confluent; scales of scutum yellowish; scutal setae dark brown; legs with mixed pale and dark scales, but without pale rings; costa, subcosta, and R_1 mostly dark-scaled, with scattered pale scales; remaining wing veins sparsely and evenly dark-scaled, without aggregations forming spots; abdominal tergites each with broad, irregular, basal, transverse band of yellowish scales.

Male: Palpus longer than proboscis by almost the length of last palpomere, with only a few short coarse setae, the longest seta at apex of third palpomere no longer than three times width of palpomere at its apex; last palpomere inflated; setae of each flagellomere of antenna fewer and shorter than is usual in male mosquitoes; posterior margin of eighth tergite rounded medially, but without median group of setae or bristles; lobe or ninth tergite studded with two irregular rows of very short, stout, blunt, peg-like spines; apical lobe of gonocoxite scarcely or not elevated, without a strongly differentiated group of setae; aedeagus strongly tapering apically, lateral margin concave, each lateral half terminating in a small laterally reflexed point similar to that of *alaskaensis* but smaller.

Larva. Antenna with fine spicules, visible only under a compound microscope or a dissecting microscope at high magnification; head seta 5-C with five or more branches, head seta 6-C with fewer (usually three or four) but distinctly longer branches than seta 5-C; prothoracic seta 1-P usually single; seta 3-P usually double or triple; metathoracic ventral seta 13-M asymmetric, its branches increasing in length and thickness ventrally; dorsolateral abdominal setae 1-III to 1-VII of segments III to VII all well developed, usually triple, each almost as long as its corresponding upper lateral (6-III to 6-VII); saddle seta rather thick and conspicuous, longer than saddle; usually no more than one precratal seta arising from saddle.

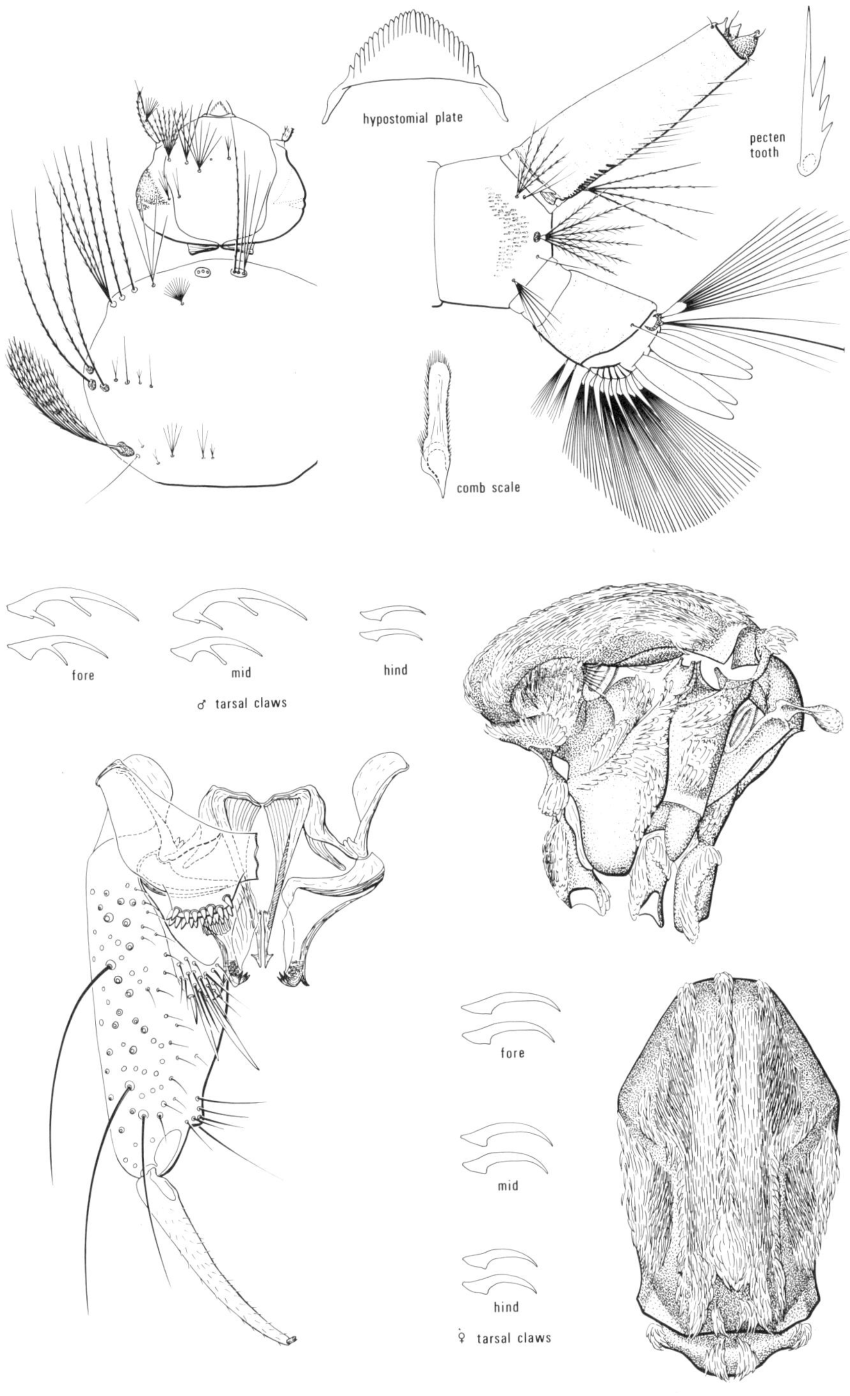

Plate 63. *Culiseta inornata*

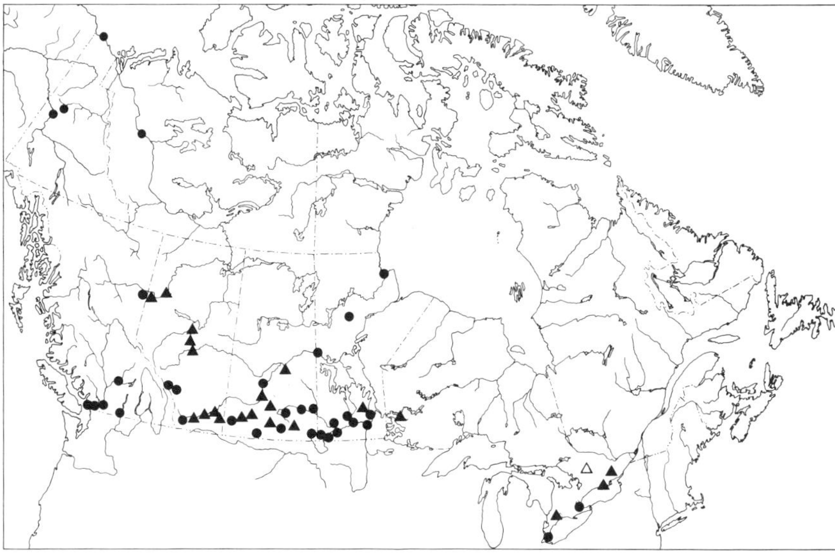

Map 63. Collection localities for *Culiseta inornata* in Canada: • specimens we examined, ▲ literature records, △ literature records based on probably misidentified specimens.

Biology. As is *alaskaensis, inornata* is widely distributed and abundant in Western Canada, but in Eastern Canada it is found only in southern Ontario. Either the species has only recently entered Ontario from Michigan or two different species are involved.

Culiseta inornata is the most common species of *Culiseta* in the Prairie Provinces. Overwintering females emerge from hibernation in early spring, and larvae appear as early as mid-May in Saskatchewan. The second generation appears by the end of May (Rempel 1953). Generations continue to develop until early October (Knab 1908).

Like *incidens*, larvae of *inornata* occur in various habitats (Rempel 1953) but not in artificial containers (Dyar 1922*b*).

Males of *inornata* do not swarm; they mate on vegetation a few centimetres above the surface of the water of their breeding sites (Rempel 1953). As a result, the species has been readily colonized (McLintock 1952). Downes (1958) noted that the antenna of the male has fewer setae than that of the males of other species of *Culiseta*.

Distribution. North America, from Yukon Territory and northern Mexico east to New Hampshire and Florida.

Subgenus *Climacura* Howard, Dyar, & Knab

Adult. Submedian stripe of scutum bare; scutal scales brown; spiracular bristles few or sometimes only one, brown; all of anepisternum including postspiracular area, anterior half of katepisternum, and lower half of mesepimeron bare of scales; underside of paratergite with one or more pale yellow hairs; scales of wing veins not aggregated; abdominal tergites entirely dark-scaled; apical lobe of gonocoxite lacking.

Larva. Antenna longer than head; antennal seta 1-A inserted near antennal apex; comb scales about 25, arranged in a vertical row, each scale elongate with scalloped, minutely fringed edges; siphon six or more times as long as maximum width, adorned with a midventral row of small tufts and several smaller dorsal and lateral scattered setae; the usual basal siphonal seta 1-S reduced, inconspicuous, and much shorter and less branched than seta 3-VIII.

Culiseta melanura (Coquillett)

Plate 64; Fig. 217; Map 64

Culex melanura Coquillett, 1902*b*:193.

Adult. Female: Integument medium brown; proboscis and palpus dark-scaled; scutum sparsely brown-scaled, with the scales never dense enough to obscure integument; submedian scutual stripe with even fewer scales than adjacent areas, sometimes bare; legs and abdominal tergites dark-scaled.

Male: Palpus about as long as proboscis, entirely dark-scaled; apex of third palpomere with tuft of about 15 long, fine setae; fourth palpomere with median and lateral fringe of moderately dense setae; scutum sparsely brown-scaled, submedian stripe usually without scales; eighth tergite truncate posteriorly, lacking setae; aedeagus obovate, with small pointed apex.

Larva. Antenna almost as long as head, with seta 1-A arising well beyond its mid length, uniformly darkly pigmented, and not especially swollen on basal portion; head seta 5-C with four or more branches; seta 6-C single, almost twice as long as seta 5-C; comb scales 18–25, each long, slender, and rounded, precisely arranged in a vertical row; siphon six or more times as long as basal width; pecten teeth small, usually only on a siphon, although one or two may arise on proximal membrane, extending over basal third of siphon; siphonal seta 1-S two- or three-branched, unusually small, shorter than basal width of siphon, and easily overlooked; siphon with an additional midventral row of more than 10 unpaired multiple tufts beginning before last pecten tooth and extending almost to apex of siphon.

Biology. *Culiseta melanura* has only recently been recorded in Canada (Chant et al. 1973). Unusual among *Culiseta*, *melanura* overwin-

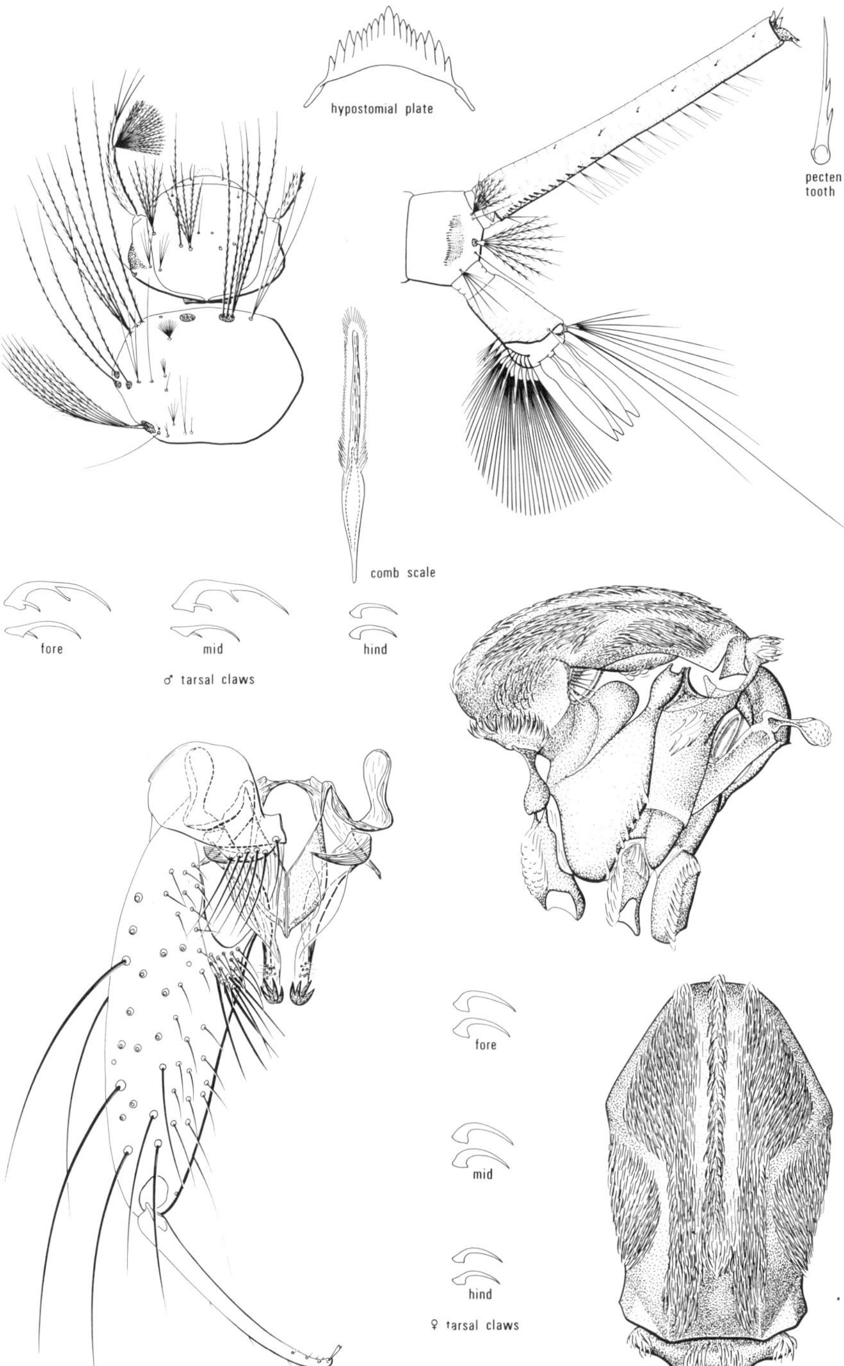

Plate 64. Culiseta melanura

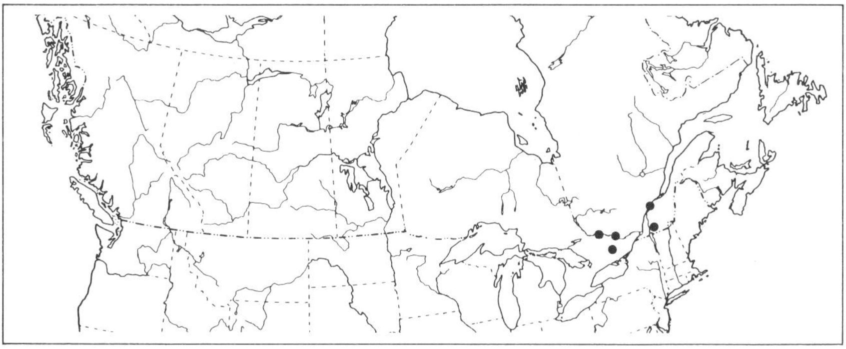

Map 64. Collection localities for *Culiseta melanura* in Canada: • specimens we examined.

ters in the larval stage (Brakeley cited by Mitchell 1907, Matheson 1944, Jamnback 1961). In deep water-filled depressions in a sphagnum bog in southern Quebec, Ellis and Wood (1974) found fully grown larvae in early May in association with the chaoborid *Eucorethra underwoodi* Underwood, another species that overwinters as a mature larva. Openings to these habitats were small, scarcely large enough to admit a dipper, and difficult to find. We assume they fill early with snow, which prevents the water therein from freezing. Larvae taken from the same habitat in November and held at about 18°C could not be induced to resume their development, even under increased photoperiod. Maloney and Wallis (1976) found that larvae of all instars developed rapidly to maturity at 28°C regardless of photoperiod. However, some larvae could be prevented from completing their development at 19°C if photoperiod was reduced to 9 hr of light. Development was arrested entirely at lower temperatures, provided that the larvae were cooled gradually for at least a week.

In the southern part of USA, this species is much more common. Larvae have been found throughout the year in Georgia (Siverly and Schoof 1962). Breeland et al. (1961) collected larvae in various small semipermanent bodies of water, including tree cavities, artificial containers, ground pools under trees, and especially in cedar swamps and basal cavities of gum trees associated with *Aedes thibaulti*. There are several generations per year in the southern USA and at least two in Canada, because larvae have been collected in early spring, midsummer, and late fall.

Although females prefer birds as hosts, they also feed on small mammals and snakes but not turtles or amphibians (Hayes 1961). They have only rarely been observed biting man (Hayes and Doane 1958, Schober 1964). Wallis (1954*b*) found gravid females in Connecticut as late as December. Eggs are laid in rafts, each containing about 100 eggs. The raft is almost circular or oval, with one side incomplete (Chamberlain et al. 1955).

The virus of eastern equine encephalitis (EEE) was first isolated from females of *melanura* by Chamberlain et al. (1955). Except under exceptional circumstances, it is difficult to imagine how such a rare and obscure species as *melanura*, which only rarely feeds on man, could serve as a vector of EEE among humans in Canada, although it may be instrumental in spreading the disease among wild birds.

Males were observed swarming on several occasions between 12 June and the end of October in Massachusetts (Hayes 1958). Swarming took place at sunset about 3–6 m above ground that was beneath the tips of overhanging branches.

Distribution. North America, from Minnesota, Colorado, and Texas east to Maine and Florida.

Subgenus *Culicella* Felt

Adult. Submedian stripe of scutum bare; spiracular bristles numerous, pale yellow; lower margin of anepisternum and underside of paratergite bare; postspiracular area bare (*morsitans*) or usually with a small group of scales (*minnesotae*); anterior half of katepisternum and lower half of mesepimeron bare of scales; scales of wing veins without spots; abdominal tergites dark-scaled, each with basal transverse band of pale scales and, in *minnesotae*, also with apical band; gonocoxite tapering uniformly to apex, with apical lobe of gonocoxite lacking.

Larva. Antenna longer than head, with antennal seta 1-A arising near its apex; siphon six or more times as long as maximum width; pecten of a few small inconspicuous teeth only at extreme base; basal siphonal seta 1-S similar in length and number of branches to seta 3-VIII.

Culiseta minnesotae Barr

Plate 65; Figs. 219, 224; Map 65

Culiseta minnesotae Barr, 1957:163.
Culiseta silvestris minnesotae Barr: Maslov, 1964:159.

Adult. Female: Integument reddish brown, usually distinctly paler than that of *morsitans*; proboscis and palpus dark-scaled; scales of scutum coppery brown anterior to transverse suture, mostly so posterior to it, with brassy yellow scales being restricted to a small round middorsocentral spot, a small supraalar spot, and a narrow postsutural dorsocentral stripe, not contiguous with the middorsocentral spot and, usually, the median prescutellar area; postspiracular area with small group of yellow scales (lacking in only 1 of 28 specimens examined); second and third and sometimes first and fourth tarsomeres each with narrow, poorly defined basal ring of pale scales; abdominal tergites each with both basal and apical bands of yellowish brown

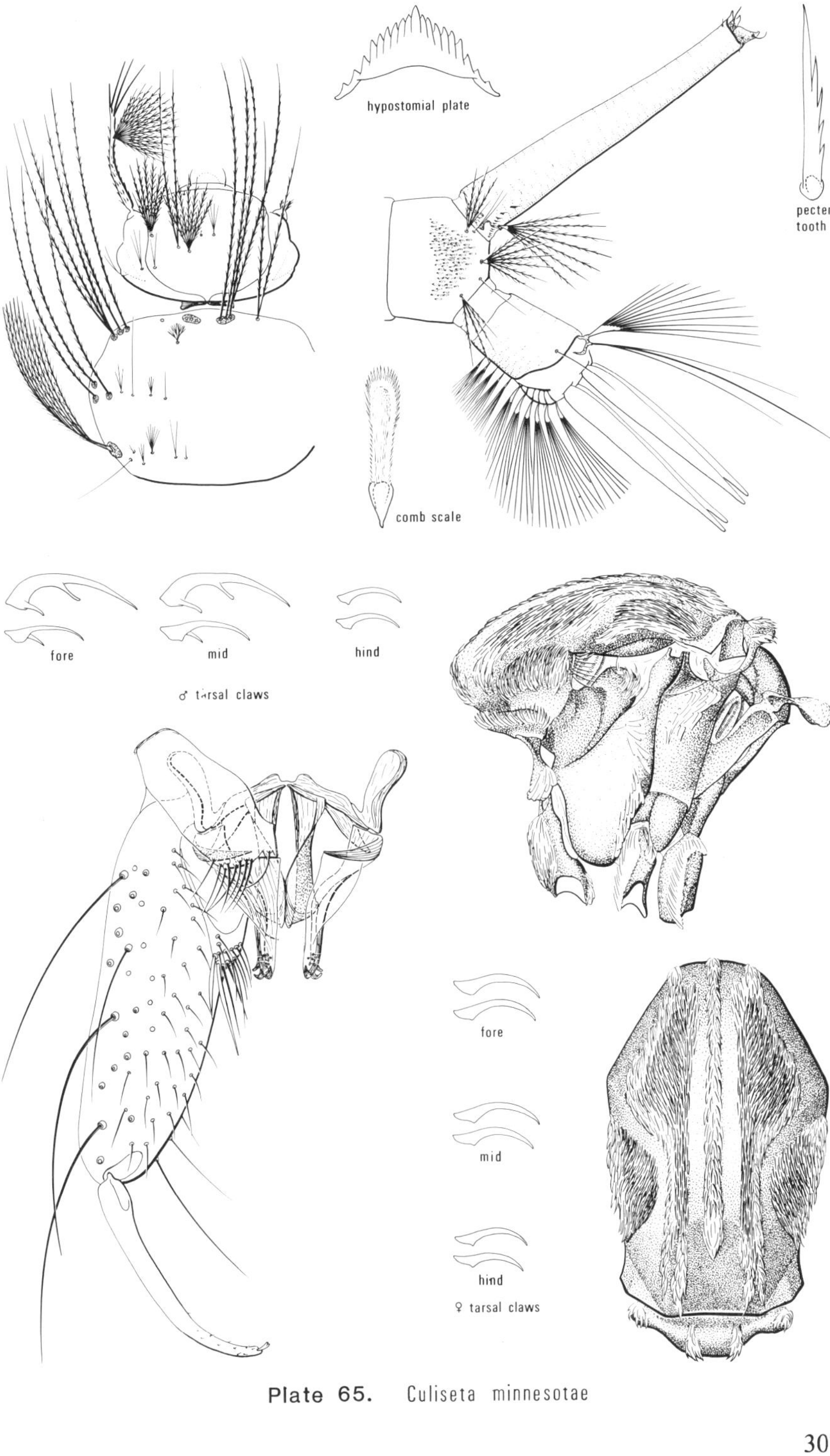

Plate 65. *Culiseta minnesotae*

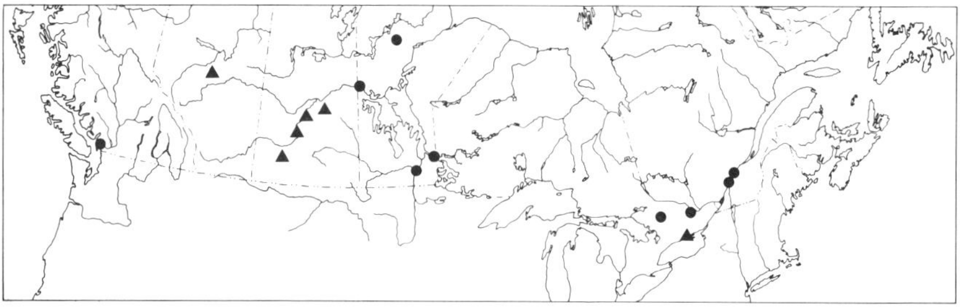

Map 65. Collection localities for *Culiseta minnesotae* in Canada: • specimens we examined, ▲ literature records.

scales separated by an irregular, poorly defined midtransverse band of darker brown scales.

Male: Palpus exceeding proboscis by combined length of last tarsomere plus more than half of second last palpomere; apical two-thirds of third palpomere and lateral and medial edges of fourth palpomere fringed with long dense setae; coloration and scaling as in female; setae on posterior margin of eighth abdominal tergite less than half as long as those of *morsitans*; aedeagus more cylindrical than that of *morsitans*, with concave lateral margins and broadly rounded apex with small median notch.

Larva. Almost identical with that of *morsitans*. According to Price (1958*a*), the number of branches of head seta 5-C is 6–11, whereas in *morsitans* it is 3–7, that of seta 7-C is 8–14 in *minnesotae* and it is 5–9 in *morsitans*, and the number of tufts of the ventral brush is 16–19 in *minnesotae* and 19–22 in *morsitans*.

Remarks. Maslov (1964) suggested that *Culiseta ochroptera* Peus, 1935 and *C. minnesotae* Barr be considered subspecies of *C. silvestris* Shingarev, 1928. Unfortunately, many North American workers began to use the combination *Culiseta silvestris minnesotae* without realizing the lack of evidence on which the proposal was based. First, the holotype of *silvestris* apparently no longer exists; Maslov saw only paratypes, which in a group such as *Culiseta* could easily have been different from the holotype. Second, Shtakelberg (1937, cited by Gutsevitch et al. 1974) concluded that the two species *ochroptera* and *silvestris* are not identical. Third, Maslov's material of *minnesotae* consisted of one male, which he discovered among *Culiseta* material that had been given to him from North America and which seemed to him indistinguishable from Asiatic specimens. Gutsevitch et al. (1974) rejected the name *silvestris*, considering it unrecognizable because the holotype was missing and the description inadequate. Instead they used *ochroptera*, a junior synonym, leaving the matter of usage of the name *minnesotae* unresolved.

The situation is analogous to the suceeding one of *morsitans* vs. *dyari*, except that here the identity of the senior name *silvestris* is in doubt. Until

Russian workers concur in choosing between *ochroptera* and *silvestris*, we are unwilling to use either name and, therefore, continue to use *minnesotae*.

Because Maslov (1964) compared so few specimens, it is difficult to know whether the slight differences he ascribed to his various subspecies in this complex were individual variations or reflected more fundamental differences.

Biology. This is one of the least known Canadian mosquitoes. Females overwinter, as in most other *Culiseta* (Barr 1957, Price 1958*a*). In central Alberta females appeared in the first week of May and disappeared in June (Graham 1969*b*, 1969*c*). Newly emerged adults reappeared in late June. Their feeding habits are similar to those of *morsitans*, preferring birds, but also feeding on small mammals and turtles (Hayes 1961). They apparently do not feed on man.

Larvae were collected in exposed semipermanent to permanent sedge–cattail marshes from mid-May to late September in Minnesota (Price 1961). They were most often found concealed among dense clumps of vegetation, occurring in association with larvae of *Culiseta morsitans* during the latter half of May. Larvae were first collected in Canada in Manitoba in a permanent cattail marsh (Trimble 1972).

Adults are seldom collected except at light or in CO_2-baited traps (Graham 1969*a*). There is probably more than one annual generation because there are adults in the Canadian National Collection from May, mid-July, and August. The adults collected in August were attracted to CO_2; they were probably not part of the overwintering generation.

Distribution. Central North America, from British Columbia and Oregon east to Quebec and Maryland.

Culiseta morsitans (Theobald)

Plate 66; Figs. 212, 218, 223; Map 66

Culex morsitans Theobald, 1901*b*:8.
Culex dyari Coquillett, 1902*b*:192.
Culex brittoni Felt, 1905*a*:79.
Culicella parodites Dyar, 1928:244.
Culiseta morsitans dyari Coquillett: Maslov, 1964:154.

Adult. Female: Integument medium brown, usually darker than that of *minnesotae*; scutum more extensively pale-scaled, with brassy scales on median stripe, presutural lateral stripe, anterior edge of transverse suture, middorsocentral spot, postsutural dorsocentral stripe, supraalar area, and periphery of prescutellar depression; postspiracular area bare; legs as in *minnesotae*; abdominal tergites dark-scaled, each with poorly defined transverse basal band of yellowish brown scales.

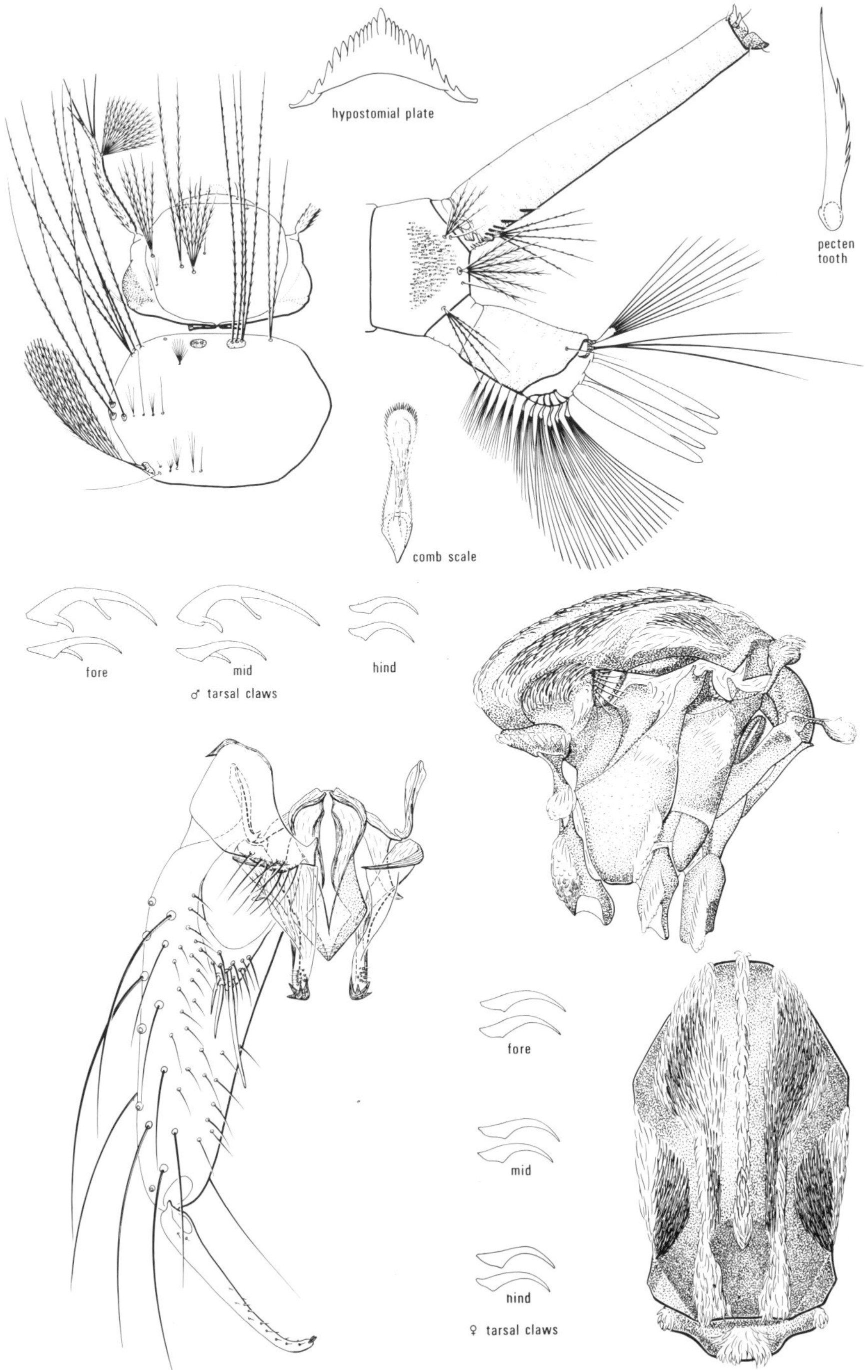

Plate 66. *Culiseta morsitans*

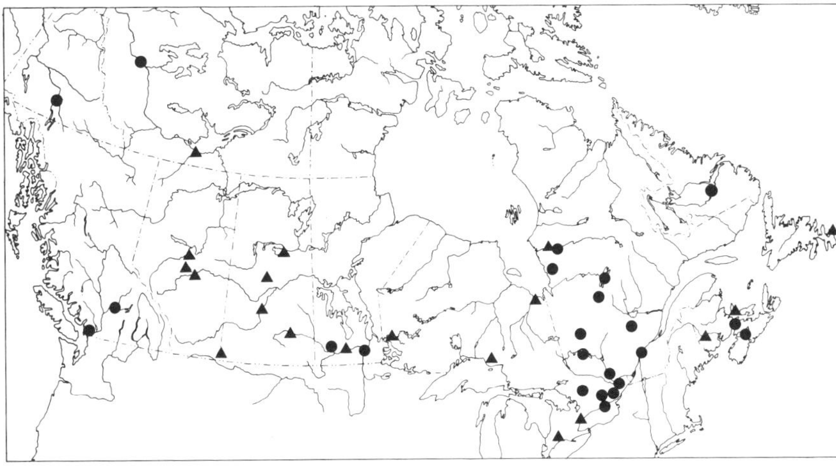

Map 66. Collection localities for *Culiseta morsitans* in Canada: • specimens we examined, ▲ literature records.

Male: Palpus longer than proboscis by slightly more than length of last palpomere; apical half of third and lateral and medial edges of fourth palpomeres with fringe of long dense setae; coloration and scaling as in female; posterior margin of eighth tergite rounded medially, with group of about 10 relatively long setae, each as long as length of tergite; apical half of aedeagus diamond-shaped with pointed apex.

Larva. Head exceptionally large in relation to body; antenna as long as head or longer, with seta 1-A arising well beyond its mid length and apex of seta extending beyond apex of antenna; base of antenna white, slightly inflated, and bowed laterally; apex of antenna darkly pigmented; head seta 5-C usually four- to six-branched; head seta 6-C double, more than half as long again as seta 5-C; thoracic setae exceptionally long; comb scales 50 or more, arranged in a large triangular patch; siphon long and slender, six or seven times as long as its basal width; pecten teeth greatly reduced in size and number, with the smaller basal ones usually arising on membrane proximal to base of siphon; row of pecten teeth barely longer than basal siphon width; siphonal seta 1-S shorter than in the subgenus *Culiseta* but conspicuous and longer than basal width of siphon; anal segment long and narrow; saddle pierced by six or more precratal setae of the ventral brush (seta 4-X).

Remarks. The name of this species is at present in a state of confusion. *Culiseta morsitans* was originally described from England and is now known from Europe and North Africa east to western Siberia (Maslov 1964, Gutsevitch et al. 1974), and in North America from central Alaska (Gjullin et al. 1961) to Newfoundland. Dyar (1928) originally maintained that differences in the terminalia between the European populations and those of North America warranted specific recognition of the latter under the name *dyari*. However, Matheson (1929) rejected *dyari* as a separate species, and North American material was subsequently called *morsitans* (with *dyari* as a synonym) until Maslov (1964) used the trinomial *C. morsitans dyari*. Unfortunately, this trinomial combination has appeared more and more often in the North American literature. Maslov used trinomials freely to denote any population that appeared to him to differ slightly from any other; he used them for almost every species of *Culiseta*. His characters for distinguishing *morsitans* from *dyari* are particularly inconsequential (for example, ratio of length of femur in all three pairs of legs to that of proboscis is usually more than one in *morsitans*, distinctly less than one in *dyari*). Perhaps the most apparently important difference is that of the overwintering stage; among North American populations, it is usually the egg stage, whereas in Europe, where the winters are less severe, the larva has been observed overwintering (Wesenberg-Lund 1920–21). Nevertheless, Wesenberg-Lund often observed total mortality as a result of a cold spell, followed by a new hatching as the weather warmed. In contrast, Shute (1933) and Callot and Dao Van Ty (1944) believed that the egg of the European form can resist desiccation during summer and can overwinter if the weather is dry. In Denmark *Aedes communis* may hatch anytime during the winter, but in Canada eggs overwinter, and hatching follows snowmelt. Thus the mode of overwintering may be a response to local conditions. We reject such a difference by itself, if in fact a real difference exists, as an indication of two separate species. More study is needed on the life cycle of *morsitans*, especially in western Siberia, where larval overwintering may not occur, and in western North America, where larval overwintering is a distinct possibility. The use of the trinomial, therefore, implies a more detailed understanding of the nature of the North American population in comparison with that of Europe than we believe exists. Because there is apparently little possibility that the two populations come into contact, the subspecific names should not be used in the sense commonly used today, for example, among birds, mammals, and butterflies, to denote anatomically different populations that interbreed freely along a zone of contact, but still maintain their anatomical distinctness. We believe that use of either the name *morsitans* or *dyari*, but not both, is more consistent with what is known. Because insufficient information is available, we prefer to use *morsitans* rather than *dyari*, for the same reason that we have used the names *Culex pipiens* and *Aedes dorsalis, sticticus*, and *pullatus* and others for species whose distributions are more or less holarctic but are not continuously distributed across both continents.

The name *parodites* Dyar deserves special mention. Matheson (1944) and Carpenter and LaCasse (1955) did not mention it, not even as a synonym. No specimens have been identified as this species, at least not in the U.S. National Museum, since the original four were described by Dyar. Maslov (1964) apparently was the first to make the synonymy, and we have examined the types and cannot find any reason to disagree with his opinion. A slide of the lectotype shows clearly the characteristic diamond-shaped aedeagus of *morsitans*. None of the four adults has any scales on the anepisternum and all have leg banding and scutal pattern similar to *morsitans*.

Biology. In most of Canada, *morsitans* overwinters in the egg stage (Morris et al. 1976). In Ontario, first instars appear in April usually in shallow, temporary pools in association with newly hatched *Aedes* spp. According to Wallis and Whitman (1968), eggs of *morsitans*, like other species of *Culiseta*, are laid in rafts. However, the rafts are not placed on the water surface, but slightly above it, secured to a moist substrate. The European form has been said to lay its eggs singly (Marshall 1938, cited by Wallis and Whitman 1968; Barr 1958). However, Marshall (1938, pp. 229–230) did not state that eggs were laid singly, but only relayed Wesenberg-Lund's (1920–21) observations that a female of *morsitans* ovipositing in a vessel without water laid eggs both singly and in small batches. Thus, the oviposition habits of the European *morsitans* appear to rest on a few tenuous observations of females living in captivity under difficult conditions. No records of this species being collected during the summer have been found, which leads us to conclude (as did Matheson 1944, Price 1961, and Morris et al. 1976) that there is only one generation per year. In this respect, the species is more like a spring *Aedes* than a *Culiseta*.

Larvae have been collected most often in shaded situations, springs, cedar swamps, sphagnum bogs, and small overflow pools caused by a rise in water level of larger woodland ponds and small lakes. Larvae often occur alone, in small numbers, or they may be associated with various species of *Aedes*.

Females prefer to feed on birds and are long lived, capable of producing at least three batches of eggs (Morris et al. 1976). Of 94 females examined by Morris et al. (1976), 22 had fed previously on birds, whereas none reacted with antimammal serum. They also bite small mammals and snakes (Hayes 1961). Horsfall (1955) reported that *morsitans* was a serious pest in eastern Europe and the Soviet Union.

Distribution. North America, Eurasia, and North Africa; Norway and North Africa east to western Siberia; Alaska and Oregon east to Newfoundland and Delaware.

Genus *Mansonia* Blanchard

Adult. Female palpus short; male palpus elongate, usually as long as proboscis or longer, with the terminal segments usually fringed with numerous long setae; postocular setae numerous, arising close to eye margin; vertex with few recumbent scales, but numerous erect forked scales; postpronotum with setae, but usually without scales, which, if present, are narrow and curled as on adjacent scutum; scutum with narrow, recumbent, often curled scales, sometimes in two sizes and colors; acrostichal, dorsocentral, and lateral setae well developed; spiracular setae absent; postspiracular setae present or (in the subgenus *Coquillettidia*) absent; anepisternum without scales; remainder of pleuron mostly bare except for three small patches of flat whitish scales; wing scales narrow or broad; legs usually with bands of paler scales; abdomen truncate in female, with the cercus small, not exserted. Gonocoxite, with small basal lobe appressed to gonocoxite, tapering to apex; gonostylus usually short, enlarged apically, bilobed or with rounded keel on distal half, and small apical spinule; claspette absent; paraproct pointed or armed with teeth apically.

Larva. Colorless, except for heavily sclerotized apex of siphon and lightly sclerotized head capsule; antenna extremely long, slender, and flexible, especially beyond insertion of antennal tuft; siphon exceptionally short, conical; valves of respiratory apparatus darkly sclerotized, folded together medially enclosing spiracles and forming a pointed device for penetrating plant tissue; posterior valve with saw-toothed apex; anterior valve reduced, but its seta greatly enlarged and directed anteriorly; saddle completely encircling a rather long, slender anal segment.

Remarks. The present classification of the Culicidae, including the concept of the genus *Mansonia* that has been in worldwide use for nearly 50 years, was the work of Edwards (1932). He placed all species whose larva shared the same peculiarly modified siphon for extracting air from underwater parts of plants in *Mansonia.* By doing this he implied, though he did not state it, that he considered this unusual larval specialization a good indication of common descent, or monophyly. All subsequent work has confirmed that Edward's assumption was correct, and *Mansonia* remains unchallenged as a monophyletic group. The peculiar siphon has always been a useful character for instantly identifying the genus.

Edwards grouped all the species of *Mansonia* known to him into four subgenera, the nominal subgenus *Mansonia*, and three others, *Mansonioides, Rhynchotaenia*, and *Coquillettidia. Mansonia* and *Rhynchotaenia* are neotropical, whereas the other two are confined to the Old World except for the North American species *Mansonia* (*Coquillettidia*) *perturbans*, which is probably a relatively recent arrival here and is clearly related to old-world stock. Edwards' descriptions and keys indicated that he was aware of a closer affinity between the subgenera *Mansonia* and *Mansonioides* and also between *Rhynchotaenia* and *Coquillettidia*, but he did not formalize this in his classification.

Without rearranging any species or adding significantly to the characters Edwards and later authors had already described, Ronderos and Bachmann (1963) elevated *Mansonia* to the rank of tribe, that is, Mansoniini, and the subgenus *Coquillettidia* to generic rank. Unfortunately, this suggestion was followed by Stone (1967), and as a result the use of *Coquillettidia* as a full genus for *perturbans* has been gaining momentum among North American workers. Adoption of *Coquillettidia* as a full genus might bring out another facet of phylogeny, but it has left most nontaxonomists confused (for example, references in technical literature to "*Coquillettidia* (*Mansonia*) *perturbans*" or "*Coquillettidia* (=*Mansonia*)"; also

Carpenter (1970*b*:44) compared with Carpenter (1974:78). We consider that Ronderos and Bachmann made changes that were unnecessary and inadvisable. Bates (1949) eloquently explained the many advantages for all culicidologists as a result of the universality and stability of Edwards' system of classification, and why this classification must be continued. Aside from the long uninterrupted universal usage that mosquito generic names, such as *Anopheles*, *Culex*, and *Aedes* have had, they are meaningful to a large segment of the population, not just to taxonomists. Anyone contemplating a system of narrow genera in which *Coquillettidia* would be one of them should consider how meaningless mosquito names would be today if Theobald's (1910) monograph had become the basis of our classification, with 21 genera of Anophelinae and well over 100 genera divided among numerous subfamilies that now constitute the present Culicinae.

We suspect that the reranking of *Coquillettidia* to the generic level serves only the taxonomist's desire to incorporate into his classification as much phylogenetic information as possible without regard for the confusion this creates for other zoologists. In some well-known groups such as the vertebrates and some of the Lepidoptera, the genera have become so fragmented and have been shifted about so many times that nonspecialists have abandoned their use in favor of vernacular names, defeating the basic premise on which zoological classification was founded. We believe that vernacular names for mosquitoes, though they have been proposed for some species, have never been necessary because their generic and specific names have been stable and therefore remained universally meaningful.

As a result of the reranking of *Coquillettidia* to the generic level, which was followed by some workers but not others, the name *Mansonia* has become ambiguous, and it is necessary to suffix it with *s. l.* (*sensu lato*) or *s. str. (sensu stricto*). Also, specimens can no longer be identified to genus at a glance; for larvae the relative lengths of the antennal portions must be analyzed. The new tribe Mansoniini joins an excess of other tribes, almost one per genus, none of which seems equivalent to another nor universally acceptable, and therefore almost never used by the majority of mosquito workers.

Most other genera, particularly *Aedes*, *Culex*, *Culiseta*, and *Psorophora*, are made up of recognizable subgenera, but to rerank any of them would be to destroy the continuity and universality of the names *Aedes*, *Culex*, *Culiseta*, and *Psorophora*. The reranking of *Coquillettidia* has set an unfortunate precedent, but it is not necessary to follow it. If enough mosquito workers continue to use *Mansonia* in the inclusive sense of Edwards, as we have done here, the use of *Coquillettidia* as a genus will eventually disappear.

Biology. All members of the genus *Mansonia* rely on air within the underwater roots of emergent plants and all have the siphon highly modified for piercing plant tissue. The genus is primarily tropical. Many of the tropical species do not burrow into the bottom ooze but attach themselves to suspended rootlets of floating plants, particularly water-lettuce, *Pistia stratiotes*. The pupae of all species also attach themselves to the roots by their pointed, well-sclerotized prothoracic respiratory trumpets. Some species attach themselves before the larval skin is fully shed and remain attached until emergence is imminent, when the tips of the trumpets break off and remain in the plant. Others are free to detach and reattach themselves at will.

Mansonia perturbans (Walker)

Plate 67; Map 67

Culex perturbans Walker, 1856:428.

Culex testaceus Wulp, 1867:128.

Culex ochropus Dyar and Knab, 1907*a*:100.

Adult. Female: Integument yellowish brown, with the base of each seta darker brown; proboscis with fairly well-defined yellow-scaled ring on middle third, the apex dark-scaled, the base predominantly so; palpus mottled with yellow and brown scales; pedicel with pale scales only; vertex with recumbent narrow yellow scales and yellow and dark erect forked scales; setae and most scales of scutum yellow, with a few smaller brown scales usually present; submedian stripe of scutum bare of scales; postpronotum with rather few narrow yellow scales in region of postpronotal setae; pleural scales flatter, paler, confined to posterior third of katepisternum and to upper third of mesepimeron; femora and tibiae mottled with dark and yellow scales, the hind tibia usually also with a well-defined ring of pale scales near but not at its apex; first tarsomere dark-scaled, with narrow basal ring and broader middle ring of white scales; all subsequent tarsomeres each with basal third to half white-scaled, the remainder dark-scaled; wing veins mottled with yellow and dark scales; abdominal tergites brown-scaled, each with irregular basal transverse band or lateral triangle of yellowish scales.

Male: Palpus longer than proboscis by length of last palpomere; apex of third and all of last two palpomeres fringed with long setae, each palpomere predominantly pale-scaled on basal two-thirds; coloration and scaling otherwise much as in female. Basal lobe of gonocoxite with single blunt club-like seta; gonocoxite with rounded keel on outer side near apex and extremely small apical spinule; apex of paraproct armed with several teeth.

Larva. With its long slender antenna, in which the apical section beyond seta 1-A is twice as long (but transparent and hard to see against a pale background) as the basal pigmented section and its conical pointed siphon is highly modified, this species is quite unlike any other in Canada.

Biology. From the time of hatching to adult emergence, larvae and pupae of *perturbans* remain buried in the mud at the bottom of permanent

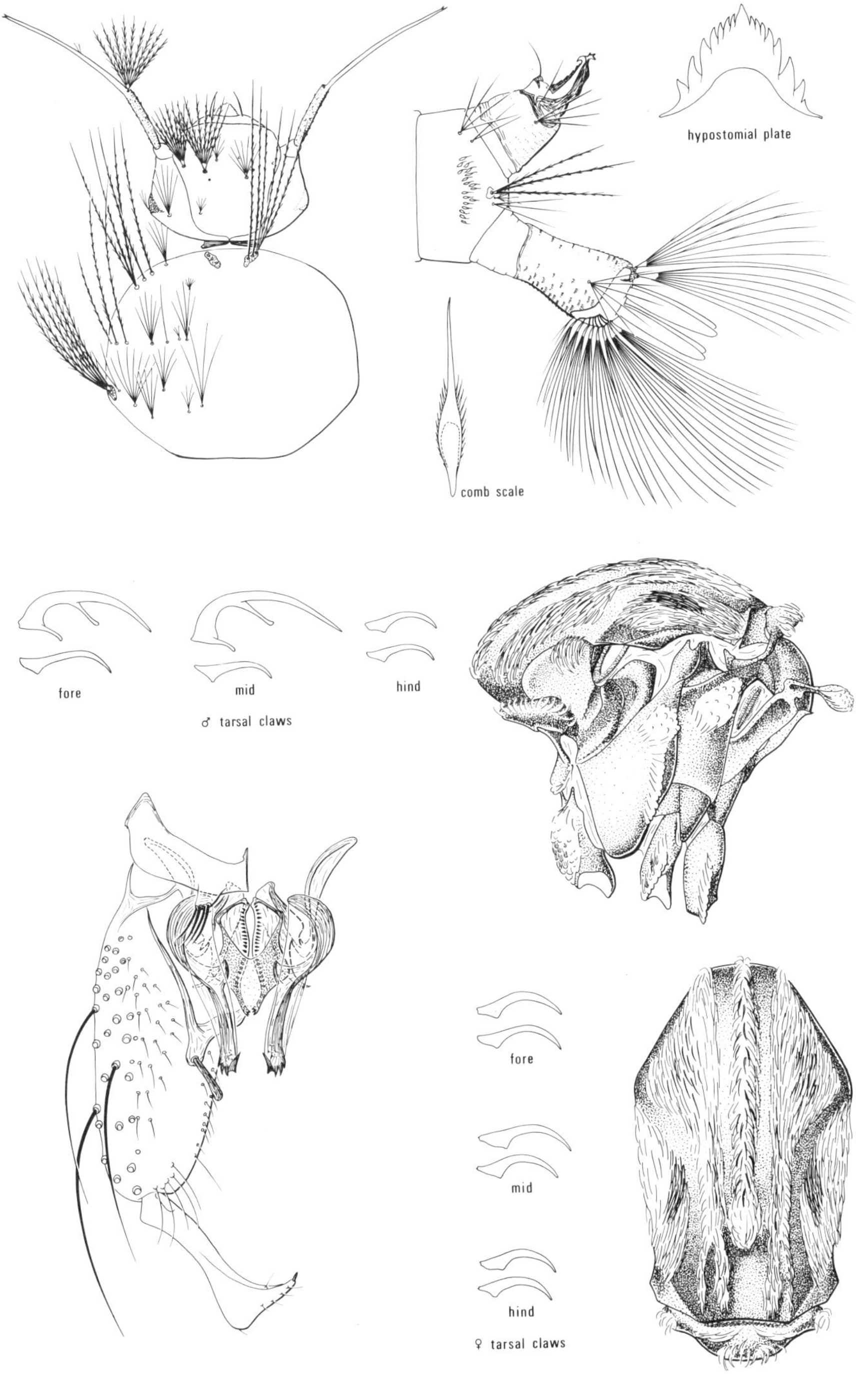

Plate 67. *Mansonia perturbans*

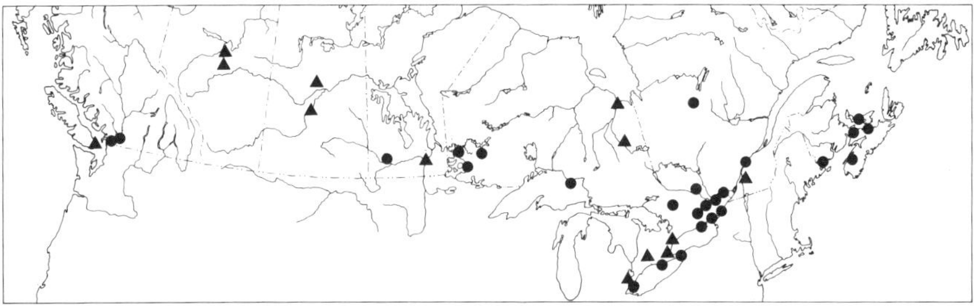

Map 67. Collection localities for *Mansonia perturbans* in Canada: • specimens we examined, ▲ literature records.

marshes, obtaining their oxygen from air tubes within the roots of emergent aquatic plants. A thick layer of soft muck or peat on the bottom of the marsh appears to be more important to *perturbans* than the species of plant, because larvae have been collected from cattails, *Typha*; arrowhead, *Sagittaria*; pickerelweed, *Pontederia*; water-lily, *Nymphaea* (Hagmann 1952); rushes, *Juncus*; reeds, *Phragmites*; sedges, *Carex*; and water-arum, *Calla*. The marsh must remain permanently wet (Brower 1953) because, although there is only one generation per year, larvae are present throughout the year and are killed by drought especially in winter. The species overwinters in the larval stage as second, third, or fourth instars, and there appears to be no larval diapause, because larvae with food in the gut were collected in winter (Hagmann 1952). The larva presumably eats the detritus surrounding it. Because larval activity is greatly reduced during winter, Hagmann recommended sampling for larvae when the water was cold by uprooting the plants, washing their roots, and screening the debris. This method was unsuccessful at temperatures above 12°C because larvae detached readily. Some larvae that Hagmann had brought back to the laboratory reattached to cattail roots in an aquarium, but when they were disturbed they quickly released their hold and sank to the bottom. Hagmann presumed that under warm conditions larvae would not spend much time attached to one plant but would move freely from plant to plant supported by oxygen in their thoracic air sacs.

Before pupation the pupal respiratory trumpets become well sclerotized (Hagmann 1953). During pupation the sharply pointed apices of these trumpets are imbedded in the root while the larval siphon is still attached. Once inserted, the trumpet apex cannot be withdrawn. Before emerging, the pupa wrenches itself free, breaking off the tips of the trumpets and leaving them behind in the plant, and rises to the water surface. Hagmann (1953) noted that, although the unbroken tips of the trumpets could not support the pupa at the surface, the broken ends did. At a water temperature of 20–22°C, the pupal stage lasted 5 days.

Emergence begins in southern Ontario in mid-June (Judd 1954) and peaks in Wisconsin in mid-July (Siverly and DeFoliart 1968*b*). Females feed rather indiscriminately on birds and mammals (Hayes 1961, Downe 1962) including man, and Downe noted a rather high proportion (15%) of multiple feedings in his precipitin tests. The species is a serious pest of man and livestock in parts of southern Canada adjacent to large permanent marshes, especially along the shores of Lakes Erie, Ontario, and Huron. Fortunately, they are not particularly active during the day, but many authors have noted a peak activity at dusk. The species is uncommon in Alberta and Saskatchewan (McLintock and Rempel 1963, Graham 1969*b*) but common in the marshy areas of southern Manitoba (Brust, personal communication). The virus of EEE was recovered from female *perturbans* in Georgia (Howitt et al. 1949).

The eggs are laid in floating rafts under overhanging stems of *Carex* (Hagmann 1952). Newly hatched larvae may come to the surface briefly, but within 24 hr they usually disappear into the mud at the bottom.

Distribution. North America, from British Columbia and Mexico east to Nova Scotia and Florida.

Genus *Orthopodomyia* Theobald

Adult. Medium in size; palpus of male slender, about as long as proboscis, and without dense tufts of long setae on apical segments; palpus of female about one-third as long as proboscis; vertex with numerous erect forked scales; scutal setae exceptionally long and numerous; postpronotal setae strong, usually two; spiracular and postspiracular setae absent; postnotum bare; calypter fringed; fourth and fifth tarsomeres of fore- and mid-legs exceptionally short, together not as long as third tarsomere; abdomen parallel-sided, truncated apically; cercus of female blunt apically, of moderate length, partly exserted, but not slender and tapering as in *Aedes* or *Psorophora*.

Larva. Most head and body setae strongly branched; lateral abdominal setae (series 6) of segments III to VI single and exceptionally long, as long as two or three abdominal segments combined; dorsolateral setae (series 1) also unusually long on segments IV and V; segments VII and VIII and in some species VI each with a sclerotized tergite that is largest on segment VIII (absent in *alba*); comb scales in North American species in two transverse rows, with scales of the hindmost row nearly twice as long as those of the preceding row; siphon without pecten and only a single seta 1-S on each side; saddle completely encircling anal segment; anal papillae unequal, with dorsal pair longer than ventral pair.

Remarks. The adults of the two species of this genus that occur in Canada are almost identical with one another but they are unlike any other genus of Canadian mosquito. Every part of the body and nearly every appendage is ornamented with lines and patches of narrow white scales, which are particularly conspicuous against the dark brown integument. The fore and mid tarsi have scarcely any white scales, but the hind tarsi are conspicuously banded, each tarsomere being banded apically as well as basally. However, the larvae of the two species are quite different, which is surprising when the adults are so similar.

Biology. *Orthopodomyia* is a small genus of the oriental and neotropical regions with a few species ranging northward into the holarctic region. The species are not found in Australia and Africa, but are present in Madagascar and Mauritius. With few exceptions, larvae of all species inhabit rot cavities in trees, broken bamboo stems, and similar "container" habitats, including the axils of bromeliads in Central America, a habitat normally occupied by *Wyeomyia* and related genera. None of the species are biting pests, and they are generally uncommon to rare.

Key to the species of *Orthopodomyia* of Canada—adults

Scales at base of R_{4+5} and on a short section of the adjacent radial sector white, these scales along with adjacent white patches of scales on M_{1+2} and CuA_1 creating an elongate white spot occupying one-third of the width of wing; R_{4+5} mostly dark-scaled except for white basal spot (Fig. 231); less than basal half of A white-scaled .... ***signifera***

Base of R_{4+5} and adjacent portion of radial sector mostly dark-scaled, leaving white patches on M_{1+2} and CuA_1 appearing together as a smaller rounded spot less than one-quarter the width of the wing; distal half to three-quarters of R_{4+5} almost or entirely white-scaled (Fig. 232); basal half or more of A white-scaled .... ***alba***

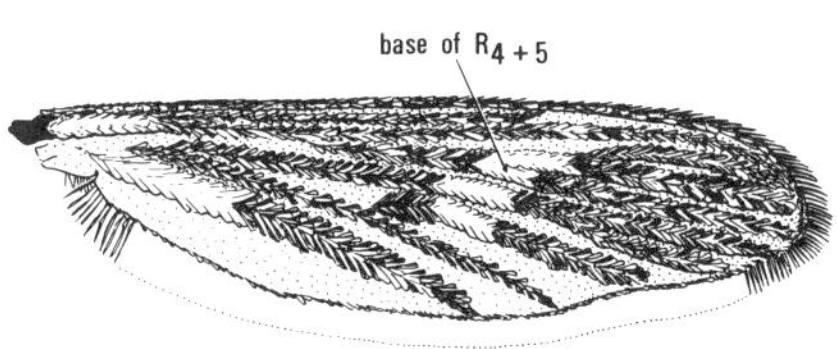

231 signifera

232 alba

Figs. 231, 232. Dorsal view of wing of *Orthopodomyia* species.

Key to the species of *Orthopodomyia* of Canada—fourth-instar larvae

Abdominal segments VII and VIII each with a sclerite partly encircling the segment dorsally; metathoracic ventral seta 13-T small, two- to three-branched (Fig. 233) .... ***signifera***

Abdominal segments VII and VIII without sclerites; metathoracic ventral seta 13-T large, five- to ten-branched (Fig. 234) .... ***alba***

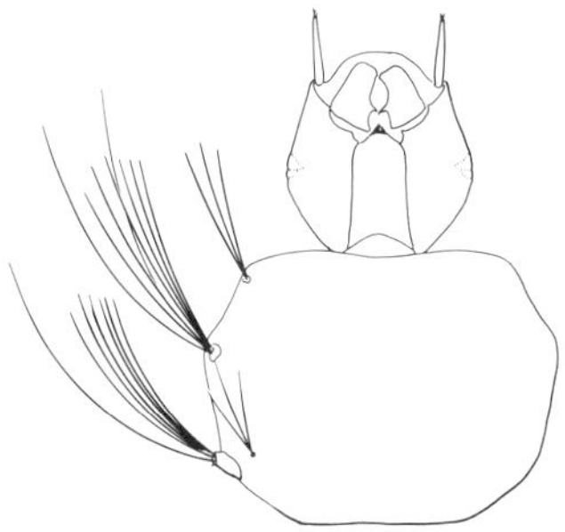

233 signifera

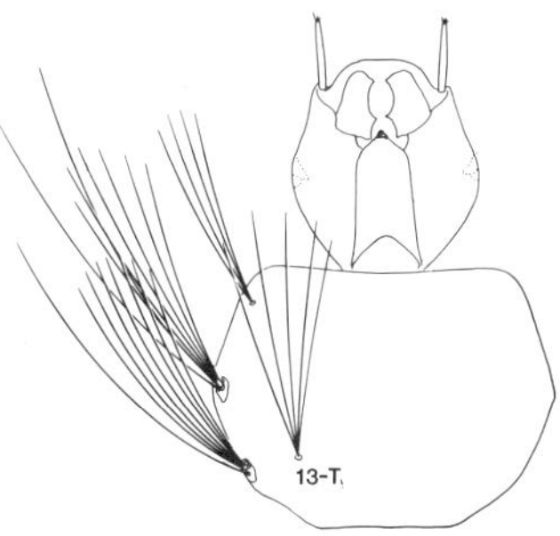

234 alba

Figs. 233, 234. Ventral view of larval head and thorax of *Orthopodomyia* species.

Orthopodomyia alba Baker

Plate 68; Figs. 232, 234; Map 68

Orthopodomyia alba Baker, 1936:1.

Adult. Female: Integument largely dark brown; proboscis dark brown scaled with a row of white scales along each dorsolateral edge; palpus dark-scaled with a row of white scales along dorsum, becoming a series of spots or dashes on distal palpomeres; pedicel with a semicircular line of white scales; basal three to five flagellomeres with one or more narrow white scales anteromedially; orbits delineated both dorsally and ventrally with a narrow line of white scales; vertex with a few recumbent white scales scattered among the dense erect forked dark brown scales; acrostichal, dorsocentral, lateral, and supraalar setae of scutum brown, long, and especially prominent; a narrow submedian line of white scales between acrostichal and dorsocentral rows of setae extending from anterior margin of scutum to prescutellar depression; a second narrow line of white scales along lateral edge of scutum from anterior margin to scutellum with interruption at transverse suture; a third similar line lateral to dorsocentral setae extending from transverse suture to and onto scutellum; acrostichal and dorsocentral stripes of brown recumbent scales among bases of setae; antepronotum with a spot of white scales; postpronotum bare except for a narrow straight line of white scales parallel to body axis; a similar broader line along upper part of katepisternum and another vertical line along posteroventral margin; postspiracular and prealar areas each with a small group of white scales; coxae white-scaled; femora and tibiae with scattered white scales that tend to be oriented into longitudinal lines; fore tarsus dark-scaled; mid tarsus dark-scaled except for a narrow ring at apex of tarsomere; hind tarsomeres each with base and apex white-scaled; wing veins with many white scales scattered among the black ones, and also white-scaled patches on stem vein, most of R_{4+5}, M_{1+2}, CuA_1, and basal half of A.

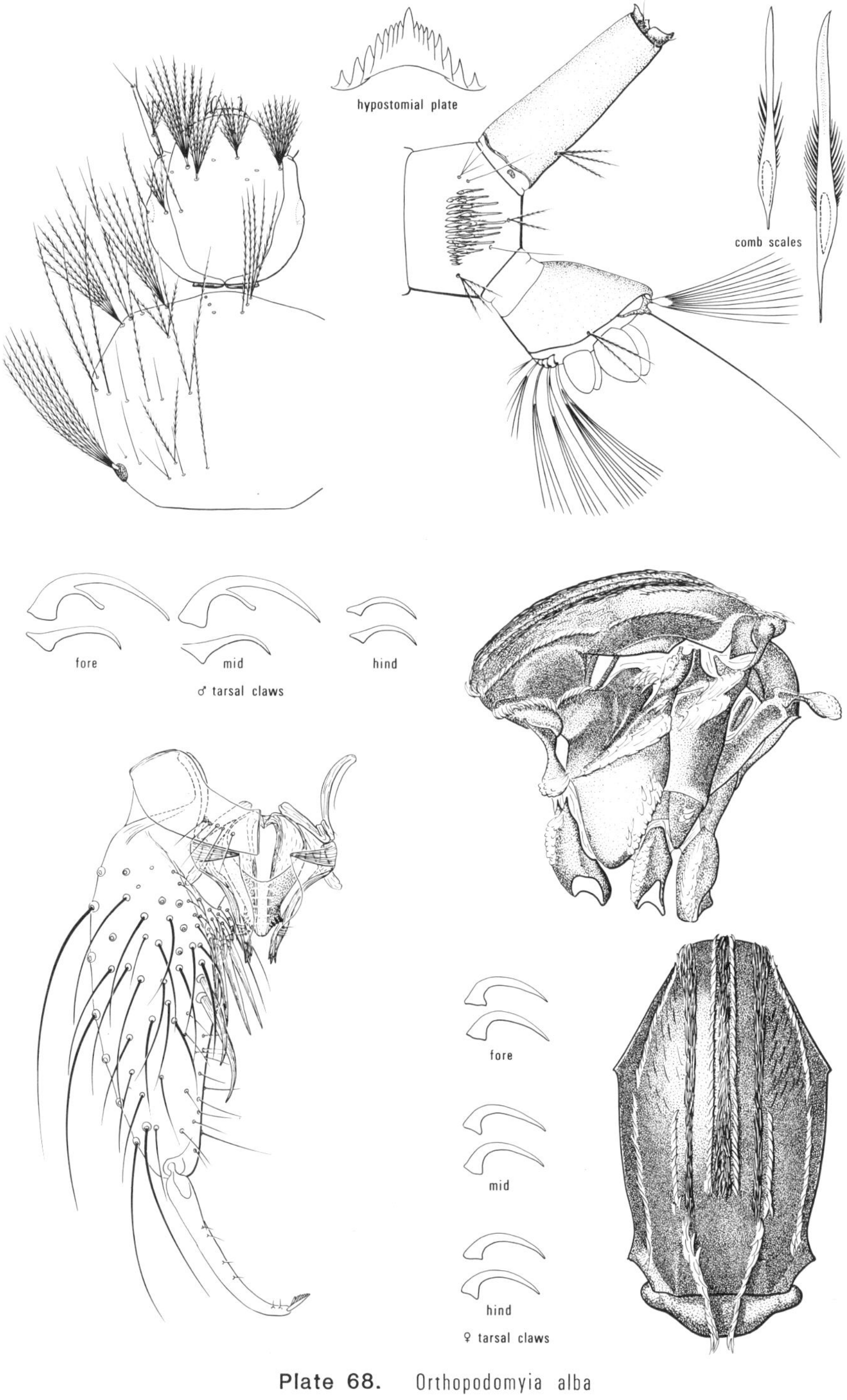

Plate 68. Orthopodomyia alba

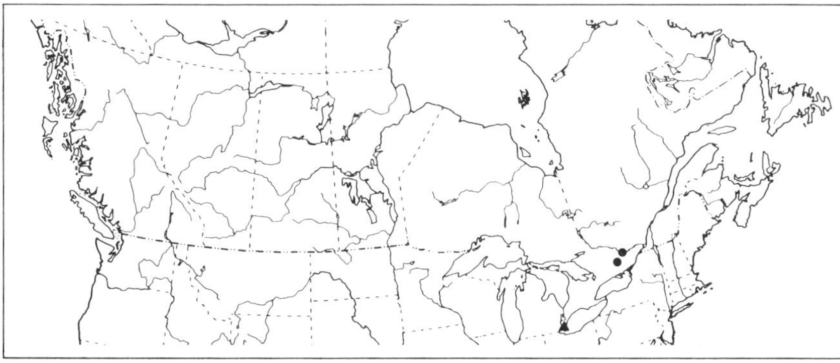

Map 68. Collection localities for *Orthopodomyia alba* in Canada: • specimens we examined, ▲ literature records.

Male: Wing narrower, with the scales fewer and paler, R_{4+5} usually white-scaled except at base, pattern otherwise as in female; gonostylus with flattened setose basal lobe, tapering evenly to apex, without apical lobe; gonostylus straight and slender, its apical spine short and peg-like, split distally into several teeth; claspette absent; paraproct toothed at apex; aedeagus divided into paired sclerites.

Larva. White, with small, rounded, lightly sclerotized head capsule; head setae all short and stiff, none longer than antenna; thoracic and abdominal setae (except for dorsolateral and lateral abdominal setae) shorter and more rigid than those of *signifera*; metathoracic ventral seta long, five- to ten-branched, abdominal segments VII and VIII not partly encircled dorsally by a sclerite; siphon more darkly sclerotized, especially toward apex, than any part of head; siphonal seta 1-S two- to four-branched, shorter than maximum width of siphon; saddle pigmented dorsally, pale ventrally.

Biology. *Orthopodomyia alba* has only recently been discovered in Canada (Smith and Trimble 1973). As in other members of the genus, larvae have been found chiefly in rot cavities in trees and rarely in artificial containers simulating tree holes (Jenkins and Carpenter 1946). Larvae seem to prefer tree cavities with a small opening and a voluminous interior containing water of pH 7.6–8.4 (Wilkins and Breland 1951). The water in such cavities is usually brown, heavily charged with organic detritus, including rotting wood and insect remains. Although cavities in various species of deciduous trees have been found to contain *alba* larvae, we have found them most often in large sugar maples, *Acer saccharum*. In certain trees, the cavity remains full of water even during drought; perhaps sap from the tree keeps it full.

Throughout most of its range, *alba* is believed to overwinter as a late-instar larva (Baker 1936, Matheson 1944, Jenkins and Carpenter

1946). Second and third instars have survived after being frozen for a week (Baker 1936). In the Ottawa area, more than 100 larvae were found in early April in about 5 L of liquid in a sugar maple; most of the larvae were in an intermediate instar (whether second or third was not determined). These larvae were alive and later they pupated and emerged in the laboratory. Only those in the last instar were dead, perhaps killed by low temperatures. In Texas, Wilkins and Breland (1951), noting that larvae were killed by freezing, suggested that *alba* overwinters as an adult. Loor and DeFoliart (1969) found a few eggs of *alba* that had overwintered (they did not state how successfully) in containers that simulated tree holes, which they had set out for *Aedes triseriatus*. In the Ottawa area, *alba* larvae have been found in many different tree holes in summer but not in spring; we know of only one cavity, mentioned previously, that served as a larval overwintering site. It may be possible that tree sap and other debris fermenting in this cavity have an ameliorating effect on the temperature of the liquid during the winter, but it seems likely that temperatures below freezing must occur, because the average minimum temperature in January is −16°C.

Larvae are also somewhat resistant to summer drought and are able to survive in the absence of free water (Breland 1947, Wilkins and Breland 1951, Breeland et al. 1961). In southern Quebec, from about 50 mL of moist sludge taken from a tree cavity that lacked free water, we recovered 205 larvae of *alba* in the last 3 instars and 35 larvae of *Anopheles barberi*.

Adults are secretive, preferring to hide in dark recesses (Wilkins and Breland 1951). Females have not been recorded biting and may presumably be autogenous, although Sudia and Gogel (1953) reported that they fed on a chick in the laboratory during the night. Eggs are laid on the water in tree cavities, where they float on the meniscus, but how much drying they can withstand is not known. In the southern USA *alba* is multivoltine, but in Canada the number of generations is unknown. Larval development is slow, even during the summer (Snow 1949, Breeland et al. 1961).

Distribution. North America, from Minnesota and Texas east to Quebec and North Carolina.

Orthopodomyia signifera (Coquillett)

Plate 69; Figs. 231, 233; Map 69

Culex signifer Coquillett, 1896:43.

Orthopodomyia californica Bohart, 1950:399.

Adults. Female and male: Apparently indistinguishable from adults of *alba* except by pattern of white scaling on wing veins as follows: in *alba* most of R_{4+5} is predominantly white-scaled, but the base is not, whereas in *signifera* most of R_{4+5} is predominantly dark-scaled, but the base is white-scaled, as is the base of adjacent R_{2+3}, thus accentuating the spots on M_{2+3} and CuA_1 by doubling the area of white scales in the center of the wing;

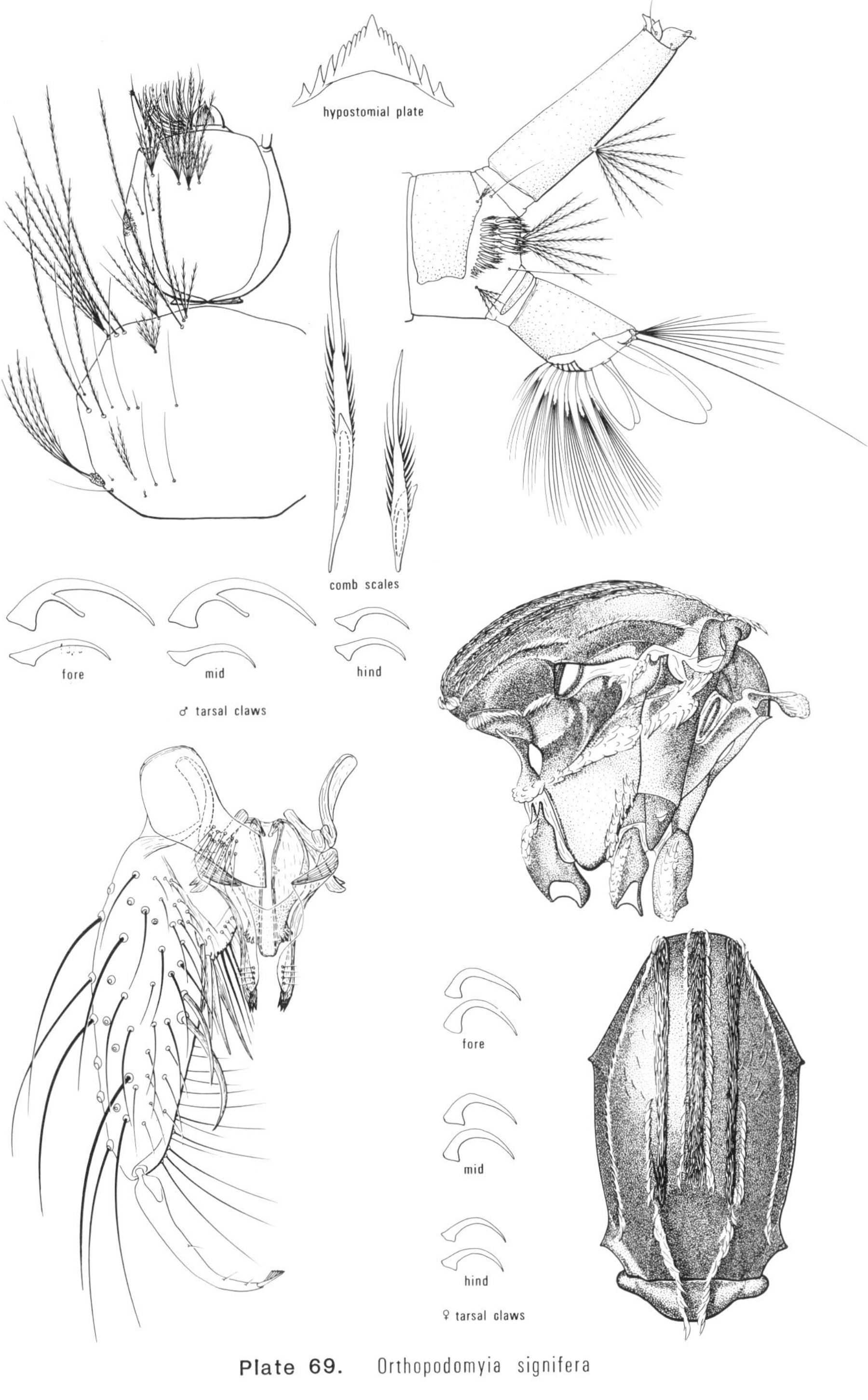

Plate 69. Orthopodomyia signifera

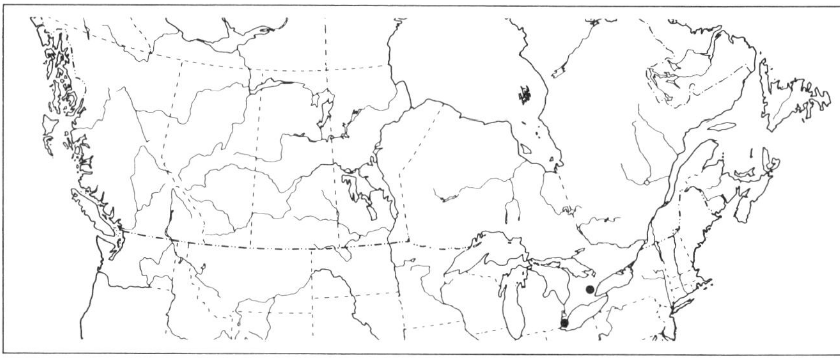

Map 69. Collection localities for *Orthopodomyia signifera* in Canada: • specimens we examined.

basal half of vein A in *alba* white-scaled, whereas in *signifera* only the basal third white-scaled.

Larva. Head capsule longer and narrower than that of *alba*, darkly pigmented except laterally around eye; head setae more profusely branched, longer than antenna; body with conspicuous reddish brown pigment dorsally; setae of thorax and abdomen longer, finer, and laxer than those of *alba*; metathoracic ventral seta short, two- to three-branched; abdominal segment VI with a small irregular "tergite"; segments VII and VIII each with a much larger sclerite; the sclerite on segment VIII encircling more than three-quarters of the segment and emarginate along its posterolateral border to accommodate comb scales; siphon longer than in *alba*, its seta 1-S more profusely branched and longer than half the length of the siphon; saddle uniformly pigmented.

Biology. *Orthopodomyia signifera* is, like the preceding species, a very recent addition to the Canadian fauna as a result of an intensive survey of tree cavities in southern Ontario (Smith and Trimble 1973). As in *alba*, larvae of *signifera* usually inhabit rot cavities in trees but they have also been reported from various shaded artificial containers that were filled with organic debris (Masters 1953, Good 1945, Michener 1947, Love and Goodwin 1961). Grimstad (1977) reported that larvae of this species were most often found in clear water in beech cavities, in contrast with larvae of *alba*, which preferred dark coffee-colored water rich in organic matter. Most authors agree that *signifera* overwinters in the larval stage and that the larvae cannot withstand freezing (Baker 1936, Matheson 1944, Snow 1949). However, Price and Abrahamsen (1958) consider that, in Minnesota, *signifera* may also overwinter in the adult or the egg stages. The overwintering stage of *signifera* in Ontario is not known, and data on the minimum temperatures reached within the liquid in various tree cavities in Ontario has not been determined. In Georgia, Jenkins and Carpenter (1946) found that larvae continued their development during the winter when exposed to as little as 1 hr of light per day, and there appeared to be no true diapause, the larvae overwintering as second, third, or fourth instars. In New York, however, larval development was curtailed by short photoperiod, but growth resumed when larvae were reared at 16 hr of light and 8 hr of darkness.

According to Brooks (1946), adults mated and oviposited successfully without any food other than the water available to them in the larval rearing pans. Because many authors have been unsuccessful in getting females of *signifera* to take a blood meal, they have concluded that the females did not feed. However, by offering them an immobilized bird during the night, Love and Goodwin (1961) succeeded in obtaining a few blood-fed females of *signifera*. Chapman (1964, as *californica*) obtained engorgement on chickens and on pads of citrated chicken and rabbit blood.

Eggs are laid singly, each covered with a gelatinous "veil marked by radial folds" (Dyar and Knab 1906*c*), on the inner surface of the tree cavity or container either on the water (Horsfall 1937) or at the water's edge (Breeland et al. 1961). Chapman (1964, as *californica*) found eggs above the waterline on partly submerged paper or leaves; newly hatched larvae had to wriggle down to the water surface.

Adults were collected while they were resting on tree trunks near their breeding sites (Dyar 1922*b*). Males apparently do not swarm.

Distribution. North America, from North Dakota and California east to New York and the West Indies.

Genus *Psorophora* Robineau-Desvoidy

Adult. Small to very large; female palpus short; male palpus exceptionally elongate, usually exceeding proboscis by length of last two palpomeres; junction of second and third palpomeres of male bare of scales, thus usually appearing paler; postocular setae numerous, in most species usually arising close to eye margin, but farther away in *ferox*, exposing a broad area of bare integument; vertex with both recumbent and erect scales; acrostichal and dorsocentral setae present but rather short and recumbent; spiracular setae present, sometimes only one or two; postspiracular area more or less covered with scales and short setae; upper margin of katepisternum scaled to anterodorsal corner (as in many *Aedes* but not other genera); legs, especially hind legs, of each of our species distinctively ornamented, either by erect scales, white bands, or both; tarsal claws simple in the subgenus *Grabhamia*, with long subbasal tooth in the other two subgenera; calypter fringed; abdomen of female tapering; cercus slender and elongate, prominently exserted; gonocoxite almost parallel-sided, with neither basal nor apical lobes; gonostyle S-shaped, with recurved apex bearing small peg-like spinule; claspette present, its stem expanded and highly elaborate apically, with more than one apical appendage or filament that may be setiform, blade-like, or flattened and convoluted.

Larva. Saddle completely encircling anal segment, with the anteriormost setae of ventral brush or precratal setae arising from within saddle; siphonal seta 1-S arising distal to pecten teeth; otherwise as in *Aedes*.

Remarks. As in *Culiseta*, the Canadian species belong to three subgenera that are reasonably distinct; the subgenera are referred to in the keys. Diagnoses of subgenera are not given because we lack the material.

Biology. This genus is found only in the New World, and its members are mostly South Temperate and Tropical. The species all appear to be associated in the larval stage with temporary summer rainpools. The eggs are deposited on the damp or dry mud after the pools have almost disappeared. The eggs remain viable until they are reflooded in warm weather, sometimes after months or years have elapsed.

Key to the subgenera and species of *Psorophora* of Canada—adults

1. Postpronotum, submedian stripe, and lateral area of scutum bare; scutum with median stripe of gold scales, sublateral area of white scales, and on its posterior half a dorsocentral stripe of black scales; palpus about one-third as long as proboscis; tibiae and apices of femora and tarsomeres with dense, long, erect, shiny scales (resembling chenille); claspette stem of male with abruptly expanded apex bearing one sickle-shaped claspette filament and numerous, recurved, unbranched setae (Fig. 235*b*); gonostylus of male slender throughout its length, with short setae along median edge (Fig. 235*a*) (subgenus *Psorophora*) .... ***ciliata***

   Postpronotum and scutum evenly clothed with scales, which tend to be uniform in color, not contrasting as in *ciliata*; palpus about one-quarter as long as proboscis; leg scales shorter and recumbent; claspette stem slightly expanded at apex, bearing feathered or flattened straight setae (Figs. 236*b*–238*b*); gonostylus slender basally and apically, strongly broadened and reticulated in middle, broadest at apical third, with one or two setae along medial edge (Figs. 236*a*–238*a*) 2

2. Proboscis, palpus, legs (except last two white-scaled hind tarsomeres), wing, and abdomen predominantly dark-scaled, with purple iridescence; all tarsal claws of both sexes with pronounced subbasal tooth; claspette stem with subapical group of medially directed, slightly sinuous, apically feathered setae, and apical group of three filaments, one flattened with pointed recurved apex, one small and spatulate, and one contorted, encircling the other two (Fig. 236*b*) (subgenus *Janthinosoma*) .... ***ferox***

   Middle third or more of proboscis pale-scaled, with contrastingly dark-scaled apex and base; tarsomeres each with basal rings of white scales; wing and abdomen with mixed pale and dark scales; all tarsal claws of female and hind claws of male lacking subbasal tooth; claspette filament with rounded, dorsally reflexed apex bearing a palmately arranged row of five to seven setae (Figs. 237*b*, 238*b*) (subgenus *Grabhamia*) 3

3. Basal hind tarsomere with two rings of white scales, one at base, the other at mid length; white wing scales randomly distributed over wing veins; apex of claspette stem bearing six or seven flattened blade-like setae and one feathered hair-like seta (Fig. 237*b*) .... ***columbiae***

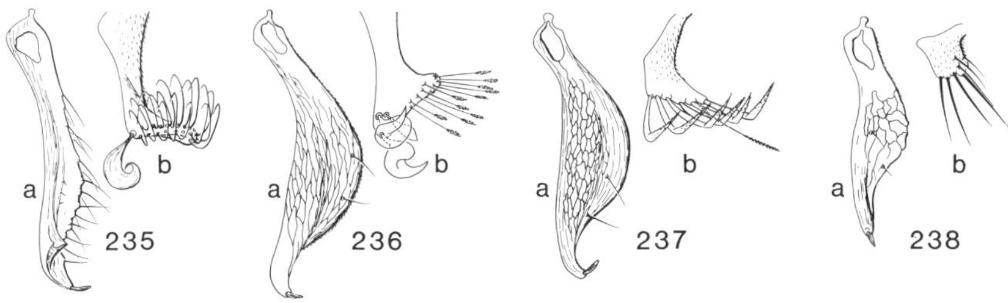

ciliata ferox columbiae signipennis

Figs. 235–238. Dorsal view of male of *Psorophora* species: *a*, gonostylus; *b*, claspette.

Basal hind tarsomere mostly pale-scaled; scales of distal half of costa, subcosta, and radius grouped into alternating transverse bands of pale and dark scales; apex of claspette stem bearing five or six slender cylindrical setae (Fig. 238*b*) .... ***signipennis***

Key to the subgenera and species of *Psorophora* of Canada—fourth-instar larvae

1. Head large, quadrate, with the labral brushes widely separated on prominent anterolateral angles of head (Fig. 34); pecten teeth 18 or more, each terminating in a hair-like apex; siphon parallel-sided; siphonal seta 1-S represented by one long hair (Fig. 37) (subgenus *Psorophora*) .... ***ciliata***
   Head smaller, rounded, with the labral brushes not elevated on anterolateral corners of head (see Fig. 36); pecten teeth six or fewer, without hair-like apex; siphon inflated; siphonal seta 1-S minute, mutliple, arising near distal third of siphon (Fig. 38) 2
2. Head setae 5-C and 6-C with four or more branches; siphonal seta 1-S about as long as apical width of siphon (subgenus *Grabhamia*) .... ***columbiae***
   Head setae 5-C and 6-C single to triple; siphonal seta 1-S minute, less than half as long as apical width of siphon 3
3. Antennae shorter than length of head (subgenus *Grabhamia*) .... ***signipennis***
   Antennae longer than length of head (subgenus *Janthinosoma*) .... ***ferox***

Psorophora ciliata (Fabricius)

Plate 70; Fig. 235; Map 70

Culex ciliatus Fabricius, 1794:401.
Culex molestus Wiedemann, 1820:7.
Culex rubidus Robineau-Desvoidy, 1827:404.
Culex boscii Robineau-Desvoidy, 1827:413.
Culex conterrens Walker, 1856:427.
Psorophora ctites Dyar, 1918:126.

Adult. Female: Large; integument yellowish brown; proboscis, palpus, apices of femora, tibiae, mid basitarsus, and most of hind tarsus with broad, dark, erect scales that greatly increase apparent width of the appendage; head rather elongate, with extensive vertex clothed with flat white scales; erect scales of vertex short, stout, and truncate, like small erect setae; scutum with median stripe of yellow scales; submedian stripe bare and shining; dorsocentral stripe of black scales narrowest anteriorly, sometimes absent, widest just in front of suture; sublateral area white-scaled; anterolateral edge of scutum in front of transverse suture bare and shining; scutal setae rather short and recumbent, numerous but inconspicuous; postpronotum bare and shiny; spiracular setae few, small, and dark; postspiracular area covered with small pale scales and hair-like setae; hypostigmal area, lower margin of subspiracular area, upper edge of katepisternum to anterodorsal angle, and upper half and anterior margin of mesepimeron with small whitish scales; wing veins dark-scaled; abdomen shiny brown, sparsely clothed with inconspicuous brown scales and hairs.

Male: Palpus longer than proboscis by almost the length of last two palpomeres; distal three-quarters of third palpomere with ventral fringe of long setae; fourth and fifth palpomeres densely fringed with long setae; coloration of scutum and leg ornamentation as in female; gonocoxite long, narrow, parallel-sided; gonostylus slender, median surface with hair-bearing tubercles, apex recurved, bearing apical spinule; claspette stem capitate, with expanded apex bearing numerous anteriorly directed hairs and a flattened sickle-shaped filament.

Larva. Large, up to 10 mm long; antenna short, less than half as long as head; head large, quadrate, labral brushes widely separated on prominent anterolateral corners of head; head setae minute, 5-C and 6-C branched beyond middle, their bases not forming a straight line with base of 7-C; clypeal margin concave anteriorly; comb scales in a broad arc; pecten teeth numerous, each with long hair-like apex; siphonal seta 1-S one long hair.

Biology. The larvae of *Psorophora ciliata* are predaceous in the second, third, and fourth instars, although the first instar appears to be equipped to filter-feed in the usual way (Shalaby 1957). The overwintering egg may not hatch until midsummer, under the same conditions of flooding by heavy rainfall that favor hatching of *Aedes vexans* and *Aedes trivittatus*. Larvae of these latter species are probably the main prey of *ciliata*, although tadpoles also serve as prey (Breeland et al. 1961). Larvae of *ciliata* feed voraciously and may consume dozens of *vexans* larvae during their development. Larval growth is even faster than that of *vexans;* Breeland and Pickard (1963) recorded hatching within 2 hr, larval maturation in 90 hr, pupation in 120 hr, and emergence in 144 hr. Breeland et al. (1961) collected *ciliata* larvae in open grassy pools, marginal woodland, and shrubby areas in association with larvae of *vexans* and various other species of *Psorophora*.

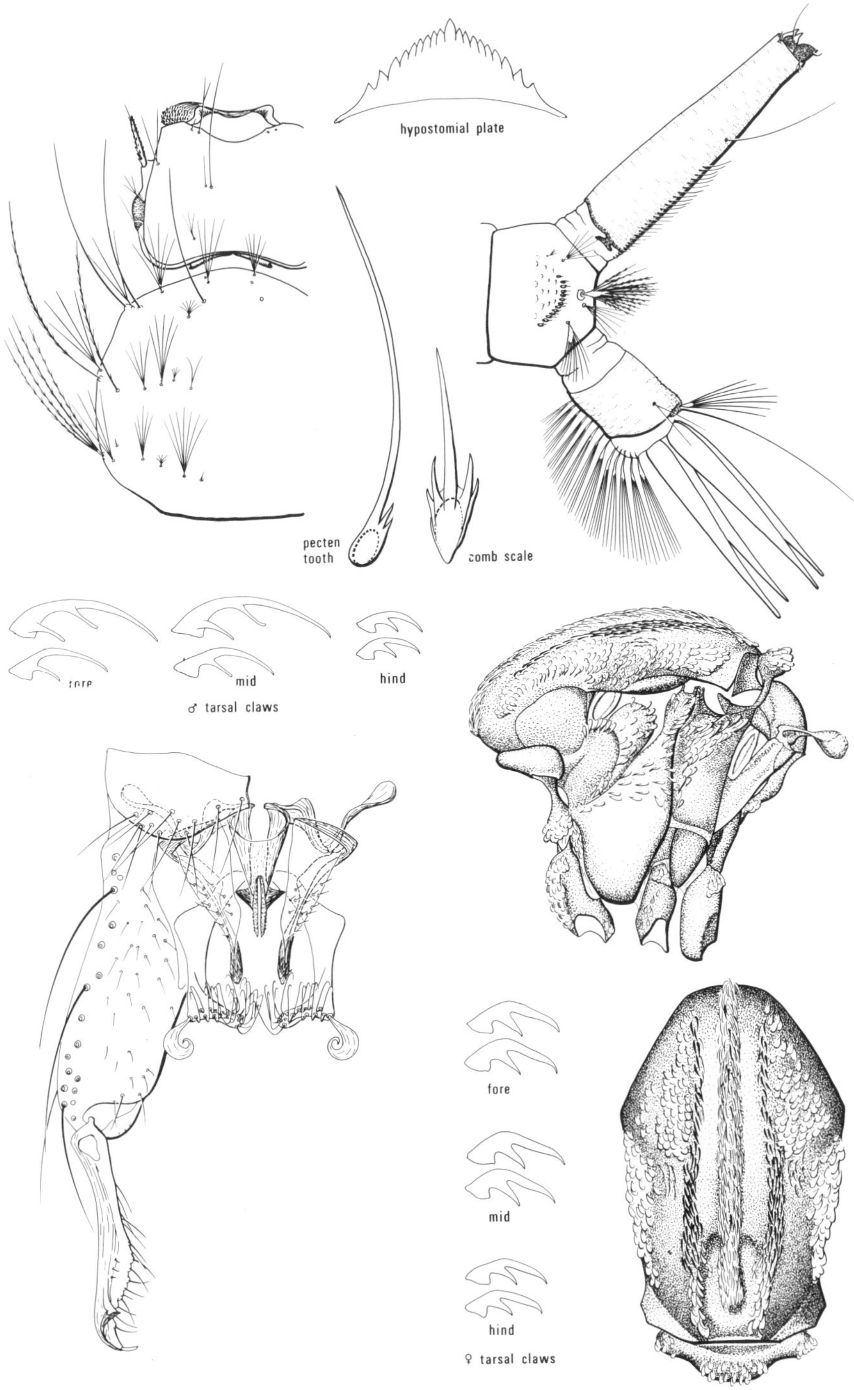

Plate 70. *Psorophora ciliata*

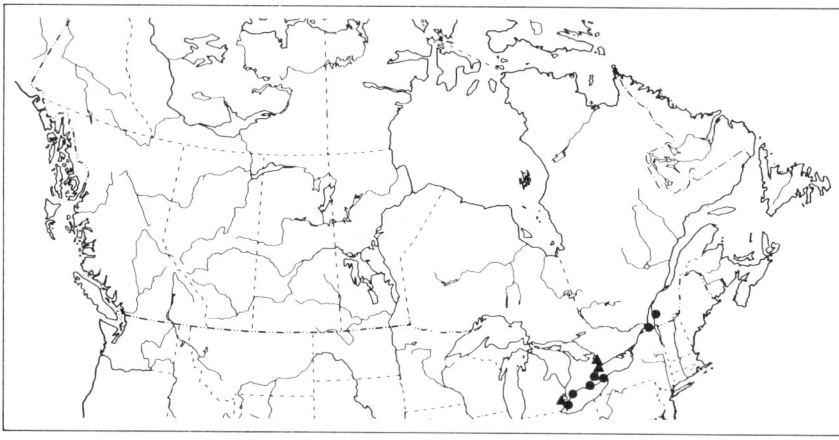

Map 70. Collection localities for *Psorophora ciliata* in Canada: • specimens we examined, ▲ literature records.

Adult females are avid biters day and night; they are said to prefer livestock (Breeland et al. 1961). They are reputed to be able to bite through heavy clothing. Few eggs are laid. Chapman and Woodward (1965) recorded an average of 31 eggs laid after the first blood meal, with a total of 90 per lifetime (Breeland and Pickard 1963). The eggs are presumably widely scattered to prevent excessive larval populations. The number of generations per year depends on rainfall.

Psorophora ciliata is a rare species in Canada. Most specimens have been collected in southwestern Ontario, and a few have been collected in the Eastern Townships of southern Quebec.

Distribution. Central and eastern North America, Central America, and South America; South Dakota and New Hampshire south to Argentina.

Psorophora columbiae (Dyar & Knab)

Plate 71; Fig. 237; Map 71

Janthinosoma columbiae Dyar and Knab, 1906*a*:135.
Janthinosoma floridense Dyar and Knab, 1906*a*:135.
Janthinosoma texanum Dyar and Knab, 1906*a*:135.
Psorophora confinnis, of authors, not Lynch Arribálzaga.

Adult. Female: Medium in size; integument dark brown; proboscis brown-scaled on apical one-quarter, remainder yellowish-scaled, with increasing number of dark scales toward base; palpus pale-scaled apically, dark basally; postocular setae close to eye margin; vertex with whitish or pinkish recumbent scales and brown erect scales; pronotum and scutum with

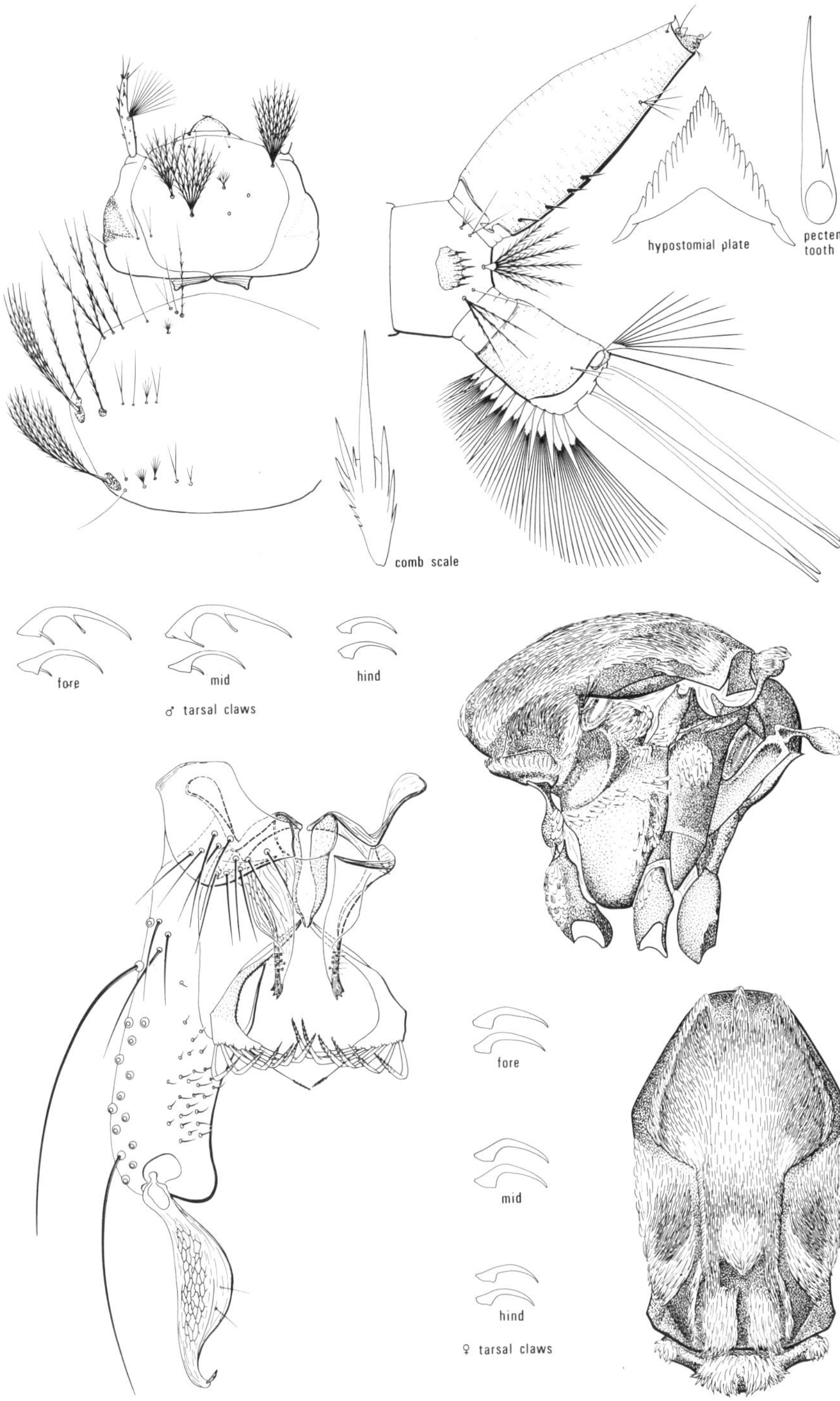

Plate 71. *Psorophora columbiae*

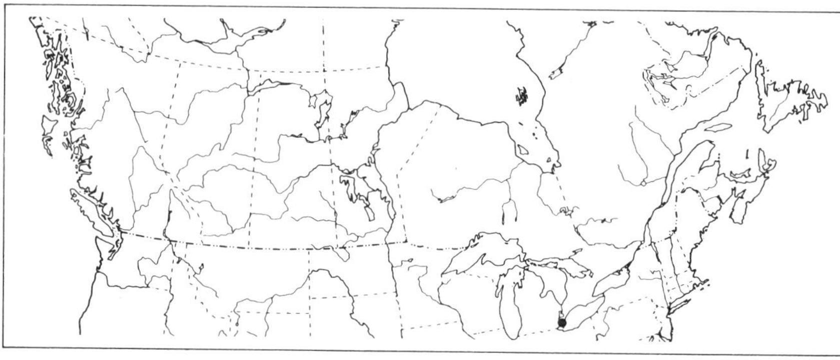

Map 71. Collection localities for *Psorophora columbiae* in Canada: • specimens we examined.

small, narrow brown scales and paler scales with a pinkish or mauve tint forming a subtle pattern, the paler scales concentrated at middorsocentral spot, along edge of scutum, and at prescutellar depression; pleural scales white; femora brown-scaled with narrow white-scaled subapical ring, an apical dorsal white spot, and often other smaller white spots; tibiae brown-scaled with prominent longitudinal row of about 10 white spots on anterodorsal surface, and scattered white scales ventrally; basitarsi with narrow basal white-scaled ring and another at mid length; remaining tarsomeres each with narrow basal white-scaled ring except last fore tarsomere; tarsal claws all simple, without subbasal tooth; wing veins with dark and pale scales intermixed, but not with aggregations of pale scales forming distinct spots; abdominal tergites dark-scaled, each with apical triangle or transverse band of paler scales.

Male: Proboscis with pale-scaled ring narrower than in female; palpus longer than proboscis by length of fourth and fifth palpomeres, brown-scaled except for a patch of pale scales at base of each of fourth and fifth segments; junction of second and third palpomeres bare of scales, exposing pale integument; most of third and all of fourth and fifth palpomeres extensively fringed with long setae; coloration and pattern of scales on scutum, legs, and wings as in female; gonocoxite rather parallel-sided, with broad truncated apex; gonocoxite S-shaped, broadest at mid length because of rounded median keel, apex reflexed, bearing a small apical spinule and one subapical hair; claspette stem expanded at apex, bearing six or seven long, flattened, pointed, anteriorly reflexed apically pubescent filaments, and one apically plumose seta.

Larva. Antenna shorter than length of head; head setae 5-C and 6-C each with four or more branches, their bases in a straight line with base of 7-C; siphon moderately inflated, with three to six widely spaced pecten teeth; siphonal seta 1-S as long as apical diameter of siphon.

Remarks. The name *columbiae* was only recently recognized for North American populations of the *confinnis* complex (Belkin et al. 1970); the name *Psorophora confinnis* is now applied to a South American species.

Biology. *Psorophora columbiae* was discovered in Canada by one of us (D.M.W.) on 18 June 1974 near the base of Fish Point at the southern end of Pelee Island, Essex Co., Ont. The few third-instar larvae that were collected, among larvae of *vexans* in a small muddy temporary puddle at the edge of a field, were reared and two adult males were obtained.

In the southern USA, *columbiae* (as *confinnis* of authors) can be an abundant and troublesome pest, especially in rice fields (Horsfall 1942), but it is not likely to be important in Canada. Like all *Psorophora*, it is a summer species. The overwintering eggs hatch in early summer wherever they are inundated with suitably warm rainwater runoff. There may be several generations per season. Unlike *ciliata, columbiae* larvae are filter feeders in all four instars; at times *columbiae* larvae make up part of the diet of *ciliata* larvae (Breeland et al. 1961).

Females seeking oviposition sites prefer short, rank vegetation; they do not oviposit on bare mud (Horsfall 1942).

Distribution. North America, from California and Mexico east to New York and Florida.

Psorophora ferox (Humboldt)

Plate 72; Fig. 236; Map 72

Culex ferox Humboldt, 1819:340.
Culex musicus Say, 1829:149 (preoccupied by *musicus* Leach).
Janthinosoma sayi Dyar and Knab, 1906*c*:181 (new name for *musicus* Say).
Janthinosoma sayi Theobald, 1907:155 (new name of *musicus* Say).

Adult. Female: Small to medium; integument dark brown; all scales of proboscis, palpus, apices of femora, tibiae, tarsi (except last two hind tarsomeres), wing veins, and abdominal tergites dark with metallic purplish iridescence; postocular setae arising some distance behind eye margin and separated from it by shiny brown bare integument; remainder of vertex behind postocular setae with both recumbent and erect yellow scales; scutum sparsely clothed with flat, broad, recumbent scales, yellowish laterally, darker medially; basal three hind tarsomeres and apices of hind tibia with long, shaggy, erect dark scales, suggestive of but not as conspicuous as in *ciliata*; tarsal claws with long subbasal tooth; last two hind tarsomeres and sometimes apex of third tarsomere white, including scales, hairs, and integument, in striking contrast with remaining tarsomeres; apical abdominal tergites with a suggestion of paler lateral triangles.

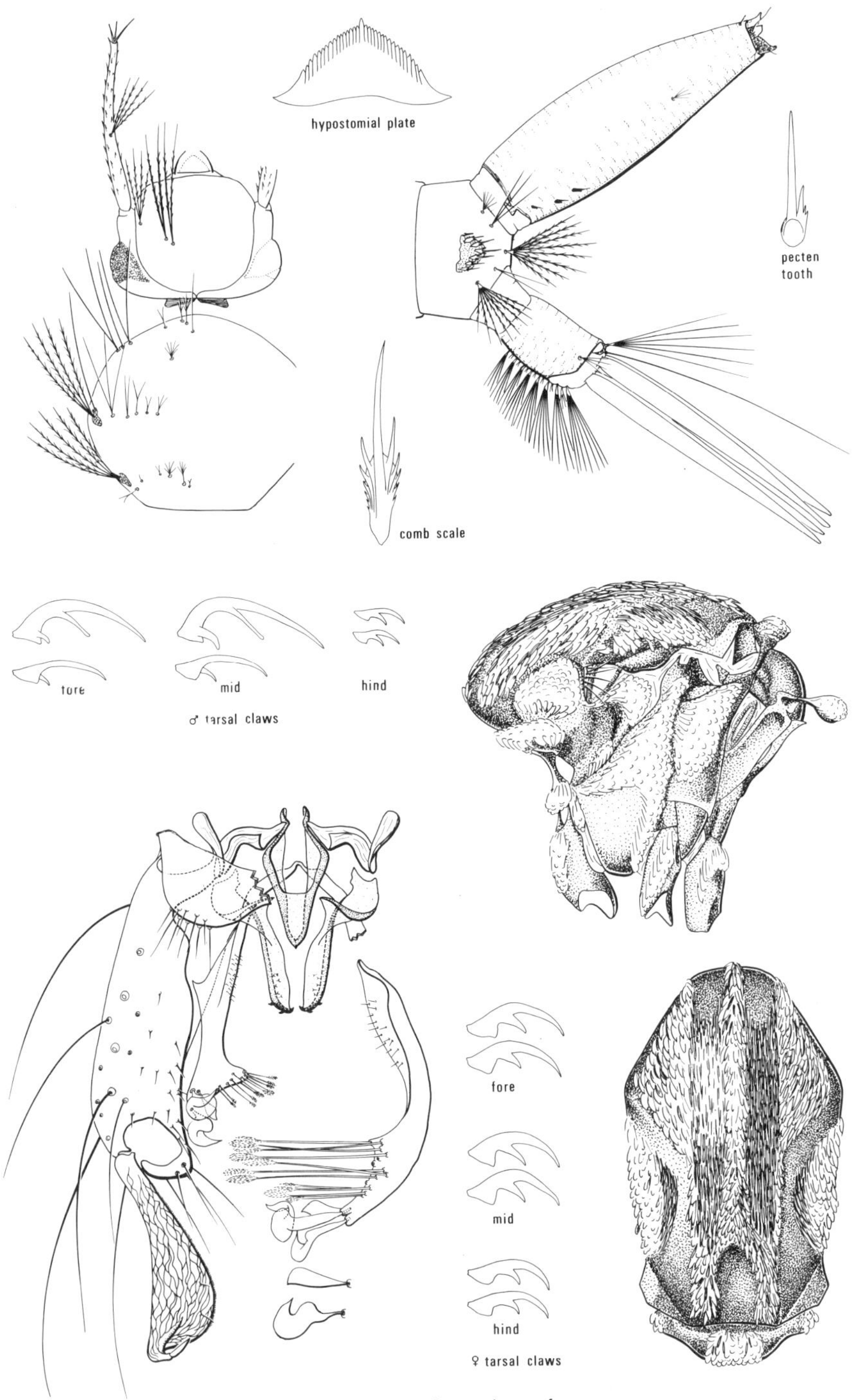

Plate 72. *Psorophora ferox*

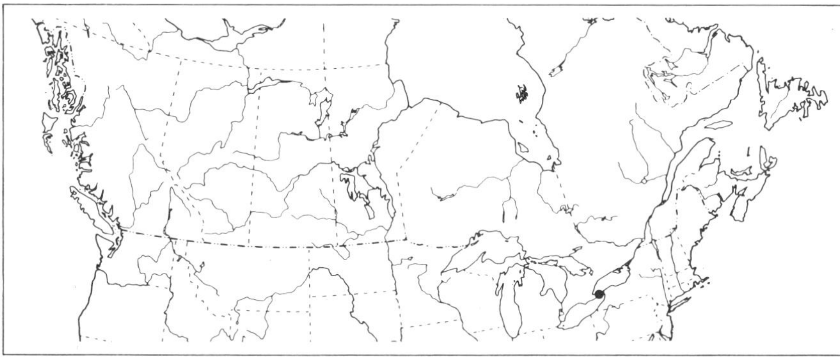

Map 72. Collection localities for *Psorophora ferox* in Canada: • specimens we examined.

Male: Palpus longer than proboscis by fifth and most of fourth palpomeres, brown-scaled, with pale bare internode between second and third palpomeres; fourth palpomere only with a sparse fringe of long setae; apex of third and fifth palpomeres with two or three setae on each; scutum sparsely yellow-scaled; hind tarsi as in female; hind claw with subbasal tooth as in female; gonocoxite narrow, parallel-sided, with prominent apicomedial angle bearing a few setae; gonostylus S-shaped, broadest near apex, the convex inner flange displaced distally almost to recurved apex; claspette stem with expanded apex bearing three flattened apically directed filaments, of which at least one is contorted apically, and a group of long straight setae with feathered apices directed medially.

Larva. Antenna noticeably longer than head (as in *Aedes diantaeus*); head setae 5-C and 6-C usually double, occasionally triple, their bases in line with base of 7-C; siphon strongly inflated, with three to five widely spaced pecten teeth; siphonal seta 1-S minute, much shorter than apical diameter of siphon.

Biology. This species has been found only once in Canada, when Hearle (1920) collected a female biting him in the woods near Jordan, Ont., on 3 August 1916. It may not even be a permanently established resident species. Like *columbiae, ferox* is more common farther south, but *ferox* is not as abundant (Dyar 1922*b*). The eggs overwinter, hatching in spring when the water temperature is above 10°C (Breeland et al. 1961). Larvae are most often collected in temporary woodland rainpools or in potholes left in stream beds rather than in fields or pastures (Lake 1963).

Females are diurnal biters, ceasing their activities after dark and preferring to remain within forested areas (Michener 1947, Breeland et al. 1961), although they have been taken in light traps (Love et al. 1963).

Distribution. Eastern North America, Central America, and South America; from Minnesota and New Hampshire south to Argentina.

Psorophora signipennis (Coquillett)

Plate 73; Fig. 238; Map 73

Taeniorhynchus signipennis Coquillett, 1904:167.

Adult. Female: Medium in size; integument medium brown; proboscis brown-scaled on apical quarter and basal quarter, but with intermediate half pale-scaled; palpus with mixed pale and brown scales; postocular setae close to eye margin; vertex, pronotum, and scutum clothed with small, narrow yellowish scales; femora, tibiae, and basitarsi mottled with pale and brown scales; remaining tarsomeres mostly pale-scaled, with apical bands of brown scales; tarsal claws simple; basal half of costa with scattered pale and dark scales, distal half with three pale-scaled spots separated by two dark-scaled areas; remaining wing veins and abdominal tergites mottled with pale and dark scales.

Male: Palpus longer than proboscis by length of last two palpomeres; second, third, and fourth palpomeres each with a narrow basal ring of pale scales and a broader middle region mottled with pale and dark scales; apex of third, fourth, and fifth palpomeres fringed with long setae; coloration of wings and legs paler but similar to those of female; claspette stem expanded apically, bearing about five seta-like filaments.

Larva. Antenna shorter than length of head; head setae 5-C and 6-C usually single, sometimes double or triple, their bases forming a straight line with base of 7-C; siphon moderately inflated, with four to six small pecten teeth, becoming progressively larger distally; siphonal seta 1-S minute, shorter than apical diameter of siphon.

Biology. Like *ferox, signipennis* is known in Canada from only one adult, a male collected near Regina, Sask., on 8 July 1942 (Rempel 1953). It is primarily a species of the Great Plains, where its larvae develop rapidly during summer in temporary rain-filled pools. There may be several generations per season, as in *columbiae* and *ferox*. Carpenter and LaCasse (1955) list collections of *signipennis* from Montana and North Dakota, and it probably occurs sporadically in southern Saskatchewan and Alberta, where adults would be easily overlooked among much larger numbers of similarly colored *Aedes dorsalis* and *Aedes campestris*.

Distribution. Central North America, from southern Saskatchewan south to Mexico.

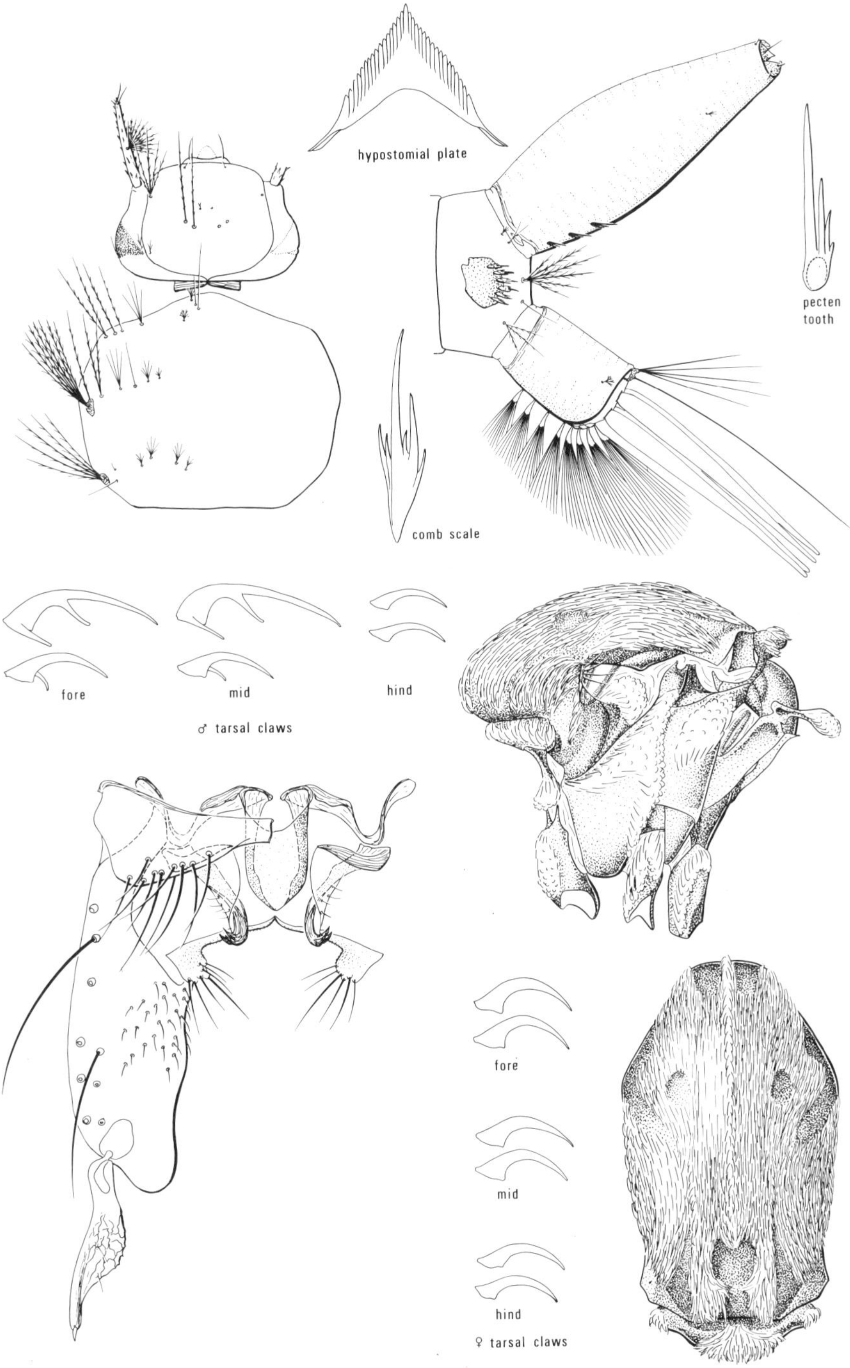

Plate 73. *Psorophora signipennis*

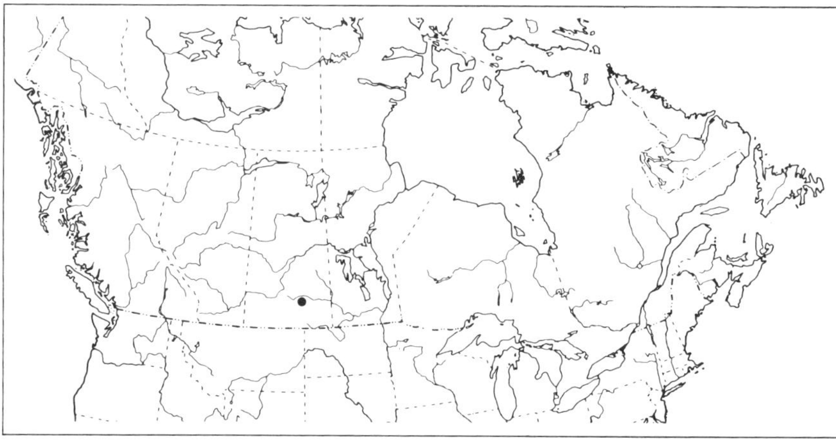

Map 73. Collection localities for *Psorophora signipennis* in Canada: • specimens we examined.

Genus *Uranotaenia* Lynch Arribálzaga

Adult. Small; palpus minute in both sexes; proboscis usually enlarged near apex; antenna of male strongly plumose; vertex covered with overlapping, flat, rounded, usually metallic scales; upright forked scales few, confined to occiput, or none; antepronotal lobes separated; scutum with both acrostichal and dorsocentral setae and usually with lines or spots of round, flat, metallic scales; usually one spiracular seta; no postspiracular setae or scales; postnotum bare; vein R_{2+3} longer than R_2 or R_3; calypter without fringe; abdomen short, parallel-sided, truncate at apex; cercus of female small, rounded, not exserted; male terminalia exserted, but not especially prominent; gonocoxite globular; gonostylus minute, short, and slender with small apical spine.

Larva. Head small in comparison with body, usually longer than broad as in anophelines; head setae 5-C and 6-C usually thickened and strongly sclerotized; comb scales usually arising along posterior edge of a lateral plate on abdominal segment VIII; pecten teeth and siphonal seta 1-S present; each pecten tooth flattened, its entire free margin fringed with long spinules; anal segment encircled by saddle, but not pierced by hairs of ventral brush (seta 4-X); two pairs of anal papillae.

Biology. A large genus without subgenera that is nearly worldwide in distribution. Most of the species are tropical and are especially numerous in Africa and the Oriental region. Most larvae inhabit ground pools; those of a few exotic species are restricted to "containers" such as pitcherplant leaves (*Nepenthes*), crabholes, rot holes, and rock pools (Horsfall 1955). Adult feeding behavior is poorly known; *Uranotaenia lowii* Theobald feeds on the blood of frogs and toads (Remington 1945), and other species have been collected that had ingested bird blood.

Uranotaenia sapphirina (Osten Sacken)

Plate 74; Map 74

Aedes sapphirinus Osten Sacken, 1868:47.
Uranotaenia coquilletti Dyar and Knab, 1906*c*:187.

Adult. Female: Integument medium brown, somewhat translucent; palpus and proboscis dull brown scaled; one long postocular seta arising from vertex at dorsomedial corner of eye and a second arising close to uppermost eye margin; vertex, especially along eye margin, covered with flat, rounded, brilliantly metallic sky blue scales; antepronotal lobe, a median longitudinal line on scutum, supraalar area of scutum, dorsal third of katepisternum, and central area of scutellum with similar metallic sky blue scales forming an unmistakable pattern quite unlike that of any other Canadian mosquito (*Orthopodomyia* also has an intricate pattern but the scales are white, not metallic blue; *Toxorhynchites* is many times larger and almost completely covered with metallic scales); remainder of scutum with sparse shiny brown recumbent scales; vein R_{2+3} more than twice as long as R_2 or R_3; wing veins, legs, and abdominal tergites covered with dull brown scales except for a small spot of white scales on tergites 3 and 5 at midposterior edge and at apex of each tibia dorsally.

Male: Coloration including the bold pattern of metallic blue scales as in female. Antennae long plumose. With *Aedes cinereus* and *Wyeomyia smithii*, the only male mosquito in Canada in which the palpus is no longer than that of the female.

Biology. This species is evidently rare in Canada. The few specimens present in the Canadian National Collection were all collected in light traps. We have not found larvae in spite of extensive searches, although they are reported as occurring with those of *Anopheles* in sunlit permanent ponds that have abundant emergent and floating vegetation, particularly *Spirogyra* or duckweed (*Lemna* sp.) (Smith 1904, Carpenter and LaCasse 1955, Barr 1958, King et al. 1960). Larvae rest parallel to the water surface and can be mistaken for *Anopheles* (Dyar 1928).

Females overwinter in caves and other shelters (Lawlor 1935, Breeland et al. 1961); they rest during the day in similar protected situations (Carpenter et al. 1946). Eggs are laid in floating rafts similar to those of *Culex* and *Mansonia* (Dyar 1922*b*). The reported feeding habits of the females are contradictory. According to Breeland et al. (1961), they land on the skin but do not attempt to bite. Mitchell (1907) claimed that "they bit but once and laid one raft, averaging 41 eggs . . . ." Females are capable of flying long distances. Some females were captured in light traps on the Atlantic Ocean, 13 km from shore (King et al. 1960).

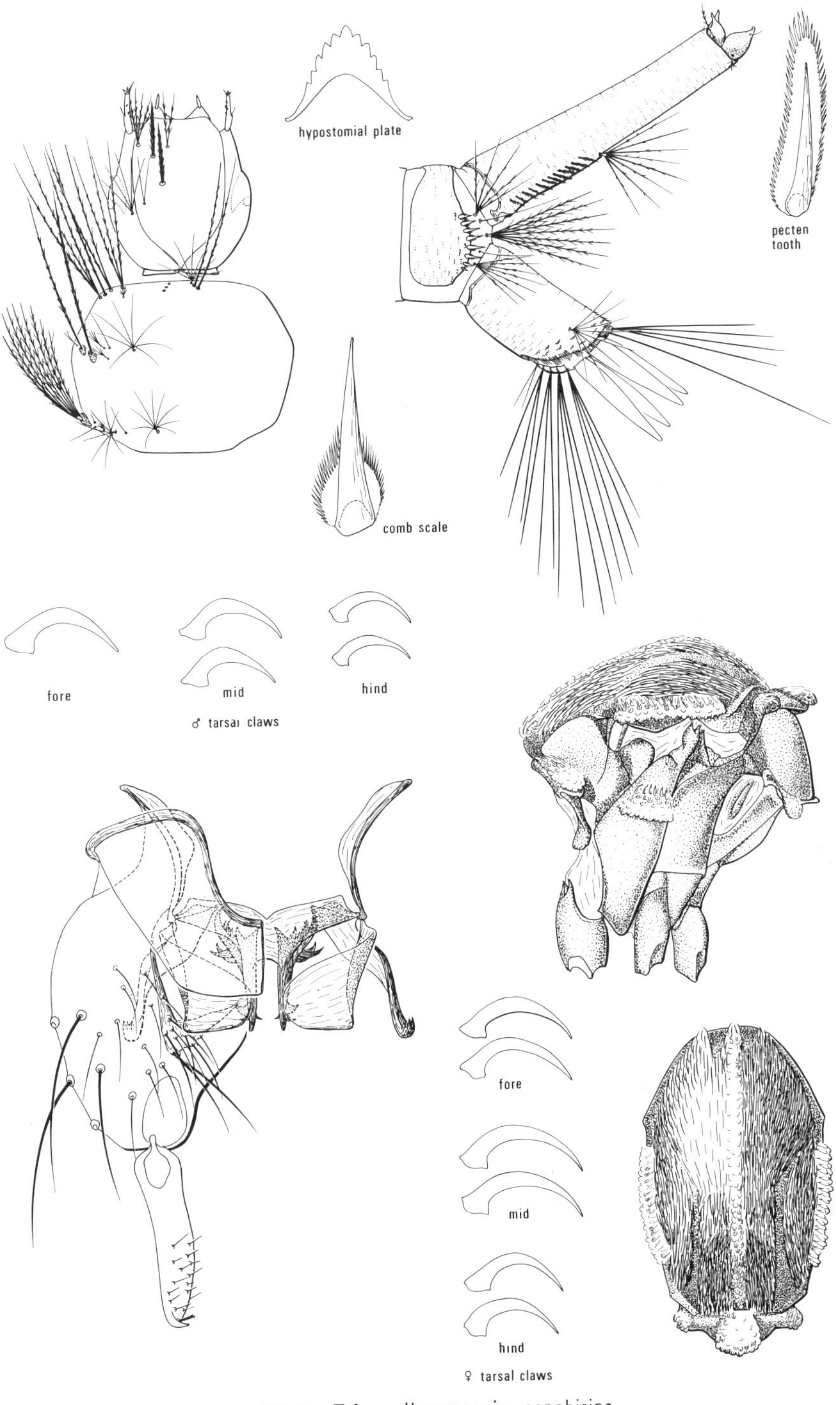

Plate 74. *Uranotaenia sapphirina*

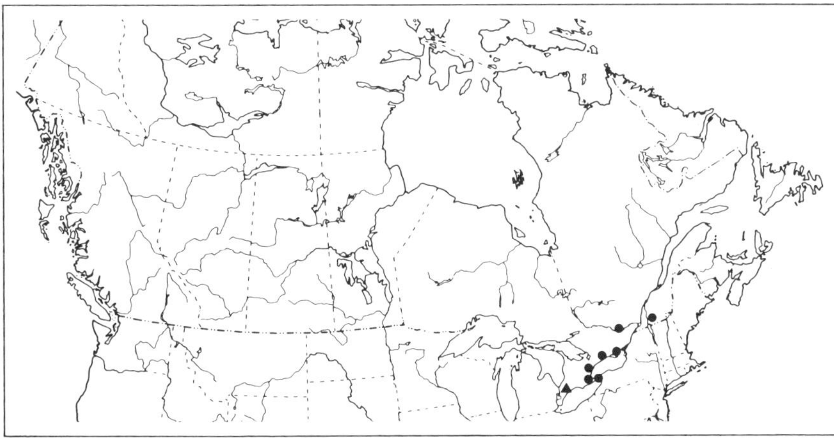

Map 74. Collection localities for *Uranoteania sapphirina* in Canada: • specimens we examined, ▲ literature records.

Distribution. North America, from North Dakota and Mexico east to southern Quebec (Twinn 1949) and the West Indies.

Genus *Wyeomyia* Theobald

Adult. Small; palpus minute in both sexes; antenna of male reduced, hardly distinguishable from that of female; one long postocular seta arising from vertex at dorsomedial corner of eye and projecting forward over eye junction; scutum uniformly covered with dull brownish gray scales, without setae except at margins; spiracular setae present; postspiracular setae absent; postnotum with a few setae in middle near posterior edge (sometimes obscured by halteres), the only Canadian species so adorned; calypter without fringe; abdomen parallel-sided, truncate at apex; male terminalia, though withdrawn into abdominal segment VIII, at least partly visible in pinned material; gonostyli of many species, including the one Canadian species, elaborately ornamented with secondary flanges and processes at apex; lobe of ninth tergite usually bearing spine-like setae, some of which may have enlarged secondarily toothed apices; claspette absent.

Larva. Comb scales usually arranged in a single transverse row, occasionally arising from a lateral sclerite; siphon lacking pecten teeth or siphonal seta 1-S, but with numerous scattered single or branched setae; saddle not encircling anal segment; lateral and dorsal setae 1-X, 2-X, and 3-X of anal segment all very long, subequal in length, single or with few branches; ventral brush (seta 4-X) reduced to a single, long, branched seta, not connected to its opposite number by a bar.

Biology. The genus is predominantly neotropical, where a large number of species are known. As larvae they inhabit the rainwater that collects in the ensheathing leaf bases of bromeliads and aroids; in the flower bracts of *Heliconia* (Musaceae), *Calathea* (Marantaceae), and other large plants; and in broken bamboo internodes, rot holes, and occasionally even artificial containers. Eggs are laid singly within the cavity. Adults fly during the day and usually do not stray far from their larval habitat. Most of the species except *smithii* take blood.

Wyeomyia smithii (Coquillett)

Plate 75; Map 75

Aedes smithii Coquillett, 1901:260.

Adult. Female: Integument of body black; proboscis, palpus, vertex, dorsal surfaces of femora, tibiae, tarsi, wing veins, and dorsum of abdomen covered with black scales, which may appear vivid metallic blue when viewed end on; scutum completely clothed with dark brownish or dark slate gray scales that may appear paler but not metallic blue when viewed end on; lower half of head behind eye, antepronotum, postpronotum, all of anepisternum, most of katepisternum, mesepimeron, coxae, undersides of femora, and venter of abdomen with round, flattened, silvery scales; cercus pale-scaled, short, rounded, and not exserted.

Male: Indistinguishable from female except by examination of terminalia; gonocoxite unornamented; gonostylus enlarged apically, with two subapical processes.

Larva. Almost transparent in life, whitish when preserved; head capsule weakly sclerotized; thorax quadrate; thoracic and abdominal setae long compared with size of body; dorsolateral and ventrolateral abdominal setae (series 1 and 13) stellately branched; comb scales 4–16, all the same length, in a single transverse row; siphon transparent or lightly sclerotized with darkened apex; siphonal setae all single; anal papillae one pair, about as long as anal segment.

Biology. Larvae have been collected only in the water-filled leaves of the pitcherplant, *Sarracenia purpurea*, where they presumably feed on the decayed remains of insects and other detritus that collect at the bottom of each leaf. The species occurs almost everywhere that the pitcherplant is found in Canada except northern Alberta and adjacent British Columbia and the Northwest Territories. They spend most of the year in the larval stage. Larvae can be found at all times of the year, even in midsummer when adults are present. Although larvae spend up to 7 months encased in ice in the frozen cores within pitcherplant leaves, Smith and Brust (1971) found that larvae could not withstand a constant laboratory temperature of –10°C for more than 2 or 3 months. Even at –5°C no larvae survived for more than 4 months; therefore Smith and Brust concluded that an adequate snow cover was essential for larval survival. During the five coldest months at Pinawa, Man., Evans and Brust (1972) measured an average ground temperature during the five coldest months of –3.7°C in a sphagnum bog under the snow; they found that even under these field conditions larval mortality was 45%.

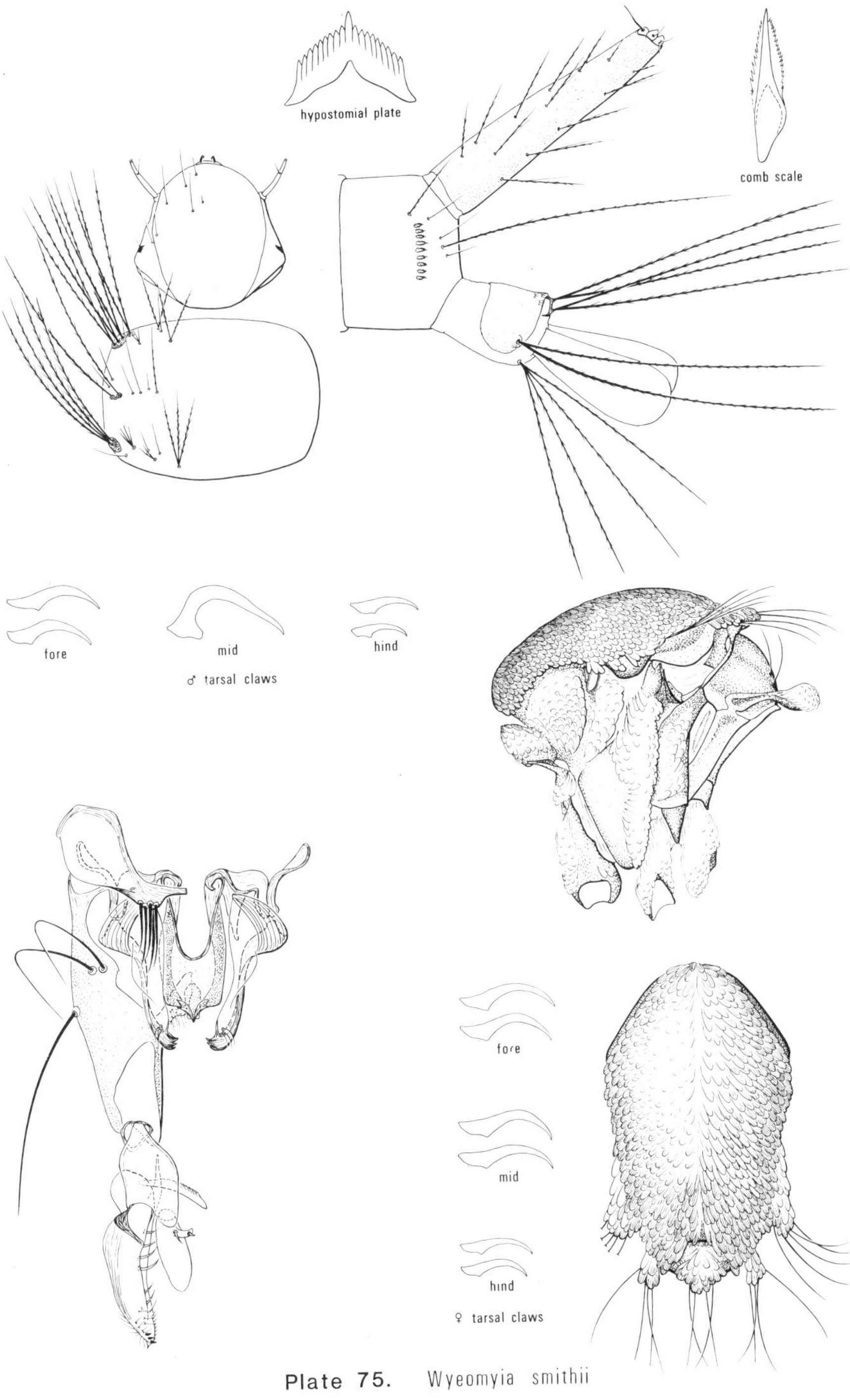

Plate 75. *Wyeomyia smithii*

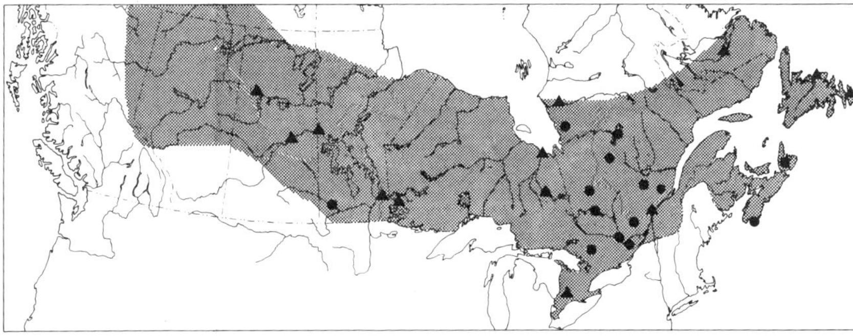

Map 75. Collection localities for *Wyeomyia smithii* in Canada: • specimens we examined; ▲ literature records; shaded area shows approximate distribution of pitcherplant, from Evans (1971).

Larvae overwinter as third instars in diapause (Smith and Brust 1971). Diapause is induced in eggs and in early instars by exposure to a daylength of 14 hr or less (Evans and Brust 1972). Development is arrested in the third instar, which ceases growth even if held at high temperatures for several months (Smith and Brust 1971). Subjected to less than 14 hr of light per day, larvae remained indefinitely in diapause. At 14 hr 30 min of light per day, 20% of the larvae proceeded to pupation. At 15 hr daylength for as short a period as 5 consecutive days, all diapausing larvae resumed development. Once diapause had been terminated by 8 days at 15 hr daylength it could not be reinstated by subjecting the larvae to shorter days again (Smith and Brust 1971). When in diapause, larvae do not ingest food, although they may appear to do so (Evans and Brust 1972).

Larval development is also curtailed by temperatures below 10°C (Smith and Brust 1971). In Manitoba, therefore, development in spring would be inhibited by such temperatures for 4–6 weeks after daylength had become long enough to terminate diapause (Evans and Brust 1972). Development from termination of diapause to pupation at 15°C requires more than 50 days. Even at 20°C, it requires an average of 16 days. In the northern part of its range, there is probably not enough time for more than one generation. Judd (1959) reported two annual generations at London, Ont. The ovaries are exceptionally well developed at emergence, being at stage IIb to IIIa with half the yolk already formed rather than stage I (Smith and Brust 1971). This first ovarian cycle may be completed on water alone, using reserves already laid down in the fat body. Carbohydrates are not required for egg development but they will prolong adult life (Price 1958*b*); they are unnecessary if the larval diet is rich (Smith and Brust 1971). Thus, females of *Wyeomyia smithii* are probably among the few fully autogenous mosquitoes known, because no blood feeding has ever been recorded. Despite this, Hudson (1970) has shown that the maxillary blades are toothed and that the mouthparts of female *smithii* appear to be fully capable of taking blood. After the first cycle is completed, the penultimate follicles do not degenerate, but contain small quantities of yolk. It is possible that females could, at this time, seek a protein meal.

In midsummer, both males and females may be found flying close to the bog surface in the vicinity of their host plants during the day. Males have no modifications for swarming, and their antennae are scarcely different from those of the females, having significantly fewer sensilla of all types than other species (McIver and Hudson 1972).

Eggs are laid singly, at night, inside dry newly opened pitcherplant leaves (Mitchell 1907, Dyar 1922*b*). They hatch when rainwater collects in them.

Distribution. North America, from Saskatchewan and Illinois east to Newfoundland and Delaware.

Species potentially occurring in or incorrectly recorded from Canada

The following is a list of species that have been recorded in states of the USA that border on Canada (except for New York, which has a distinct southern and coastal element in its flora and fauna that is unlikely to occur in Canada) but are not known to be represented by authentic Canadian specimens or, as in the case of *idahoensis*, are not recognized by us as separate species.

Anopheles crucians Wiedemann. Venard and Mead (1953) recorded this species as rare in Ohio. Adults have pale spots on their wings as in *punctipennis* but none on the costa except at its apex.

Anopheles occidentalis Dyar & Knab. As explained under *Anopheles earlei* previously, we believe that all Canadian records of *occidentalis* and possibly those of Washington State as well are *earlei*.

Aedes idahoensis (Theobald). See *Aedes spencerii*.

Aedes niphadopsis Dyar & Knab. Though recorded from Canada by Pucat (1965) and Steward (1968), the specimen on which this was based was a misidentified example of *dorsalis*. Gjullin and Eddy (1972) show no records of this species north of southern Idaho.

Aedes nevadensis Chapman & Barr. Ellis and Brust (1973) suspected that a few larvae from southern Manitoba belonged to this species, but these specimens were lost. These authors also recorded specimens from southern Washington State. The species is closely related to *communis*. The comb scales of the larva usually have one, often two, long apical spines that are more than twice as long as the shorter adjacent spinules, whereas those of *communis* are rounded and fringed apically with subequal spinules. Adults are scarcely distinguishable from those of *communis* except by population analysis (see Ellis and Brust 1973).

Aedes punctodes Dyar. Although Gjullin et al. (1961) treated this as a synonym of *punctor*, it is usually regarded as a separate species (Knight 1951, Frohne 1953*a*, Carpenter and LaCasse 1955). It is closely related to *punctor* from which it can be distinguished only in the larval stage, being smaller, with short anal papillae, and lacking a completely encircling saddle in the last instar. It occurs in Pacific coastal salt marshes of Alaska but has not yet been collected on the coast of British Columbia.

Aedes tormentor Dyar & Knab. This distinctively marked, dark-legged species reaches its northern limit in Ohio (Venard and Mead 1953). Adults have a broad, pale-scaled middorsal stripe on the scutum flanked by brown scales. The last-instar larva is found in summer rainpools; it resembles that of *trivittatus* except that the pecten teeth extend distal to the siphonal seta 1-S.

Aedes ventrovittis Dyar. This species has been recorded from Chelan Co., in the high Cascades of northern Washington, south of Manning Provincial Park, B.C. It should be expected to be found in British Columbia. This small, dark alpine species seems to be rather close to *impiger*. Adult females resemble and could easily be mistaken for *impiger* except that *ventrovittis* lacks lower mesepimeral setae. Recently, Bickley (1976) identified as *ventrovittis* a larva collected at Cantwell, Alaska, by Frohne in 1954. We have not seen specimens of larvae but the description suggests that of *impiger* except that the distal pecten teeth are more widely and unevenly spaced (Carpenter and LaCasse 1955).

Culex boharti Brookman & Reeves. Gjullin and Eddy (1972) listed the most northerly record for this species as Lincoln Co., in east central Washington, but it is not impossible that *boharti* may be discovered in the Okanagan Valley of central British Columbia. It is closest to *territans* in all respects and can be readily distinguished from that species only by the male terminalia. According to Carpenter and LaCasse (1955), the abdominal segment IV of the larva is less pigmented than the adjoining segments, the pale scales on the female vertex are yellowish rather than ashy white as in *territans,* and, in the male, the bridge connecting the two halves of the aedeagus is broader and closer to the apex, and the distal appendages of the subapical lobe are setiform, not flattened, and lack serrated edges.

Culex erraticus (Dyar & Knab). Carpenter and LaCasse (1955) recorded this species from Michigan and Ohio. The larva with its head seta 6-C single or double would key to *territans* in our key, but each comb scale has a long spine-like apex and short subapical spinules, whereas that of *territans* is rounded and apically fringed with subequal spinules. The larval body integument is covered with minute hairs. In the male, the subapical lobe of the gonocoxite is split into three long arms, each bearing one or more apical appendages. The scales of wing vein R_s in the female are broadened, the vertex usually having broad appressed scales dorsally and pale scales on the abdomen forming basal rather than apical bands or spots.

Culex peccator Dyar & Knab. This species is similar to *erraticus* and, like it, has been recorded from Michigan. The larva of *peccator* also has minute integumental hairs over the body, like *erraticus,* and rounded fringed comb scales, like *territans*. However, the head seta 5-C is very short, less than half as long as seta 6-C. In the male, the gonostylus is rounded, with bifurcate apex, unlike that of *erraticus* or *territans*. The female is apparently inseparable from *erraticus,* but may be easily distinguished from *territans* by the presence of basal rather than apical pale-scaled abdominal bands or spots.

Culex peus Speiser (syn. *Culex stigmatosoma* Dyar). Not found north of southern Washington State (Gjullin and Eddy 1972), adults of this species would be most easily mistaken for *tarsalis,* because the tarsi and proboscis are similarly banded with white scales. However, the anterodorsal surface of each femur and tibia lacks the longitudinal line of white scales characteristic of *tarsalis.* The larva appears to be scarcely distinguishable from that of *pipiens* except that in *peus* the spicules along the distal margin of the saddle are larger and more conspicuous.

Culex salinarius Coquillett. Twinn (1945) recorded this species from Nova Scotia but there are no specimens of *salinarius* from Canada in the Canadian National Collection, Ottawa, and we suspect that Twinn misidentified specimens of *pipiens* as this species. For distinguishing characteristics of the larva, see under *Culex pipiens.* Adult females of *salinarius* are more reddish than *pipiens,* and the transverse basal abdominal pale-scaled bands are yellowish and poorly differentiated (they may be absent also in female *pipiens*). The best distinguishing character is the lateral arm of the male paraproct, which is long and curved distally, even more pronounced than in *restuans.*

Culiseta particeps (Adams) (syn. *Culiseta maccrackenae* Dyar & Knab). Larvae presumed to be this species were recorded from southeastern Alaska by Frohne and Sleeper (1951). However, Gjullin et al. (1961) pointed out that this identification may be doubtful, because the species is otherwise known only from southwestern Oregon. More recently, however, the species was recorded from Jefferson Co., northwestern Washington State, by Myklebust (1966, cited in Gjullin and Eddy 1972). A larva from Ketchikan, Alaska, collected in 1949 by Frohne was also identified as *particeps* by Bickley (1976).

Although this species belongs to the subgenus *Culiseta* and the adults would probably key to *alaskaensis* in our key, the femora have a narrow subapical ring of pale scales. The saddle of the larva has a group of conspicuous spicules dorsally at its apex, and the mesothoracic seta 1-M is elongate and two- to four-branched, unlike any other larvae of *Culiseta.*

Psorophora spp. Venard and Mead (1953) recorded four species of *Psorophora: cyanescens, discolor, horrida,* and *varipes* from Ohio. All were rare in that state and are unlikely to occur in Canada. Also *P. horrida* has recently been recorded from southern Michigan (Zaim et al. 1977).

Glossary

abdomen The cylindrical portion of the body behind the thorax composed of 10 segments in adults (Figs. 9, 10) and larvae; each segment is numbered in Roman numerals, beginning with the anteriormost. Abdominal setae of the larva are also numbered (see p. 41).

acrostichal setae A rather weak, irregular, longitudinal row of setae arising on the scutum along the midline (Fig. 5).

acus A small sclerite at or near the proximal margin of the siphon (Fig. 180).

aedeagus A single median sclerite partly or completely encircling the male genital duct and serving as an intromittent organ (Figs. 10, 11).

allopatric Species or populations whose distributions do not overlap.

anal papillae Paired outgrowths of the rectum of the larva that project externally through the anus (Fig. 16) and serve as regulators of water balance in the blood (osmoregulation); generally their length is inversely correlated with the salinity of the larval medium.

anal vein (abbr., **A**) The posteriormost vein of the wing (Fig. 7); of taxonomic importance only in *Orthopodomyia* (Figs. 231, 232).

anautogeny (adj., **anautogenous**) The inability of an adult female to produce eggs without a blood meal, that is, the state of requiring a blood meal for egg production.

anepisternal cleft A narrow membrane tapering ventrally, extending from the wing base to the anterodorsal corner of the katepisternum (Fig. 4).

anepisternum The upper half of the episternum divided into an anterior and a posterior portion by the membranous anepisternal cleft; the anterior portion is further subdivided into a subspiracular area and a postspiracular area (Fig. 4).

antenna (pl., **antennae**) A pair of 15-segmented sensory appendages arising on the front of the head between the eyes and the clypeus (see also scape, pedicel, and flagellum) (Figs. 1, 2).

anteprocoxal membrane The membrane connecting the fore coxa and the probasisternum (Fig. 3).

antepronotum The anterior division of the pronotum separated from the latter by a transverse groove, forming a setose lobe behind the head and anterior to the postpronotum (Figs. 3, 4).

apical lobe of gonocoxite A rounded flange-like extension of the apical dorsomedial corner of the gonocoxite; usually present in *Aedes* and often armed with specialized setae (Fig. 11).

apotome The diamond-shaped or shield-shaped dorsum of the larval head capsule, outlined laterally and posteriorly by ecdysal lines, bearing externally head setae 4-C to 8-C and supporting internally the origins of the large muscles that operate the labral brushes.

autogeny (adj., **autogenous**) The ability in an adult female to produce eggs without a blood meal. A species that does not take blood is obligatorily autogenous, whereas a species, or members thereof, that can produce a few eggs without blood, though blood is normally required, are said to be facultatively autogenous (Corbet 1964).

band Running transversely or encircling a segment, in contrast with a stripe, which runs longitudinally; usually referring to the abdominal or leg segments.

basal lobe of gonocoxite A variously shaped process, usually bearing long setae, on the proximal medioventral corner of the gonocoxite (Fig. 11).

bivoltine Capable of having two generations per year (see also univoltine and multivoltine).

bifurcation Separation or branching into two parts.

bog A water-holding depression that has become partly or completely filled in with peat (which is largely undecomposed sphagnum moss) that supports a carpet of living sphagnum, a variety of ericaceous shrubs (such as leatherleaf, laurel, blueberry), and in drier areas, black spruce and tamarack. Water-filled depressions are usually acid and support a characteristic mosquito fauna. In the north, most of the flat, low-lying terrain is covered with this type of vegetation.

calypter (pl., **calypters**) A scale-like lobe behind the base of each wing, folded longitudinally when the wing is folded forming the upper and the lower calypters (Fig. 7); the lower calypter is scarcely developed in mosquitoes, but the upper one is much larger and its periphery is fringed in most genera (Fig. 26).

cercus (pl., **cerci**) A paired process at the apex of the abdomen; in mosquitoes, present in the adult female only, lost in the larva and in the adult male; elongate and slender in *Psorophora* and most *Aedes*, rounded and pad-like in other genera (Fig. 9).

cibarium A cavity best developed in the adult that lies between the bases of the mouthparts, in front of the true mouth, and serves as a sucking pump for drawing liquids up the canal formed by the labrum; it is activated by muscles inserted on its dorsal surface that originate on the clypeus (Snodgrass 1959, p. 60).

claspette An appendage of the gonocoxite, arising from its medial anteroventral corner, that may be lobular (in *Anopheles*) or slender (in *Aedes* and *Psorophora*), usually crowned apically with setae or one or more filaments (Fig. 11).

clypeus A plate or shield on the anterior median aspect of an insect's head that bears the labrum on its anterior margin; internally the clypeus supports the muscles of the cibarial pump.

comb or **comb scales** A patch or transverse row of flattened, thorn-like outgrowths of the integument on each side of abdominal segment VIII of the larva; their shape and number is often species-specific (Figs. 41, 42, vertical arrows).

costa (abbr., **C**) The thickened anterior margin of the wing, extending from base to apex (Fig. 7).

coxa The basal segment of the leg, with which the leg articulates with the thorax (Fig. 4).

cranium (adj., **cranial**) The sclerotized capsule forming the external skeleton of the head of adult, pupa, and larva.

crossvein A short transverse vein connecting two major longitudinal veins (Fig. 7), for example, the r-m crossvein connects the posteriormost branch of the radius (R_{4+5}) with the media, and the m-cu crossvein (sometimes called the base of M_4, see under media) connects the media with CuA_1. Crossveins lack scales except in *Culiseta alaskaensis*.

cubitus (abbr., **Cu**) The fourth of the five major wing veins, lying between the media and the anal vein; branching at its base into a strong anterior branch, CuA, which bifurcates again into CuA_1 and CuA_2, and an insignificant scaleless posterior branch, CuP, closely paralleling CuA (Fig. 7).

diapause A physiological state in which development is temporarily arrested and normal metabolic processes are greatly curtailed that enables the egg, larva, or adult to survive an unfavorably cold or dry period; diapause apparently always takes place at a particular species-specific stage in the life cycle in response to changing environmental conditions; it is almost invariably obligatory.

distal (or **apical**) In reference to an appendage or extremity, the part that is remote from the body of an organism; the opposite of proximal (or basal).

dorsal Pertaining to the dorsum, or upper surface; in special cases, such as in the male terminalia that become inverted after emergence, the lower surface of the apex of the abdomen is still referred to as dorsal in spite of its apparently ventral position (see also ventral).

dorsocentral setae A usually well-defined paired longitudinal row of setae arising on the scutum about midway between the midline and the lateral edge (Fig. 5).

dorsocentral stripe A narrow stripe of scales lying on the scutum coincident with the dorsocentral setae and lateral to the submedian stripe.

dorsolateral abdominal seta Arising from a point about midway between middorsal and midlateral, on the abdominal segments of the larva; this seta is exceptionally well-developed on some segments in several species of *Aedes* (Fig. 16).

ecdysal line Line of weakness in the larval cuticle where splitting normally occurs before molting.

encephalitis (also called **encephalomyelitis**, pl., **encephalitides**) One of several related diseases of the central nervous system caused by arboviruses, that is, arthropod-borne viruses.

epimeron The posterior portion of a thoracic pleural sclerite, separated from the anterior portion (the episternum) by the pleural suture; hence proepimeron, mesepimeron, and metepimeron for each of the three thoracic segments.

episternum The anterior portion of a thoracic pleural sclerite separated from the posterior portion (the epimeron) by the pleural suture; hence, proepisternum, mesepisternum, and metepisternum for the three thoracic segments.

epizootic A rapid spread of disease among an animal population (analogous to an epidemic in the human population).

exserted Projecting beyond an enclosing organ or part.

femur (pl., **femora**) The thickest segment of the leg, proximal to the tibia.

fenestrated Having one or more transparent spots.

filament The apical seta-like or blade-like portion(s) of the claspette (Fig. 11).

flagellomere One of the segments of the flagellum (Figs. 1, 2).

flagellum The elongate apical portion of the antenna, beyond the pedicel, subdivided into 13 flagellomeres; in most mosquitoes much more elaborately setose in the male than in the female (compare Figs. 1 and 2).

follicle One of the many elongate tubular subdivisions of an ovary made up of a series of cells, each of which is capable when stimulated of developing into an egg. Only the most distal cell develops at one time, typically following a blood meal.

fossa A cavity or depression.

frons The upper part of the front of the head between the antennae and vertex; in mosquitoes, largely obliterated medially by the compound eyes.

generation The complete cycle of egg, four larval instars, pupa, and adult.

gonocoxal apodeme An anteriorly directed internal strut for supporting the paramere, arising from the proximal dorsomedial corner of the gonocoxite (Figs. 10, 11); called "parameral apodeme" by Knight and Laffoon (1971), although it is not an apodeme of the paramere.

gonocoxite The basal segment of the gonopod; always the largest element of the male terminalia; usually cylindrical or conical in shape (Figs. 10, 11).

gonopod The two-segmented paired appendage that comprises the major portion of the male terminalia, consisting of a basal gonocoxite and a distal gonostylus.

gonostylus The apical segment of the gonopod, or clasper; usually long, curved, and slender, bearing a peg-like seta at its apex (Figs. 10, 11).

hair(s), hairy Fine delicate seta(e).

halter (pl., **halteres**) A club-shaped appendage, representing the reduced metathoracic wing, arising from the posterodorsal corner of the thorax (Fig. 7).

holarctic Distributed in both North America and Eurasia.

hypopharynx In the adult mosquito, the "tongue," a slender median stylet extending the length of the proboscis, bearing the opening of the salivary duct at its apex; in the larva rudimentary, fused with the labium.

hypostigmal area A minute portion of the subspiracular area at its dorsal extremity, just below the mesothoracic spiracle (Fig. 4).

hypostomial plate The strongly sclerotized, toothed, triangular plate forming the anterior midventral margin of the head capsule of the larva, which serves as an anvil against which the apical mandibular teeth strike when the mandibles are closed (adducted).

instar One of the stages of larval development, between molts, that is, the first instar is that larval stage between hatching and the first molt, the fourth (and final in the mosquito) is that between the third instar and the pupa.

integument The outer covering of an organism, which in insects is usually in the form of relatively rigid (sclerotized) plates, or sclerites, interconnected by more flexible (membranous) areas.

isotherm A line on a map connecting points having the same mean temperature for a given period.

katepisternal setae (upper and lower) A row of setae more or less interrupted (hence upper and lower) across the dorsal end of the katepisternum and extending down along its posterior margin (the mesopleural suture) (Fig. 4).

katepisternum A large convex sclerite, the ventral subdivision of the mesepisternum, lying between the fore and mid coxae below and the anepisternum above, bordered anteriorly by the postprocoxal membrane and posteriorly by the mesopleural suture (Fig. 4).

labium The lower lip formed by the fusion of the pair of head appendages behind the maxillae (that is, the appendages of the last head segment); in the adult, the outer trough-like part of the proboscis holding the remaining mouthparts; in the larva, reduced and concealed between the much larger maxillae.

labral brush A large mass of long setae arranged in tightly packed rows on either side of the labrum of the larva that when adducted (folded medially) capture food particles from the surrounding water (Fig. 13).

labrum The upper lip, a median unpaired flap above the mouth and mandibles; in the adult, the uppermost stylet of the proboscis, whose edges are rolled ventrally and medially to form a tube-like food canal; in the larva, bearing the labral brushes and associated sclerites that are the principal food-gathering mechanisms in filter-feeding (that is, nonpredaceous) species.

lateral Pertaining to either side of a bilaterally symmetrical organism.

lateral abdominal seta, lower A paired seta, series 7, arising just ventral to the upper lateral seta.

lateral abdominal seta, upper A paired seta, series 6, usually the longest of all abdominal setae, arising midlaterally from each of the first seven abdominal segments of the larva (see discussion p. 42).

lateral stripe The scales along the lateral edge of the scutum; in some *Aedes* these scales are darker than those of the adjacent sublateral area (Fig. 6).

line A longitudinal row of scales, usually one or two scales wide (that is, a narrow stripe).

marsh An unshaded stagnant body of water, sometimes surrounded with shrubs, and usually supporting such emergent vegetation as cattails (*Typha*), bur-reed (*Sparganium*), or sedge (*Carex*).

maxilla (pl., **maxillae**) The paired appendage behind the mandible; in the adult female, sharp, stylet-like, and apically toothed; in the larva, pad-like and fringed with setae.

media (abbr., **M**) The third major wing vein, lying between the radius and the cubitus (Fig. 7), branching beyond the middle of the wing into M_1 and M_2 (some authors have interpreted CuA_1 as containing a branch (M_4) of the media, others have denied the existence of M_4 in the Diptera).

medial Direction toward the midline (in contrast with lateral).

median Pertaining to the midline of a bilaterally symmetrical organism.

median stripe A narrow middorsal longitudinal stripe of scales coincident with the acrostichal setae, usually paler in color than the scales of the adjacent submedian stripes (Fig. 5).

mesepimeral setae, lower A row or group of well-developed setae arising on the lower half of the mesepimeron just behind and parallel to the mesopleural suture (Figs. 4, 83 *a*).

mesepimeral setae, upper A group of well-developed setae occupying the dorsal portion of the mesepimeron, lying between the wing base and the metathoracic spiracle.

mesepimeron The prominent sclerite of the mesothorax lying posterior to the mesopleural suture and above the mid coxa (Fig. 4).

mesepisternum The entire pleural area of the mesothorax lying anterior to the mesopleural suture, and further subdivided into a dorsal anepisternum and a ventral katepisternum (Fig. 4).

mesomeron A small triangular sclerite posterodorsal to the mid coxa (Fig. 4).

mesonotum The dorsal surface of the mesothorax, which because of the extensive development of this segment at the expense of the prothorax and metathorax forms virtually the entire dorsum of the thorax.

mesothorax The middle thoracic segment, which because of extensive development of the wing muscles is many times larger than the prothorax or the metathorax.

metathorax The third segment of the thorax, lying just anterior to the abdomen and bearing the halter (second pair of wings), hind leg, and metathoracic spiracle (Fig. 7).

microtrichium (pl., **microtrichia**) One of the minute fixed hairs on the integument.

middorsocentral spot A paired group of pale-colored scales on the scutum at the junction of the transverse suture and the dorsocentral row of setae; most noticeable in *Culiseta* and *Culex* (Fig. 5).

multivoltine Capable of having more than one generation per year (see also univoltine and bilvoltine).

nulliparous Pertaining to a female that has not yet laid eggs (see also parous).

occiput The convex posterior surface of the head (Figs. 1, 2).
oocyte A cell that gives rise to an egg.
osmoregulation Regulation of osmotic pressure.
ovarian cycle The cycle initiated by a blood meal, which stimulates the production of a hormone resulting in egg development, ending in oviposition.
ovarian development The formation of an egg in each ovarian follicle following a blood meal and the subsequent release of hormone.
oviposition The act of depositing eggs.
ovoid Egg-shaped with the large end toward the point of attachment.

palpomere One of the three to five segments of the palpus (Figs. 1, 2).
palpus (pl., **palpi**) A paired, slender, segmented appendage of the adult head, arising on both sides of the base of the proboscis; in mosquitoes, almost always longer and more elaborate in the male than in the female (Figs. 1, 2).
paramere A paired appendage articulating on either side of the aedeagus with the gonocoxal apodeme and supporting the aedeagus; reduced in mosquitoes to a short bar connecting the aedeagus to an anterodorsal apodeme (gonocoxal apodeme) of the gonocoxite (Figs. 10, 11).
paratergite A small oblong sclerite just below the lateral edge of the scutum, anterior to the wing base and behind the mesothoracic spiracle (Fig. 4).
parous Pertaining to a female that has laid at least one batch of eggs.
pecten A regular row of dagger-like rigidly attached spines, the pecten teeth along each side of the siphon near its base in Culicinae (Figs. 16, 37) or on either side below the spiracular openings in Anophelinae (Fig. 30).
pedicel The second and widest antennal segment, usually doughnut-shaped (Figs. 1, 2); also called torus by some authors.
permanent Referring to a mosquito larval habitat that does not normally dry up in summer.
pharynx The first part of the digestive tube behind the true mouth, equipped in the larva with a complex filtering apparatus for engulfing food particles stuffed into it by the mouthparts, then retaining these particles among rows of fine setae while water is expelled, analogous to the baleen of a whale; in the adult, forming a muscular pump, for imbibing nectar or blood (see also cibarium).
phenology The study of the sequence of appearance of each species in the course of a season.
pleural suture A reinforcement of the pleuron in winged insects, extending from the wing base to the coxa and appearing externally as a vertical depression.
pleuron The side of a thoracic segment, between the dorsum above and the venter below.

plumose Finely branched or feathered, or with densely arranged whorls of long setae, as in the male antenna.

pollinose Covered with submicroscopic outgrowths of the integument that together form a "bloom" or pruinosity.

polyphagus Feeding on or utilizing many kinds of food.

postnotum The convex posteriormost part of the mesonotum, behind the scutellum and in front of the narrow metanotum, bare of scales except in *Wyeomyia* (Fig. 4).

postocular setae A row of setae arising on the vertex parallel to and just behind the upper hind margin of the eye.

postprocoxal membrane The membrane connecting the fore coxa to the katepisternum, bearing in some *Aedes* a taxonomically important group of scales (Figs. 4, 75).

postpronotum The posterior division of the pronotum, bordered anteriorly by the antepronotum, dorsally by the scutum, and posteriorly by the mesothoracic spiracle and the subspiracular area (Fig. 4).

postspiracular setae A taxonomically important group of erect well-developed setae on the postspiracular area (not to be confused with scales that may also occur there) that are characteristic of *Aedes* and *Psorophora* (Figs. 4, 22, and 23 *ps*).

postsutural portion of scutum That portion of the scutum behind the transverse suture and in front of the scutellum.

prealar setae A group of setae on the convex upper extremity of the posterior portion of the anepisternum, directly in front of the wing base and behind the anepisternal cleft (Fig. 4).

prescutellar depression A concave area in the middle of the posterior part of the scutum, usually without scales (Fig. 5).

presutural portion of scutum That portion of the scutum between the pronotum and the transverse suture.

probasisternum A midventral sclerite between and anterior to the fore coxae and below the neck (Fig. 3).

proboscis The long slender tubular bundle of stylets formed from the mouthparts (labrum, mandibles, maxillae, hypopharynx, and labium) of the adult (Figs. 1, 2).

proclinate Directed forward.

proctiger The tenth segment of the male bearing the anus apically; in Culicinae, sclerotized and variously modified as an accessory copulatory apparatus partly enveloping the aedeagus (Figs. 10, 11).

prothorax The first segment of the thorax, bearing the first pair of legs (Figs. 3, 4).

proximal (or **basal**) In reference to an appendage or extremity, that which is nearer the body of an organism; the opposite of distal (or apical).

pseudopod Leg-like appendage, absent in mosquito larvae but present in larvae of some related families (such as Chironomidae, Simuliidae).

radial sector (abbr., $\mathbf{R_s}$) The posterior branch of the radius, which bifurcates again to become R_{2+3} and R_{4+5}; R_{2+3} again bifurcates into R_2 and R_3; R_{4+5} remains unbranched in mosquitoes (Fig. 7).

radius (abbr., **R**) The second and thickest major longitudinal wing vein, branching near the basal third of the wing into an anterior branch (R_1) and a posterior branch, the radial sector (R_s) (Fig. 7).

recumbent Of scales or hairs, lying flat, appressed to the surface of the integument.

row Pertaining to the setae of the scutum, distinct from a stripe of scales.

saddle The tergite of the last (tenth) abdominal segment of the larva, oval in the first three instars (and the fourth in *Anopheles* and in most *Aedes*), but usually encircling the segment in the fourth instar like a cylinder (Fig. 16).

saddle seta Numbered 1-X, that is, the first seta on segment X of the larva; the position of this seta, whether arising on the saddle or on the membrane near its edge, is important in instar determination (Fig. 16 and discussion, p. 43)

scale A minute flattened seta arising from a microscopic socket; scales may assume various shapes and colors and are sometimes iridescent; they may be erect or recumbent (appressed to the integument).

scape The first segment of the antenna; in mosquitoes, greatly reduced and almost concealed by the swollen second segment (pedicel) (Figs. 1, 2).

sclerite A region of the integument that is thickened and relatively rigid.

sclerotized Pertaining to thickened parts of the integument.

scutal fossa A slightly depressed area on each side of the scutum lateral to the dorsocentral setae and anterior to the transverse suture (Fig. 6).

scutellum A transverse sclerite, part of the mesonotum, lying between the scutum and the postnotum (Figs. 4, 5).

scutum The large, convex, rounded sclerite forming most of the dorsal surface of the mesonotum, ornamented with taxonomically important stripes or patches of scales (Figs. 3, 4).

semipermanent Referring to a mosquito larval habitat that remains wet until early summer, but that eventually dries up under normal conditions; not a readily distinguishable habitat because it may appear temporary in dry years and may remain permanent in exceptionally wet ones.

seta (pl., **setae**) Any hair-like or bristle-like outgrowth of the integument that arises from a basal socket.

setose Covered with setae.

setula (pl., **setulae**) A minute seta.

siphon A conical or semicylindrical, gradually tapering, strongly sclerotized tube projecting from the dorsum of abdominal segment VIII of the larva, bearing the paired spiracles at its apex (Fig. 16).

siphonal seta Numbered 1-S; this usually conspicuous, branched, paired seta arises either at the base of the siphon, as in *Culiseta*, or nearer the middle, as in *Aedes*; the siphon may also be adorned with unnumbered additional paired setae or tufts, as in *Culex*.

slough A semipermanent prairie marsh; when the content of dissolved salts, mostly carbonates, is high, it is called an alkaline slough.

spicule A minute rod-like outgrowth of the integument of the larvae.

spine(s), spinose A pointed conical outgrowth of the integument that does not arise from a basal socket and hence is immovable.

spinule A minute spine.

spiracle An opening in the integument, connecting internally with a tracheal tube for exchange of gases; larvae have a single pair at the apex of the siphon; in adults, the anterior (mesothoracic) and posterior (metathoracic) spiracles of the thorax are the most conspicuous (Fig. 4).

spiracular setae A row or small group of posteriorly directed setae that arise on a narrow sclerite along the anterior border of the mesothoracic spiracle, behind the postpronotum (the postpronotal setae arise just in front of them and can be easily mistaken for spiracular setae) (Figs. 4, 22 *s*).

sternite The sclerite or sclerotized plate of the sternum, or the ventral portion of a segment.

stripe Running longitudinally, in contrast with a band that runs transversely or encircles a segment; usually used to describe the scales that form the scutal pattern.

subapical lobe of gonocoxite In *Culex*, the basal lobe, which has been shifted apically, usually armed with several characteristically shaped setae that are often called "appendages" (Figs. 204–207).

subcosta (abbr., **Sc**) The first major longitudinal wing vein, lying directly behind and parallel to the anterior margin (costa) (Fig. 7).

sublateral area or stripe The scales forming a somewhat triangular area between the dorsocentral setae and the lateral edge of the scutum (Fig. 5); in some *Aedes*, the presutural part of the sublateral area is pale-scaled, while the postsutural part forms a stripe on either side of the prescutellar depression (the postsutural sublateral stripe).

submedian stripe A longitudinal stripe of scales, usually darker than the scales of adjacent bands, on the scutum between the acrostichal setae and the dorsocentral setae (Fig. 5).

subspiracular area That portion of the anepisternum ventral to the mesothoracic spiracle, bounded anteriorly by the postpronotum and posteriorly by the postspiracular area (Fig. 4).

supraalar setae A group of well-developed erect setae arising from the lateral edge of the scutum above the wing base.

swarm Pertaining to male Diptera (and some other orders, for example, Ephemeroptera), the assembly, usually in flight, of one or more males over a species-specific object or site (the swarm marker), such as a stone in a river, a bridge, a clearing, or a hilltop, to await the arrival of females.

tarsal claw Arising from the apex of each leg, almost always paired, therefore 12; in the male, one claw on the fore- and mid-legs is always larger than the other member of its pair for grasping the female; in the female, all claws are usually about the same shape, but those of the forelegs are larger.

tarsomere Any one of the five segments of the tarsus.

tarsus (pl., **tarsi**) The apical portion of the leg divided into five segments or tarsomeres.

temporary A mosquito larval habitat that normally dries up in late spring or early summer; in exceptionally wet seasons, such a habitat may refill with water and then produce a quite different mosquito fauna from that which it supported in the spring.

tergite The sclerite or sclerotized plate of the tergum or dorsal portion of the segment.

tibia (pl., **tibiae**) The longest segment of the leg, distal to the femur and proximal to the tarsus.

transverse suture A faint lyre-shaped impression on the scutum, scarcely visible in itself, but usually accentuated by a scaleless line or by a row of differently colored usually paler scales (Fig. 5).

truncate Having the end square.

univoltine Having one, and only one, generation per year (see also bivoltine and multivoltine).

ventral Pertaining to the venter or lower surface (see also dorsal).

ventral brush A longitudinal row of long closely packed setae on abdominal segment X of the larva, each laterally braced at its base to prevent sideways movement, which serves as a paddle as the abdomen is lashed from side to side (Fig. 16).

ventrolateral abdominal seta A well-developed seta on the third, fourth, and fifth abdominal segments of the larva of many species of *Aedes*, arising midway between midlateral and midventral points and usually considered to belong to setal series 13 (Fig. 16).

vertex A rather general, imprecise term for the uppermost part of the head, bordered anteriorly by the eyes and the frons and posteriorly by the occiput (Figs. 1, 2).

References

Abdel-Malek, A. 1949. A study of the morphology of the immature stages of *Aedes trivittatus* (Coquillett) (Diptera: Culicidae). Ann. ent. Soc. Am. 42:19-37.

Addison, E.M. 1973. Life cycle of *Dipetalonema sprenti* Anderson (Nematoda: Filarioidea) of beaver (*Castor canadensis*). Can. J. Zool. 51:403-416.

Aitken, T.H.G. 1939. The *Anopheles maculipennis* complex of western North America. Pan-Pacif. Ent. 15:191-192.

Aitken, T.H.G. 1945. Studies on the anopheline complex of western North America. Univ. Calif. Publs. Ent. 7:273-364.

Amin, O.M., and A.G. Hageman. 1974. Mosquitoes and tabanids in southeast Wisconsin. Mosquito News 34:170-177.

Bailey, S.F., D.A. Eliason, and B.L. Hoffman. 1965. Flight and dispersal of the mosquito *Culex tarsalis* Coquillett in Sacramento Valley of California. Hilgardia 37(3):73-113.

Baker, F.C. 1935. The effect of photoperiodism on resting, treehole, mosquito larvae. Can. Ent. 67:149-153.

Baker, F.C. 1936. A new species of *Orthopodomyia*, *O. alba* sp. n. (Diptera, Culicidae). Proc. ent. Soc. Wash. 38:1-7.

Barr, A.R. 1955. The resurrection of *Aedes melanimon* Dyar. Mosquito News 15:170-172.

Barr, A.R. 1957. A new species of *Culiseta* (Diptera: Culicidae) from North America. Proc. ent. Soc. Wash. 59:163-167.

Barr, A.R. 1958. The mosquitoes of Minnesota (Diptera: Culicidae: Culicinae). Univ. Minn. agric. Exp. Stn Tech. Bull. 228. 154 pp.

Basham, E.H., J.A. Mulrennan, and A.J. Obermuller. 1947. The biology and distribution of *Megarhinus* Robineau-Desvoidy in Florida. Mosquito News 7:64-66.

Bates, M. 1949. The natural history of mosquitoes. Macmillan Publishing Co., Inc., New York. 379 pp.

Beck, D.E. 1961. Central Utah County, Utah, mosquito survey studies. Mosquito News 21:6-11.

Beckel, W.E. 1953. Preliminary observation on a hatching stimulus for *Aedes* eggs (Culicidae). Science (Wash. D.C.) 118(3062):279-280.

Beckel, W.E. 1954. The identification of adult female *Aedes* mosquitoes (Diptera, Culicidae) of the black-legged group taken in the field at Churchill, Manitoba. Can. J. Zool. 32:324-330.

Beckel, W.E. 1955. Oviposition site preference of *Aedes* mosquitoes (Culicidae) in the laboratory. Mosquito News 15:224-228.

Beckel, W.E. 1958. Observations on the rearing of larvae, pupae and adults of some *Aedes* mosquitoes of northern Canada. Can. J. Zool. 36:797-808.

Beckel, W.E., and C.A. Barlow. 1953. Studies of the biology of the *Aedes* of northern Canada (Culicidae). III. Field oviposition of *Aedes communis* (DeGeer) with a method of separating the eggs from the substrate. Tech. Pap. Def. Res. Board Can. 8.

Beckel, W.E., and T.P. Copps. 1953. Studies of the biology of the *Aedes* of northern Canada (Culicidae). IV. Laboratory rearing of northern *Aedes* mosquito larvae and pupae. Tech. Pap. Def. Res. Board Can. 8.

Belkin, J.N. 1950. A revised nomenclature for the chaetotaxy of the mosquito larva (Diptera: Culicidae). Am. Midl. Nat. 44:678-698.

Belkin, J.N. 1968. Mosquito studies (Diptera, Culicidae). IX. The type specimens of New World mosquitoes in European museums. Contrib. Am. ent. Inst. 3(4):1-69.

Belkin, J.N., S.J. Heinemann, and W.A. Page. 1970. Mosquito studies (Diptera, Culicidae). XXI. The Culicidae of Jamaica. Contrib. Am. ent. Inst. 6(1):1-458.

Belkin, J.N., and W.A. McDonald. 1956. *Aedes sierrensis* (Ludlow, 1905), a change in name for the western tree-hole mosquito of the Pacific Slope (Diptera, Culicidae). Proc. ent. Soc. Wash. 58(6):344.

Belkin, J.N., and W.A. McDonald. 1957. A new species of *Aedes* (*Ochlerotatus*) from tree holes in southern Arizona and a discussion of the *varipalpus* complex (Diptera: Culicidae). Ann. ent. Soc. Am. 50:179-191.

Bellamy, R.E., and E.H. Kardos. 1958. A strain of *Culex tarsalis* Coq. reproducing without blood meals. Mosquito News 18:132-134.

Belton, P., and D.E. French. 1967. A specimen of *Aedes thibaulti* collected near Belleville, Ontario. Can. Ent. 99:1336.

Belton, P., and M.M. Galloway. 1966. Light-trap collections of mosquitoes near Belleville, Ontario, in 1965. Proc. ent. Soc. Ont. (1965) 96:90-96.

Benach, J.L., and W.J. Crans. 1973. Larval development and transmission of *Foleyella flexicauda* Schacher and Crans, 1973 (Nematoda: Filarioidea) in *Culex territans*. J. Parasit. 59:797-800.

Benedict, W.C. 1962. Mosquitoes in and about Windsor, Ontario. Proc. ent. Soc. Ont. 93:82-84.

Bennington, E.E., J.S. Blackmore, and C.A. Sooter. 1958*a*. Soil temperature and the emergence of *Culex tarsalis* from hibernation. Mosquito News 18:297-298.

Bennington, E.E., C.A. Sooter, and H. Baer. 1958*b*. The diapause in adult female *Culex tarsalis* Coquillett (Diptera: Culicidae). Mosquito News 18:299-304.

Bickley, W.E. 1976. Notes on the distribution of Alaskan mosquitoes. Mosquito Syst. 8(3):232-236.

Bickley, W.E., and J.T. Whitlan. 1956. Notes on some freshwater *Aedes* mosquitoes in Maryland. Mosquito News 16:138-139.

Bidlingmayer, W.L. 1954. Description of a trap for *Mansonia* larvae. Mosquito News 14:55-58.

Blanchard, R. 1905. Les moustiques. Histoire naturelle et médicale. Paris. 673 pp., 316 figs.

Boddy, D.W. 1948. An annotated list of the Culicidae of Washington. Pan-Pacif. Ent. 24(2):85-94.

Bohart, R.M. 1950. A new species of *Orthopodomyia* from California (Diptera, Culicidae). Ann. ent. Soc. Am. 43:399-404.

Bohart, R.M. 1954. Identification of first stage larvae of California *Aedes* (Diptera, Culicidae). Ann. ent. Soc. Am. 47:355-366.

Bohart, R.M., and R.K. Washino. 1978. Mosquitoes of California. 3rd ed. Univ. of Calif., Div. Agric. Sci. Publ. 4804. 153 pp.

Bradley, G.H., and R.F. Fritz. 1945. Observations of the seasonal occurrence and abundance of *Anopheles quadrimaculatus* Say. J. natn. Malar. Soc. 4:251-262.

Breeland, S.G., and E. Pickard. 1963. Life history studies on artificially produced broods of floodwater mosquitoes in the Tennessee Valley. Mosquito News 23:75-85.

Breeland, S.G., and E. Pickard. 1964. Insectary studies on longevity, blood feeding, and oviposition behavior of four floodwater mosquito species in the Tennessee Valley. Mosquito News 24:186-196.

Breeland, S.G., and E. Pickard. 1965. The Malaise trap—an efficient and unbiased mosquito collecting device. Mosquito News 25:19-21.

Breeland, S.G., W.E. Snow, and E. Pickard. 1961. Mosquitoes of the Tennessee Valley. J. Tenn. Acad. Sci. 36:249-319.

Breland, O.P. 1947. Variations in the larvae of the mosquito *Orthopodomyia alba* Baker (Diptera, Culicidae). Bull. Brooklyn ent. Soc. 42:81-86.

Breland, O.P. 1948. Notes on some carnivorous mosquito larvae. Ent. News 59:156-157.

Breland, O.P. 1949. The biology and the immature stages of the mosquito, *Megarhinus septentrionalis* Dyar and Knab. Ann. ent. Soc. Am. 42:38-47.

Breland, O.P. 1960. Restoration of the name *Aedes hendersoni* Cockerell, and its elevation to full specific rank (Diptera: Culicidae). Ann. ent. Soc. Am. 53:600-606.

Britton, W.E. 1912. The mosquito plaque of the Connecticut coast region and how to control it. Bull. Conn. agric. Exp. Stn 173:1-14.

Brooks, I.C. 1946. Tree-hole mosquitoes in Tippecanoe County, Indiana. Proc. Indiana Acad. Sci. 56:154-156.

Brower, L.P. 1953. The distribution of *"Mansonia perturbans"* (Walker) in Morris County. Proc. 40th A. Meeting New Jers. Mosq. Exterm. Ass. 58:147-149.

Brust, R.A. 1968. Temperature-induced intersexes in *Aedes* mosquitoes: comparative study of species from Manitoba. Can. Ent. 100:879-891.

Brust, R.A. 1971. Laboratory mating in *Aedes diantaeus* and *Aedes communis* (Diptera: Culicidae). Ann. ent. Soc. Am. 64:234-237.

Brust, R.A. 1974. Reproductive isolation within the *Aedes atropalus* group, and description of eggs. J. med. Ent. 11:459-466.

Brust, R.A., and R.A. Costello. 1969. Mosquitoes of Manitoba. II. The effect of storage temperature and relative humidity on hatching of eggs of *Aedes vexans* and *Aedes abserratus* (Diptera: Culicidae). Can. Ent. 101:1285-1291.

Brust, R.A., and W.R. Horsfall. 1965. Thermal stress and anomalous development of mosquitoes (Diptera: Culicidae). IV. *Aedes communis*. Can. J. Zool. 43:17-53.

Brust, R.A., and K.S. Kalpage. 1967. New records for *Aedes* species in Manitoba. Mosquito News 27:117-118.

Burgess, L. 1957. Note on *Aedes melanimon* Dyar, a mosquito new to Canada (Diptera: Culicidae). Can. Ent. 89:532.

Burgess, L., and W.O. Haufe. 1960. Stratification of some prairie and forest mosquitoes in the lower air. Mosquito News 20:341-346.

Burgess, R.W. 1948. The experimental hybridization of *Anopheles quadrimaculatus* Say and *Anopheles maculipennis freeborni* Aitken. Am. J. Hyg. 48:171-172.

Burgess, R.W. 1955. Experiments in hybridizing *Anopheles freeborni* Aitken and *Anopheles punctipennis* (Say). Ann. ent. Soc. Am. 48:229-231.

Callot, J., and Dao Van Ty. 1944. Contribution à l'étude des moustiques français culicides de Richelieu (Indre-et-Loire). Annls Parasit. hum. comp. 20:43-66.

Cameron, A.E. 1918. Blood sucking flies in Saskatchewan. Agric. Gaz. Can. 5:556-561.

Carpenter, M.J., and L.T. Nielsen. 1965. Ovarian cycles and longevity in some univoltine *Aedes* species in the Rocky Mountains of western United States. Mosquito News 25:127-134.

Carpenter, S.J. 1962. Observations on the distribution and ecology of mountain *Aedes* mosquitoes in California. VI. *Aedes increpitus* Dyar. Calif. Vector Views 9(8):39-43.

Carpenter, S.J. 1968. Review of recent literature on mosquitoes of North America. Calif. Vector Views 15(8):71-98.

Carpenter, S.J. 1970*a*. Observations on the distribution and ecology of mountain *Aedes* mosquitoes in California. XIV. Mosquito problems in the Bishop Creek recreational region. Calif. Vector Views 17(3):13-17.

Carpenter, S.J. 1970*b*. Review of recent literature on mosquitoes of North America. Supplement I. Calif. Vector Views 17(6):39-65.

Carpenter, S.J. 1970*c*. Observations on the distribution and ecology of mountain *Aedes* mosquitoes in California. XV. Mosquito problems in the Yosemite National Park recreational region. Calif. Vector Views 17(11):101-107.

Carpenter, S.J. 1974. Review of recent literature on mosquitoes of North America. Supplement II. Calif. Vector Views 21(12):73-99.

Carpenter, S.J. and W.J. LaCasse. 1955. Mosquitoes of North America (north of Mexico). Univ. of California Press, Berkeley and Los Angeles. vi, 360 pp., 127 pls.

Carpenter, S.J., W.W. Middlekauft, and R.W. Chamberlain. 1946. The mosquitoes of the southern United States east of Oklahoma and Texas. Am. Midl. Nat. Monogr. No 3. 292 pp.

Chamberlain, R.W., W.D. Sudia, and D.B. Nelson. 1955. Laboratory observations on a mosquito, *Culiseta melanura* (Coquillett). Mosquito News 15:18-21.

Chant, G.D., W.F. Baldwin, and L. Forster. 1973. Occurrence of *Culiseta melanura* (Diptera: Culicidae) in Canada. Can. Ent. 105:1359.

Chao, J. 1958. An autogenous strain of *Culex tarsalis* Coq. Mosquito News18:134-136.

Chapman, H.C. 1959. Overwintering of larval populations of *Culex erythrothorax* in Nevada. Mosquito News 19:244-246.

Chapman, H.C. 1961. Additional records and observations on Nevada mosquitoes. Mosquito News 21:136-138.

Chapman, H.C. 1962. The bio-ecology of *Culex erythrothorax* Dyar. Mosquito News 22:130-134.

Chapman, H.C. 1964. Observations on the biology and ecology of *Orthopodomyia californica* Bohart (Diptera: Culicidae). Mosquito News 24:432-439.

Chapman, H.C., and A.R. Barr. 1969. Techniques for successful colonization of many mosquito species. Mosquito News 29:532-535.

Chapman, H.C., and D.B. Woodward. 1965. Blood feeding and oviposition of some floodwater mosquitoes in Louisiana: Laboratory studies. Mosquito News 25:259-262.

Cockerell, T.D.A. 1918. The mosquitoes of Colorado. J. econ. Ent. 11:195-200.

Cook, F.E., M.M. Bodine, and E.P. Wermerskirchen. 1974. Some further notes on *Aedes* (*Ochlerotatus*) *sticticus* (Meigen) in Minnesota. Mosquito News 34:239.

Coquillett, D.W. 1896. New Culicidae from North America. Can. Ent. 28:43-44.

Coquillett, D.W. 1901. Three new species of Culicidae. Can. Ent. 33:258-260.

Coquillett, D.W. 1902*a*. Three new species of *Culex*. Can. Ent. 34:292-293.

Coquillett, D.W. 1902*b*. New forms of Culicidae from North America. Jl N.Y. ent. Soc. 10:191-194.

Coquillett, D.W. 1902*c*. New Diptera from North America. Proc. U.S. natn. Mus. (1903) 25:83-126.

Coquillett, D.W. 1903*a*. Four new species of *Culex*. Can. Ent. 35:255-257.

Coquillett, D.W. 1903*b*. A new *Anopheles* with unspotted wings. Can. Ent. 35:310.

Coquillett, D.W. 1904. New North American Diptera. Proc. ent. Soc. Wash. 6:166-192.

Coquillett, D.W. 1906. A new *Culex* near *curriei*. Ent. News 17:109.

Corbet, P.S. 1964. Autogeny and oviposition in arctic mosquitoes. Nature (Lond.) 203(4945):668.

Corbet, P.S. 1965. Reproduction in mosquitoes of the High Arctic. Proc. int. Congr. Ent. (Lond.) 12:817-818.

Corbet, P.S. 1966. Diel patterns of mosquito activity in a high arctic locality: Hazen Camp, Ellesmere Island, N.W.T. Can. Ent. 98:1238-1252.

Corbet, P.S. 1967. Facultative autogeny in arctic mosquitoes. Nature (Lond.) 215(5101): 662-663.

Corbet, P.S., and H.V. Danks. 1973. Seasonal emergence and activity of mosquitoes (Diptera: Culicidae) in a high-arctic locality. Can. Ent. 105:837-872.

Corbet, P.S., and H.V. Danks. 1975. Egg-laying habits of mosquitoes in the high arctic. Mosquito News 35:8-14.

Corbet, P.S., and A.E.R. Downe. 1966. Natural hosts of mosquitoes in northern Ellesmere Island. Arctic 19:153-161.

Crampton, G.C. 1942. The external morphology of the Diptera. *In* Guide to the insects of Connecticut. Part VI. The Diptera or true flies of Connecticut. First fasc. Bull. Conn. State Geol. Nat. Hist. Surv. 64:10-165.

Crane, G.T., R.E. Elbel, and C.H. Calisher. 1977. Transovarial transmission of California encephalitis virus in the mosquito *Aedes dorsalis* at Blue Lake, Utah. Mosquito News 37:479-482.

Crans, W.J. 1969. Preliminary observations of frog filariasis in New Jersey. Bull. Wildlife Disease Ass. 5:342-347.

Crans, W.J. 1970. The blood feeding habits of *Culex territans* Walker. Mosquito News 30:445-447.

Cupp, E.W., and W.R. Horsfall. 1969. Biological bases for placement of *Aedes sierrensis* (Ludlow) in the subgenus *Finlaya* Theobald. Mosquito Syst. News Lett. 1:51-52.

Curtis, L.C. 1967. The mosquitoes of British Columbia. B.C. Provincial Mus. Occasional Paper No. 15. 90 pp.

Danilov, V.N. 1974. Restoration of the name *Aedes* (*O.*) *mercurator* Dyar to the mosquito known in the U.S.S.R. as *Aedes riparius ater* Gutsevich (Diptera, Culicidae) [in Russian]. Parazitologiya (Leningr.) 8(4):322-328.

Danks, H.V., and P.S. Corbet. 1973. A key to all stages of *Aedes nigripes* and *A. impiger* (Diptera: Culicidae) with a description of first-instar larvae and pupae. Can. Ent. 105:367-376.

Darsie, R.F., Jr. 1949. Pupae of the anopheline mosquitoes of the northeastern United States (Diptera, Culicidae). Revta Ent. 20(1-3):509-530.

Darsie, R.F., Jr. 1951. Pupae of the culicine mosquitoes of the northeastern United States (Diptera, Culicidae, Culicini). Memoir 304. Cornell Univ. Agric. Exp. Stn. Ithaca, N.Y. 67 pp.

Darsie, R.F., Jr. 1955. Notes on American mosquito pupae, I. Description of *Aedes riparius* and *Aedes pionips* (Diptera, Culicidae). Proc. ent. Soc. Wash. 57:23-29.

Darsie, R.F., Jr. 1957. Notes on American mosquito pupae, II. The *Aedes* (*Ochlerotatus*) *punctor* subgroup, with key to known Nearctic *Aedes* pupae (Diptera, Culicidae). Ann. ent. Soc. Am. 50:611-620.

Darsie, R.F., Jr. 1973. A record of changes in mosquito taxonomy in the United States of America, 1955–1972. Mosquito Syst. News Lett. 5:187-193.

Darsie, R.F. 1978. Additional changes in mosquito taxonomy in North America, north of Mexico, 1972-1977. Mosquito Syst. 10:246-248.

Darsie, R.F., Jr., E.E. Tindall, and A.R. Barr. 1962. Description of the pupa of *Culiseta melanura* (Diptera: Culicidae). Proc. ent. Soc. Wash. 64:167-170.

Davis, H.G., and H.H. Stage. 1959. Mosquito control activities in the Columbia Basin, Washington, for 1958. Mosquito News 19:62-63.

DeGeer, C. 1776. Mémoires pour servir à l'histoire des Insectes. Vol. 6. Stockholm. 523 pp., 30 pls.

Detinova, T.S. 1968. Age structure of insect populations of medical importance. A. Rev. Ent. 13:427-450.

Dixon, R.D., and R.A. Brust. 1972. Mosquitoes of Manitoba. III. Ecology of larvae in the Winnipeg area. Can. Ent. 104:961-968.

Dodge, H.R. 1963. Studies on mosquito larvae. I. Later instars of eastern North American species. Can. Ent. 95:786-813.

Dodge, H.R. 1966. Studies on mosquito larvae. II. The first-stage larvae of North American Culicidae and of world Anophelinae. Can. Ent. 98:337-393.

Dorsey, C.K. 1944. Mosquito activities at Camp Peary, Virginia. Ann. ent. Soc. Am. 37:376-387.

Downe, A.E.R. 1960. Blood-meal sources and notes on host preferences of some *Aedes* mosquitoes (Diptera: Culicidae). Can. J. Zool. 38:689-699.

Downe, A.E.R. 1962. Some aspects of host selection by *Mansonia perturbans* (Walk.) (Diptera: Culicidae). Can. J. Zool. 40:725-732.

Downes, J.A. 1958. Assembly and mating in the biting Nematocera. Proc. 10th int. Congr. Ent. (1956) 2:425-434.

Dyar, H.G. 1904*a*. The larva of *Culex punctor* Kirby, with notes on an allied form. Jl N.Y. ent. Soc. 12:169-171.

Dyar, H.G. 1904*b*. Brief notes on mosquito larvae. Jl N.Y. ent. Soc. 12:243-246.

Dyar, H.G. 1905. Remarks on genitalic genera in the Culicidae. Proc. ent. Soc. Wash. 7:42-49.

Dyar, H.G. 1916. New *Aedes* from the mountains of California (Diptera, Culicidae). Insecutor Inscit. menstr. 4:80-90.

Dyar, H.G. 1917*a*. The mosquitoes of the mountains of California (Diptera, Culicidae). Insecutor Inscit. menstr. 5:11-21.

Dyar, H.G. 1917*b*. The mosquitoes of the Pacific Northwest (Diptera, Culicidae). Insecutor Inscit. menstr. 5:97-102.

Dyar, H.G. 1917*c*. Notes on the *Aedes* of Montana (Diptera, Culicidae). Insecutor Inscit. menstr. 5:104-121.

Dyar, H.G. 1917*d*. A new *Aedes* from the Rocky Mountain region (Diptera, Culicidae). Insecutor Inscit. menstr. 5:127-128.

Dyar, H.G. 1918. New American mosquitoes (Diptera, Culicidae). Insecutor Inscit. menstr. 6:120-129.

Dyar, H.G. 1919*a*. The mosquitoes collected by the Canadian Arctic Expedition, 1913–18 (Diptera, Culicidae) *in* R.M. Anderson, ed. Report of the Canadian Arctic Expedition, 1913–18. Vol. 3. Insects. Pt. C: Diptera. Ottawa. 90 pp.

Dyar, H.G. 1919*b*. Westward extension of the Canadian mosquito fauna. Insecutor Inscit. menstr. 7:11-39.

Dyar, H.G. 1920*a*. The mosquitoes of British Columbia and Yukon Territory, Canada (Diptera, Culicidae). Insecutor Inscit. menstr. 8:1-27.

Dyar, H.G. 1920*b*. The American mosquitoes of the *stimulans* group (Diptera, Culicidae). Insecutor Inscit. menstr. 8:106-120.

Dyar, H.G. 1920*c*. The *Aedes* of the mountains of California and Oregon (Diptera, Culicidae). Insecutor Inscit. menstr. 8:165-173.

Dyar, H.G. 1921*a*. The mosquitoes of Canada. Trans. R. Can. Inst. 13:71-120.

Dyar, H.G. 1921*b*. The American *Aedes* of the *punctor* group (Diptera, Culicidae). Insecutor Inscit. menstr. 9:69-80.

Dyar, H.G. 1922*a*. New mosquitoes from Alaska (Diptera, Culicidae). Insecutor Inscit. menstr. 10:1-3.

Dyar, H.G. 1922*b*. The mosquitoes of the United States. Proc. U.S. natn. Mus. 62:1-119.

Dyar, H.G. 1923. Note on the swarming of *Aedes cinereoborealis* Felt and Young. Insecutor Inscit. menstr. 11:56-57.

Dyar, H.G. 1924*a*. Two new mosquitoes from California (Diptera, Culicidae). Insecutor Inscit. menstr. 12:125-127.

Dyar, H.G. 1924*b*. Note on *Aedes aloponotum* and other species of its region (Diptera, Culicidae). Insecutor Inscit. menstr. 12:176-179.

Dyar, H.G. 1924*c*. The American forms of *Aedes cinereus* Meigen (Diptera, Culicidae). Insecutor Inscit. menstr. 12:179-180.

Dyar, H.G. 1928. The mosquitoes of the Americas. Carnegie Inst. Wash. Publ. 387:1-616.

Dyar, H.G. 1929. A new species of mosquito from Montana with annotated list of the species known from the state. Proc. U.S. natn. Mus. 75(23):1-8.

Dyar, H.G., and F. Knab. 1906*a*. Diagnoses of new species of mosquitoes. Proc. biol. Soc. Wash. 19:133-141.

Dyar, H.G., and F. Knab. 1906*b*. Notes on some American mosquitoes with descriptions of new species. Proc. biol. Soc. Wash. 19:159-172.

Dyar, H.G., and F. Knab. 1906*c*. The larvae of Culicidae classified as independent organisms. Jl N.Y. ent. Soc. 14:169-230.

Dyar, H.G., and F. Knab. 1906*d*. The species of mosquitoes in the genus *Megarhinus*. Smithson. misc. Collns (1907) 48:241-258.

Dyar, H.G., and F. Knab. 1907*a*. New American mosquitoes. Jl N.Y. ent. Soc. 15:100-101.

Dyar, H.G., and F. Knab. 1907*b*. Descriptions of three new North American mosquitoes. Jl N.Y. ent. Soc. 15:213-214.

Dyar, H.G., and F. Knab. 1908. Descriptions of some new mosquitoes from tropical America. Proc. U.S. natn. Mus. 35:53-70.

Dyar, H.G., and F. Knab. 1909*a*. Mosquito comment. Can. Ent. 41:101-102.

Dyar, H.G., and F. Knab. 1909*b*. Descriptions of some new species and a new genus of American mosquitoes. Smithson. misc. Collns 52:253-266.

Dyar, H.G., and F. Knab. 1910. Description of three new American mosquitoes (Diptera, Culicidae). Proc. ent. Soc. Wash. (1909) 11:173-174.

Dyar, H.G., and F. Knab. 1918. New American mosquitoes (Diptera, Culicidae). Insecutor Inscit. menstr. (1917) 5:165-169.

Dyar, H.G., and R.C. Shannon. 1925. The mosquitoes of Peary's North Pole expedition of 1908 (Diptera, Culicidae). J. Wash. Acad. Sci. 15:77-78.

Eads, R.B. 1965. Biological notes on *Culex tarsalis* in the Lower Rio Grande Valley of Texas. Mosquito News 25:61-63.

Edman, J.D., and A.E.R. Downe. 1964. Host-blood sources and multiple feeding habits of mosquitoes in Kansas. Mosquito News 24:154-160.

Edmunds, L.R. 1955. Notes on the biology and seasonal abundance of the larval stages of *Culex tarsalis* Coquillett in irrigated areas of Scotts Bluff County, Nebraska (Diptera: Culicidae). Mosquito News 15:157-160.

Edwards, F.W. 1932. Diptera. Fam. Culicidae. Genera Insectorum. Fasc. 194. P. Wytsman, Brussels. 258 pp.

Ellis, R.A., and R.A. Brust. 1973. Sibling species delimitation in the *Aedes communis* (DeGeer) aggregate (Diptera: Culicidae). Can. J. Zool. 51:915-959.

Ellis, R.A., and D.M. Wood. 1974. First Canadian record of *Corethrella brakeleyi* (Diptera: Chaoboridae). Can. Ent. 106:221-222.

Enfield, M.A. 1977. Additions and corrections to the records of *Aedes* mosquitoes in Alberta. Mosquito News 37:82-85.

Evans, K.W. 1971. Bionomics of *Wyeomyia smithii* (Coquillett) the pitcherplant mosquito (Diptera; Culicidae; Sabethini). M.Sc. Thesis, Univ. of Man. 279 pp.

Evans, K.W., and R.A. Brust. 1972. Induction and termination of diapause in *Wyeomyia smithii* (Diptera: Culicidae), and larval survival studies at low and subzero temperatures. Can. Ent. 104:1937-1950.

Fabricius, J.C. 1794. Entomologica systematica emendata et aucta. Vol. 4. Hafniae, Copenhagen. 472 pp.

Felt, E.P. 1904. Mosquitoes or Culicidae of New York State. N. Y. State Mus. Bull. 79:239-400, 113 figs., 57 pls.

Felt, E.P. 1905*a*. *Culex brittoni* n. sp. Ent. News 16:79-80.

Felt, E.P. 1905*b*. Studies in Culicidae. Pages 442-497 *in* E.P. Felt, 20th report of the State Entomologist on injurious and other insects of the State of New York, 1904. N. Y. State Mus. Bull. 97:359-564.

Felt, E.P., and D.B. Young. 1904. Importance of isolated rearings from culicid larvae. Science (Wash. D.C.) 20:312-313.

Fitch, A. 1847. Winter insects of eastern New York. Am. J. Agric. Sci. 5:274-284.

Freeborn, S.B. 1926. The mosquitoes of California. Univ. Calif. Publ. Ent. 3:333-460.

Freeborn, S.B., and R.M. Bohart. 1951. The mosquitoes of California.

Freeman, T.N. 1952. Interim report of the distribution of the mosquitoes obtained in the Northern Insect Survey. Defense Research Board of Ottawa. Tech. Rep. 1. 2 pp., 43 maps.

Frohne, W.C. 1951. Seasonal incidence of mosquitoes in the Upper Cook Inlet, Alaska. Mosquito News 11:213-216.

Frohne, W.C. 1952. Mosquito news from Alaska. Mosquito News 12:263.

Frohne, W.C.1953*a*. Mosquito breeding in Alaskan salt marshes, with especial reference to *Aedes punctodes* Dyar. Mosquito News 13:96-103.

Frohne, W.C. 1953*b*. Natural history of *Culiseta impatiens* (Walker) (Diptera, Culicidae) in Alaska. Trans. Am. microsc. Soc. 72:103-118.

Frohne, W.C. 1954*a*. Mosquito distribution in Alaska with especial reference to a new type of life cycle. Mosquito News 14:10-13.

Frohne, W.C. 1954*b*. Biology of an Alaskan mosquito, *Culiseta alaskaensis* (Ludlow). Ann. ent. Soc. Am. 47:9-24.

Frohne, W.C. 1956. The biology of northern biting flies. U.S. public Health Serv. Rep. 71:616-621.

Frohne, W.C. 1957. Reconnaissance of mountain mosquitoes in the McKinley Park Region, Alaska. Mosquito News 17:17-22.

Frohne, W.C. 1959. Predation of dance flies (Diptera: Empididae) upon mosquitoes in Alaska, with especial reference to swarming. Mosquito News 19:7-11.

Frohne, W.C., and R.G. Frohne. 1952. Mating swarms of males of the mosquito *Aedes punctor* (Kirby), in Alaska. Mosquito News 12:248-251.

Frohne, W.C., and R.G. Frohne. 1954. Diurnal swarms of *Culex territans* Walker, and the crepuscular swarming of *Aedes* about a small glade in Alaska. Mosquito News 14:62-64.

Frohne, W.C., and D.A. Sleeper. 1951. Reconnaissance of mosquitoes, punkies and blackflies in southeast Alaska. Mosquito News 11:209-213.

Furlow, B.M., and K.L. Hays. 1972. Some influences of aquatic vegetation of the species and number of Culicidae (Diptera) in small pools of water. Mosquito News 32:595-599.

Gebhardt, L.P., G.J. Stanton, D.W. Hill, and G.C. Collett. 1964. Natural overwintering hosts of the virus of western equine encephalitis. New Engl. J. Med. 4:172-177.

Gerberg, E.J. 1970. Manual for mosquito rearing and experimental techniques. Bull. Am. Mosq. Control Ass. 5:1-109.

Gibson, A. 1926. Anti-mosquito work in Canada. Proc. New Jers. Mosq. Exterm. Ass. 13:54-59.

Gibson, A. 1928. Mosquito suppression in Canada in 1927. Proc. New Jers. Mosq. Exterm. Ass. 15:136-146.

Gibson, A. 1929. Mosquito suppression in Canada in 1928. Proc. New Jers. Mosq. Exterm. Ass. 16:102-108.

Gibson, A. 1932. Mosquito investigations in Canada during 1931. Proc. Que. Soc. Prot. Plants, 1931. 12 pp.

Gibson, A. 1933. Mosquito suppression work in Canada in 1932. Proc. New Jers. Mosq. Exterm. Ass. 20:92-102.

Gibson, A. 1935. Mosquito suppression work in Canada in 1934. Proc. New Jers. Mosq. Exterm. Ass. 22:77-92.

Gibson, A. 1936. Mosquito suppression work in Canada in 1935. Proc. New Jers. Exterm. Ass. 23:88-98.

Gibson, A. 1937. Mosquito suppression work in Canada in 1936. Proc. New Jers. Mosq. Exterm. Ass. 24:96-108.

Giles, G.M. 1900. A handbook of the gnats or mosquitoes giving the anatomy and life history of the Culicidae. London. 374 pp., 16 figs., 7 pls.

Gjullin, C.M., and G.W. Eddy. 1972. The mosquitoes of the northwestern United States. U.S. Dep. Agric. Tech. Bull. No. 1447. 111 pp.

Gjullin, C.M., L.F. Lewis, and D.M. Christenson. 1968. Notes on the taxonomic characters and distribution of *Aedes aloponotum* Dyar and *Aedes communis* (DeGeer) (Diptera: Culicidae). Proc. ent. Soc. Wash. 70:133-136.

Gjullin, C.M., R.I. Sailer, A. Stone, and B.V. Travis. 1961. The mosquitoes of Alaska. U.S. Dep. Agric. Agric. Handb. 182:1-98.

Gjullin, C.M., and W.W. Yates. 1945. *Anopheles* and malaria in the northwestern States. Mosquito News 5:121-127.

Gjullin, C.M., W.W. Yates, and H.H. Stage. 1950. Studies on *Aedes vexans* (Meig.) and *Aedes sticticus* (Meig.), flood water mosquitoes in the lower Columbia River Valley. Ann. ent. Soc. Am. 43:262-275.

Good, N.E. 1945. A list of the mosquitoes of the District of Columbia. Proc. ent. Soc. Wash. 47:168-179.

Graham, P. 1969*a*. A comparison of sampling methods for adult mosquito populations in central Alberta, Canada. Quaest. Ent. 5:217-261.

Graham, P. 1969*b*. Observations on the biology of the adult female mosquitoes (Diptera: Culicidae) at George Lake, Alberta, Canada. Quaest. Ent. 5:309-339.

Graham, P. 1969*c*. *Culiseta silvestris minnesotae* Barr and *C. morsitans dyari* (Coquillett) (Diptera: Culicidae) in Alberta. Mosquito News 29:261-262.

Grimstad, P.R. 1977. Occurrence of *Orthopodomyia alba* Baker and *Orthopodomyia signifera* (Coquillett) in Michigan. Mosquito News 37:129-130.

Grimstad, P.R., C.E. Garry, and G.R. DeFoliart. 1974. *Aedes hendersoni* and *Aedes triseriatus* (Diptera: Culicidae) in Wisconsin: characterization of larvae, larval hybrids and comparison of adult and hybrid mesoscutal patterns. Ann. ent. Soc. Am. 67:797-804.

Grossbeck, J.A. 1904*a*. Description of a new *Culex*. Can. Ent. 36:332.

Grossbeck, J.A. 1904*b*. Description of two new species of *Culex*. Ent. News 15:332-333.

Grossbeck, J.A. 1905. New species of Culicidae. Can. Ent. 37:359-360.

Grossbeck, J.A. 1906. Notes on *Culex squamiger* Coq., with description of a closely allied species. Can. Ent. 38:129-131.

Gutsevitch, A.V., A.S. Monchadskii, and A.A. Shtakelberg. 1974. Fauna of the U.S.S.R. Diptera. Vol. 3, No. 4. Mosquitoes Family Culicidae [translation from Russian]. Keter Press, Jerusalem. 408 pp.

Haeger, J.S. 1955. The non-blood feeding habits of *Aedes taeniorhynchus* (Diptera, Culicidae) on Sanibel Island, Florida. Mosquito News 15:21-26.

Hagmann, L.E. 1952. *Mansonia perturbans* recent studies in New Jersey. Proc. New Jers. Mosq. Exterm. Ass. 30:60-65.

Hagmann, L.E. 1953. Biology of "*Mansonia perturbans*" (Walker). Proc. 40th A. Meeting New Jers. Mosq. Exterm. Ass. 58:141-147.

Hammon, W.McD., W.C. Reeves, S.R. Benner, and B. Brookman. 1945. Human encephalitis in the Yakima Valley, Washington, 1942, with forty-nine virus isolations (western equine and St. Louis types) from mosquitoes. J. Am. med. Ass. 128:1133-1139.

Happold, D.C.D. 1965*a*. Mosquito ecology in Central Alberta. I. The environment, the species, and studies of the larvae. Can. J. Zool. 43:795-819.

Happold, D.C.D. 1965*b*. Mosquito ecology in Central Alberta. II. Adult populations and activities. Can. J. Zool. 43:821-846.

Harbach, R.E., and K.L. Knight. 1978. A mosquito taxonomic glossary. XIV. The larval body (except chaetotaxy). Mosquito Syst. 10:53-105.

Harmston, F.C. 1969. Separation of the females of *Aedes hendersoni* Cockerell and *Aedes triseriatus* (Say) (Diptera: Culicidae) by the tarsal claws. Mosquito News 29:490-491.

Harmston, F.C., and F.A. Lawson. 1967. Mosquitoes of Colorado. U.S. Dep. Health Educ. Welfare. 140 pp.

Harrison, R.J., and G. Cousineau. 1973. Les moustiques au Québec, leur importance médicale, vétérinaire, économique et la nécessité d'un programme de démonstration. Ann. ent. Soc. Queb. 18:138-146.

Harwood, R.F. 1962. Trapping overwintering adults of the mosquitoes *Culex tarsalis* and *Anopheles freeborni*. Mosquito News 22:26-31.

Haufe, W.O. 1952. Observations on the biology of mosquitoes (Diptera: Culicidae) at Goose Bay, Labrador. Can. Ent. 84:254-263.

Haufe, W.O., and L. Burgess. 1956. Development of *Aedes* (Diptera: Culicidae) at Fort Churchill, Manitoba, and prediction of dates of emergence. Ecology 37:500-519.

Hayes, R.O. 1958. Observations on the swarming of *Culiseta melanura* Coquillett. Mosquito News 18:70-73.

Hayes, R.O. 1961. Host preferences of *Culiseta melanura* and allied mosquitoes. Mosquito News 21:179-187.

Hayes, R.O., and O.W. Doane, Jr. 1958. Primary record or *Culiseta melanura* biting man in nature. Mosquito News 18:216-217.

Hayes, R.O., L.C. Lamotte, L.A. White, and L.D. Beadle. 1960. Isolation of eastern encephalitis virus from the mosquito *Culex restuans* collected in New Jersey during 1959. Mosquito News 20:190.

Hayes, R.O., and H.B. Morlan. 1957. Notes on *Aedes triseriatus* egg incubation and colonization. Mosquito News 17:33-36.

Headlee, T.J. 1916. Some recent advances in knowledge of the natural history and the control of mosquitoes. Bull. New Jers. agric. Exp. Stn 306:1-26.

Headlee, T.J. 1945. The mosquitoes of New Jersey and their control. Rutgers Univ. Press, New Brunswick, N.J. 326 pp.

Hearle, E. 1920. Notes on some mosquitoes new to Canada. Can. Ent. 52:114-116.

Hearle, E. 1923. A new mosquito from British Columbia (Culicidae, Diptera). Can. Ent. 55:4-5.

Hearle, E. 1926. The mosquitoes of the Lower Fraser Valley, British Columbia, and their control. Natn. Res. Counc. Can. Rep. 17:1-94.

Hearle, E. 1927*a*. List of mosquitoes of British Columbia. Proc. ent. Soc. Br. Columb. 24:11-19.

Hearle, E. 1927*b*. Notes on the occurrence of *Aedes (Ochlerotatus) nearcticus* Dyar in the Rocky Mountain Park, Alberta (Culicidae, Diptera). Can. Ent. 59:61-63.

Hearle, E. 1927*c*. A new Canadian mosquito. Can. Ent. 59:101-103.

Hearle, E. 1929. The life history of *Aedes flavescens* Müller. Trans. R. Soc. Can., Third Ser. 23:85-101.

Hedeen, R.A. 1953. The biology of the mosquito *Aedes atropalpus* Coquillett. J. Kans. ent. Soc. 26:1-10.

Helson, B.V., G.A. Surgeoner, R.E. Wright, and S.A. Allan. 1978. *Culex tarsalis, Aedes sollicitans, Aedes grossbecki*: new distribution records from southwestern Ontario. Mosquito News 38:137-138.

Hennig, W. 1973. Diptera (Zweiflügler). *In* Handbuch der Zoologie. 2 Aufl., Bd. 4, Hfte. 2, T. 2/31. 337 pp., 20 figs.

Hocking, B. 1952. Autolysis of flight muscles in a mosquito. Nature (Lond.) 169:1101.

Hocking, B. 1954. Flight muscle autolysis in *Aedes communis* (DeGeer). Mosquito News 14:121-123.

Hocking, B., W.R. Richards, and C.R. Twinn. 1950. Observations on the bionomics of some northern mosquito species (Culicidae: Diptera). Can. J. Res., D, 28: 58-80.

Holmberg, R.G., and D. Trofimenkoff. 1968. *Aedes melanimon* in Saskatchewan. Mosquito News 28:651-652.

Horsfall, W.R. 1937. Mosquitoes of southeastern Arkansas. J. econ. Ent. 30:743-748.

Horsfall, W.R. 1939. Habits of *Aedes thibaulti* Dyar and Knab (Diptera, Culicidae). J. Kans. ent. Soc. 12:70-71.

Horsfall, W.R. 1942. Breeding habits of a rice field mosquito. J. econ. Ent. 35:478-482.

Horsfall, W.R. 1955. Mosquitoes their bionomics and relation to disease. Ronald Press Co., New York. 723 pp.

Horsfall, W.R., and H.W. Fowler. 1961. Eggs of floodwater mosquitoes. VIII. Effect of serial temperatures on conditioning of eggs of *Aedes stimulans* Walker (Diptera: Culicidae). Ann. ent. Soc. Am. 54:664-666.

Howard, L.O., H.G. Dyar, and F. Knab. 1913. The mosquitoes of North and Central America and the West Indies. Vol. 2. Carnegie Inst. Wash. Publ. (1912) 159(2):1-10, 150 pls.

Howard, L.O., H.G. Dyar, and F. Knab. 1915. The mosquitoes of North and Central America and the West Indies. Systematic description (in two parts). Vol. 3. Part I. Carnegie Inst. Wash. Publ. 159(3):523 pp.

Howard, L.O., H.G. Dyar, and F. Knab. 1917. The mosquitoes of North and Central America and the West Indies. Systematic description. Vol. 3, Part II. Carnegie Inst. Wash. Publ. 159(4):524-1064.

Howden, H.F., J.E.H. Martin, E.L. Bousfield, and D.E. McAllister. 1970. Fauna of Sable Island and its zoogeographic affinities—a compendium. Natn. Mus. Nat. Sci. (Ottawa) Publ. Zool. 4:1-45.

Howitt, B.F., H.R. Dodge, L.K. Bishop, and R.H. Gorrie. 1949. Recovery of the virus of eastern equine encephalomyelitis from mosquitoes (*Mansonia perturbans*) collected in Georgia. Science (Wash. D.C.) 110:141-142.

Hsaio, T., and R.M. Bohart. 1946. The mosquitoes of Japan and their medical importance. Bull. U.S. Nav. med. Wash. 1095:1-44.

Hubert, A.A. 1953. Observations on the continuous rearing of *Culiseta incidens* (Thomson). Mosquito News 13:207-208.

Hudson, A. 1970. Notes on the piercing mouthparts of three species of mosquitoes (Diptera: Culicidae) viewed with the scanning electron microscope. Can. Ent. 102:501-509.

Hudson, A., and J. McLintock. 1967. A chemical factor that stimulates oviposition by *Culex tarsalis* Coquillett (Diptera, Culicidae). Anim. Behav. 15:336-341.

Huffaker, C.B. 1944. The temperature relations of the immature stages of the malarial mosquito, *Anopheles quadrimaculatus* Say, with a comparison of the developmental power of constant and variable temperatures in insect metabolism. Ann. ent. Soc. Am. 37:1-27.

Humboldt, F.H.A. von. 1819. Voyage aux régions équinoxiales du Nouveau Continent, fait en 1799, 1800, 1801, 1802, 1803 et 1804 par A. de Humboldt et A. Bonpland. Vol. 2. 722 pp. *In* F.H.A. Humboldt, ed. Voyage de M.M. Alexandre de Humboldt et Aimé Bonpland. Pt. I: Relation historique. Paris.

Hurlbut, H.S. 1938. Further notes on the overwintering of the eggs of *Anopheles walkeri* Theobald with a description of the eggs. J. Parasit. 24:521-526.

Intermill, R.W. 1973. Development of *Dirofilaria immitis* in *Aedes triseriatus* Say. Mosquito News 33:176-181.

Irwin, W.H. 1941. A preliminary list of the Culicidae of Michigan. Part 1. Culicinae (Diptera). Ent. News 52:101-105.

James, H.G. 1960. Note on effects of drought on larvae of *Aedes vexans* Meigen (Diptera: Culicidae). Mosquito News 20:320-321.

James, H.G. 1964. Insect and other fauna associated with the rock pool mosquito *Aedes atropalpus* (Coq.). Mosquito News 24:325-329.

Jamnback, H. 1961. *Culiseta melanura* (Coq.) breeding on Long Island, N.Y. Mosquito News 21:140-141.

Jenkins, D.W. 1948. Ecological observations on the mosquitoes of central Alaska. Mosquito News 8:140-147.

Jenkins, D.W. 1950. Bionomics of *Culex tarsalis* in relation to western equine encephalomyelitis. Am. J. trop. Med. 30:909-916.

Jenkins, D.W., and S.J. Carpenter. 1946. Ecology of the tree hole breeding mosquitoes of nearctic North America. Ecol. Monogr. 16:33-48.

Jenkins, D.W., and K.L. Knight. 1950. Ecological survey of the mosquitoes of Great Whale River, Quebec (Diptera, Culicidae). Proc. ent. Soc. Wash. 52:209-223.

Jenkins, D.W., and K.L. Knight. 1952. Ecological survey of the mosquitoes of southern James Bay. Am. Midl. Nat. 47:456-468.

Judd, W.W. 1950. Mosquitoes collected in the vicinity of Hamilton, Ontario, during the summer of 1948. Mosquito News 10:57-59.

Judd, W.W. 1954. Results of a survey of mosquitoes conducted at London, Ontario, in 1952 with observations on the biology of the species collected. Can. Ent. 86:101-108.

Judd, W.W. 1959. Studies of the Byron Bog in southwestern Ontario. X. Inquilines and victims of the pitcher-plant, *Sarracenia purpurea* L. Can. Ent. 91:171-180.

Kalpage, K.S.P. 1970. The effect of daylength and temperature on the induction and termination of diapause in *Aedes atropalpus* (Coquillett), and field and laboratory studies of autogeny and hibernation in some mosquitoes from Manitoba. Ph.D. Thesis, University of Manitoba, Winnipeg, Man.

Kalpage, K.S.P., and R.A. Brust. 1968. Mosquitoes of Manitoba. I. Descriptions and a key to *Aedes* eggs (Diptera: Culicidae). Can. J. Zool. 46:699-718.

Kalpage, K.S.P., and R.A. Brust. 1973. Oviposition attractant produced by immature *Aedes atropalpus*. Environ. Ent. 2:729-730.

Kalpage, K.S.P., and R.A. Brust. 1974. Studies on diapause and female fecundity in *Aedes atropalpus*. Environ. Ent. 3:139-145.

Khelevin, N.W. 1958. The effect of environmental factors on the embryonic diapause and on the number of generations during one season in *Aedes caspius dorsalis* Mg. (Diptera: Culicidae) [in Russian, with English summary]. Ent. Obozr. 37:24-46.

King, W.V., G.H. Bradley, C.N. Smith, and W.C. McDuffie. 1960. A handbook of the mosquitoes of the southeastern United States. U.S. Dep. Agric., Agric. Handb. 173:1-188.

Kirby, W. 1837. The insects. Pt. 4. 325 pp., 8 pls. *In* J. Richardson, Fauna Boreali-Americana. Norwick, London.

Klassen, W., and B. Hocking. 1964. The influence of a deep river valley system on the dispersal of *Aedes* mosquitoes. Bull. ent. Res. 55:289-304.

Knab, F. 1908. Observation on the mosquitoes of Saskatchewan. Smithson. misc. Collns 50:540-547.

Knab, F. 1916. A new mosquito from the eastern United States. Proc. biol. Soc. Wash. 29:161-163.

Knight, K.L. 1951. The *Aedes* (*Ochlerotatus*) *punctor* subgroup in North America (Diptera, Culicidae). Ann. ent. Soc. Am. 44:87-99.

Knight, K.L. 1965. Some physical and chemical characteristics of coastal soils underlying mosquito breeding areas. Mosquito News 25:154-159.

Knight, K.L., and T.E. Baker. 1962. The role of the substrate moisture content in the selection of oviposition sites by *Aedes taeniorhynchus* (Wied.) and *A. sollicitans* (Walk.). Mosquito News 22:247-254.

Knight, K.L., and J.L. Laffoon. 1970*a*. A mosquito taxonomic glossary. III. Adult thorax. Mosquito Syst. Newslett. 2:132-148.

Knight, K.L., and J.L. Laffoon. 1970*b*. A mosquito taxonomic glossary. IV. Adult thoracic appendages. Mosquito Syst. Newslett. 2:165-176.

Knight, K.L., and J.L. Laffoon. 1971. A mosquito taxonomic glossary. VIII. The larval chaetotaxy. Mosquito Syst. Newslett. 3:160-194.

Knight, K.L., and A. Stone. 1977. A catalog of the mosquitoes of the world (Diptera: Culicidae). Ent. Soc. Am. (Thomas Say Found.), Wash. 6(2nd ed.): 1-611.

LaCasse, W.J., and S. Yamaguti. 1955. Mosquito fauna of Japan and Korea. Off. Surgeon, 8th U.S. Army, Kyoto, Honshu.

Laffoon, J.L., and K.L. Knight. 1973. A mosquito taxonomic glossary. IX. The larval cranium. Mosquito Syst Newslett. 5:31-96.

Lake, R.W. 1954. Some biological observations on tree hole mosquitoes in Passaic County, New Jersey. Proc. New Jers. Mosq. Exterm. Ass. 41:193-198.

Lake, R.W. 1963. The occurrence of *Aedes dupreii* (Coquillett) and *Psorophora howardii* Coquillett in Delaware. Mosquito News 23:160.

Lawlor, W.K. 1935. Hibernation of *Uranotaenia sapphirina* (Osten Sacken). Bull. Brooklyn ent. Soc. 30:14.

Linnaeus, C. 1758. Systema naturae per regna tria naturae. Ed. 10, Vol. 1. Holmiae [Stockholm] 824 pp.

Loor, K.A., and G.R. DeFoliart. 1969. An oviposition trap for detecting the presence of *Aedes triseriatus* (Say). Mosquito News 29:487-488.

Loor, K.A., and G.R. DeFoliart. 1970. Field observations on the biology of *Aedes triseriatus*. Mosquito News 30:60-64.

Lord, J.K. 1866. A naturalist in Vancouver Island and British Columbia. Vol. 2. London. 375 pp., 4 figs.

Love, G.J., and M.M. Goodwin, Jr. 1961. Notes on the bionomics and seasonal occurrence of mosquitoes in southwestern Georgia. Mosquito News 21:195-215.

Love, G.J., R.B. Platt, and M.H. Goodwin, Jr. 1963. Observations on the spatial distribution of mosquitoes in southwestern Georgia. Mosquito News 23:13-22.

Ludlow, C.S. 1905*a*. A new North American *Taeniorhynchus*. Can. Ent. 37:231-232.

Ludlow, C.S. 1905*b*. Mosquito notes. No. 4. Can. Ent. 37:385-388.

Ludlow, C.S. 1906*a*. Mosquito notes. No. 4 (concl.). Can. Ent. 38:132-134.

Ludlow, C.S. 1906*b*. An Alaskan mosquito. Can. Ent. 38:326-328.

Ludlow, C.S. 1906*c*. A new American mosquito. George Washington Univ. Bull. 5(4):83-84.

Ludlow, C.S. 1907. Mosquito notes. No. 5 (contd.). Can. Ent. 39:129-131, 266-268.

Ludlow, C.S. 1911. A new Alaskan mosquito. Can. Ent. 43:178-179.

Mail, G.A. 1934. The mosquitoes of Montana. Bull. Mont. agric. Exp. Stn 288:1-72.

Maire, A., and A. Aubin. 1976. Inventaire et classification écologiques des biotopes à larves de moustiques (Culicides) de la région de Radisson (territoire de la Baie de James, Québec). Can. J. Zool. 54:1979-1991.

Maire, A., and Y. Mailhot. 1978. A new record of *Aedes cantator* from the tidal zone of southeastern James Bay, Quebec. Mosquito News 38:207-209.

Maloney, J.M., and R.C. Wallis. 1976. Response of colonized *Culiseta melanura* to photoperiod and temperature. Mosquito News 36:190-196.

Marshall, J.F. 1938. The British Mosquitoes. British Museum of Natural History, London. 341 pp.

Maslov, A.V. 1964. Bloodsucking mosquitoes of the subtribe Culisetina (Diptera, Culicidae) of the world fauna [in Russian]. Akad. Nauk. S.S.S.R., Zool. Inst., Leningr. 182 pp.

Masters, C.A. 1953. Seasonal succession of mosquito species and the relationship existing between dissolved minerals in mosquito-breeding waters and species. Mosquito News 13:159-161.

Matheson, R. 1929. A handbook of the mosquitoes of North America. 1st ed. Charles C. Thomas, Pubs., Springfield, Ill., and Baltimore, Md. 268 pp.

Matheson, R. 1933. A new species of mosquito from Colorado (Diptera, Culicidae). Proc. ent. Soc. Wash. 35:69-71.

Matheson, R. 1944. A handbook of the mosquitoes of North America. 2nd ed. Comstock Publishing Associates, Ithaca, N.Y. 314 pp.

Matheson, R., and H.S. Hurlbut. 1937. Notes on *Anopheles walkeri* Theobald. Am. J. trop. Med. 17:237-242.

Mattingly, P.F., L.E. Rozeboom, K.L. Knight, H. Laven, F.H. Drummond, S.R. Christophers, and P.G. Shute. 1951. The *Culex pipiens* complex. Trans. R. ent. Soc. Lond. 102:331-382.

McAlpine, J.F., and D.D. Munroe. 1968. Swarming of lonchaeid flies and other insects, with descriptions of four new species of Lonchaeidae (Diptera). Can. Ent. 100:1154-1178.

McClure, H.E. 1943. Aspection in the biotic communities of the Churchill area, Manitoba. Ecol. Monogr. 13:1-35.

McIver, S., and A. Hudson. 1972. Sensilla on the antennae and palps of selected *Wyeomyia* mosquitoes. J. med. Ent. 9:337-345.

McLean, D.M. 1975. Arboviruses and human health in Canada. Natn. Res. Counc. Can. Publ. 14106. 35 pp.

McLintock, J. 1944. The mosquitoes of the Greater Winnipeg area. Can. Ent. 76:89-104.

McLintock, J. 1952. Continuous laboratory rearing of *Culiseta inornata* (Will.) (Diptera: Culicidae). Mosquito News 12:195-201.

McLintock, J. 1976. The arbovirus problem in Canada. Can. J. Public Health 67 (Suppl. 1):8-12.

McLintock, J., P.S. Curry, R.J. Wagner, M.K. Leung, and J.O. Iversen. 1976. Isolation of Snowshoe Hare virus from *Aedes implicatus* larvae in Saskatchewan. Mosquito News 36:233-237.

McLintock, J., and J. Iversen. 1975. Mosquitoes and human disease in Canada. Can. Ent. 107:695-704.

McLintock, J., and J.G. Rempel. 1963. Mid-summer mosquito abundance in southern Saskatchewan, 1962. Mosquito News 23:242-249.

Means, R.G. 1965. *Culex territans* Walker biting man in nature. Mosquito News 25:489.

Meigen, J.W. 1830. Systematische Beschreibung der bekannten europäischen zweiflügeligen Insekten. Vol. 6. Hamm. iv, 401 pp., pls. 55-66.

Meigen, J.W. 1838. Systematische Beschreibung der bekannten europäischen zweiflügeligen Insekten. Vol. 7. Hamm. xii, 434 pp., pls. 67-74.

Meredith, J., and J.E. Phillips. 1973. Ultrastructure of anal papillae from a seawater mosquito larva (*Aedes togoi* Theobald). Can J. Zool. 51:349-353.

Meredith, J., and J.E. Phillips. 1973. Ultrastructure of anal papillae from a seawater mosquito larva (*Aedes togoi* Theobald). Can J. Zool. 51:349-353.

Michener, C.D. 1947. Mosquitoes of a limited area in southern Mississippi. Am. Midl. Nat. 37:325-374.

Mitchell, E.G. 1907. Mosquito life. G.P. Putnam's Sons, New York. 281 pp.

Miura, T., and R.M. Takahashi. 1973. Observations on the hatching of *Aedes nigromaculis* (Ludlow) eggs: seasonal variation. Proc. Pap. 41st A. Conf. Calif. Mosq. Control Ass. pp. 111-114.

Morris, C.D., R.H. Zimmerman and L.A. Magnarelli. 1976. The bionomics of *Culiseta melanura* and *Culiseta morsitans dyari* in Central New York State (Diptera: Culicidae). Ann. ent. Soc. Am. 69:101-105.

Müller, O.F. 1764. Fauna insectorum Fridrichsdalina, sive methodica descriptio insectorum agri Fridrichsdalinensis cum characteribus genericis et specificis, nominibus trivialibus, locis natalibus, iconibus allegatis, novisque pluribus speciebus additis. Hafniae et Lipsiae [Copenhagen and Leipzig]. xxiv, 96 pp.

Munroe, E. 1956. Canada as an environment for insect life. Can. Ent. 88:372-476.

Myklebust, R.J. 1966. Distribution of mosquitoes and chaoborids in Washington State, by counties. Mosquito News 26:515-519.

Natvig, L.R. 1948. Contributions to the knowledge of the Danish and Fennoscandian mosquitoes: Culicini. Norsk ent. Tidsskr., Suppl. 1. 567 pp.

Niebanck, L. 1957. Observations on mosquito species in Nassau County, Long Island, N.Y., based on larval and adult collections during the ten-year period 1947-1956. Mosquito News 17:315-318.

Nielsen, L.T., and D.M. Rees. 1959. The mosquitoes of Utah—a revised list. Mosquito News 19:45-47.

Nielsen, L.T., and D.M. Rees. 1961. An identification guide to the mosquitoes of Utah. Univ. Utah biol. Ser. 12(3):1-58.

Norris, M. 1946. Recovery of a strain of western equine encephalitis virus from *Culex restuans* (Theo.) (Diptera: Culicidae). Can. J. Res., E, 24:63-70.

O'Meara, G.F., and G.B. Craig. 1970. Geographical variation in *Aedes atropalpus* (Diptera: Culicidae). Ann. ent. Soc. Am. 63:1392-1400.

Omori, N. 1962. A review of the role of mosquitoes in the transmission of Malayan and Bancroftian filariasis in Japan. Bull. Wld Hlth org. 27:585-594.

O'Rourke, F.J. 1959. *Anopheles* and the problem of malaria in Canada. Can. Ent. 91:346-358.

Osten Sacken, C.R. 1868. Description of a new species of Culicidae. Trans. Am. ent. Soc. 2:47-48.

Osten Sacken, C.R. 1877. Western Diptera: Descriptions of new genera and species of Diptera from the region west of the Mississippi and especially from California. Bull. U.S. Geol. and Geog. Surv. of the Territ. 3:189-354.

Owen, W.B. 1937. The mosquitoes of Minnesota, with special reference to their biologies. Tech. Bull. Univ. Minn. agric. Exp. Stn 126:1-75.

Ozburn, R.H. 1945. Preliminary report on anopheline mosquito survey in Canada. Pt. 1. A report on light trap collections. A. Rep. ent. Soc. Ont. (1944) 75:37-44.

Parker, D.J. 1977. The biology of the tree-holes of Point Pelee National Park, Ontario. II. First record of *Toxorhynchites rutilus septentrionalis* (Dyar and Knab) in Canada (Diptera: Culicidae). Can. Ent. 109:93-94.

Parkin, W.E., W.M. Hammon, and G.E. Sather. 1972. Review of current epidemiological literature on viruses of the California arbovirus group. Am. J. trop. Med. Hyg. 21:964-978.

Parsons, M.A., R.L. Berry, M. Jalil, and R.A. Masterson. 1972. A revised list of the mosquitoes of Ohio with some new distribution and species records. Mosquito News 32:223-226.

Penn, G.H. 1938. Larval development of *Aedes triseriatus* Say. Tulane Biologist 2:5-8.

Penn, G.H. 1949. Pupae of the nearctic anopheline mosquitoes north of Mexico. J. natn. Malar. Soc. 8:50-69.

Peters, H.T. 1943. Studies on the biology of *Anopheles walkeri* Theobald (Diptera: Culicidae). J. Parasit. 29:117-122.

Petersen, J.J., H.C. Chapman, and O.R. Willis. 1969. Predation of *Anopheles barberi* Coquillett on first instar mosquito larvae. Mosquito News 29:134-135.

Peterson, A.G., and W.W. Smith. 1945. Occurrence and distribution of mosquitoes in Mississippi. J. econ. Ent. 38:378-383.

Peus, F. 1972. The subgenus *Aedes* sensu stricto in Germany (Diptera, Culicidae). Zeit. angew. Ent. 72:177-194.

Pickavance, J.R., G.F. Bennett, and J. Phipps. 1970. Some mosquitoes and blackflies from Newfoundland. Can. J. Zool. 48:621-624.

Portman, R.F. 1954. Control of mosquitoes in California rice fields. J. econ. Ent. 47:818-824.

Pratt, H.D. 1952. Notes on *Anopheles earlei* and other American species of the *Anopheles maculipennis* complex. Am. J. trop. Med. Hyg. 1:484-493.

Price, R.D. 1958*a*. A description of the larva and pupa of *Culiseta* (*Culicella*) *minnesotae* Barr (Diptera, Culicidae). J. Kans. ent. Soc. 31:47-53.

Price, R.D. 1958*b*. Notes on the biology and laboratory colonization of *Wyeomyia smithii* (Coquillett) (Diptera: Culicidae). Can. Ent. 90:473-478.

Price, R.D. 1960. Identification of first-instar aedine mosquito larvae of Minnesota (Diptera: Culicidae). Can. Ent. 92:544-560.

Price, R.D. 1961. Biological notes on *Culiseta minnesotae* Barr and *Culiseta morsitans* (Theobald) (Diptera: Culicidae). J. Kans. ent. Soc. 34:22-26.

Price, R.D. 1963. Frequency of occurrence of spring *Aedes* (Diptera: Culicidae) in selected habitats in northern Minnesota. Mosquito News 23:324-329.

Price, R.D., and L.R. Abrahamsen. 1958. The discovery of *Orthopodomyia signifera* (Coquillett) and *Anopheles barberi* Coquillett in Minnesota (Diptera, Culicidae). J. Kans. ent. Soc. 31:92.

Price, R.D., T.A. Olson, M.E. Rueger, and L.L. Schlottman. 1960. A survey on potential overwintering sites of *Culex tarsalis* Coquillett in Minnesota. Mosquito News 20:306-311.

Pucat, A. 1964. Seven new records of mosquitoes in Alberta. Mosquito News 24:419-421.

Pucat, A. 1965. List of mosquito records from Alberta. Mosquito News 25:300-302.

Rapp, W.F., Jr. 1960. Notes on the chemical conditions in the mosquito larval environment. Mosquito News 20:29-30.

Rapp, W.F., Jr., and F.C. Harmston. 1965. Notes on the mosquitoes (Culicinae) of northwestern Nebraska. Mosquito News 25:302-306.

Reeves, W.C., and W.M. Hammon. 1944. Feeding habits of the proven and possible mosquito vectors of western equine and St. Louis encephalitis in the Yakima Valley, Washington. Am. J. trop. Med. 24:131-134.

Remington, C.L. 1945. The feeding habits of *Uranotaenia lowii* Theobald (Diptera: Culicidae). Ent. News 56:32-37; 64-68.

Rempel, J.G. 1950. A guide to the mosquito larvae of western Canada. Can. J. Res., D, 28:207-248.

Rempel, J.G. 1953. The mosquitoes of Saskatchewan. Can. J. Zool. 31:433-509.

Robineau-Desvoidy, J.B. 1827. Essai sur la tribu des Culicides. Mém. Soc. Hist. nat. Paris 3:390-413.

Rohlf, F.J. 1963. Classification of *Aedes* by numerical taxonomic methods (Diptera: Culicidae). Ann. ent. Soc. Am. 56:798-804.

Ronderos, R.A., and A.O. Bachmann. 1963. A proposito del complejo *Mansonia* (Diptera-Culicidae). Revta Soc. ent. Argent. (1962) 25:43-51.

Rowe, J.S. 1972. Forest regions of Canada. Can. For. Serv. Publ. 1300. 172 pp. Dep. of the Environment, Ottawa.

Rozeboom, L.E. 1952. The significance of *Anopheles* species complexes in problems of disease transmission and control. J. econ. Ent. 45:222-226.

Rudolfs, W. 1929. Environmental factors and mosquito breeding. Trans. 4th int. Congr. Ent. 2(1928):945-959.

Rudolfs, W., and J.B. Lackey. 1929. The composition of water and mosquito breeding. Am. J. Hyg. 10:160-180.

Rueger, M.E. 1958. *Aedes* (*Ochlerotatus*) *barri*, a new species of mosquito from Minnesota (Diptera, Culicidae). J. Kans. ent. Soc. 31:34-46.

Rush, W.A. 1962. Observations on an overwintering population of *Culex tarsalis* with notes on other species. Mosquito News 22:176-181.

Rush, W.A., J.M. Brennan, and C.M. Eklund. 1958. A natural hibernation site of the mosquito *Culex tarsalis* Coquillett in the Columbia River basin, Washington. Mosquito News 18:288-293.

Ryckman, R.E., and K.Y. Arakawa. 1951. *Anopheles freeborni* hibernating in wood rats' nests (Diptera: Culicidae). Pan-Pacif. Ent. 27:172.

Sandholm, H.A., and R.D. Price. 1962. Field observations on the nectar feeding habits of some Minnesota mosquitoes. Mosquito News 22:346-349.

Sasa, M., S. Hayashi, R. Kano, K.I. Komine, and S. Ishu. 1952. Studies on filariasis due to *Wuechereria malayi* (Brug 1927) discovered from Hachijo-Koshima Island, Japan. Jap. J. exp. Med. 22:357-390.

Say, T. 1823. Descriptions of dipterous insects of the United States. J. Acad. nat. Sci. Philad. 3:9-54, 73-104.

Say, T. 1824. Appendix. Part 1. Natural history. 1. Zoology. E. Class Insecta. Pages 268-378 *in* W.H. Keating, Major Long's second expedition. Vol. 2. Philadelphia. 459 pp., pls. 6-15.

Say, T. 1829. Descriptions of North American dipterous insects. J. Acad. nat. Sci. Philad. (1829-30) 6:149-178.

Schober, H. 1964. Notes on the behavior of *Culiseta melanura* (Coq.) with three instances of its biting man. Mosquito News 24:67.

Schremmer, F. 1949. Morphologische und funktionelle Analyse der Mundteile und des Pharynx der Larve von *Anopheles maculipennis* Meig. Öst. zool. Z. 2:173-222.

Shalaby, A.M. 1957. The mouthparts of the larval instars of *Psorophora ciliata* (Fabricius) (Diptera: Culicidae). Bull. Soc. ent. Egypte 41:429-455.

Shaw, F.R., and S.A. Maisey. 1961. The biology and distribution of the rockpool mosquito, *Aedes atropalpus* (Coq.). Mosquito News 21:12-16.

Shemanchuk, J.A. 1965. On the hibernation of *Culex tarsalis* Coquillett, *Culiseta inornata* Williston, and *Anopheles earlei* Vargas (Diptera: Culicidae) in Alberta. Mosquito News 25:456-462.

Shemanchuk, J.A., A.E.R. Downe, and L. Burgess. 1963. Hosts of mosquitoes (Diptera: Culicidae) from the irrigated areas of Alberta. Mosquito News 23:336-341.

Shields, S.E., and J.B. Lackey. 1938. Conditions affecting mosquito breeding with special reference to *Aedes thibaulti* Dyar and Knab (Diptera, Culicidae). J. econ. Ent. 31:95-102.

Shute, P.G. 1933. The life-history and habits of British mosquitoes in relation to their control by antilarval operations. J. trop. Med. Hyg. 36:83-88.

Siverly, R.E., and G.R. DeFoliart. 1968*a*. Mosquito studies in northern Wisconsin. I. Larval studies. Mosquito News 28:149-154.

Siverly, R.E., and G.R. DeFoliart. 1968*b*. Mosquito studies in northern Wisconsin. II. Light trapping studies. Mosquito News 28:162-167.

Siverly, R.E., and H.F. Schoof. 1962. Biology of *Culiseta melanura* (Coq.) in southeast Georgia. Mosquito News 22:274-282.

Smith, J.B. 1904. The common mosquitoes of New Jersey. Bull. New Jers. agric. Exp. Stn 171:1-40.

Smith, L.W., Jr. 1969. The relationship of mosquitoes to oxidation lagoons in Columbia, Missouri. Mosquito News 29:556-563.

Smith, M.E. 1952. A new northern *Aedes* mosquito, with notes on its close ally, *Aedes diantaeus* H., D., and K. (Diptera, Culicidae). Bull. Brooklyn ent. Soc. 47:19-28, 29-40.

Smith, M.E. 1965. Laval differences between *Aedes communis* (DeG.) and *A. implicatus* Vock. (Diptera: Culicidae) in a Colorado community. Mosquito News 25:187-191.

Smith, M.E. 1969*a*. The *Aedes* mosquitoes of New England (Diptera: Culicidae). II. Larvae: keys to instars, and to species exclusive of first instar. Can. Ent. 101:41-51.

Smith, M.E. 1969*b*. The *Aedes* mosquitoes of New England. III. Saddle hair position in 2nd and 3rd instar larvae, with particular reference to instar recognition and species relationships. Mosquito Syst. Newslett. 1(3):57-62.

Smith, S.M., and R.A. Brust. 1970. Autogeny and stenogamy of *Aedes rempeli* (Diptera: Culicidae) in arctic Canada. Can. Ent. 102:253-256.

Smith, S.M., and R.A. Brust. 1971. Photoperiodic control of the maintenance and termination of larval diapause in *Wyeomyia smithii* (Coq.) (Diptera: Culicidae) with notes on oogenesis in the adult female. Can. J. Zool. 49:1065-1073.

Smith, S.M., and R.M. Trimble. 1973. The biology of tree-holes of Point Pelee National Park, Ontario. 1. New mosquito records for Canada (Diptera: Culicidae). Can. Ent. 105:1585-1586.

Snodgrass, R.E. 1959. The anatomical life of the mosquito. Smithson. misc. Collns 139(8):1-87.

Snow, W.E. 1949. Studies on portable resting stations for *Anopheles quadrimaculatus* in the Tennessee Valley. J. natn. Malar. Soc. 8:336-343.

Snow, W.E., and G.E. Smith. 1956. Observations on *Anopheles walkeri* Theobald in the Tennessee Valley. Mosquito News 16:294-298.

Sommerman, K.M. 1964. Notes on activities of Alaskan *Culiseta* adults (Diptera: Culicidae). Mosquito News 24:60-64.

Sommerman, K.M. 1977. Biting fly – arbovirus probe in interior Alaska (Culicidae) (Simuliidae)—(SSH: California complex) (Northway: Bunyamwera group). Mosquito News 37:90-103.

Stage, H.H. 1943. Four days in the muskeg. Mosquito News 3:127-130.

Stage, H.H., C.M. Gjullin, and W.W. Yates. 1952. Mosquitoes of the northwestern States. U.S. Dep. Agric., Agric. Handb. 46:95 pp.

Steward, C.C. 1968. Numerical classification of the Canadian species of the genus *Aedes* (Diptera: Culicidae). Syst. Zool. 17:426-437.

Stone, A. 1961. A synoptic catalog of the mosquitoes of the world. Suppl. I. Proc. ent. Soc. Wash. 63:29-52.

Stone, A. 1963. A synoptic catalog of the mosquitoes of the world. Suppl. II. Proc. ent. Soc. Wash. 65:117-140.

Stone, A. 1965. Family Culicidae. *In* A. Stone, et al. A catalog of the Diptera of America north of Mexico. U.S. Dep. Agric., Agric. Handb. 276:105-120.

Stone, A. 1967. A synoptic catalog of the mosquitoes of the world. Suppl. III (Diptera, Culicidae). Proc. ent. Soc. Wash. 69(3):197-224.

Stone, A., K.L. Knight, and H. Starcke. 1959. A synoptic catalog of the mosquitoes of the world (Diptera, Culicidae). Ent. Soc. Am. (Thomas Say Found.), Wash. 6:1-358.

Strickland, E.H. 1938. An annotated list of the Diptera (flies) of Alberta. Can. J. Res., D, 16:175-219.

Sudia, W.D., and R.H. Gogel. 1953. The occurrence of *Orthopodomyia alba* Baker in Georgia (Diptera: Culicidae). Bull. Brooklyn ent. Soc. 48(5):129-131.

Tauthong, P. 1975. The biology and systematics of *Aedes campestris* Dyar and Knab (Diptera: Culicidae) and related species in Manitoba and Saskatchewan. Ph.D. Thesis, Univ. Man., Winnipeg, Man. 109 pp.

Tauthong, P., and R.A. Brust. 1977*a*. The effect of photoperiod on diapause induction, and temperature on diapause termination in embryos of *Aedes campestris* Dyar and Knab (Diptera: Culicidae). Can. J. Zool. 55:129-134.

Tauthong, P., and T.[sic] A. Brust. 1977*b*. The effect of temperature and survival of two populations of *Aedes campestris* Dyar and Knab (Diptera: Culicidae). Can. J. Zool. 55:135-137.

Tawfik, M.S., and R.H. Gooding. 1970. Observations of mosquitoes during 1969 control operations at Edmonton, Alberta. Quaest. Ent. 6(4):307-310.

Taylor, B.W., and R.A. Brust. 1974. Laboratory mating of *Aedes vexans* (Diptera: Culicidae). Ann. ent. Soc. Am. 67:137-138.

Theobald, F.V. 1901*a*. A monograph of the Culicidae or mosquitoes. Vol. 1. British Museum of Natural History, London. 424 pp.

Theobald, F.V. 1901*b*. A monograph of the Culicidae or mosquitoes. Vol. 2. British Museum of Natural History, London. 391 pp.

Theobald, F.V. 1901*c*. The classification of mosquitoes. J. trop. Med. 4:229-235.

Theobald, F.V. 1903*a*. A monograph of the Culicidae or mosquitoes. Vol. 3. British Museum of Natural History, London. 359 pp.

Theobald, F.V. 1903*b*. Description of a new North American *Culex*. Can. Ent. 35:211-213.

Theobald, F.V. 1903*c*. Notes on Culicidae and their larvae from Pecos, New Mexico, and description of a new *Grabhamia*. Can. Ent. 35:311-316.

Theobald, F.V. 1907. A monograph of the Culicidae or mosquitoes. Vol. 4. British Museum of Natural History, London. 639 pp.

Theobald, F.V. 1910. A monograph of the Culicidae or mosquitoes. Vol. 5. British Museum of Natural History, London. 646 pp.

Thibault, J.K. 1910. Notes on the mosquitoes of Arkansas. Proc. ent. Soc. Wash. 12:13-26.

Thien, L.B. 1969*a*. Mosquito pollination of *Habenaria obtusata* (Orchidaceae). Am. J. Bot. 56:232-237.

Thien, L.B. 1969*b*. Mosquitoes and *Habenaria obtusata* (Orchidaceae). Mosquito News 29:252-255.

Thomson, C.G. 1869. 6. Diptera. Species nova descripsit. Pages 443-614, pl. 9 (Heft 12, No. 2) *in* K. Svenska Vetenskaps-Akademien, Konglia svenska fregatten Eugenies resa omkring jorden. Pt. 2: Zoologie, Sect. 1: Insekter. Stockholm, 1868. 617 pp., 9 pls.

Tremblay, H.L. 1947. Biological characteristics of laboratory reared *Aedes atropalpus*. J. econ. Ent. 40:244-250.

Tremblay, H.L. 1955. Mosquito culture techniques and experimental procedures. Bull. Am. Mosq. Control Ass. 3:1-73.

Trimble, R.M. 1972. Occurrence of *Culiseta minnesotae* and *Aedes trivittatus* (Diptera: Culicidae) in Manitoba, including a list of mosquitoes from Manitoba. Can. Ent. 104:1535-1537.

Truman, J.W., and G.B. Craig, Jr. 1968. Hybridization between *Aedes hendersoni* and *Aedes triseriatus*. Ann. ent. Soc. Am. 61:1020-1025.

Twinn, C.R. 1929. Report of the mosquito control committee of the Ottawa District, 1928. Ottawa. 12 pp.

Twinn, C.R. 1931. Notes on the biology of mosquitoes of eastern Canada. Proc. New Jers. Mosq. Exterm. Ass. 18:10-22.

Twinn, C.R. 1945. Report of a survey of anopheline mosquitoes in Canada in 1944. Proc. New Jers. Mosq. Exterm. Ass. 32:242-251.

Twinn, C.R. 1949. Mosquitoes and mosquito control in Canada. Mosquito News 9:35-41.

Vargas, L. 1943*a*. *Anopheles earlei* Vargas, 1942, n. sp. norteamericana del grupo *maculipennis*. Boln Of. sanit. pan-am. 22(1):8-12.

Vargas, L. 1943*b*. El "grupo *maculipennis*" del nuevo mundo y el *Anopheles earlei*. Revta Inst. Salubr. Enferm. trop., Mex. 4:279-286.

Vargas, L. 1944. Algunas consideraciones sobre *Anopheles occidentalis* Dyar y Knab, 1906. Revta Inst. Salubr. Enferm. trop., Mex. 5:215-220.

Vargas, L., and R. Matheson. 1948. Estado actual del *Anopheles earlei* Vargas, 1943, y *Anopheles occidentalis* Dyar and Knab, 1906, con claves para larvas, pupas y adultos del llamodo complejo *maculipennis* de Norteamerica. Revta Inst. Salubr. Enferm. trop., Mex. 9:27-33.

Venard, C.E., and F.W. Mead. 1953. An annotated list of Ohio mosquitoes. Ohio J. Sci. 53:327-331.

Vockeroth, J.R. 1950. Specific characters in tarsal claws of some species of *Aedes* (Diptera, Culicidae). Can. Ent. 82:160-162.

Vockeroth, J.R. 1954*a*. Notes on Northern Species of *Aedes*, with descriptions of two new species (Diptera, Culicidae). Can. Ent. 86:109-116.

Vockeroth, J.R. 1954*b*. Notes on the identities and distributions of *Aedes* species of northern Canada, with a key to the females (Diptera: Culicidae). Can. Ent. 86:241-255.

Vockeroth, J.R. 1966. A method of mounting insects from alcohol. Can. Ent. 98:69-70.

Wada, Y. 1965. Population studies on Edmonton mosquitoes. Quaest. Ent. 1:187-222.

Walker, F. 1848. List of the specimens of dipterous insects in the collection of the British Museum. Vol. 1. London. 229 pp.

Walker, F. 1856. Diptera (concl.). Vol. 1. Pages 415-474. *in* W.W. Saunders, ed. Insecta Saundersiana, London.

Wallis, R.C. 1954*a*. A study of oviposition activity of mosquitoes. Am. J. Hyg. 60:135-168.

Wallis, R.C. 1954*b*. Notes on the biology of *Culiseta melanura* (Coquillett). Mosquito News 14:33-34.

Wallis, R.C., and L. Whitman. 1968. Oviposition of *Culiseta morsitans* (Theobald) and comments of the life cycle of the American form. Mosquito News 28:198-200.

Watts, D.M., S. Pantuwatana, G.R. DeFoliart, T.M. Yuill, and W.H. Thompson. 1973. Transovarial transmission of LaCrosse virus (California encephalitis group) in the mosquito, *Aedes triseriatus*. Science (Wash. D.C.) 182:1140-1141.

Weathersby, A.B. 1962. Colonization of six species of mosquitoes in Japan. Mosquito News 22:31-34.

Weathersby, A.B. 1963. Harvesting mosquito pupae with cold water. Mosquito News 23:249-251.

Wesenberg-Lund, C. 1920–1921. Contributions on the biology of the Danish Culicidae. A.F. Host & Son, Copenhagen. 210 pp.

West, A.S., and A. Hudson. 1960. Notes on mosquitoes of eastern Ontario. Proc. New Jers. Mosq. Exterm. Ass. 47:68-74.

West, A.S., and D.W. Jenkins. 1951. Plant feeding habits of northern mosquitoes studied with radioisotopes. Mosquito News 11:217-219.

Wiedemann, C.R.W. 1820. Diptera exotica. Pt. 1. Kiliae [Kiel]. xix, 42 pp.

Wilkins, O.P., and O.P. Breland. 1951. The larval stages and the biology of the mosquito, *Orthopodomyia alba* Baker (Diptera: Culicidae). Jl N.Y. ent. Soc. 59:225-240.

Williston, S.W. 1893. List of Diptera of the Death Valley Expedition. Pages 253-259 *in* C.V. Riley. The Death Valley Expedition. A biological survey of parts of California, Nevada, Arizona, and Utah. Pt. II, 4. Report on a small collection of insects made during the Death Valley Expedition. North American Fauna 7:235-268.

Wishart, G., and H.G. James. 1946. Notes on the anopheline mosquitoes of the Kingston, Trenton and Peterborough, Ontario, areas. 76th A. Rep. ent. Soc. Ont. 1945:39-48.

Wood, D.M. 1977. Notes on the identities of some common nearctic *Aedes* mosquitoes (Diptera, Culicidae). Mosquito News 37:71-81.

Woodward, D.B., and H.C. Chapman. 1965. Blood volumes ingested by various pest mosquitoes. Mosquito News 25:490-491.

Wray, F.C. 1946. Six generations of *Culex pipiens* without a blood meal. Mosquito News 6:71-72.

Wulp, F.M. van der. 1867. Eenige Noord-Americaansche Diptera. Tijdschr. Ent. 10(Ser. 2, No. 2):125-164, pls. 3-5.

Yamanda, S. 1927. An experimental study on twenty-four species of Japanese mosquitoes regarding their suitability as intermediate hosts for *Filaria bancrofti* Cobbold. Sci. Rep. Inst. infectious Dis., Tokyo 6:559-622.

Yates, W.W. 1953. Notes on the ecology of *Culiseta* mosquitoes found in the Pacific Northwest. Mosquito News 13:229-232.

Zaim, M., H.D. Newson, and G.D. Dennis. *Psorophora horrida* in Michigan. Mosquito News 37:763.

Zavortink, T.J. 1972. Mosquito studies (Diptera, Culicidae). XXVIII. The New World species formerly placed in *Aedes* (Finlaya). Contr. Am. ent. Inst. 8(3):1-206.

Index

(Page numbers of principal entries are in boldface; synonyms are in italic type.)